# FLOWERING
# PLANTS
## OF THE WORLD

| | |
|---|---|
| *Consultant Editor:* | Professor V. H. Heywood PhD DSc FLS FIBiol |
| *Advisory Editors:* | Professor D. M. Moore PhD FLS<br>I. B. K. Richardson BSc<br>Professor W. T. Stearn Hon DSc Leiden, Hon ScD<br>Cambridge, Hon DPhil Uppsala, FLS FIBiol |
| *Artists:* | Victoria Goaman<br>Judith Dunkley<br>Christabel King |

# FLOWERING
# PLANTS
## OF THE WORLD

MAYFLOWER BOOKS

# CONTRIBUTORS

Project Editor: Graham Bateman PhD
Editor: Bernard Dod
Maps Editor: Bill MacKeith
Production: Elizabeth Digby Firth
Managing Editor: Ben Lenthall
Design: Richard Brookes

MAYFLOWER BOOKS, INC.,
575 Lexington Avenue,
New York, NY 10022

First American Edition 1978

Planned and produced by Elsevier
International Projects Ltd, Oxford. © 1978
Elsevier Publishing Projects SA, Lausanne

ISBN 0-8317-3400-0

Origination by Art Color Offset, Rome, Italy,
and City Engraving Ltd, Hull, Humberside
Filmset by Keyspools Ltd, Golborne,
Lancashire
Printed by Jolly & Barber Ltd, Rugby

Great Britain

**S.C.H.B.** S. C. H. Barrett BSc PhD
Toronto University, Canada

**B.N.B.** B. N. Bowden BSc DPhil
Chelsea College
University of London

**D.B.** D. Bramwell BSc MSc PhD
Jardín Botánico
Tafira Alta, Las Palmas de Gran
Canaria, Canary Islands

**S.R.C.** S. R. Chant BSc PhD Dip Ag Sci DJA
Chelsea College
University of London

**W.D.C.** W. D. Clayton PhD BSc ARCS
Royal Botanic Gardens, Kew

**C.D.C.** Professor C. D. K. Cook BSc PhD FLS
Botanic Gardens and Institute for
Systematic Botany
University of Zurich, Switzerland

**J.C.** J. Cullen BSc PhD
Royal Botanic Garden, Edinburgh

**D.C.** D. F. C. Cutler BSc PhD DIC
Royal Botanic Gardens, Kew

**J.E.D.** J. E. Dandy MA
British Museum (Natural History)
Tring

**M.C.D** M. C. Doggett BSc
University of Reading

**J.M.E.** J. M. Edmonds BSc PhD MA
University of Cambridge

**T.T.E.** T. T. Elkington BSc PhD
University of Sheffield

**B.S.F.** B. S. Field BSc MPhil
University College
University of London

**S.A.H.** S. A. Heathcote BSc
British Museum (Natural History),
London

**I.H.** I. C. Hedge BSc
Royal Botanic Garden, Edinburgh

**V.H.H.** Professor V. H. Heywood PhD DSc
FLS FIBiol
University of Reading

**F.B.H.** F. B. Hora MA DPhil
University of Reading

**C.J.H.** C. J. Humphries BSc PhD
British Museum (Natural History),
London

**P.F.H.** P. F. Hunt MSc MIBiol
Frome, Somerset

**C.J.** C. Jeffrey BA
Royal Botanic Gardens, Kew

**S.W.J.** S. W. Jones BSc
British Museum (Natural History)

**S.L.J.** S. L. Jury BSc FLS
University of Reading

**M.K.** M. Kovanda PhD
Czechoslovak Academy of Sciences,
Pruhonice, Czechoslovakia

**F.K.K.** F. K. Kupicha BSc PhD
British Museum (Natural History),
London

**D.J.M.** D. J. Mabberley MA DPhil
University of Oxford

**B.F.M.** B. F. Matthew FLS
Royal Botanic Gardens, Kew

**A.F.M.** A. F. Mitchell BA BAg(For) VMH
Forest Research Station
Farnham, Surrey

**D.M.M.** Professor D. M. Moore BSc PhD
University of Reading

**B.M.** B. Morley PhD
The Botanic Garden of Adelaide,
Australia

**B.P.** B. Pickersgill BSc PhD
University of Reading

**M.C.F.P.** M. C. F. Proctor MA PhD
University of Exeter

**A.R.-S.** A. Radcliffe-Smith BSc
Royal Botanic Gardens, Kew

**I.B.K.R.** I. B. K. Richardson BSc
Royal Botanic Gardens, Kew

**R.H.R.** R. H. Richens MA PhD
Commonwealth Bureau of Plant
Breeding and Genetics, Cambridge

**N.K.B.R.** N. K. B. Robson BSc PhD
British Museum (Natural History),
London

**G.D.R.** G. D. Rowley BSc
University of Reading

**N.W.S.** N. W. Simmonds ScD AICTA FRSE
FIBiol
University of Edinburgh

**C.A.S.** C. A. Stace BSc PhD FLS
University of Leicester

**W.T.S.** Professor W. T. Stearn FLS DSc ScD
FilD
British Museum (Natural History),
London

**R.F.T.** Professor R. F. Thorne PhD ABMS
Rancho Santa Ana Botanic Garden,
California, USA

**D.A.W.** Professor D. A. Webb ScD
Trinity College
University of Dublin, Eire

**T.C.W.** T. C. Whitmore MA PhD ScD
University of Oxford

**H.P.W.** H. P. Wilkinson BSc PhD FLS
Royal Botanic Gardens, Kew

**T.J.W.** T. J. Wright BSc
Wye College (University of London)

# PREFACE

Since the ancient beginnings of botany as a science, its practitioners have attempted to classify plants into logical and coherent groups. Botanists recognized natural classes through the centuries quite by intuition, because they had no established conceptual means by which to measure affinities or relationships until Darwin published his theory of organic evolution in 1859. As early as the 17th century, John Ray had published a surprisingly perceptive classification; yet a century later, Linnaeus found himself inhibited by his total belief that all plants were unchanging and unchangeable because they had arisen as the result of divine creation. Finally, he was forced to admit that his very widely used and workable system of classification was wholly artificial, and he never attained the natural system of which he had dreamed.

Once Darwin had formally proposed his theory of evolution, by which all modern plants could presumably have arisen from a common ancestry, then the way was opened immediately for fruitful speculation on the causes of coherence, uniqueness or naturalness among plant families, as well as their arrangement in natural phylogenetic (evolutionary) systems. The isolating individuality, on the one hand, and the relating affinities, on the other, of any given flowering plant family were at first measured and identified largely by flower morphology and, to a lesser degree, by other obvious morphological characteristics, following closely the practices of earlier pre-Darwinian writers. Today, more than a century later, the arrangement of plant families into a natural phylogenetic system has become one of the most sophisticated of all botanical disciplines, since it at the same time draws from and transcends them all in its search for information of all kinds with which to reach logical conclusions. Natural families possess whole constellations of characteristics which in the aggregate give them coherence. However, we still need to know much more about the evolutionary mechanisms by which whole large groups of characteristics are linked genetically and inherited more or less as a unit, generation after generation.

In a nutshell, then, most of the disciplines of botany have been brought together in the one transcendental field of scientific classification of flowering plant families. This clearly written and beautifully illustrated book therefore provides an excellent introduction to the whole science of botany, which should afford much pleasure to the interested layman and amateur botanist, as well as to the serious student and professional botanist. Most of our standard treatises dealing with the arrangement of flowering plant families and the reasoning for such positioning of them have been couched in technical botanical language to such an extent that they are limited in their major usefulness only to the initiated. This new book, on the other hand, with its attention to the useful members of each family, as well as to other matters of general interest, offers much utterly fascinating and easily accessible information for any intelligent reader.

Above all, perhaps, the illustrations set this book apart from its contemporaries. Their immediate impact as superb examples of the art of modern flower painting is balanced by a soundly based scientific accuracy. The reader is not simply introduced to the vast array of gross forms found in the flowering plants but is also shown details of their innermost structure—secrets not normally visible to the untrained eye.

William Campbell Steere, PhD, DSc.
(President Emeritus, The New York Botanical Garden)

# CONTENTS

# FLOWERING PLANTS

The flowering plants, known scientifically as the class Angiospermae, are the dominant group of vascular plants on earth today. They arose in the early Cretaceous period, some 120 million years ago or more, and by the end of this period, about 80—90 million years ago, they had become the characteristic plants dominating most parts of the world.

Not only are the flowering plants the largest and most successful plant group today but they are of fundamental importance to the life and survival of Man. He is, in fact, dependent on them for his major sources of food and sustenance, either directly through agricultural or horticultural crops such as cereals, legumes and fruits, or indirectly through their ability to provide pasture or feed for the animals he eats. They also provide a source of raw materials for building and shelter, for the manufacture of paper, fabrics and plastics, and for oils, fibers, waxes, spices, herbs, resins, drugs, medicines, tannins, intoxicants, beverages—the list seems endless.

Because of their dominance, flowering plants form the major vegetation in the landscape as well as providing habitats for most terrestrial animal life. Ecologically they are important to Man by affording windbreaks, protection against erosion, defense against the encroachment of the sea, cover for spoil tips, as well as a source of pleasure and recreation by their use in constructing gardens, parks, sports grounds, street plantings, house decorations, and so on.

The flowering plants have also played a major role in developing the cultural consciousness of Man and have occupied and continue to occupy a conspicuous place in religions and ceremonies as well as providing great symbolism in diverse cultures throughout the world. The shapes and form of stems, leaves and flowers have served as an inspiration in art, architecture and design in many parts of the world.

The angiosperms are characterized by the possession of true flowers which are more advanced and complex than the reproductive structures of the gymnosperms (eg conifers) from which they have almost certainly been ultimately derived. The common name flowering plants, or even the loose use of the term flower to refer to the whole plant, testifies to the importance of the flower as their most significant feature. Indeed the origin of the angiosperms has been traditionally regarded as synonymous with the origin of the flower.

Morphologically the flower is usually interpreted as an abbreviated and highly modified spore-bearing shoot, possibly derived from a gymnosperm-like reproductive structure called a strobilus (or cone). Basically the flower consists of four series of elements arranged centrifugally on an axis: (i) an outer series of modified bracts or leaves, the *sepals*, which are usually, although by no means always, green and serve a protective function, and which make up the *calyx*; (ii) an inner series of modified bracts or leaves, the *petals*, very often colored and serving an attractant role, making up the *corolla*; (iii) one or more series of male organs, the *stamens*, which bear the pollen and are collectively known as the *androecium*; and (iv) a series of female structures, the *carpels*, which contain the ovules (later becoming the seeds), collectively forming the ovary or *gynoecium*. (Plate I, IV.)

A major characteristic of the angiosperms is that the ovules are *enclosed* in an ovary which is crowned by a style and stigma, the latter receiving the pollen grains. This contrasts with the gymnosperms which have the ovules exposed and unprotected.

Biologically, the flower's prime function is the production of seed. Seed may be produced as a result of self-pollination and fertilization or by cross-pollination, or in some cases by a non-sexual process called *apomixis*. There is a major evolutionary advantage in cross-fertilization and many flowering plants possess adaptations or modifications to favor or ensure this. Outcrossing is what Darwin called Nature's urge, leading to vigor and maintaining fertility.

From its earliest origins, the flower has been intimately associated with pollinators – beetles being probably the earliest pollinators of primitive flowers in the early Cretaceous period. The evolution of the flower in all its complexity of form, color and scent has gone hand in hand with the evolution of the pollinating insects or other animals. There is, in fact, a close association between the form of the flower and the sensory perception of different classes of pollinators, and there has been a parallel co-evolution of the insect–flower relationship. Insect pollination was most likely associated with the closure of the primitive leaf-like carpel by infolding and fusion of the edges or sutures, so as to protect the ovules from the visits of marauding insects. The closure of the carpel not only served to protect the ovules but set in train and utilized the insect visitations as a pollen-transfer mechanism.

Associated with the evolution of the flower as such was the reduction of the female gametophyte (the stage in the life cycle from which the female gametes develop) which at maturity consists of an eight-nucleate embryo sac (see Plate X); this then undergoes a process of double fertilization: one male nucleus (gamete) produced by the pollen grain fertilizes the egg which then develops into the embryo, while the other male nucleus fuses with two female nuclei, which then develop into the food-storing endosperm tissue of the seed. The apparent uniqueness of the eight-nucleate embryo sac and the double fertilization is interpreted as evidence of the evolutionary coherence of the flowering plants and their distinctness from other plant groups.

Not only have the angiosperms developed highly complex and diverse reproductive organs but they possess a very advanced level of cell structure and differentiation, especially in the water conducting tissues (xylem) which contain tube-like cells called vessels, absent only in some primitive members. Their high level of physiological efficiency and their wide range of vegetative plasticity and floral diversity have allowed the flowering plants to occupy nearly every conceivable habitat extreme. This in turn is reflected in the almost bewildering array of growth habits found throughout the group. They grow in, and indeed characterize, every major ecological formation in the world – forests, grasslands, deserts, and many aquatic habitats. Within their major phytogeographical regions the flowering plants form a wide array of ecological communities, often characterized by the predominance of particular families, genera or species.

Most angiosperm families are tropical in their distribution and about two-thirds of the species are confined to the tropics or adjacent regions. Unfortunately, the floras of tropical regions are much less well-known than those of temperate areas, partly because of the imbalance of scientific manpower and institutions between developed and underdeveloped or developing countries, partly because of the sheer numbers of species involved. Despite two centuries of scientific study of plant classification, we are still largely ignorant about the detailed taxonomy and biology of the majority of flowering plants. Moreover, the situation has now become highly critical because the tropical countries which house the rich floras are now under great pressure. Population growth, energy problems, economic difficulties and inflation, encourage them to consume, and in effect destroy, their plant resources, in a headlong race for development or even survival. It is only within recent years that there has been a dawning appreciation of the extent to which Man has been consuming, destroying and modifying his global environment, especially in the tropics. The pressure on plant resources is enormous. It stems not only from the increased need for land to feed expanding populations but also from the increasing consumption of forests for newsprint, construction materials and furniture. Urban

Plate I. Comparison of primitive and advanced floral features.

PRIMITIVE

Parts many
Parts indefinite in number
Parts free
Spiral arrangement of parts
Sepals, petals, stamens,
carpels all present
Bisexual
Ovary superior
Symmetry regular
(actinomorphic)

*Magnolia campbellii*

*Nymphaea elegans*

development, road building and industrialization also decrease the areas available for plants to grow. Large-scale devastation is caused by uninformed and irrational methods of land usage. Slash and burn techniques in tropical forests allowing the cultivation of crops for a year or two in fact leave behind exhausted and eroded soils without adequate plant cover, and the use of heavy machinery such as bulldozers, earthmovers and prairie busters is directed with "intelligence' to achieve the maximum detrimental effect on natural vegetation.

Official estimates by UNESCO suggest that 10,000,000 hectares of tropical forest are being felled each year – comparable to an area the size of the Royal Botanic Gardens, Kew (126 hectares) every $6\frac{1}{2}$ minutes – and it is calculated that at the present rate of destruction there will be no undisturbed tropical humid lowland forest left anywhere in the world by the end of this century. Many of the plants in these tropical forests and other plant communities will be destroyed before they have even been collected and described for the first time and many others will disappear before we have had a chance to study them thoroughly and learn how best to use them.

Readers of this account of the flowering plant families can scarcely fail to be impressed by the enormous wealth and diversity of the plant-types described, adapted to an enormous range of habitats. Striking examples are the tiny green specks of the aquatic *Wolffia* species, the smallest known flowering plants; the Australian gum trees (*Eucalyptus*) some of which attain 150m (500ft) in height – the tallest flowering plants known; the bizarre baobab tree (*Adansonia digitata*) with its grotesquely swollen soft-wooded trunk; the enormous banyan tree (*Ficus benghalensis*) which sends down roots from its branches that extend the growth of the tree indefinitely; the moss-like aquatic herbs of the Podostemaceae; the majestic smooth-trunked palms of *Roystonea*, which appear to be made of concrete; or the remarkable parasite *Rafflesia*, whose vegetative body is reduced to a few filaments growing in the roots or stems of its host plant, but which produces gigantic fleshy flowers that stink of putrid flesh. It is this almost incredible diversity which it is the purpose of this book to describe.

The economic uses made of flowering plants are manifest, as can be seen in the brief summary given for each family. The total number of species of economic value is difficult to estimate. One standard reference work lists over 6,000 species which are known to be used in agriculture, forestry, fruit and vegetable growing and pharmacognosy. Some of these are used in world trade and some are of strictly local value such as those which are of purely ethnological interest in producing local foods, medicines and other commodities of value in the daily life of native communities. In addition there are more than twice as many more species grown as purely ornamental garden plants.

Adopting a more strict definition of plants that fit into Man's economic activity, the numbers can be reduced to 1,000–2,000, and of these relatively few, about 100–200, are of major importance in world trade, while only 15 provide the bulk of the world's food crops – rice, wheat, corn, sorghum, barley, sugar cane and sugar beet, potato, sweet potato, cassava, beans, soya beans and peanut, coconut and banana. When one considers the thousands of species of flowering plants that can be exploited for human benefit, it comes as a surprise to note how incredibly restricted is the range of staple crops. Attempts are being made to see if this range can be extended but there are many technical, agronomic and sociological problems involved. The range of minor, often local crops remains vast and indeed there is a tendency for these to become more widely known as a result of travel, advances in food packaging and processing and improved transport facilities. There is renewed interest today in seeking new sources of oils, fibers, drugs and medicines from plants, and modern chemical screening techniques encourage rapid surveys of potentially valuable species. Modern medicine and chemotherapy have made enormous progress through the introduction of new plant-based drugs. It may well be that countless surprises are in store for

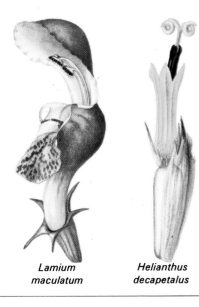

ADVANCED

Parts few

Parts definite in number

Parts fused

Whorled arrangement of parts

Loss of sepals, petals, stamens, or carpels

Unisexual

Ovary inferior

Symmetry irregular (zygomorphic)

*Lamium maculatum*          *Helianthus decapetalus*

us. The richest source of valuable new plants is, however, in the tropics and it is precisely there where plant communities are being destroyed.

## Classification of Flowering Plants

There are about 250,000 species of flowering plants. Each species when recognized, is given a formal scientific name in Latin, and the application of these names is governed by an international code of nomenclature. These species, the basic units of classification, are in turn grouped into higher or more-inclusive units which again are laid down by the international code. The main units are shown in the accompanying table.

The flowering plants (or Angiospermae) are variously treated as constituting a single division, sub-division or class in modern systems of classification. They are subdivided into two subclasses (or classes) which in turn are further subdivided into orders and families. It is the family which is the most frequently employed of these higher categories, both by botanists and by laymen and which constitute the basis of this volume.

Because of their distinctive features some families accepted today such as the palms (Palmae or Arecaceae), grasses (Gramineae or Poaceae) and umbels (Umbelliferae or Apiaceae) were recognized long before the family was formally accepted as a taxonomic category. Indeed the Umbelliferae was the first family to receive general recognition and members of it were known to several prehistoric peoples. The Greek scientist Theophrastus (about 372–287 BC) recognized the Umbelliferae as one of several natural plant families, and several umbellifers were included in the classic Chinese *Materia Medica* compiled in the late Han dynasty about the second or third century BC and apparently based on an earlier oral tradition.

The numbers of families recognized in recent systems of classification are relatively small, ranging from 179 in the British Bentham and Hooker's treatment of the flowering plants published last century (1862–1883) to the 354 recognized by the American Cronquist in his 1968 system, 321 by the American Thorne in his 1976 phylogenetic classification of the Angiospermae and 418 by the Russian Takhtajan in his 1969 book. These discrepancies reflect differences of opinion about the status of groups. For example, some authors recognize the three major components of the Leguminosae as either subfamilies (Papilionoideae, Caesalpinioideae and Mimosoideae) or as separate families (Papilionaceae, Caesalpiniaceae and Mimosaceae). Different viewpoints also reflect the ability to distinguish particular groups, and some authors prefer either larger or smaller groupings as better indicating their interpretations of evolutionary status, especially the derivation of all members from a common origin. Numerous examples of such differences of opinion will be noted in the discussion of classification and relationships given in the family treatments in this volume.

The flowering plant families vary very widely in their content, from those containing only a single genus and species (monotypic) as in the Trochodendraceae, Brunoniaceae and Adoxaceae – to the vast polytypic

| CLASSIFICATION OF THE BEAN, SHOWING THE HIERARCHY OF CATEGORIES AND THEIR NAMES | | |
|---|---|---|
| CATEGORY | SCIENTIFIC NAME OF TAXONOMIC GROUPS (TAXA) | VERNACULAR NAME |
| Class | Angiospermae | Angiosperms; Flowering Plants |
| Subclass | Dicotyledoneae | Dicotyledons, Dicots |
| Superorder | Ros**idae** | Rose Superorder |
| Order | Fab**ales** | Legume Order |
| Family | Fab**aceae** (Leguminosae) | Legumes; Legume Family |
| Subfamily | Papilion**oideae** | Pea Subfamily |
| Tribe | Phaseol**eae** | Bean Tribe |
| Subtribe | Phaseol**inae** | Bean Subtribe |
| Genus | *Phaseolus* | Bean |
| Species | *Phaseolus vulgaris* | French or Kidney Bean |
| Variety | *Phaseolus vulgaris* var *humilis* | Bush Bean |
| Cultivar | *Phaseolus vulgaris* 'Canadian Wonder' | |
| Endings in **heavy type** are standardized indications of rank | | |

| | Dicotyledons | Monocotyledons | |
|---|---|---|---|
|  | EMBRYO: Two cotyledons (seed leaves) present; endosperm present or lacking in the seed. | EMBRYO: One cotyledon (seed leaf), present; endosperm frequently present in the seed. | 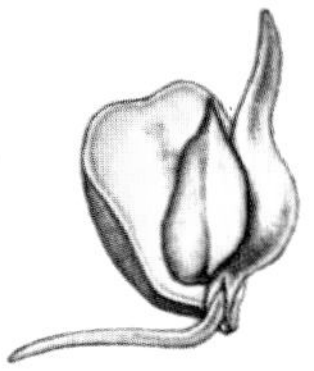 |
| 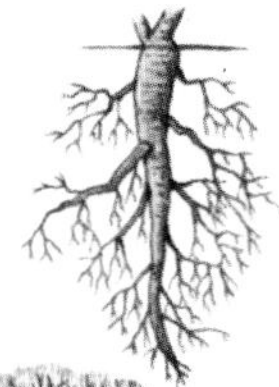 | ROOTS: The primary root often persists and becomes a strong taproot, with smaller secondary roots. | ROOTS: The primary root is of short duration and is soon replaced by adventitious roots, which form a fibrous root system or sometimes a bundle of fleshy roots. | 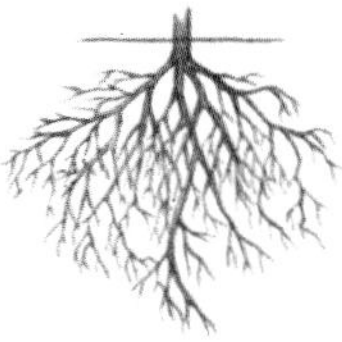 |
|  | GROWTH FORM: May be either herbaceous or woody. | GROWTH FORM: Mostly herbaceous; a few are arborescent (tree-like). |  |
| 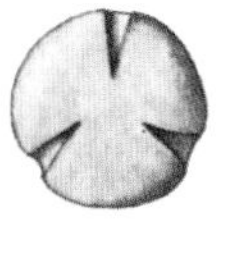 | POLLEN: Basically tricolpate (having three furrows or pores). | POLLEN: Basically monocolpate (having one furrow or pore). | 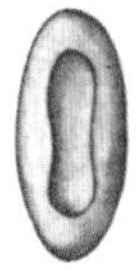 |
| 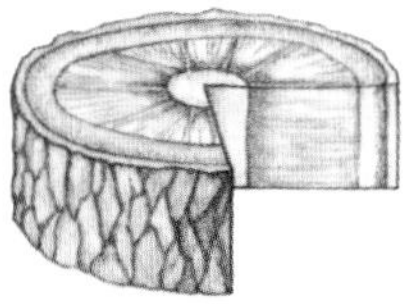 | VASCULAR SYSTEM: Usually consists of a ring of primary bundles with a cambium, and secondary growth in diameter of the stem; stem differentiated into cortex and stele. | VASCULAR SYSTEM: Consists of numerous scattered bundles, without definite arrangement and in a ground parenchyma; cambium only exceptionally present; no differentiation into cortical and stelar regions in stems. | 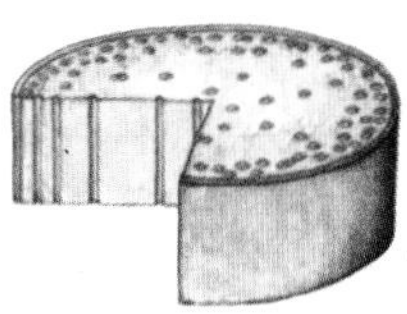 |
|  | LEAVES: Usually net-veined (pinnate or palmate), usually broad in shape and seldom sheathing at the base; petiole (stalk) commonly developed and often bearing stipules. | LEAVES: Usually parallel-veined, commonly oblong or linear in shape and often sheathing at the base; petiole (stalk) seldom developed and stipules absent. | 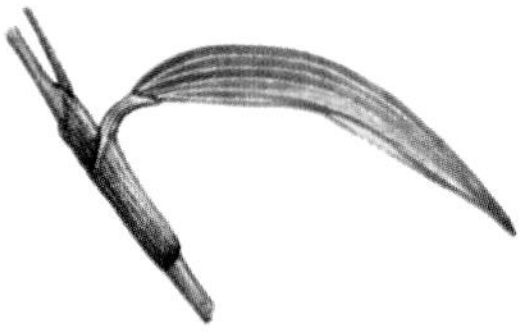 |
| 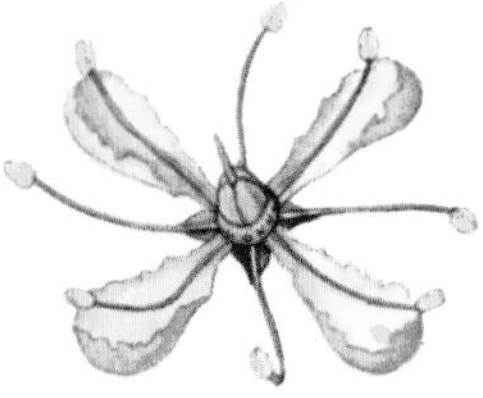 | FLOWERS: Parts are usually in fours or fives. | FLOWERS: Parts are usually in threes or multiples of three. |  |

Plate II. Comparison of features diagnostic of Dicotyledons and Monocotyledons.

assemblages containing hundreds of genera and thousands of species such as the Compositae, Euphorbiaceae and Rubiaceae.

Whilst there is a basic agreement between different authors as to the 200 or so "core" families to be recognized, there is no such consensus as regards the next higher level, the order, and the different systems of classification differ widely not only in the number of orders (and even superorders) recognized but also in their names and content. Thus the same order may appear under different names in different systems, and different families may be included in the same order and the same family in different orders in different systems! Fortunately, orders are not widely used except in formal presentations of classification systems. However they are occasionally referred to informally. Many botanists favor the use of informal groupings above the family level.

There is general acceptance that there are two subclasses of flowering plants (dicotyledons and monocotyledons) which differ from each other on a number of characters (see Plate II).

Current systems of classification of the flowering plants are concerned not so much with which group should be recognized and at which rank but with the relationships between the different families and how these relationships might best be expressed in a scheme or arrangement. The relationships expressed are intended to be a reflection of the phylogenetic or evolutionary status and origins of the families. In the absence of a satisfactory fossil record, such evolutionary relationships have to be deduced from a comparison of the present-day representatives of the families and from attempts to extrapolate backwards what their past evolutionary history might have been. Various principles have been suggested for evolutionary trends in characters and for the relative degrees of advancement or primitiveness. On the basis of such arguments conclusions are drawn as to the evolutionary states and position of the families possessing such characters. Unfortunately, while there is general agreement on some of these principles, there is much dispute as to their application and much depends on which characters are regarded as having greater phyletic importance. Moreover, there is a steady stream of new information published every year, from fields such as anatomy, embryology, palynology (the study of pollen), chemistry and morphology, all of which has to be assessed and interpreted.

Because of the complex nature of the task and the high degree of subjectivity involved, no fully documented, consistent system of classification has yet been produced. On the contrary, nearly every recent system has been published with quite inadequate documentation. In addition there are the technical problems of attempting to present multi-dimensional relationships in a two-dimensional arrangement. Earlier phylogenetic systems were often presented in the form of a tree but more recently the metaphor of a phylogenetic shrub, with numerous main stems and branches rising through geological time to the present, is gaining popularity. More realistic are two-dimensional transections of imaginary phylogenetic trees whose stems and branches are unknown, indicating in some cases the degree of specialization of the branches by their radial distance from the center of the diagram which is the position of the extinct ancestral complex. Such diagrams (Plate III) have the advantage that they can be interpreted as essentially a set of overall relationships without any necessary evolutionary implications.

It would have been inappropriate to present a new system here although some system had to be followed for practical purposes. We have, therefore, followed the sequence of families and order given by G. L. Stebbins in his *Flowering Plants – Evolution Above the Species Level* (1974) because, though based largely on Cronquist's *The Evolution and Classification of Flowering Plants* (1968), it combines, to some extent, views from several of the current systems. It is used here mainly for convenience and its adoption in no way implies acceptance of its correctness or otherwise. Indeed it is likely that future systems may be radically different. In some cases we have departed from the system by including small segregate families in their nearest "parent" family.

The family accounts have been written by a team of scientists who are specialists in the field of plant taxonomy, and who in many cases have spent considerable research time on the families they describe. Because of this, the individual authors have been free to express their own comments on the relationships of the families they have contributed, within the restraints of the nomenclature of the orders and families used in the Stebbins' system. I have occasionally added further suggested relationships from recent literature. The overall classification may be seen by referring to the Classification section of this book (pages 14–15).

It was always our aim to reach as wide a public as possible. The difficulty with Botany is that, far more than Zoology, there exists a daunting array of unfamiliar technical terms which are essential to a precise description. Most of these cannot, and should not be avoided and to help with the understanding of these, a fully illustrated glossary is included at the end of this Introduction. To further improve the readability and accessibility of the

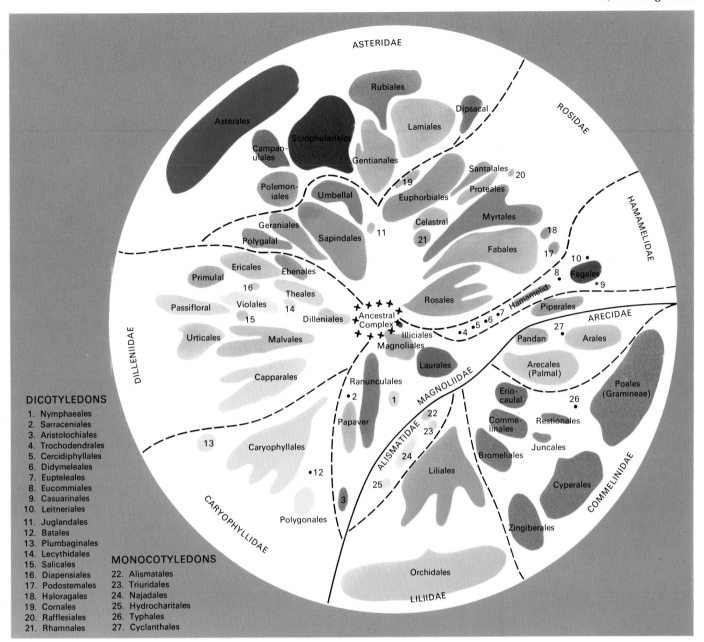

ASTERIDAE

ROSIDAE

HAMAMELIDAE

ARECIDAE

COMMELINIDAE

LILIIDAE

CARYOPHYLLIDAE

DILLENIIDAE

MAGNOLIIDAE

ALISMATIDAE

Rubiales
Dipsacal
Lamiales
Scrophulariales
Campan-uiales
Asterales
Gentianales
Santalales 20
Proteales
Polemon-iales
Umbellal
19
Euphorbiales
Myrtales
Geraniales
Sapindales
11
Celastral
21
18
Polygalal
Fabales
17
Ericales
Ebenales
Primulal
16
Theales
Rosales
10
8
Fagales
9
Violales
14
Passifloral
15
Dilleniales
Hamamelid
Piperales
Urticales
Malvales
Illiciales
Magnoliales
Ancestral Complex
4 5 6 7
27
Pandan
Arales
Capparales
Laurales
Arecales (Palmal)
Ranunculales
2
1
Erio-caulal
26
Restionales
Papaver
22
Commel-inales
13
Caryophyllales
23
24
Juncales
12
Liliales
Bromeliales
Cyperales
3
25
Zingiberales
Polygonales
Orchidales

**DICOTYLEDONS**

1. Nymphaeales
2. Sarraceniales
3. Aristolochiales
4. Trochodendrales
5. Cercidiphyllales
6. Didymeleales
7. Eupteleales
8. Eucommiales
9. Casuarinales
10. Leitneriales
11. Juglandales
12. Batales
13. Plumbaginales
14. Lecythidales
15. Salicales
16. Diapensiales
17. Podostemales
18. Haloragales
19. Cornales
20. Rafflesiales
21. Rhamnales

**MONOCOTYLEDONS**

22. Alismatales
23. Triuridales
24. Najadales
25. Hydrocharitales
26. Typhales
27. Cyclanthales

Plate III. Evolutionary diagram showing the relative degree of specialization of the orders of Angiosperms (after Stebbins, 1974).

information we have used popular name subtitles wherever it was practicable to do so. Family names have been treated as singular nouns throughout, and distribution maps and information summary panels have been included for every family. There can be no doubt about the uniqueness and beauty of the illustrations. Each subject has been carefully selected by the three highly experienced botanical artists so that within each panel the main features of the family concerned are illustrated. The captions indicate many of the features of diagnostic interest, but the panels should also be studied in close conjunction with the text. Most panels illustrate a single family, but exceptionally, where families were particularly closely related, we have illustrated two together.

Finally it is a pleasure to acknowledge the enormous help the publishers and I have received from the Director and Staff of the Royal Botanic Garden, Kew.

V. H. Heywood

# CLASSIFICATION

## DICOTYLEDONS

### MAGNOLIIDAE

**Magnoliales**
Magnoliaceae
*Magnolias and Tulip Tree*
Winteraceae
*Winter's Bark*
Himantandraceae
Canellaceae
*White Cinnamon*
Annonaceae
*Sweetsop and Soursop*
Myristicaceae
*Nutmeg and Mace*
Degeneriaceae

**Illiciales**
Illiciaceae
Schisandraceae

**Laurales**
Austrobaileyaceae
Lactoridaceae
Eupomatiaceae
Gomortegaceae
Monimiaceae
Calycanthaceae
*Carolina Allspice and Chimonanthus*
Chloranthaceae
Lauraceae
*Avocado, Bay Laurel and Cinnamon*
Hernandiaceae

**Piperales**
Piperaceae
*Pepper and Kava*
Saururaceae
Peperomiaceae
*Pepper Elders*

**Aristolochiales**
Aristolochiaceae
*Aristolochias and Asarums*
Nepenthaceae
*Pitcher Plants*

**Nymphaeales**
Ceratophyllaceae
Nymphaeaceae
*Water Lilies*

**Ranunculales**
Berberidaceae
*Barberry, Sacred Bamboo and May Apple*
Ranunculaceae
*The Buttercup Family*
Lardizabalaceae
*Akebias and Holboellias*
Menispermaceae
*Curare*

**Papaverales**
Papaveraceae
*Poppies*
Fumariaceae
*Dicentras and Fumitory*

**Sarraceniales**
Sarraceniaceae
*Pitcher Plants*

### HAMAMELIDAE

**Trochodendrales**
Trochodendraceae

**Hamamelidales**
Cercidiphyllaceae
Platanaceae
*Plane and Buttonwood Trees*
Hamamelidaceae
*Witch Hazels and Sweet Gums*

**Eucommiales**
Eucommiaceae

**Leitneriales**
Leitneriaceae
*Florida Corkwood*

**Myricales**
Myricaceae

**Fagales**
Betulaceae
*Birches, Alders and Hornbeams*
Fagaceae
*Beeches, Oaks and Sweet Chestnuts*
Balanopaceae

**Casuarinales**
Casuarinaceae
*She Oak and River Oak*

### CARYOPHYLLIDAE

**Caryophyllales**
Cactaceae
*The Cactus Family*
Aizoaceae
*The Mesembryanthemum Family*
Caryophyllaceae
*The Carnation Family*
Nyctaginaceae
*Bougainvilleas*
Amaranthaceae
*Cockscombs and Celosias*
Phytolaccaceae
*Pokeweed and Bloodberry*
Chenopodiaceae
*Sugar Beet, Beetroot and Spinach*
Didiereaceae
Portulacaceae
*Lewisias and Portulacas*
Basellaceae
*Madeira Vine*

**Batales**
Batidaceae
*Saltwort*

**Polygonales**
Polygonaceae
*Buckwheats, Rhubarb and Sorrels*

**Plumbaginales**
Plumbaginaceae
*Sea Lavender and Thrift*

### DILLENIIDAE

**Dilleniales**
Dilleniaceae
*Dillenias and Hibbertias*
Paeoniaceae
*Peonies*
Crossosomataceae

**Theales**
Theaceae
*Tea, Camellias and Franklinia*
Ochnaceae
*African Oak and Ochnas*
Dipterocarpaceae
Guttiferae
*Mangosteen and Mammey Apple*
Elatinaceae
Quiinaceae
Marcgraviaceae

**Malvales**
Scytopetalaceae
Elaeocarpaceae
*Crinodendrons and Aristotelias*
Tiliaceae
*Limes, Lindens, Basswoods and Jute*
Sterculiaceae
*Cocoa and Kola*
Bombacaceae
*Baobab and Balsa*
Malvaceae
*Cotton, Mallows and Hollyhocks*
Sphaerosepalaceae
Sarcolaenaceae

**Urticales**
Ulmaceae
*Elms and Hackberries*
Moraceae
*Fig, Hemp and Mulberries*
Urticaceae
*Stinging Nettles and Ramie Fiber*

**Lecythidales**
Lecythidaceae
*Brazil Nuts and Cannonball Tree*

**Violales**
Violaceae
*Violets and Pansies*
Flacourtiaceae
*Chaulmoogra Oil and West Indian Boxwood*
Lacistemataceae
Passifloraceae
*Passion Flowers and Granadillas*
Turneraceae
Malesherbiaceae
Fouquieriaceae
*Ocotillo*
Caricaceae
*Papaw*
Bixaceae
*Annatto*
Cochlospermaceae
*Rose Imperial*
Cistaceae
*Rockroses*
Tamaricaceae
*Tamarisks*
Ancistrocladaceae
Frankeniaceae
Achariaceae
Begoniaceae
*Begonias*
Loasaceae
*Blazing Star and Rock Nettle*
Datiscaceae
*Datiscas*
Cucurbitaceae
*The Gourd or Pumpkin Family*

**Salicales**
Salicaceae
*Aspens, Poplars and Willows*

**Capparales**
Capparaceae
*Capers*
Tovariaceae
Cruciferae
*The Mustard Family*
Resedaceae
*Mignonette*
Moringaceae

**Ericales**
Clethraceae
*Lily-of-the-valley Tree*
Grubbiaceae
Cyrillaceae
*Leatherwood and Buckwheat Trees*
Ericaceae
*The Heath Family*
Epacridaceae
Empetraceae
*Crowberries*
Pyrolaceae
*Wintergreens*

**Diapensiales**
Diapensiaceae
*Shortia and Galax*

**Ebenales**
Sapotaceae
*Chicle, Gutta-percha and Sapodilla*
Ebenaceae
*Persimmons and Ebonies*
Styracaceae
*Silverbell and Snowbell Trees*

**Primulales**
Primulaceae
*The Primrose Family*
Myrsinaceae

### ROSIDAE

**Rosales**
Cunoniaceae
Pittosporaceae
*Parchment-bark*
Droseraceae
*Sundews and Venus' Fly Trap*
Brunelliaceae
Eucryphiaceae
*Eucryphias*
Bruniaceae
Rosaceae
*The Rose Family*
Crassulaceae
*Stonecrops and Houseleeks*
Cephalotaceae
*Flycatcher Plant*
Saxifragaceae
*Currants, Hydrangeas and Saxifrages*
Chrysobalanaceae
*Coco Plum*

**Fabales**
Leguminosae
*The Pea Family*

**Podostemales**
Podostemaceae

**Haloragales**
Theligonaceae
Haloragaceae
*Gunneras, Water Milfoil and Haloragis*
Hippuridaceae
*Mare's Tail*

**Myrtales**
Sonneratiaceae
Trapaceae
*Water Chestnuts*
Lythraceae
*Cupheas, Lythrums and Henna*
Rhizophoraceae
*Mangroves*
Penaeaceae
Thymelaeaceae
Myrtaceae
*Myrtles, Eucalyptus and Cloves*
Punicaceae
*Pomegranate*
Onagraceae
*Clarkias, Fuchsias and Evening Primroses*
Oliniaceae
Melastomataceae
*Dissotis and Medinilla*
Combretaceae
*Terminalia, Combretum and Quisqualis*

**Cornales**
Nyssaceae
*Handkerchief and Tupelo Trees*
Garryaceae
Alangiaceae
Cornaceae
*Dogwoods*

**Proteales**
Elaeagnaceae
*Oleaster and Sea Buckthorn*
Proteaceae
*Proteas, Banksias and Grevilleas*

**Santalales**
Santalaceae
*Sandalwood*
Medusandraceae
Olacaceae
*American Hog Plum*
Loranthaceae
*Mistletoes*
Misodendraceae
Cynomoriaceae
Balanophoraceae

**Rafflesiales**
Rafflesiaceae
Hydnoraceae

**Celastrales**
Geissolomataceae
Celastraceae
*Spindle Tree*
Stackhousiaceae
Salvadoraceae
Corynocarpaceae
Icacinaceae
Aquifoliaceae
*Hollies and Yerba Maté*
Dichapetalaceae

**Euphorbiales**
Buxaceae
*Box*
Pandaceae
Euphorbiaceae
*The Spurge Family*

**Rhamnales**
Rhamnaceae
*Buckthorn and Jujube*
Vitaceae
*Grapevine and Virginia Creeper*

**Sapindales**
Staphyleaceae
*Bladder Nuts*
Melianthaceae
Connaraceae
*Zebra Wood*
Sapindaceae
*Akee, Litchi and Rambutan*
Sabiaceae
Julianiaceae
Hippocastanaceae
*Horse Chestnuts and Buckeyes*
Aceraceae
*Maples*
Burseraceae
*Frankincense and Myrrh*
Anacardiaceae
*Cashew, Mango, Sumacs and Poison Ivy*
Simaroubaceae
*Quassia and Tree of Heaven*
Coriariaceae
*Coriarias*
Meliaceae
*Mahoganies*
Cneoraceae
*Spurge Olive*
Rutaceae
*The Citrus Fruit Family*

Zygophyllaceae
*Lignum Vitae*

**Juglandales**
Juglandaceae
*Walnuts, Hickories and Pecan Nuts*

**Geraniales**
Houmiriaceae
*Bastard Bullet Tree*
Linaceae
*Flax and Linseed*
Geraniaceae
*Geraniums and Pelargoniums*
Oxalidaceae
*Wood Sorrel and Bermuda Buttercup*
Erythroxylaceae
*Coca/Cocaine*
Limnanthaceae
*Poached Egg Flower*
Balsaminaceae
*Balsams*
Tropaeolaceae
*Nasturtiums and Canary Creeper*

**Polygalales**
Malpighiaceae
Trigoniaceae
Tremandraceae
Vochysiaceae
Polygalaceae
Krameriaceae

**Umbellales**
Araliaceae
*Ivies and Ginseng*
Umbelliferae
*The Carrot Family*

---

**ASTERIDAE**

**Gentianales**
Loganiaceae
*Buddleias and Strychnine*
Gentianaceae
*Gentians*
Apocynaceae
*Periwinkles and Oleanders*
Asclepiadaceae
*Milkweeds and Wax Plant*
Oleaceae
*Olive, Ashes and Lilacs*

**Polemoniales**
Nolanaceae
Solanaceae
*The Potato Family*
Convolvulaceae
*Bindweeds, Morning Glory and Sweet Potato*
Menyanthaceae
*Bogbean*
Lennoaceae
Polemoniaceae
*Phlox*
Ehretiaceae
Hydrophyllaceae
Boraginaceae
*Forget-me-not and Alkanet*

**Lamiales**
Verbenaceae
*Teaks and Verbenas*
Labiatae
*The Mint Family*
Tetrachondraceae
Callitrichaceae
Phrymaceae

**Plantaginales**
Plantaginaceae
*Plantains*

**Scrophulariales**
Columelliaceae
Myoporaceae
*Emu Bushes*
Scrophulariaceae
*The Foxglove Family*
Globulariaceae
Gesneriaceae
*African Violets and Gloxinias*
Orobanchaceae
*Broom rapes and Toothworts*
Bignoniaceae
*Catalpa*
Acanthaceae
*Black-eyed Susan and Sea Holly*
Pedaliaceae
*Sesame*
Hydrostachydaceae
Martyniaceae
*Unicorn Plant*
Lentibulariaceae
*Bladderworts and Butterworts*

**Campanulales**
Campanulaceae
*The Bellflower Family*
Lobeliaceae
*Lobelias*
Stylidiaceae
*Trigger Plant*
Brunoniaceae
Goodeniaceae
*Leschenaultias and Scaevolas*

**Rubiales**
Rubiaceae
*Gardenias, Coffee and Quinine*

**Dipsacales**
Adoxaceae
Caprifoliaceae
*Elders and Honeysuckles*
Valerianaceae
*Spikenard and Valerians*
Dipsacaceae
*Teasel and Scabious*
Calyceraceae

**Asterales**
Compositae
*The Sunflower Family*

# MONO-COTYLEDONS

**ALISMATIDAE**

**Alismatales**
Butomaceae
*Flowering Rush*
Limnocharitaceae
*Water Poppy*
Alismataceae
*Water Plantains*

**Hydrocharitales**
Hydrocharitaceae
*Canadian Waterweed and Frog's Bit*

**Najadales**
Aponogetonaceae
*Water Hawthorn*
Scheuchzeriaceae
*Arrowgrass*
Juncaginaceae
Lilaeaceae
Najadaceae
Potamogetonaceae
*Pondweeds*
Zannichelliaceae
*Horned Pondweed*
Ruppiaceae
*Ditch Grasses*
Zosteraceae
*Eel Grasses*
Posidoniaceae
Cymodoceaceae

**Triuridales**
Triuridaceae

---

**COMMELINIDAE**

**Commelinales**
Xyridaceae
*Yellow-eyed Grasses*
Rapateaceae
Mayacaceae
Commelinaceae
*The Spiderwort Family*

**Eriocaulales**
Eriocaulaceae

**Restionales**
Flagellariaceae
Centrolepidaceae
Restionaceae

**Poales**
Gramineae
*The Grass Family*

**Juncales**
Juncaceae
*Rushes*
Thurniaceae

**Cyperales**
Cyperaceae
*Reeds and Sedges*

**Typhales**
Typhaceae
*Reedmace Bulrush and Cattails*

Sparganiaceae
*Bur-reeds*

**Bromeliales**
Bromeliaceae
*The Pineapple Family*

**Zingiberales**
Musaceae
*Bananas and Manila Hemp*
Strelitzaceae
*Bird of Paradise Flower*
Zingiberaceae
*Ginger, Cardamom and Turmeric*
Cannaceae
*Queensland Arrowroot*
Marantaceae
*Arrowroot*

---

**ARECIDAE**

**Arecales**
Palmae
*The Palm Family*

**Cyclanthales**
Cyclanthaceae
*Panama Hat Plant*

**Pandanales**
Pandanaceae
*Screw Pines*

**Arales**
Lemnaceae
*Duckweeds*
Araceae
*The Aroids*

---

**LILIIDAE**

**Liliales**
Pontederiaceae
*Water Hyacinth and Pickerel Weed*
Philydraceae
Iridaceae
*The Iris Family*
Liliaceae
*The Lily Family*
Amaryllidaceae
*The Daffodil Family*
Agavaceae
*Sisal Hemp, Pulque and Dragon Tree*
Xanthorrhoeaceae
*Grass Trees*
Velloziaceae
Haemodoraceae
*Kangaroo Paw*
Taccaceae
*East Indian and Hawaiian Arrowroot*
Stemonaceae
Cyanastraceae
Smilacaceae
*Smilax and Sarsaparilla*
Dioscoreaceae
*Yams*

**Orchidales**
Burmanniaceae
Orchidaceae
*The Orchid Family*

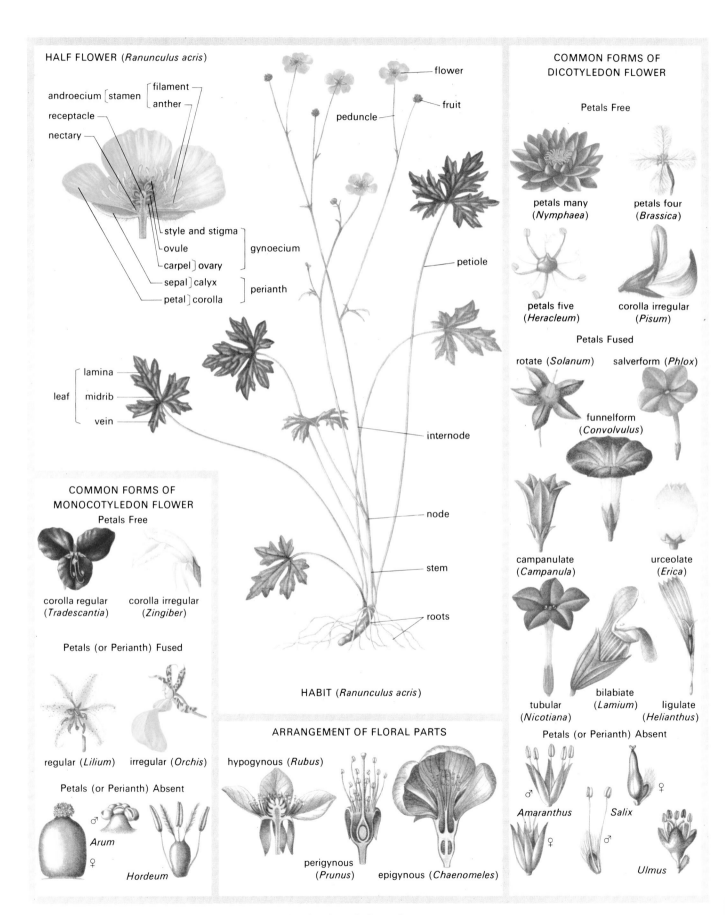

HALF FLOWER (*Ranunculus acris*)

androecium [stamen [filament / anther
receptacle
nectary
flower
fruit
peduncle

style and stigma
ovule
carpel ] ovary } gynoecium
sepal ] calyx
petal ] corolla } perianth

petiole

leaf [ lamina / midrib / vein

internode

node

stem

roots

HABIT (*Ranunculus acris*)

COMMON FORMS OF MONOCOTYLEDON FLOWER

Petals Free

corolla regular (*Tradescantia*)     corolla irregular (*Zingiber*)

Petals (or Perianth) Fused

regular (*Lilium*)     irregular (*Orchis*)

Petals (or Perianth) Absent

♂ *Arum* ♀

*Hordeum*

ARRANGEMENT OF FLORAL PARTS

hypogynous (*Rubus*)

perigynous (*Prunus*)     epigynous (*Chaenomeles*)

COMMON FORMS OF DICOTYLEDON FLOWER

Petals Free

petals many (*Nymphaea*)     petals four (*Brassica*)

petals five (*Heracleum*)     corolla irregular (*Pisum*)

Petals Fused

rotate (*Solanum*)     salverform (*Phlox*)

funnelform (*Convolvulus*)

campanulate (*Campanula*)     urceolate (*Erica*)

tubular (*Nicotiana*)     bilabiate (*Lamium*)     ligulate (*Helianthus*)

Petals (or Perianth) Absent

♂ *Amaranthus* ♀ *Salix*

♀ ♂

*Ulmus*

Plate IV. Structure of the flowering plant and flower, also showing the main flower forms.

# GLOSSARY

**Abaxial**  On the side facing away from the stem or axis.

**Acaulescent**  Stemless or nearly so.

**Achene**  A small, dry, single-seeded fruit that does not split open (Plate VI).

**Acuminate**  Narrowing gradually to a point (Plate IX).

**Acute**  Having a sharp point (Plate IX).

**Adaxial**  On the side facing the stem or axis.

**Adnate**  Joined to or attached to; applied to unlike organs, eg stamens adnate to the perianth; cf connate.

**Adventitious**  Arising from an unusual position, eg roots from a stem or leaf.

**Aerial root**  A root that originates above ground level.

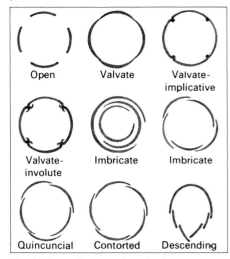

Plate V. The main types of aestivation.

**Aestivation**  The arrangement of the parts of a flower within the bud, usually referring to sepals and petals (Plate V).

**Alternate** (of leaves)  One leaf at each node of the stem (Plate IX); (of stamens) between the petals.

**Amphitropous** (of ovules)  Attached near its middle, half-inverted (Plate X).

**Anatropous** (of ovules)  Bent over through 180 degrees to lie alongside its stalk (funicle) (Plate X).

**Androecium**  All the male reproductive organs of a flower; the stamens; cf gynoecium (Plate XII).

**Androgynophore**  A column on which stamens and carpels are borne.

**Angiosperm**  A plant producing seeds enclosed in an ovary. A flowering plant.

**Annual**  A plant that completes its life cycle from germination to death within one year.

**Annular**  Ring-like.

**Anther**  The terminal part of the male organs (stamen), usually borne on a stalk (filament) and developing to contain pollen (Plate XII).

**Anthesis**  The period of flowering; from the opening of the flower bud to the setting of the seed.

**Anthocyanin**  The pigment usually responsible for pink, red, purple, violet and blue colors in flowering plants.

**Antipetalous**  Occurring opposite the petals, on the same radius, as distinct from alternating with the petals.

**Antisepalous**  Occurring opposite the sepals, on the same radius, as distinct from alternating with the sepals.

**Aperturate** (of pollen)  Having one or more apertures.

**Apetalous**  Without petals.

**Apex**  The tip of an organ; the growing point.

**Apical**  Pertaining to the apex.

**Apocarpous**  With carpels free from each other (Plate VII).

**Apomixis** (adj. apomictic)  Reproduction by seed formed without sexual fusion.

**Aquatic**  Living in water.

**Aril**  A fleshy or sometimes hairy outgrowth from the hilum or funicle of a seed.

**Asepalous**  Without sepals.

**Auricle** (adj. auriculate)  Small ear-like projections at the base of a leaf or leaf blade or bract (Plate IX).

**Awn**  A stiff, bristle-like extension to an organ, usually at the tip.

**Axil**  The upper angle formed by the union of a leaf with the stem (Plate IV).

**Axile placentation**  A type of placentation in which the ovules are borne on placentas on the central axis of an ovary that has two or more locules (Plate XI).

**Axillary**  Pertaining to the organs in the axil, eg the buds, flowers or inflorescence.

**Axis**  The main or central stem of a herbaceous plant or of an inflorescence.

**Baccate**  Berry-like.

**Basal**  Borne at or near the base.

**Basal placentation**  Having the placenta at the base of the ovary (Plate XI).

**Basifixed** (of anthers)  Attached at the base to the filament, and therefore lacking independent movement; cf dorsifixed (Plate XII).

**Berry**  A fleshy fruit without a stony layer, usually containing many seeds (Plate VI).

**Betalains**  Red and yellow alkaloid pigments present in members of the Caryophyllales.

**Bi-**  A prefix meaning two or twice.

**Bicarpellate** (of ovaries)  Derived from two carpels.

**Biennial**  A plant that completes its life cycle in more than one, but less than two years and which usually flowers in the second year.

**Bifid**  Forked; having a deep fissure near the center.

**Bilabiate**  Two-lipped (Plate IV).

**Bipinnate** (of leaves)  A pinnate leaf with the primary leaflets themselves divided in a pinnate manner; cf pinnate (Plate IX).

**Biseriate**  In two rows.

**Bisexual** (of flowers)  Containing both male and female reproductive organs in a single flower; cf unisexual (Plate IV).

**Blade**  The flattened part of a leaf; the lamina (Plate IX).

**Bostryx** (of inflorescences)  A cymose inflorescence with successive branches on one side only; normally coiled like a spring (Plate VIII).

**Bract**  A leaf, often modified or reduced, which subtends a flower or inflorescence in its axil (Plate VIII).

**Bracteole**  A small leaf-like organ, occurring along the length of a flower stalk, between a true subtending bract and the calyx (Plate VIII).

**Bulb**  An underground organ comprising a short disk-like stem, bearing fleshy scale leaves, buds and surrounded by protective scale leaves; it acts as a perennating organ and is a means of vegetative reproduction; cf corm, tuber.

**Bulbil**  A small bulb or bulb-like organ often produced on above-ground organs.

**Caducous**  Falling off prematurely or easily.

**Calcicole**  A plant that favors soil containing lime.

**Calcifuge**  A plant that avoids soil containing lime.

**Calyculus**  A group of leaf-like appendages below the calyx.

**Calyx**  Collective term for all the sepals of a flower (Plate IV).

**Cambium**  A layer of cells that occurs within the stem and roots which divides to form secondary permanent tissues.

**Campanulate**  Bell-shaped (Plate IV).

**Campylotropous** (of ovules)  Bent over through 90 degrees so that the stalk (funicle) appears to be attached to the side of the ovule (Plate X).

**Capitate**  Head-like.

**Capitulum**  An inflorescence consisting of a head of closely packed stalkless flowers (Plate VIII).

**Capsule**  A dry fruit which normally splits open to release its seeds (Plate VI).

**Carnivorous plant**  A plant that is capable of catching and digesting small animals such as insects.

**Carpel**  One of the flower's female reproductive organs, comprising an ovary and a stigma, and containing one or more ovules (Plate IV).

**Caruncle** (adj. carunculate)  A fleshy, sometimes colored, out-growth near the hilum of some seeds.

**Caryopsis**  A dry fruit (achene) typical of grasses (Plate VI).

**Catkin**  A pendulous inflorescence of simple, usually unisexual flowers (Plate VIII).

**Ciliate** (of margins) Fringed with small hairs (Plate IX).

**Cincinnus** A monochasial, cymose inflorescence with branches alternating from one side of the vertical axis to the other; normally curved to one side (Plate VIII).

**Circumscissile** Opening all round by a transverse split (Plate VI).

**Cladode** A flattened stem which has assumed the form and function of a leaf.

**Claw** The narrow basal part of some petals and sepals.

**Cleistogamic (cleistogamous)** (of flowers) Self-pollinating, without the flower ever opening.

**Clone** A group of plants that have arisen by vegetative reproduction from a single parent, and which therefore all have identical genetic material.

**Colpate** (of pollen) Having one or more colpi (oblong-elliptic apertures in the pollen-wall).

**Column** (of a flower) The combined style and stigma, typically of orchids.

**Compound** Consisting of several parts, eg a leaf with several leaflets or an inflorescence with more than one group of flowers (Plates VIII, IX).

**Connate** Joined or attached to; applied to similar organs fused during development, eg stamens fused into a tube; cf adnate.

**Connective** (of stamens) The tissue connecting the pollen sacs of an anther (Plate XII).

**Contorted** (of sepals and petals) Twisted in the bud so that they overlap on one side only; spirally twisted.

**Convolute** Rolled together.

**Cordate** (of leaves) Heart-shaped (Plate IX).

**Coriaceous** Leathery.

**Corm** A bulbous, swollen, underground, stem base, bearing scale leaves and adventitious roots; it acts as a perennating organ and is a means of vegetative propagation; cf bulb, tuber.

**Corolla** All the petals of a flower; it is normally colored (Plate IV).

**Corona** A series of petal-like structures in a flower, either outgrowths from the petals, or modified from the stamens, eg a daffodil "trumpet".

**Corymb** A rounded or flat-topped inflorescence of racemose type, in which the lower (outer) flower stalks (pedicels) are longer than the upper (inner) ones, so that all the flowers are at about the same level (Plate VIII).

**Corymbose** Arranged in a corymb; corymb-like.

**Cotyledon** The first leaf, or pair of leaves of an embryo within the seed. See Dicotyledon, Monocotyledon.

**Crenate** (of leaf margins) Round-toothed (Plate IX).

**Crenulate** Finely crenate (Plate IX).

**Cross-fertilization, cross-pollination** See Fertilization, Pollination.

**Cupule** (adj. cupulate) A cup-shaped sheath, surrounding some fruits.

**Cyme** An inflorescence in which each terminal growing point produces a flower. Subsequent growth is therefore from a lateral growing point, the oldest flowers being at the apex, or center, if flat (Plate VIII).

**Cymose** Arranged in a cyme; cyme-like (Plate VIII).

**Cypsela** A single-seeded fruit derived from a unilocular, inferior ovary (Plate VI).

**Cystolith** A crystal or deposit of lime, within a cell.

**Deciduous** The shedding of leaves seasonally.

**Declinate** (of stamens) Curving downwards.

**Decussate** (of leaves) Arranged in opposite pairs on the stem, with each pair at 90 degrees to the preceding pair (Plate IX).

**Dehiscence** The method or act of opening or splitting (Plates VI, XII).

**Dehiscent** Opening to shed pollen or seeds.

**Dehiscing** In the act of shedding pollen or seeds.

**Dentate** Having a toothed margin (Plate IX).

**Denticulate** Having a finely toothed margin (Plate IX).

**Derived** Originating from an earlier form or group.

**Di-** A prefix meaning two.

**Dichasium** (of inflorescences) A form of cymose inflorescence with each branch giving rise to two other branches; cf monochasium (Plate VIII).

**Dicotyledon** One of two subclasses of angiosperms; a plant whose embryo has two cotyledons; cf monocotyledon (Plate II).

**Didymous** In pairs.

**Didynamous** Having two stamens longer than others (Plate XII).

**Dimorphism** (adj. dimorphic) Having two distinct forms.

**Dioecious** Having male and female flowers borne on separate plants.

**Disk** The fleshy outgrowth developed from the receptacle at the base of the ovary or from the stamens surrounding the ovary; it often secretes nectar.

**Distichous** Arranged in two vertical rows (Plate IX).

**Dorsal** Upper.

**Dorsifixed** (of anthers) Attached at the back to the filament; cf basifixed (Plate XII).

**Drepanium** (of inflorescences) A cymose inflorescence with successive branches on one side only; normally flattened in one plane and curved to one side (Plate VIII).

**Drupe** A fleshy fruit containing one or more seeds, each of which is surrounded by a stony layer (Plate VI).

**Elaiosome** A fleshy outgrowth on a seed with oily substances attractive to ants.

**Elliptic** (of leaves) Oval-shaped, with narrowed ends (Plate IX).

**Embryo** The rudimentary plant within the seed.

**Embryo sac** The central portion of the ovule; a thin-walled sac within the nucellus containing the egg nucleus (Plate X).

**Endocarp** The innermost layer of the ovary wall (pericarp) of a fruit. In some fruits it becomes hard and "stony"; cf drupe (Plate VI).

**Endosperm** Fleshy tissue, containing stored nutritive material, found in some seeds (Plate VI).

**Entire** (of leaves) With an undivided margin (Plate IX).

**Epicalyx** A whorl of sepal-like appendages resembling the calyx but outside the true calyx.

**Epidermis** Usually a single layer of living cells forming the protective covering to many plant organs, particularly leaves,

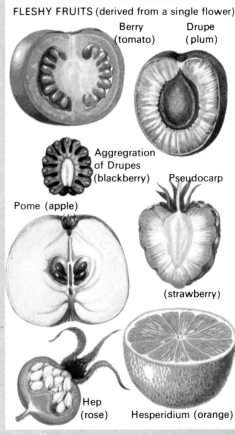

FLESHY FRUITS (derived from a single flower)

Berry (tomato)

Drupe (plum)

Aggregation of Drupes (blackberry)

Pseudocarp

Pome (apple)

(strawberry)

Hep (rose)

Hesperidium (orange)

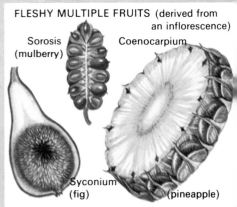

FLESHY MULTIPLE FRUITS (derived from an inflorescence)

Sorosis (mulberry)

Coenocarpium

Syconium (fig)

(pineapple)

petals, sepals and herbaceous stems.

**Epigynous** (of flowers) With the sepals, petals and stamens inserted near the top of the ovary (Plate IV).

**Epipetalous** Attached to the petals or corolla (Plate XII).

**Epiphyte** A plant that grows on the surface of another, without deriving food from its host.

**Erect** (of an ovule) Upright, with its stalk at the base (Plate X).

**Exine** The outer layer of the wall of a pollen grain.

**Exocarp** The outermost layer of the fruit wall (Plate VI).

**Exserted** Protruding.

**Exstipulate** Without stipules.

**Extrorse** (of anthers) Opening away from

the axis of growth towards the corolla (Plate XII).

**Fascicle** A cluster or bundle.

**Female flower** A flower containing functional carpels, but not stamens (Plate IV).

**Fertilization** The fusion of male and female reproductive cells (gametes) in the ovary after pollination. *Cross-fertilization* occurs between flowers from separate plants; *self-fertilization* occurs between flowers on the same plant or within the same flower.

**Filament** The anther-bearing stalk of a stamen (Plate XII).

**Filiform** Thread-like.

**Fimbriate** (of margins) Fringed, usually with hairs.

**Flower** The structure concerned with

sexual reproduction in the Angiosperms. Essentially it consists of the male organs (androecium) comprising the stamens, and the female organs (gynoecium) comprising the ovary, style(s) and stigma(s), usually surrounded by a whorl of petals (the corolla) and a whorl of sepals (the calyx). The male and female parts may be in the same flower (bisexual) or in separate flowers (unisexual) (Plate IV).

**Follicle** A dry fruit which is derived from a single carpel and which splits open along one side only (Plate VI).

**Free** (of petals, sepals, etc.) Not joined to each other or to any other organ (Plate I, Plate IV).

**Free central placentation** A type of placentation in which the ovules are borne on pla-

---

Plate VI. The main types of fruit and their structure.

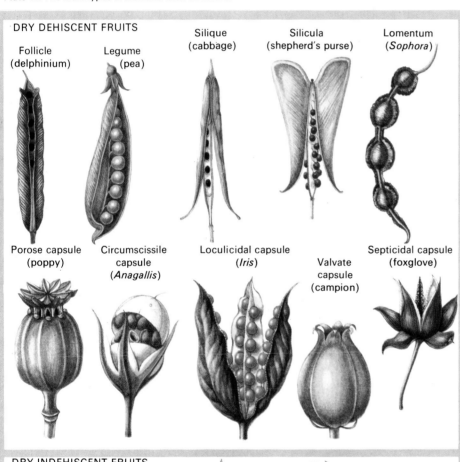

DRY DEHISCENT FRUITS

Follicle (delphinium) · Legume (pea) · Silique (cabbage) · Silicula (shepherd's purse) · Lomentum (*Sophora*)

Porose capsule (poppy) · Circumscissile capsule (*Anagallis*) · Loculicidal capsule (*Iris*) · Valvate capsule (campion) · Septicidal capsule (foxglove)

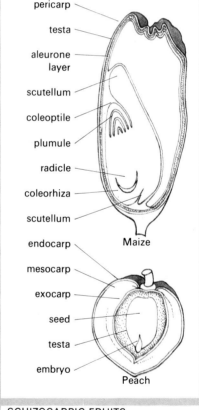

CROSS SECTION OF FRUITS

pericarp, testa, aleurone layer, scutellum, coleoptile, plumule, radicle, coleorhiza, scutellum — Maize

endocarp, mesocarp, exocarp, seed, testa, embryo — Peach

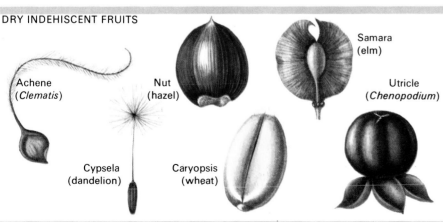

DRY INDEHISCENT FRUITS

Achene (*Clematis*) · Nut (hazel) · Samara (elm) · Utricle (*Chenopodium*) · Cypsela (dandelion) · Caryopsis (wheat)

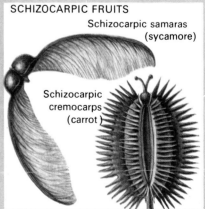

SCHIZOCARPIC FRUITS

Schizocarpic samaras (sycamore) · Schizocarpic cremocarps (carrot)

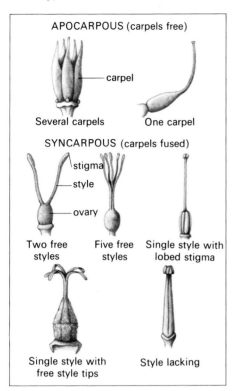

APOCARPOUS (carpels free)

carpel

Several carpels          One carpel

SYNCARPOUS (carpels fused)

stigma
style
ovary

Two free          Five free          Single style with
styles            styles             lobed stigma

Single style with          Style lacking
free style tips

Plate VII. The main types of gynoecium.

centas on a free, central column within an ovary that has only one locule (Plate XI).

**Fruit** Strictly the ripened ovary of a seed plant and its contents. Loosely, the whole structure containing ripe seeds, which may include more than the ovary; cf achene, berry, capsule, drupe, follicle, nut, samara (Plate VI).

**Funicle** The stalk of an ovule (Plate X).

**Gamopetalous** With petals fused, at least at the base.

**Gamosepalous** With sepals fused, at least at the base.

**Glabrous** Without hairs or projections.

**Gland** (adj. glandular) Secreting organ producing oil, resin, nectar, water, etc.; cf hydathode, nectary.

**Glaucous** With a waxy, grayish-blue bloom.

**Globose** Spherical, rounded.

**Gynobasic style** A style that arises near the base of a deeply-lobed ovary (Plate XIV).

**Gynoecium** All the female reproductive organs of a flower, comprising one or more free or fused carpels (Plate VII).

**Gynophore** Stalk of a carpel or gynoecium.

**Habit** The characteristic mode of growth or occurrence; the form and shape of a plant.

**Halophyte** A plant that tolerates salty conditions.

**Hardy** Able to withstand extreme conditions, usually of cold.

**Haustorium** A peg-like fleshy outgrowth from a parasitic plant, usually embedded in the host plant and drawing nourishment from it.

**Head** A dense inflorescence of small,

crowded, often stalkless, flowers – a capitulum (Plate VIII).

**Helicoid** (of cymose inflorescences) Coiled like a spring.

**Herb** (adj herbaceous) A plant that does not develop persistent woody tissue above ground and either dies at the end of the growing season or overwinters by means of underground organs, eg bulbs, corms, rhizomes.

**Heterophylly** Having leaves of more than one type on the same plant.

**Heterostyly** Having styles (and usually stamens) of two or more lengths in different flowers within a species.

**Hilum** The scar left on a seed marking the point of attachment to the stalk of the ovule.

**Hirsute** Covered in rough, coarse hairs.

**Honey guide** Markings (eg lines or dots) on the perianth which direct insects to the nectar.

**Hydathode** A specialized gland, usually found in leaves that exude water.

**Hydrophyte** An aquatic plant.

**Hypanthium** A cup-shaped enlargement of the floral receptacle or the bases of the floral parts, which often enlarges and surrounds the fruits, eg the fleshy tissue in rose-hips.

**Hypha** The thread-like unit of construction of fungi.

**Hypogynous** (of flowers) With the sepals, petals and stamens attached to the receptacle or axis, below and free from the ovary (Plate IV).

**Imbricate** (of sepals and petals) Overlapping, as in a tiled roof.

**Imparipinnate** (of leaves) A pinnate leaf with an unpaired terminal leaflet occurring centrally (Plate IX).

**Inaperturate** (of pollen grains) Without an aperture; without any pores or furrows.

**Incised** (of leaves) Sharply and deeply cut (Plate IX).

**Incompatible** Of plants between which hybrids cannot be formed.

**Indefinite** (of flower parts) Of a number large enough to make a precise count difficult.

**Indehiscent** Fruits not opening to release seeds; cf dehiscent.

**Indumentum** A covering, usually of hairs.

**Inferior** (of ovaries) An ovary with the sepals, petals and stamens attached to its apex (Plate IV).

**Inflorescence** Any arrangement of more than one flower, eg bostryx, capitulum, corymb, cyme, dichasium, fascicle, panicle, raceme, rhipidium, spadix, spike, thyrse and umbel (Plate VIII).

**Infructescence** A cluster of fruits, derived from an inflorescence.

**Inserted** Growing out of another organ.

**Integument** (of ovules) The outer protective covering of the ovule; usually two are found in angiosperms (Plate X).

**Internode** The length of stem that lies

between two leaf-joints (nodes) (Plate IV).

**Introrse** Directed and opening inwards toward the center of the flower; cf extrorse (Plate XII).

**Involucel** A whorl of bracteoles; cf bracteole.

**Involucre** A whorl of bracts beneath an inflorescence; cf bract.

**Irregular** (of flowers) Not regular; not divisible into halves by an indefinite number of longitudinal planes; zygomorphic.

**Lacerate** (of leaves) Irregularly cut (Plate IX).

**Lamina** The thin, flat blade of a leaf or petal (Plate IV).

**Lanceolate** Narrow, as a lance, with tapering ends (Plate IX).

**Lateral** Arising from the side of the parent axis or attached to the side of another organ.

**Latex** A milky and usually whitish fluid that is produced by the cells of various plants and is the source of, eg rubber, gutta percha, chicle and balata.

**Laticiferous** Producing a milky juice (latex).

**Leaf** An aerial and lateral outgrowth from a stem which makes up the foliage of a plant. Its prime function is the manufacture of food by photosynthesis. It typically consists of a stalk (petiole) and a flattened blade (lamina) (Plate IX).

**Leaflet** Each separate lamina of a compound leaf (Plate IX).

**Liana** A woody, climbing vine.

**Ligulate** Strap-shaped or tongue-shaped (Plate IV).

**Ligule** (of leaves) A scale-like membrane on the surface of a leaf; (of flowers) the strap-shaped corolla in some Compositae.

**Limb** The upper, expanded portion of a calyx or corolla with fused parts; cf tube.

**Linear** (of leaves) Elongated, and with parallel sides (Plate IX).

**Lithophyte** A plant which grows on stones and not in the soil.

**Lobe** (of leaves or perianths) A curved or rounded part.

**Lobed** (of leaves) With curved or rounded edges (Plate IX).

**Locule** The chamber or cavity of an ovary which contains the ovules, or of an anther which contains the pollen (Plate XI).

**Loculicidal** Splitting open longitudinally along the dorsal suture (mid-rib) of each segment of the wall (Plate VI).

**Male flower** A flower containing functional stamens, but no carpels.

**Marginal placentation** A type of placentation in which the ovules are borne along the fused margins of a single carpel, eg pea seeds in a pod (Plate XI).

**Membranous** Resembling a membrane; thin, dry and semi-transparent.

**Mericarp** A one-seeded portion of a fruit which splits up when the fruit is mature, eg the fruits of the Umbelliferae (Plate VI).

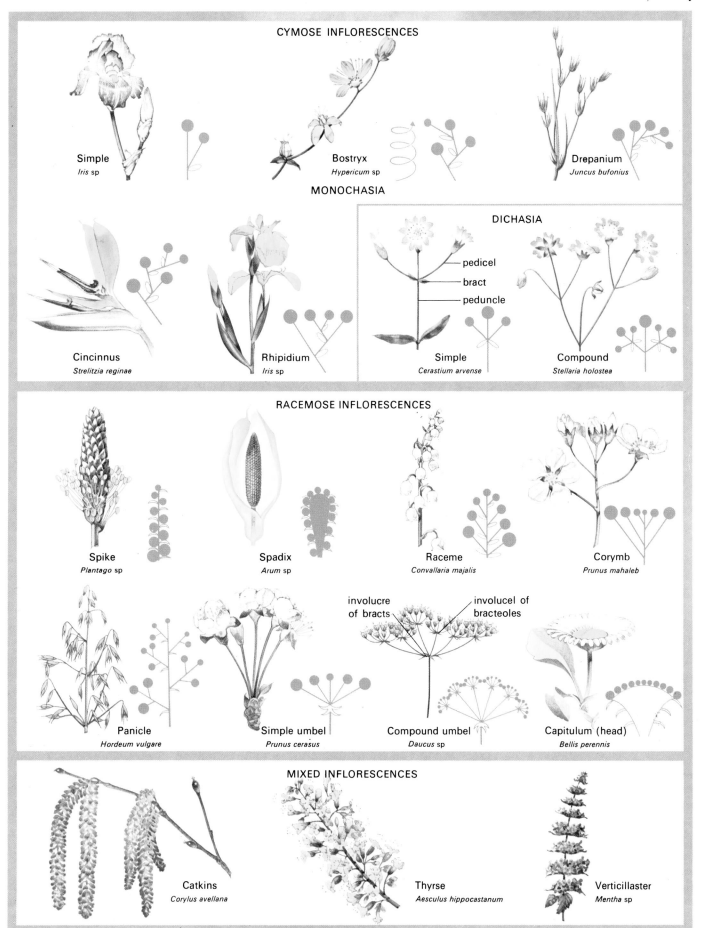

CYMOSE INFLORESCENCES

Simple
*Iris* sp

Bostryx
*Hypericum* sp

Drepanium
*Juncus bufonius*

MONOCHASIA

Cincinnus
*Strelitzia reginae*

Rhipidium
*Iris* sp

DICHASIA

pedicel
bract
peduncle

Simple
*Cerastium arvense*

Compound
*Stellaria holostea*

RACEMOSE INFLORESCENCES

Spike
*Plantago* sp

Spadix
*Arum* sp

Raceme
*Convallaria majalis*

Corymb
*Prunus mahaleb*

Panicle
*Hordeum vulgare*

Simple umbel
*Prunus cerasus*

involucre
of bracts

involucel of
bracteoles

Compound umbel
*Daucus* sp

Capitulum (head)
*Bellis perennis*

MIXED INFLORESCENCES

Catkins
*Corylus avellana*

Thyrse
*Aesculus hippocastanum*

Verticillaster
*Mentha* sp

Plate VIII. The main types of inflorescence. Both the theoretical structure (the largest balls indicate the oldest flowers) and actual examples are shown.

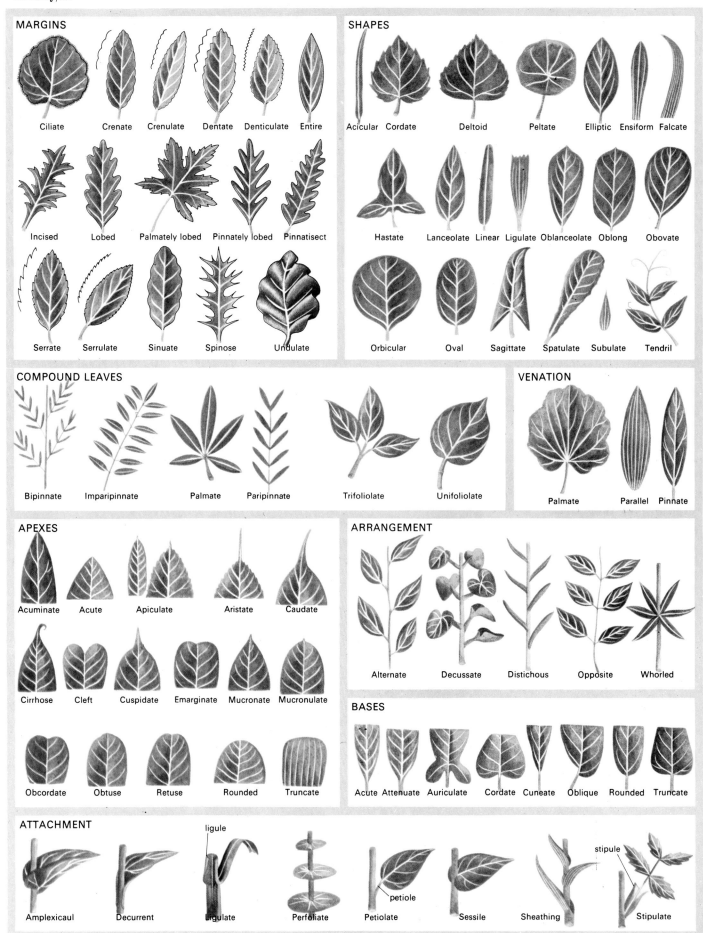

**MARGINS**

Ciliate Crenate Crenulate Dentate Denticulate Entire

Incised Lobed Palmately lobed Pinnately lobed Pinnatisect

Serrate Serrulate Sinuate Spinose Undulate

**SHAPES**

Acicular Cordate Deltoid Peltate Elliptic Ensiform Falcate

Hastate Lanceolate Linear Ligulate Oblanceolate Oblong Obovate

Orbicular Oval Sagittate Spatulate Subulate Tendril

**COMPOUND LEAVES**

Bipinnate Imparipinnate Palmate Paripinnate Trifoliolate Unifoliolate

**VENATION**

Palmate Parallel Pinnate

**APEXES**

Acuminate Acute Apiculate Aristate Caudate

Cirrhose Cleft Cuspidate Emarginate Mucronate Mucronulate

Obcordate Obtuse Retuse Rounded Truncate

**ARRANGEMENT**

Alternate Decussate Distichous Opposite Whorled

**BASES**

Acute Attenuate Auriculate Cordate Cuneate Oblique Rounded Truncate

**ATTACHMENT**

ligule

stipule

Amplexicaul Decurrent Ligulate Perfoliate Petiolate petiole Sessile Sheathing Stipulate

Plate IX. Leaf form, arrangement and attachment (partly after Radford *et al.,* 1974).

**Meristem** A group of cells capable of dividing indefinitely.

**Mesocarp** The middle layer of the fruit wall (pericarp). It is usually fleshy, as in a berry (Plate VI).

**Mesophyte** A plant having moderate moisture requirements.

**Micropyle** The opening through the integuments of an ovule, through which the pollen-tube grows after pollination (Plate X).

**Midrib** The central or largest vein of a leaf or carpel (Plate IV).

**Mono-** A prefix meaning single, one or once.

**Monocarpic** Fruiting only once and then dying.

**Monochasium** A cymose inflorescence in which there is a single terminal flower with below it a single branch bearing flower(s) (Plate VIII).

**Monocolpate** (of pollen) Having a single colpus (an oblong-elliptic aperture).

**Monocotyledon** One of two subclasses of angiosperms; a plant whose embryo has one cotyledon; cf dicotyledon (Plate II).

**Monoecious** Having separate male and female flowers on the same plant.

**Monogeneric** (of a family) Containing only one genus.

**Monopodial** (of stems or rhizomes) With branches or appendages arising from a simple main axis; cf sympodial.

**Monotypic** A genus or a family containing a single species.

**Mucilage** A slimy secretion, which swells on contact with water.

**Multiseriate** (of flower parts) Borne in many series or whorls.

**Mycorrhiza** The symbiotic association of the roots of some seed plants with fungi.

**Naked** (of flowers) Lacking a perianth (Plate IV).

**Nectar** A sugary liquid secreted by some plants; it forms the principle raw material of honey.

**Nectary** The gland in which nectar is produced.

**Node** The point on a stem where one or more leaves are borne (Plate IX).

**Nucellus** The central nutritive tissue of the ovule, containing the embryo sac and surrounded by, in angiosperms, two integuments (Plate X).

**Numerous** (of floral parts) Usually meaning more than ten; cf indefinite.

**Nut** A dry, single-seeded and non-opening (indehiscent) fruit with a woody pericarp (Plate VI).

**Obligate** Parasite unable to grow on its own; entirely dependent on a host for nutrition.

**Obovate** (of leaves) Having the outline of an egg, with the broadest part above the middle and attached at the narrow end (Plate IX).

**Ochrea** A cup-shaped structure formed by the joining of stipules or leaf bases around a stem.

**Opposite** (of leaves) Occurring in pairs on opposite sides of the stem; (of stamens) inserted in front of the petals (Plate IX).

**Orbicular** More or less circular (Plate IX).

**Orthotropous** (of ovules) Borne on a straight stalk (funicle); not bent over (Plate X).

**Ovary** The hollow basal region of a carpel, containing one or more ovules and surmounted by the style(s) and stigma(s). It is made up of one or more carpels which may fuse together in different ways to form one or more chambers (locules). The ovary is generally above the perianth parts (superior) or below them (inferior) (Plate IV).

**Ovate** (of leaves) Having the outline of an egg with the narrow end above the middle (Plate IX).

**Ovoid** (of leaves) Egg-shaped (Plate IX).

**Ovule** The structure in the chamber (locule) of an ovary containing the egg cell within the embryo sac which is surrounded by the nucellus. It is enclosed by two integuments. The ovule develops into the seed after fertilization (Plate X).

**Palmate** (of leaves) With more than three segments or leaflets arising from a single point, as in the fingers of a hand (Plate IX).

**Panicle** (of inflorescences) Strictly a branched raceme, with each branch bearing a further raceme of flowers. More loosely, it applies to any complex, branched inflorescence (Plate VIII).

**Paniculate** Arranged in a panicle.

**Parasite** A plant that obtains its food from another living plant to which it is attached.

**Parenchyma** Tissue made up of thin-walled living photosynthetic or storage cells which is capable of division even when mature.

**Parietal placentation** A type of placentation in which the ovules are borne on placentas on the inner surface of the outer wall of the ovary (Plate XI).

**Paripinnate** A pinnate leaf with all leaflets in pairs; cf imparipinnate.

**Pedate** (of leaves) A palmately divided compound leaf, having three main divisions, and having the outer division one or more times. There may be a free central leaflet.

**Pedicel** The stalk of a single flower (Plate IV).

**Peduncle** The stalk of an inflorescence (Plate VIII).

**Peltate** (of leaves) More or less circular and flat with the stalk inserted in the middle (Plate IX).

**Pendulous** Hanging down (Plate XI).

**Perennating** Living over from season to season.

**Perennial** A plant that persists for more than two years and normally flowers annually.

**Perfect flower** A flower with functional male and female organs.

**Perianth** The floral envelope whose segments are usually divisible into an outer whorl (calyx) of sepals, and an inner whorl (corolla) of petals. The segments of either or both whorls may fuse to form a tube (Plate IV).

**Pericarp** The wall of a fruit that encloses the seeds and which develops from the ovary wall (Plate IV).

**Perigynous** (of flowers) Having the stamens, corolla and calyx inserted around the ovary, their bases often forming a disk (Plate IV).

**Perisperm** The nutritive storage tissue in some seeds, derived from the nucellus.

**Persistent** Remaining attached, not falling off.

**Petal** A non-reproductive (sterile) part of the flower, usually conspicuously colored; one of the units of the corolla (Plate IV).

**Petaloid** Petal-like.

**Petiole** The stalk of a leaf (Plate IX).

**Phloem** That part of the tissue of a plant

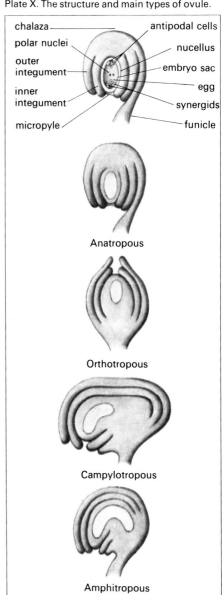

Plate X. The structure and main types of ovule.

chalaza
polar nuclei
outer integument
inner integument
micropyle
antipodal cells
nucellus
embryo sac
egg
synergids
funicle

Anatropous

Orthotropous

Campylotropous

Amphitropous

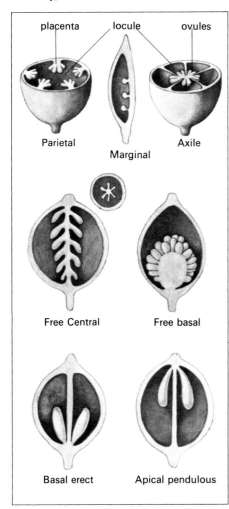

placenta    locule    ovules

Parietal            Axile

Marginal

Free Central        Free basal

Basal erect        Apical pendulous

Plate XI. The main forms of placentation.

which is concerned with conducting food material. In woody stems it is the innermost layer of the bark; cf xylem.

**Photosynthesis**   The process by which green plants manufacture sugars from water and carbon dioxide by converting the energy from light into chemical energy with the aid of the green pigment chlorophyll.

**Phyllode**   A flattened leaf stalk (petiole) which has assumed the form and function of a leaf blade.

**Pinnate**   (of leaves)   Compound, with leaflets in pairs on opposite sides of the midrib; cf imparipinnate and paripinnate (Plate IX).

**Pinnatisect** (of leaves)   Pinnately divided, but not as far as the midrib (Plate IX).

**Pistil**   The female reproductive organ consisting of one or more carpels, comprising ovary, style and stigma; the gynoecium as a whole.

**Pistillate**   A flower that has only female organs.

**Pistillode**   A sterile, often reduced pistil.

**Placenta**   Part of the ovary wall to which the ovules are attached.

**Placentation**   The arrangement and distribution of the ovule-bearing placentas within the ovary; cf axile, basal, free central, marginal and parietal (Plate XI).

**Plumule**   The rudimentary shoot in an embryo.

**Pollen**   Collective name for the pollen grains, ie the minute spores (microspores) produced in the anthers.

**Pollen sac**   The chamber (locule) in an anther where the pollen is formed (Plate XII).

**Pollination**   The transfer of pollen grains from stamen to stigma. *Cross-pollination* occurs between flowers of different plants of the same species; *self-pollination* occurs between flowers of the same plant, or within one flower.

**Pollinium**   A mass of pollen grains produced by one anther-lobe, cohering together and transported as a single unit during pollination, as in the orchids.

**Polygamodioecious**   Having male and bisexual flowers on one individual plant and female and bisexual flowers on another.

**Polygamous**   Having separate male, female and bisexual flowers on the same plant.

**Polypetalous**   With petals free from each other.

**Pore**   A small hole.

**Protandrous** (of flowers)   The maturing of stamens and the consequent release of their pollen before the stigmas of the same flower become receptive.

**Protogynous** (of flowers)   The receptiveness of the stigmas before the stamens of the same flower mature and release their pollen.

**Pseudocopulation**   The attempted copulation by male insect visitors with a part of a flower which resembles the female of the insect species, as in the orchids.

**Pseudo-whorled** (of leaves)   Arising close together and so appearing to arise at the same level, although not in fact doing so.

**Pubescent**   Covered in soft, short hairs.

**Punctate**   Shallowly pitted or dotted, often with glands.

**Raceme**   An inflorescence consisting of a main axis, bearing single flowers alternately or spirally on stalks (pedicels) of approximately equal length. The apical growing point continues to be active so there is usually no terminal flower and the youngest branches or flowers are nearest the apex. This mode of growth is known as monopodial (Plate VIII).

**Racemose**   Arranged like a raceme; in general any inflorescence capable of indefinite prolongation, having lateral and axillary flowers (Plate VIII).

**Radical** (of leaves)   Arising from the base of a stem or from a rhizome; basal.

**Radicle**   The rudimentary root in an embryo.

**Raphe**   A ridge or tissue visible on the testa of seeds developed from ovules which are bent over through 180 degrees (anatropous). It results from the fusion of the stalk (funicle) with the rest of the bent-over ovules.

**Ray** (of wood)   Radial strands of living cells concerned with the transport of water and food.

**Receptacle**   Flat, concave or convex part of the stem from which all parts of a flower arise; the floral axis (Plate IV).

**Recurved**   Curved backwards.

**Reflexed**   Bent sharply backwards at an angle.

**Regular** (of flowers)   Radially symmetrical, with more than one plane of symmetry; actinomorphic.

**Reticulate**   Marked with a network pattern, usually of veins.

**Rhachis**   The major axis of an inflorescence.

**Rhipidium** (of inflorescences)   A cymose inflorescence with branches alternating from one side of the vertical axis to the other; normally flattened in one plane and fan-shaped (Plate VIII).

**Rhizome**   A horizontally creeping underground stem which lives over from season to season (perennates) and which bears roots and leafy shoots.

**Root**   The lower, usually underground, part of a plant. It anchors the plant in the soil and absorbs water and mineral nutrients by means of the root hairs (Plate IV).

**Rosette**   A group of leaves arising closely together from a short stem, forming a radiating cluster on or near the ground.

**Rotate** (of corollas)   Wheel-shaped; with the petals or lobes spreading out from the axis of a flower (Plate IV).

**Ruminate**   (of endosperm in seeds) Irregularly grooved and ridged; having a "chewed" appearance.

**Sagittate** (of leaves)   Shaped like an arrow head; with two backward-directed barbs (Plate IX).

**Samara**   A dry fruit that does not split open and has part of the fruit wall extended to form a flattened membrane or wing (Plate VI).

**Saponins**   A toxic, soap-like group of compounds which is present in many plants.

**Saprophyte**   A plant that cannot live on its own, but which needs decaying organic material as a source of nutrition.

**Scale**   A small, often membranous, reduced leaf frequently found covering buds and bulbs.

**Scaled**   Covered by scale leaves.

**Scape**   A leafless flower-stalk.

**Scarious**   Dry and membranous, with a dried-up appearance.

**Schizocarp**   A fruit derived from a simple or compound ovary in which the locules separate at maturity to form single-seeded units (Plate VI).

**Sclerenchyma**   Tissue with thickened cell walls, often woody (lignified), and which give mechanical strength and support.

**Scorpioid**   (of cymose inflorescences) Curved to one side like a scorpion's tail.

**Scrambler**   A plant with a spreading, creeping habit usually anchored with the help of hooks, thorns or tendrils.

**Seed**   The unit of sexual reproduction developed from a fertilized ovule; an embryo

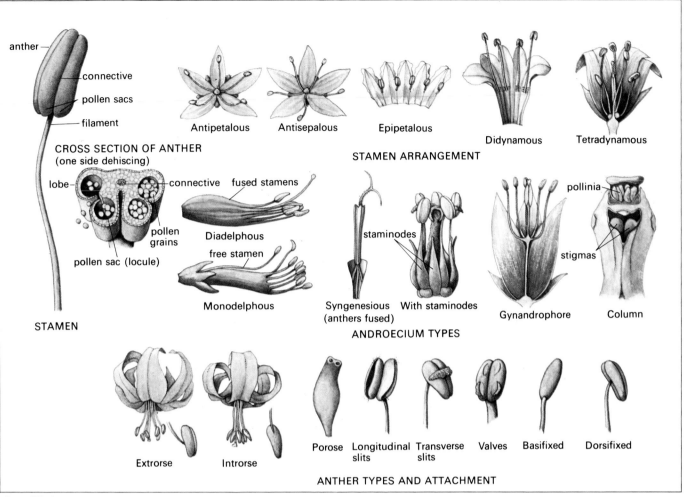

anther

connective

pollen sacs

filament

Antipetalous  Antisepalous  Epipetalous

Didynamous  Tetradynamous

STAMEN ARRANGEMENT

CROSS SECTION OF ANTHER
(one side dehiscing)

lobe  connective  fused stamens

pollen
grains

Diadelphous

free stamen

staminodes

pollinia

stigmas

pollen sac (locule)

Monodelphous

STAMEN

Syngenesious  With staminodes
(anthers fused)

Gynandrophore  Column

ANDROECIUM TYPES

Extrorse  Introrse

Porose  Longitudinal  Transverse  Valves  Basifixed  Dorsifixed
slits  slits

ANTHER TYPES AND ATTACHMENT

Plate XII. Stamen and androecium forms and structure.

enclosed in the testa which is derived from the integument(s).

**Seedling**  The young plant that develops from a germinating seed.

**Self-fertilization**  See Fertilization.

**Self-incompatible**  Of plants incapable of self-fertilization, usually because the pollen-tube cannot germinate or grows very slowly.

**Self-pollination**  See Pollination.

**Semi-parasite**  A plant which, although able to grow independently, is much more vigorous if it establishes a parasitic relationship on another plant.

**Sepal**  A floral leaf or individual segment of the calyx of a flower, usually green (Plate IV).

**Septate** (of ovaries)  Divided into locules by walls.

**Septicidal** (of fruits)  Splitting open longitudinally through the septa so that the carpels are separated (Plate VI).

**Septum** (of ovaries)  The wall between two chambers (locules) of an ovary made up of two or more fused carpels (syncarpous ovary).

**Seriate**  Arranged in a row.

**Serrate** (of margins)  Toothed, like a saw (Plate IX).

**Serrulate** (of margins)  Finely toothed, like a saw (Plate IX).

**Sessile**  Without a stalk, eg leaves without petioles or stigmas without a style.

**Sheath** (of leaves)  The base of a leaf or leaf-stalk (petiole) which encases the stem.

**Sheathing** (of leaves)  With a sheath that encases the stem (Plate IX).

**Shoot**  The above-ground portions of a vascular plant, such as the stems and leaves; the part of a plant which develops from the plumule of the embryo.

**Shrub**  A perennial woody plant with well developed side-branches that appear near the base, so that there is no trunk. They are less than 10 metres (30 feet) high.

**Silicule** or **Silicula**  A dry fruit that opens along two lines and has a central persistent partition; it is as broad as, or broader, than it is long, as in the Cruciferae (Plate VI).

**Silique** or **Siliqua**  A silicule-type of fruit that is longer than it is broad, as in the Cruciferae (Plate VI).

**Simple** (of leaves)  Not divided or lobed in any way (Plate IX).

**Simple umbel** (of inflorescences)  An umbel in which the stalks (pedicels) arise directly from the top of the main stalk (Plate VIII).

**Sinuate** (of margins)  Divided into wide irregular teeth or lobes which are separated by shallow notches (Plate IX).

**Solitary** (of flowers)  Occurring singly in each axil.

**Spadix**  A spike of flowers on a swollen, fleshy axis (Plate VIII).

**Spathe**  A large bract subtending and often ensheathing an inflorescence. Applied only in the monocotyledons.

**Spatulate** or **Spathulate** (of leaves)  Shaped like a spoon (Plate IX).

**Spicate**  Spike-like.

**Spike**  An inflorescence of simple racemose type in which the flowers are stalkless (sessile) (Plate VIII).

**Spikelet**  A small spike, as in the grasses (Plate VIII).

**Spine**  The hard and sharply-pointed tip of a branch or leaf, usually round in cross section.

**Spinose**  Spiny (Plate IX).

**Spur**  A hollow, usually rather conical, projection from the base of a sepal, petal, or fused corolla.

**Stamen**  The male reproductive organ of a flower. It consists of a usually bilobed anther borne on a stalk (filament) (Plate XII).

**Staminate**  Having stamens (male organs), but no carpels (female organs); cf pistillate.

**Staminode**  A sterile, often reduced or modified stamen (Plate XII).

**Stellate**  Star-shaped.

**Stem**  The main supporting axis of a plant. It bears leaves with buds in their axils. Usually aerial, it can however be subterranean (Plate IV).

**Sterile**   Unable to reproduce sexually.

**Stigma**   The receptive part of the female reproductive organs on which the pollen grains germinate; the apical part of the carpel (Plate XIII).

**Stipitate**   Having a stalk or stipe.

**Stipulate**   Having stipules (Plate IX).

**Stipule**   A leafy appendage, often paired, and usually at the base of the leaf stalk.

**Stomata**   The pores that occur in large numbers in the epidermis of plants and through which gaseous exchange takes place.

**Stooling** (of plants)   Having several stems arising together at the base.

**Style**   The elongated apical part of a carpel or ovary bearing the stigma at its tip (Plate XIV).

**Subapical**   Below the apex.

**Succulent**   With fleshy or juicy organs containing reserves of water.

**Suffrutescent** (of herbaceous plants) Having a persistent woody stem base.

**Superior** (of ovaries)   Occurring above the level at which the sepals, petals and stamens are borne; cf inferior (Plate IV).

**Suture**   A line of union; the line along which dehiscence often takes place in fruits.

**Symbiosis**   The non-parasitic relationship between living organisms to their mutual benefit.

**Sympetalous**   With the petals united along their margins, at least at the base.

**Sympodial** (of stems or rhizomes)   With the apparent main stem consisting of a series of usually short axillary branches; cf cyme and monopodial.

**Syncarpous** (of ovaries)   Made up of two or more fused carpels.

**Tendril**   Part or all of a stem, leaf or petiole modified to form a delicate, thread-like appendage; a climbing organ with the ability to coil around objects (Plate IX).

**Tepal**   A perianth-segment that is not clearly distinguishable as being either a sepal or a petal.

**Terminal**   Situated at the apex.

**Ternate** (of leaves)   Compound, divided into three parts more or less equally. Each part may itself be further sub-divided.

**Tessellated** (of leaves)   Marked with a fine chequered pattern, like a mosaic.

**Testa**   The outer protective covering of a seed (Plate VI).

**Thallus**   A type of plant body that is not differentiated into root, stem or leaf.

**Theca**   One half of an anther containing two pollen sacs.

**Throat**   The site in a calyx or corolla of united parts where the tube and limbs meet.

**Thyrse** (of inflorescences)   Densely branched, broadest in the middle and in which the mode of branching is cymose.

**Tomentose**   Densely covered in short hairs.

**Tree**   A large perennial plant with a single branched and woody trunk and with few or no branches arising from the base.

**Tri-**   A prefix meaning three.

**Trichome**   A hair-like outgrowth.

**Tricolpate** (of pollen)   Having three colpi.

**Trifoliolate** (of leaves)   Having three leaflets (Plate IX).

**Tube**   The united, usually cylindrical part of the calyx or corolla made up of united parts; cf limb.

**Tuber**   An underground stem or root that lives over from season to season (perennates) and which is swollen with food reserves; cf bulb, corm, rhizome.

**Turion**   A short, scaly branch produced from a rhizome.

**Umbel**   An umbrella-shaped inflorescence with all the stalks (pedicels) arising from the top of the main stem. Umbels are sometimes compound, with all the stalks (peduncles) arising from the same point and giving rise to several terminal flower stalks (pedicels) (Plate VIII).

**Undershrub**   A perennial plant with lower woody parts, but herbaceous upper parts that die back seasonally.

**Understory**   A layer of shrubs, usually in deciduous woodland or tropical rain forests.

**Undulate** (of leaves)   With wavy margins.

**Unifoliolate**   With a single leaflet that has a stalk distinct from the stalk of the whole leaf (Plate IX).

**Unilocular** (of ovaries)   Containing one chamber (locule) in which the ovules or seeds occur.

**Uniseriate**   Arranged in a single row, series or layer, eg perianth-segments.

**Unisexual** (of flowers)   Of one sex.

**Utricle**   A small bladder-like, single-seeded dry fruit (Plate VI).

**Valvate** (of perianth-segments)   With the margins adjacent without overlapping; cf imbricate.

**Vascular**   Possessing vessels; able to conduct water and nutrients.

**Vascular bundle**   A strand of tissue involved in water and food transport.

**Vein**   Any of the visible strands of conducting and strengthening tissues running through a leaf (Plate IX).

**Venation**   The arrangement of the veins of a leaf (Plate IX).

**Ventral**   On the lower side.

**Vernation** (of leaves)   The manner and pattern of arrangement within the bud.

**Verticillate**   Arranged in whorls.

**Verticillaster** (of inflorescences)   Whorled dichasia at the nodes of an elongate rachis.

**Vessels**   Tube-like cells arranged end to end in the wood of flowering plants and which form the principal pathway in the transport of water and mineral salts.

**Whorl**   The arrangement of organs, such as leaves, petals, sepals and stamens so that they arise at the same level on the axis in an encircling ring.

**Xeromorphic**   Possessing characteristics such as reduced leaves, succulence, dense hairiness or a thick cuticle which are adaptations to conserve water and so withstand extremely dry conditions.

**Xerophyte**   A plant which is adapted to withstand extremely dry conditions.

**Xylem**   The woody fluid-conveying (vascular) tissue concerned with the transport of water about the plant; cf phloem; vessels.

Plate XIII. The main forms of stigma (after Radford *et al.*, 1974).

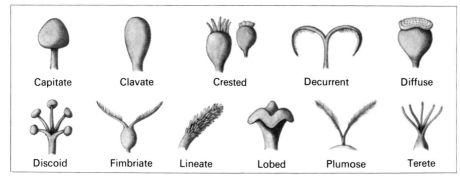

Capitate   Clavate   Crested   Decurrent   Diffuse

Discoid   Fimbriate   Lineate   Lobed   Plumose   Terete

Plate XIV. The main forms of style (after Radford *et al.*, 1974).

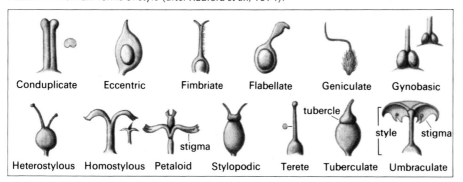

Conduplicate   Eccentric   Fimbriate   Flabellate   Geniculate   Gynobasic

Heterostylous   Homostylous   Petaloid   Stylopodic   Terete   Tuberculate   Umbraculate

# DICOTYLEDONS

# MAGNOLIIDAE

## MAGNOLIALES

### MAGNOLIACEAE
*Magnolias and Tulip Tree*

The Magnoliaceae is a family of about 220 species of trees and shrubs native to Asia and America. Its best-known representatives are the horticulturally important species of the genus *Magnolia*.

**Distribution.** Approximately four-fifths of the species are distributed in temperate and tropical Southeast Asia from the Himalayas eastwards to Japan and southeastwards through the Malay Archipelago to New Guinea and New Britain; the remaining one-fifth are found in America, from temperate southeast North America through tropical America to Brazil. All the American species belong to the three genera *Magnolia*, *Talauma* and *Liriodendron*, which occur also in Asia and thus have independent discontinuous distributions. The other genera are confined to Asia. Fossil records indicate that the family was formerly much more widely distributed in the Northern Hemisphere, eg in Greenland and Europe.

**Diagnostic features.** The family is morphologically very distinct and easily recognizable. The leaves are alternate, simple, petiolate, with large stipules which at first surround the stem but fall off as the leaf expands and leave a characteristic scar around the node. The flowers are bisexual (rarely unisexual), often large and showy, pedunculate, solitary at the ends of the branches or in the axils of the leaves (sometimes paired when axillary); the peduncle bears one or more spathaceous bracts which enclose the young flower but fall off as it expands. The perianth is composed of two or more (usually three) whorls of free tepals which are petaloid; the outer tepals are sometimes reduced and sepal-like. The stamens are numerous, free, spirally arranged, with stout filaments; the anthers have two locules opening by longitudinal slits.

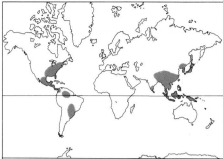

**Number of genera:** 12
**Number of species:** about 220
**Distribution:** centered in temperate and tropical SE Asia.
**Economic uses:** valued ornamentals (*Magnolia* and *Liriodendron*) and some useful timbers.

Magnoliaceae. 1 *Magnolia stellata* leaf and flowering shoot showing bracts on flower stalks ( $\times \frac{2}{3}$ ). 2 *M. heptapeta* half flower showing two whorls of perianth segments, numerous spirally arranged stamens and numerous free carpels on an elongate receptacle ( $\times \frac{1}{2}$ ). 3 *M. grandiflora* dehisced fruits with arillate seeds attached by silky thread ( $\times \frac{2}{3}$ ). 4 *Liriodendron tulipifera* (a) flower and leaf ( $\times \frac{1}{2}$ ); (b) vertical section of carpel ( $\times 1\frac{1}{3}$ ); (c) fruiting head ( $\times \frac{2}{3}$ ). 5 *Talauma ovata* fruit with upper portions of carpels falling away to reveal one or two seeds in each carpel locule ( $\times \frac{1}{3}$ ).

The carpels are numerous or few (rarely single), spirally arranged, free or partly fused. Each carpel has two or more ventrally placed ovules. The fruit is composed of separate or united carpels, which are longitudinally dehiscent or circumscissile or indehiscent. The seeds (except in *Liriodendron*) are large, with an arilloid testa free from the endocarp, but attached by a silky thread; in *Liriodendron* they adhere to the endocarp, without an arilloid testa. The seeds have copious endosperm and a minute embryo.

**Classification.** There are 12 genera, forming two tribes. The genus *Liriodendron*, which differs from the others in its characteristic lobed leaves, extrorse anthers and deciduous winged indehiscent fruiting carpels, forms the small tribe LIRIODENDREAE. All the other genera belong to the tribe MAGNOLIEAE, with inward or laterally dehiscing anthers and seeds with an arilloid testa which is free from the endocarp. The Magnolieae with terminal flowers include the two largest genera, *Magnolia* (over 80 species) and *Talauma* (over 50 species), both of which are common to Asia and America, as well as the Asian genera *Manglietia* (25 species), *Alcimandra* (one species), *Aromadendron* (four species), *Pachylarnax* (two species) and *Kmeria* (two species). The genera with axillary flowers are all Asian and include the genera *Michelia* (about 40 species), *Elmerrillia* (six species), *Paramichelia* (three species) and *Tsoongiodendron* (one species).

The Magnoliaceae is an extremely natural group, very distinct from any other known family. They are the type of an order Magnoliales under which have been placed a number of families, including the Winteraceae and Annonaceae, which have been traditionally associated with the Magnoliaceae because of a similarity in floral structure. The Magnoliaceae, however, is an isolated family with no really close allies. There are good grounds for regarding it as overall the most primitive living family of flowering plants although many specialists would not agree.

**Economic uses.** The wood of the North American tulip tree (*Liriodendron tulipifera*) is a valuable timber product of the eastern United States of America. Some other species such as *Magnolia hypoleuca* and *Michelia champaca* (sapu) yield serviceable timbers which are used locally. The bark and flower buds of *Magnolia officinalis* and other species yield a valuable drug exported from China for medicinal use. It is in the horticultural field, however, that the Magnoliaceae are best known to Man, the genus *Magnolia* being one of the most popular genera of trees and shrubs in cultivation, with numerous species and hybrids now available. Species of *Manglietia*, *Michelia* and *Liriodendron* are also cultivated in temperate countries, while *Michelia champaca* and *M. figo* are widely grown in warmer regions of the world.          J.E.D.

## WINTERACEAE
*Winter's Bark*

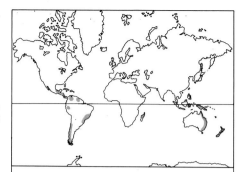

**Number of genera:** 7 or 8
**Number of species:** 60–120
**Distribution:** montane to cool temperate forests of islands in and continents bordering S Pacific.
**Economic uses:** some horticultural and medicinal uses.

The small aromatic family Winteraceae is of great evolutionary interest to botanists because its members, whose wood lacks vessels, are possibly the least specialized of all living flowering plants.

**Distribution.** Most of the perhaps 120 species of Winteraceae are characteristic of wet tropical montane to cool temperate rain forests in the continents and islands fringing the southern half of the Pacific Ocean. The four American species of *Drimys* range from the tropical highlands of Mexico southward to Guayana, southeastern Brazil, Cape Horn and the Juan Fernández Islands. The other genera are abundantly represented from New Zealand and Tasmania north through New Caledonia and eastern Australia to the Solomon Islands and New Guinea, thence as one wide-ranging species to Borneo and the Philippines. A little known species has also been discovered in Madagascar.

**Diagnostic features.** Although the winteraceous trees and shrubs with their alternate, leathery, entire-margined, simple leaves without stipules are typical of rain forest plants, they are readily spotted by the appearance of the under surface of the leaves, which is glaucous owing to the waxy deposits plugging the stomata. In more technical details family members are distinguished by the wood without vessels, but with long cambial initials and tracheids, and heterocellular rays. The flowers are mostly in fascicles or cymes, bisexual or functionally unisexual, with parts mostly separate, spirally arranged, and indefinite in number. The sepals and petals are mildly differentiated, the former being valvate or capped, the latter imbricate. The stamens consist of short, broad sporophylls with lateral to apical sporangia; the pollen grains are distally monoporate in coherent tetrads. The carpels are mostly stalked and apparently folded longitudinally with partly

free margins of the ventral suture bearing papillae and forming double stigmatic crests; anatropous ovules are many to few on marginal to submarginal placentas. The fruits are mostly indehiscent berries of obvious follicular origin; and the seeds have a rudimentary embryo in abundant endosperm.

The primitive, vesselless wood; mostly bisexual flowers with largely undifferentiated stamens producing distally monoporate pollen grains; apocarpous, styleless carpels with partly unsealed stigmatic margins; and anatropous ovules developing into primitive seeds are a combination of preserved, ancient features that mark the Winteraceae as perhaps the closest relative to the extinct angiospermous forebears.

**Classification.** There are seven or eight closely related genera with 60–120 species, with *Drimys* in America; *Belliolum*, *Bubbia*, *Exospermum*, *Pseudowintera*, *Tasmannia* and *Zygogynum* in Australasia to western Malaysia; and an undescribed genus from Madagascar.

The Winteraceae is distinct enough from other related families to warrant its own suborder Winterineae, yet there is no doubt about its inclusion with other primitive relicts belonging to the orders Magnoliales, Illiciales, Laurales, Piperales and Aristolochiales (a group of orders brought together in most modern systems as a single order the Annonales). Its closest relatives are the Illiciaceae and Schizandraceae. The Magnoliaceae, Degeneriaceae, Annonaceae and Myristicaceae are also closely related.

**Economic uses.** The peppery, aromatic leaves and bark of some species are said to have medicinal qualities as astringents and stimulants. Winter's bark, from the South American *Drimys winteri*, was once used by mariners as a scurvy preventative. A number of species, especially *D. winteri*, are cultivated in arboreta and parks.          R.F.T.

## HIMANTANDRACEAE

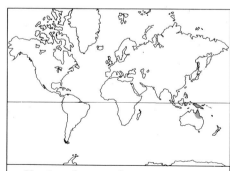

**Number of genera:** 1
**Number of species:** 3
**Distribution:** NE Australia and parts of New Guinea and the Moluccas.
**Economic uses:** none.

The Himantandraceae is a family of aromatic trees consisting of a single genus, *Galbulimima*, with only three species.

**Distribution.** The family is found in northeast Australia, and parts of the islands of New Guinea and the Moluccas immediately to the north of Australia.

**Diagnostic features.** All are aromatic trees with a downy covering of numerous minute scales each attached by a central stalk. The leaves are alternate, simple and without stipules. The flowers are regular, bisexual and solitary. The flower bud has two unusual leathery sepals which form a cap round the bud. The stamens (about 40) are very much like the petals (four to seven) in their shape and texture, the anthers being at the edges of the flat, lanceolate blades. There are seven to ten free superior carpels, each with one locule and one or two pendulous ovules. The carpels become fused to form a globose fruit. The seeds have oily endosperm and a small embryo.

**Classification.** The family is interesting because it is primitive – a botanical relict. Its primitive features include petal-like stamens, simple form of the female parts, vessels with porous perforations and ladder-like pits, and pollen grains with a single aperture. The close relatives of the Himantandraceae are not clear. The family appears to belong to the Annonaceae–Magnoliaceae– Winteraceae–Degeneriaceae complex. In characters such as the stamens, pollen and the unusual calyx it is similar to the Degeneriaceae. It shares the character of peltate scales with the Annonaceae, although it differs in other respects.

**Economic uses.** The family is not important economically. S.W.J.

## CANELLACEAE
*White Cinnamon*

The Canellaceae is a small family of aromatic trees, some with medicinal properties.

**Distribution.** The family is restricted to the tropics, but discontinuously distributed there, three genera occurring in Central America and the West Indies, the other two in East Africa and Madagascar.

**Diagnostic features.** The leaves are leathery, entire and gland-dotted and without stipules. The flowers are bisexual with four to five sepals, usually four up to 12 petals, united or free. The stamens are arranged in a tube and the ovary is superior, composed of two to five united carpels with a single locule and two to many ovules on parietal placentas. The fruit is a berry with two to many seeds which have a straight or slightly curved embryo and oily endosperm. The secretory

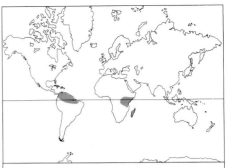

**Number of genera:** 5
**Number of species:** 16 or 17
**Distribution:** discontinuous in tropics.
**Economic uses:** source of the tonic white cinnamon and several local uses.

cells which account for the aromatic features of the wood are found in the parenchyma also.

**Classification.** *Cinnamosma*, characterized by its united petals, has three species restricted to Madagascar; *Canella*, with two species in Florida and the West Indies, has free petals, but a terminal inflorescence, the other genera having axillary flowers; *Pleodendron*, with one or two species restricted to Puerto Rico, has 12 petals, while

Winteraceae. 1 *Drimys axillaris* leafy shoot with small axillary flowers ($\times\frac{2}{3}$). 2 *D. winteri* (a) leafy shoot with bundles of conspicuous flowers ($\times\frac{2}{3}$); (b) half flower showing free sepals and petals, numerous short stamens and free carpels containing ovules on marginal placentas ($\times 2$); (c) fruits— berries ($\times 2$). Illiciaceae. 3 *Illicium floridanum* leafy shoot and axillary flowers ($\times\frac{2}{3}$). 4 *I. anisatum* dehiscing fruits—follicles ($\times 1\frac{1}{3}$).

*Cinnamodendron*, with seven species in tropical South America and the West Indies, has ten petals with many stamens, and *Warburgia*, with three species in tropical East Africa, has ten petals with ten stamens.

The family is usually considered to be allied to Annonaceae or Myristicaceae, but the structure of the seeds indicates a closer relationship with Winteraceae, from which it differs in having wood that contains vessel elements.

**Economic uses.** Canella bark, or white cinnamon, is derived from the aromatic bark of *Canella winterana* and is used as a tonic and stimulant or as a condiment. The orange-colored bark is also used as a fish poison in Puerto Rico. The violet flowers, when dried and resoftened in water, will release a musky scent. The bark of *Cinnamodendron corticosum* of the West Indies is also used as a tonic. The aromatic wood of *Cinnamosma fragrans* of Madagascar is exported via Zanzibar to Bombay, where it is used in religious ceremonies. *Warburgia ugandensis*, a tree from Uganda, yields a resin used in fixing handles to tools, while its leaves are sometimes used in curries. The bark is used as a purgative.      D.J.M

# ANNONACEAE
*Sweetsop and Soursop*

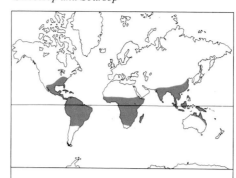

**Number of genera:** about 120
**Number of species:** about 2,000
**Distribution:** pantropical, but centered in the Old World tropics.
**Economic uses:** cultivated for fruits (sweetsop, soursop) and aromatic oils.

The Annonaceae is a large family of chiefly tropical trees and shrubs, some of which are used for their fruit.

**Distribution.** Found throughout the tropics, the Annonaceae species are especially characteristic of lowland evergreen forest in several parts of the Old World tropics.

**Diagnostic features.** All but one species are trees or shrubs, sometimes climbing, usually evergreen, with resin canals and septate pith in the stems. The entire leaves, arranged in two ranks, without stipules, are often recognizable in the field by a glaucous or metallic sheen. The fragrant flowers, often nodding, frequently open before all the parts are fully developed. They are regular and bisexual (rarely unisexual). The perianth is usually in three whorls of three and the stamens are usually numerous and spirally arranged, rarely few and whorled. The carpels are usually separate, the fruit being an aggregate of berries, though the fruits of *Annona* species are worth eating because the berries coalesce with an edible, fleshy receptacle. The fruits are attractive to bats, squirrels and monkeys, but those of *Anaxagora javanica* burst open explosively. The seeds have ruminate endosperm, and are of complicated construction. Some develop an aril (third integument) after fertilization; others are pachychalazal, ie the apical growth of the integuments is halted and an intercalary growth beneath them gives rise to the bulk of the seed, eg in *Cananga*; others

Annonaceae. 1 *Annona squamosa* (a) shoot with leaves in two ranks and axillary flower ($\times\frac{2}{3}$); (b) young fruit ($\times 1$); (c) vertical section of fruit (sweetsop) comprising an aggregate of numerous berries with the fleshy receptacle ($\times\frac{2}{3}$). 2 *Monodora myristica* (a) leafy shoot and flower ($\times\frac{2}{3}$); (b) androecium of numerous short-stalked stamens and gynoecium comprising united carpels, shown entire (left) and in vertical section (right) ($\times 3$); (c) vertical section of fruit with many seeds ($\times\frac{2}{3}$). 3 *Asimina triloba* (a) flowering shoot ($\times\frac{2}{3}$); (b) gynoecium of five free carpels ($\times 2$); (c) stamen with anther surmounted by swollen extension of connective ($\times 5$); (d) young fruit ($\times\frac{2}{3}$).

have a new integument intercalated between the inner and outer ones. The flower in many species acts as a trap for pollinating beetles.

**Classification.** Although its limits are well defined, the family is notoriously difficult to divide into "natural" groupings of genera. Two genera, *Monodora* and *Isolona* of Africa, have syncarpous ovaries and are separable as the subfamily MONODOROIDEAE. The other subfamily, ANNONOIDEAE, is divided into six tribes by Sinclair. The larger genera include the pantropical *Xylopia* (about 150 species), the Old World genera *Uvaria* (about 100 species), *Polyalthia* (about 100 species), *Friesodielsia* (about 50 species) and *Artabotrys* (about 100 species of climbing or erect shrubs with inflorescences borne on hooklike peduncles), the American *Annona* (about 100 species, a few in Africa) and *Monanthotaxis* with 56 species in Africa and Madagascar.

The Annonaceae is one of the group of primitive families of flowering plants sometimes termed the Annoniflorae, often with indefinite numbers of free floral parts and spirally arranged stamens. It is allied to the Magnoliaceae from which it differs in having ruminate seeds and no stipules, and growing at lower altitudes and latitudes, Winteraceae from which it differs in having vessels, Myristicaceae in its free stamens and Lauraceae and Calycanthaceae in having hypogynous flowers.

**Economic uses.** South American species of *Annona* were early introduced to the Old World for their fruits (notably soursop and sweetsop), though those of the Old World *Artabotrys* are edible too. The aromatic oils characteristic of the family give ylang-ylang from the flowers of *Cananga odorata*, and the perfume of *Mkilua fragrans* is used by Arab and Swahili women. The spicy fruits of the West African *Xylopia aethiopica* are the so-called "Negro pepper" used as a condiment, and those of *Monodora myristica* of the same area are used as a nutmeg substitute. Although some *Xylopia* woods are used, the timber is of little importance, except for pliable lancewoods.                         D.J.M.

# MYRISTICACEAE
*Nutmeg and Mace*

The Myristicaceae is a family of tropical trees. *Myristica*, with about 125 species, including the nutmeg, is the largest genus, centered in New Guinea. Species of the

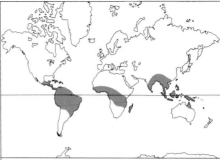

**Number of genera:** 16
**Number of species:** 380
**Distribution:** tropical, mainly in lowland rain forests.
**Economic uses:** source of the nutmeg and mace.

South American genus *Virola* are an important constituent of the Amazonian forests.

**Distribution.** The family is exclusively tropical, nearly all members inhabiting lowland rain forests. They are found in the Malaysian region, especially in New Guinea, in tropical America, especially the Amazonian basin, and in Africa and Madagascar.

**Diagnostic features.** The family comprises dioecious (ie male and female flowers on

Myristicaceae. 1 *Virola glaziouii* (a) leaf and male inflorescence ($\times\frac{2}{3}$); (b) male flower ($\times 4$); (c) half female flower ($\times 4$); (d) vertical section of fruit ($\times\frac{2}{3}$). 2 *Knema pectinata* shoot with male flowers ($\times\frac{2}{3}$). 3 *Myristica fragrans* (a) shoot with flowers and fruit (nutmeg) split open to show seed covered by the red aril ($\times\frac{2}{3}$); (b) fruit cut open ($\times\frac{2}{3}$); (c) vertical section of seed ($\times\frac{2}{3}$); (d) half male flower ($\times 2$); (e) androecium with stamens in a column ($\times 4$); (f) vertical section of column ($\times 4$); (g) half female flower ($\times 2$); (h) female flower opened out ($\times 2$). 4 *Horsfieldia macrocoma* (a) male inflorescence ($\times\frac{2}{3}$); (b) female and (c) male flowers with one sepal removed ($\times 4$); (d) fruit cut open ($\times\frac{2}{3}$); (e) cross section of seed ($\times\frac{2}{3}$).

separate individuals) trees with alternate entire leaves which often have glandular dots containing aromatic oil. Many species have a distinctive growth pattern with the branches whorled and almost horizontal. When wounded, the wood exudes a red sap. The flowers are small and inconspicuous and are borne in capitate, fascicled or corymbose inflorescences. There are no petals, but a calyx usually of three partly united sepals. Male flowers have two–30 stamens fused in a column; the anthers shed their pollen through two longitudinal slits. The female flower has a unilocular ovary containing one basal ovule; there is disagreement about whether the ovary is formed from one or two carpels. As the fruit develops it becomes fleshy and on maturity splits into two or four valves, disclosing the large seed. (This is one of the rare examples of a dehiscent berry.) The seed contains a small embryo and much endosperm, rich in fat and starch. The endosperm in most species is penetrated by ingrowing branches of the perisperm (a layer derived from the nucellus) and thus appears ruminate in cross-section. The seed is enveloped by a coarse, brightly-colored network of tissue, the aril, which is derived from the funicle.

**Classification.** The Myristicaceae has about 380 species. For a long time all members were classified in the single genus *Myristica*, probably because the female flower and the fruit show such uniformity of structure throughout the family. More detailed study later resulted in its division into numerous genera, and 16 are recognized today including *Knema*, *Virola*, *Pycnanthus* and *Horsfieldia*. The genera, each of which is confined to Malaysia, Africa or America, are separated mainly by differences in the structures of the staminal column, aril and inflorescence.

The family is very distinct, and opinions differ as to where its taxonomic affinities lie. The groups most frequently proposed as nearest relatives are the Monimiaceae (in turn related to the Lauraceae) and the Annonaceae; both of these comprise tropical rain forest trees with aromatic leaves. The Monimiaceae resembles the Myristicaceae in having mainly unisexual flowers with no petals, and carpels with a single ovule. The Annonaceae are similar to the Myristicaceae in their three-part perianth and large seeds with ruminate endosperm.

**Economic uses.** *Myristica fragrans* is the nutmeg tree, from which the spices nutmeg and mace are obtained. The pericarp of the fruit is also used to make a jelly preserve and inferior seeds are pressed to make "nutmeg butter" used in perfumery. Some species, eg the Indian *Gymnacranthera farquhariana* and the Brazilian *Virola surinamensis* (ucahuba), have waxy seeds used as a source of "butter" for consumption and candlemaking. The wood of this family makes poor-quality lumber with a high moisture content and is little used.                F.K.K.

## DEGENERIACEAE

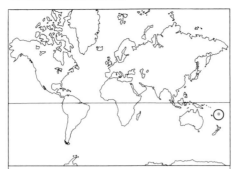

**Number of genera:** 1
**Number of species:** 1
**Distribution:** Fiji.
**Economic uses:** none.

The Degeneriaceae consists of a single Fijian tree species, *Degeneria vitiensis*.

The leaves are alternate, simple, entire and without stipules. The solitary bisexual flowers are considered primitive in that there are only three sepals, and the 12–13 petals and numerous stamens are arranged in several series. The stamens are fleshy and are associated with staminodes. The gynoecium consists of a single, superior carpel with a ventral, elongate, spreading stigma, open when young and containing numerous ovules set in a double row on two placentas. The fruit is large, indehiscent and leathery, containing numerous flattened seeds with copious endosperm and a minute embryo. It is closely related to the Magnoliaceae and the Winteraceae with which it shares many unspecialized characters. No economic uses have been reported.                S.R.C.

# ILLICIALES

## ILLICIACEAE

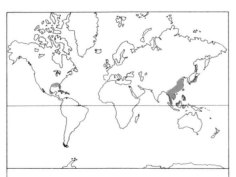

**Number of genera:** 1
**Number of species:** about 40
**Distribution:** W Indies, N America, Japan, China and SE Asia.
**Economic uses:** source of anise.

The Illiciaceae contains a single genus of shrubs and small trees (*Illicium*) that are native to the West Indies, North America, Japan, China and Southeast Asia.

The leaves are alternate, simple, entire, lacking stipules and often aromatic. The solitary, bisexual flowers have a multiseriate perianth of seven to numerous free segments and often numerous stamens. The outermost perianth segments are small and more sepaloid, the inner gradually becoming more petaloid and smaller. The gynoecium consists of a whorl of five–20 superior free carpels, each with a single ovule. The fruits are follicular, with glossy seeds containing a minute embryo and copious endosperm. It is related to the Magnoliaceae, Winteraceae and Schisandraceae. Oils used for flavoring are obtained from the bark of *Illicium parviflorum* (yellow star anise) and the fruits and seeds of *I. verum* (Chinese star anise).
S.R.C.

## SCHISANDRACEAE

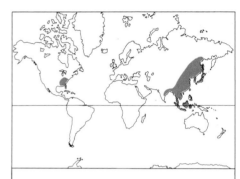

**Number of genera:** 2
**Number of species:** 47
**Distribution:** India, E and SE Asia, SE USA.
**Economic uses:** garden ornamentals.

The Schisandraceae is a family of two genera of trailing or twining woody shrubs, *Kadsura* and *Schisandra* native to India and eastern and Southeast Asia; one species of *Schisandra* is native in the south-eastern United States and Southeast Asia.

The leaves are alternate, without stipules, simple and frequently dotted with pellucid glands. The flowers are solitary, generally small, unisexual (male and female being borne on the same or different plants) and regular. The perianth of 9–15 free parts is not clearly differentiated into sepals and petals, but the innermost are more distinctly petaloid. The stamens are numerous to few with short bilocular anthers on short filaments, more or less fused at the base in a fleshy mass. The gynoecium consists of 12 to numerous superior free carpels on an elongated or obovoid torus. There are two (but rarely more) ovules in each carpel and the fruit is an aggregate of numerous sessile drupes each with one to five flattened seeds with a small embryo surrounded by copious endosperm. The Schisandraceae is related to the Magnoliaceae but differs in lack of stipules. Some *Schisandra* species are cultivated as garden ornamentals.                S.R.C.

# LAURALES

## AUSTROBAILEYACEAE

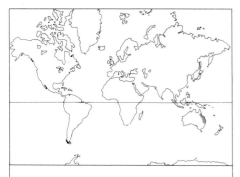

**Number of genera:** 1
**Number of species:** 2
**Distribution:** Queensland.
**Economic uses:** none.

The Austrobaileyaceae is a family of large climbing shrubs to high climbing lianas containing a single genus (*Austrobaileya*) which is native to Queensland, Australia.

The leaves are opposite, entire and possess small stipules. The bisexual flowers are solitary, axillary and possess a perianth of about 12 greenish, free segments undifferentiated into sepals and petals. There are numerous (12–25) petaloid stamens, the innermost of which are sterile. The anthers are bilocular with longitudinal dehiscence. The gynoecium consists of about eight superior free carpels with unsealed styles each containing several ovules in two rows and bearing a two-lobed style. Fruits are not known. The Austrobaileyaceae is a primitive family placed in the order Laurales, but it shows no clear relationship with other families in this order and its affinities are obscure. No economic uses are known.

S.R.C.

## LACTORIDACEAE

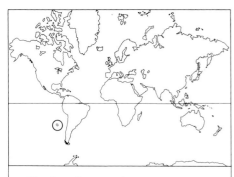

**Number of genera:** 1
**Number of species:** 1
**Distribution:** Juan Fernández Islands.
**Economic uses:** none.

The Lactoridaceae is represented by one shrubby species (*Lactoris fernandeziana*) which is endemic to the Juan Fernández Islands.

The leaves are small, stipulate, alternate and dotted with tiny glands. The flowers are bisexual or unisexual, with both sexes on the same plant. The floral parts are in threes; there are three sepals, no petals, six stamens in two rows, and three free carpels each with six to eight ovules. The follicular fruit has either four or six seeds, each containing much oily endosperm and a small embryo. The Lactoridaceae occupies an isolated position in the angiosperms. With its trimerous flowers it is excluded from the order Magnoliales which it otherwise resembles. Here, it is ascribed to a more advanced order, the Laurales, perhaps derived from the stipulate family, the Chloranthaceae. Beyond this its relationships are doubtful. No economic uses are known.

S.W.J.

## EUPOMATIACEAE

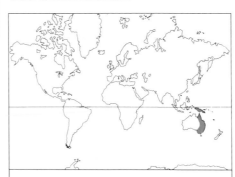

**Number of genera:** 1
**Number of species:** 2
**Distribution:** E Australia and New Guinea.
**Economic uses:** timber is used locally.

The Eupomatiaceae comprises a single genus (*Eupomatia*) of shrubs or small trees.

**Distribution.** The family is native only to eastern Australia and New Guinea, mainly in rain forests.

**Diagnostic features.** The leaves are alternate, simple and lack stipules. The flowers are showy, perigynous, solitary and bisexual without clear differentiation between petals and sepals. The perianth is fused into one mass, the upper part falling off as a conical lid, the lower part filled with the thickened receptacle on the edge of which the stamens arise. The innermost stamens are sterile and petal-like. the others prong-shaped with pointed anthers. The carpels are sunk into the top-shaped receptacle, each carpel containing a few ovules and all maturing to form a flat-topped berry with oil glands, the several locules each containing one or two angular seeds. The seeds contain copious endosperm and a very small embryo.

**Classification.** The single genus contains only two species. *Eupomatia bennettii* from Queensland is a shrub with fleshy roots, flowers to 2.5cm (1in) diameter, and having yellow central staminodes, orange marginal staminodes, and fertile stamens reflexed onto the flower stalk. The flowers are beetle-pollinated, which is considered to be a primitive trait in evolutionary terms. The other species, *E. laurina* from Victoria, New South Wales, Queensland and New Guinea, has white flowers and spreading instead of reflexed stamens. It forms a tree growing to about 10m (33ft).

The Eupomatiaceae was formerly included in the Annonaceae, to which it seems to be allied, apart from exhibiting certain more highly developed characters. The family is also related to the Magnoliaceae.

**Economic uses.** The prettily marked wood of *E. laurina* is used locally for its decorative qualities and *E. bennettii* has limited uses as an ornamental.

B.M.

## GOMORTEGACEAE

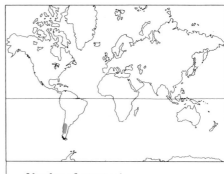

**Number of genera:** 1
**Number of species:** 1
**Distribution:** Chile.
**Economic uses:** none.

The Gomortegaceae consists of a single evergreen tree species, *Gomortega keule* (*G. nitida*) which possesses aromatic oil glands.

**Distribution.** *Gomortega keule* is native to South central Chile.

**Diagnostic features.** The tree bears simple, pinnately veined leaves, which are shiny, without stipules and borne in opposite pairs on the stem. The flowers are regular and bisexual, borne in axillary or terminal racemes. They are subtended by two bracts and have a perianth of six to 10 spirally arranged segments, which are probably best regarded as sepals. There are between two and 11 stamens also arranged spirally. The very short stamen filaments are free; the innermost possess two short-stalked oil glands at the base of each filament. The anthers dehisce by means of valves. The ovary is inferior of two or three fused carpels, and has two or three locules with a single pendulous ovule in each. The single style terminates in two or three stigmatic lobes. The fruit is drupe-like with a very hard endocarp and fleshy exocarp. The seed contains a large embryo surrounded by a quantity of oily endosperm.

**Classification.** The family shares a number of features with other families in the order Laurales, where it is usually placed. Perhaps the closest relationship is with the laurel family (Lauraceae).

**Economic uses.** No economic uses are known for this family.

S.R.C.

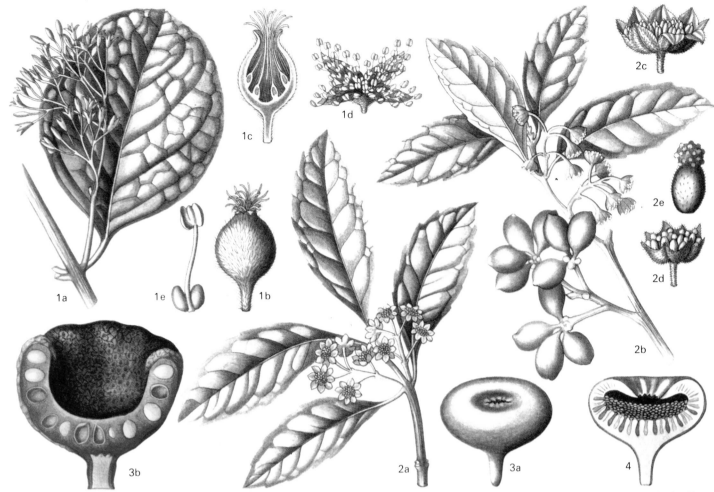

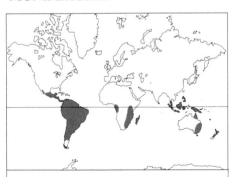

Monimiaceae. 1 *Monimia rotundifolia* (a) leaf and axillary inflorescence ( ×⅔); (b) female flower with hairy receptacle and many separate styles ( ×4); (c) vertical section of female flower showing free carpels ( ×4); (d) male flower with many stamens ( ×4); (e) stamen with basal glands ( ×18). 2 *Hedycarya arborea* (a) leafy shoot bearing axillary inflorescences of male flowers ( ×⅔); (b) leafy shoot with female flowers and fruits ( ×⅔); (c) male flower with uniform perianth segments ( ×2); (d) female flower with numerous free carpels ( ×2); (e) carpel ( ×6). 3 *Tambourissa elliptica* (a) flower ( ×2); (b) vertical section of fruit showing achenes deeply embedded in the receptacle ( ×2). 4 *Tambourissa* sp fruit ( ×⅔).

## MONIMIACEAE

**Number of genera:** 30
**Number of species:** about 450
**Distribution:** tropics—Polynesia, Australasia, Madagascar, America.
**Economic uses:** local uses for timber, as fruits, and in perfumes, medicines and dyes

The Monimiaceae comprises 30 genera and some 450 species of trees and shrubs many of which produce aromatic oils.

**Distribution.** The family is native to austral regions such as Polynesia, Australasia, Madagascar and tropical America, being noticeably absent from India and poorly represented in Africa. *Hortonia* is endemic to Sri Lanka.

**Diagnostic features.** The leaves are opposite without stipules, evergreen, leathery and in many species contain aromatic oils. The flowers are perigynous, solitary, in racemes or cymes, often unisexual and regularly symmetrical. The perianth consists of four to many segments in two or more overlapping series, sometimes with the outer sepal-like and the inner petal-like, or rarely lacking altogether. The stamens are few to numerous in one or two series with short filaments, the anthers opening by slits. The ovary is superior and consists of many free carpels (sometimes few or one), each locule containing one erect, basal or rarely pendulous ovule. In fruit the carpels become dry achenes often more or less enclosed within or on the fleshy receptacle. The seeds contain a straight embryo and copious endosperm.

**Classification.** The family's numerous monotypic (single-species) genera include *Lauterbachia* from New Guinea, *Hedycaryopsis* and *Schrameckia* from Madagascar, *Amborella* (sometimes placed in the segregate family Amborellaceae) and *Carnegiea* (*Carnegieodoxa*) from New Caledonia, *Hennecartia* from Paraguay and Brazil, *Macropeplus* and *Macrotorus* from eastern Brazil, and *Peumus* from Chile

The larger genera include: *Mollinedia* with about 75 species in the American tropics, *Kibara* with 30 Indo-malaysian and tropical Australasian species, *Tambourissa* with 25 species in Madagascar, Mauritius and the Mascarenes, *Hedycarya* with 15 species in Polynesia and Australasia, and the five species of *Wilkiea*, all restricted to eastern Australia.

Among the seven genera placed either in a separate family, the Atherospermataceae, or in the Monimiaceae itself are *Siparuna* of tropical America, *Laurelia*, the two species of *Atherosperma* from Tasmania and Victoria, and the monotypic *Doryphora sassafras* from New South Wales, the Australian sassafras. Also placed in the Monimiaceae are the genera *Trimenia* and *Piptocalyx*, which are sometimes segregated as the family Trimeniaceae.

The Monimiaceae constitutes an interesting link with the Magnoliales, especially the Magnoliaceae, to which it is related through the Calycanthaceae, and the Lauraceae, with

Calycanthaceae. 1 *Chimonanthus praecox* (a) shoot bearing solitary flowers, surrounded by numerous bracts in bud, which open before the leaves appear (×⅔); (b) half flower with bract-like outer and petaloid inner perianth segments (×3); (c) half false fruit comprising a fleshy receptacle surrounding the true fruits – achenes (×⅔); (d) leaf (×⅔). 2 *Calycanthus occidentalis* (a) shoot bearing leaves and axillary flowers with numerous petaloid perianth segments (×1); (b) half flower showing numerous perianth segments, stamens with short filaments and the ovary comprising free carpels with free styles and stigmas, all inserted on a cup-like receptacle (×2).

which there is a relationship via *Atherosperma*. The monimiaceous flower structurally suggests magnolia-like flowers with a concave floral axis, the stamens and carpels depressed within the axis.

**Economic uses.** Local use is made of the wood of some species. Aromatic oils extracted from leaves and bark are widely used locally to prepare perfumes (as with *Doryphora sassafras*) and medicines (as with *Peumus boldus* which also yields dye bark, edible fruit and a hardwood, boldo).   B.M.

# CALYCANTHACEAE
## Carolina Allspice and Chimonanthus

The Calycanthaceae is a small family of hardy deciduous or evergreen shrubs.

**Distribution.** The three genera have a discontinuous distribution: *Calycanthus* (four species) in the southwest and eastern United States of America, *Chimonanthus* (three species) in eastern Asia, and *Idiospermum* in the rain forests of northeastern Queensland.

**Diagnostic features.** The leaves are simple, entire, without stipules and opposite. The fragrant, bisexual, actinomorphic flowers are axillary and solitary. The perianth is in several overlapping series with a gradual transition from sepals to petals. The outermost in the series are bract- or scale-like, the remainder distinctly showy and petaloid. These, and the numerous stamens, are inserted on the rim of a thick cup-like receptacle. The stamens have short, free filaments and anthers with two locules which dehisce longitudinally by means of slits. The innermost stamens are sometimes sterile (staminodes). There are few to numerous (up to 20) free carpels inserted on the inside of the receptacle. The single locule of each carpel contains one or two superposed anatropous ovules. The fruits are one-seeded achenes enclosed in the enlarged fleshy receptacle. The seeds have no endosperm, but a large embryo with two folded leafy or three or four massive fleshy cotyledons.

**Classification.** The three genera show a number of distinctive features: *Calycanthus* are deciduous shrubs with reddish or purple flowers with both perianth and stamens in several series. *Chimonanthus* are deciduous or evergreen shrubs with small yellowish flowers with the perianth in several series but the stamens in only two series, the inner one sterile. *Idiospermum* is a large evergreen tree with laminar stamens and an ovary of one or

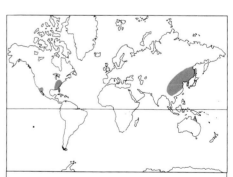

**Number of genera:** 3
**Number of species:** 8
**Distribution:** N America, E Asia, NE Australia.
**Economic uses:** ornamental shrubs, medicines and spice (Carolina allspice).

two, free, carpels with subsessile stigmas.

The family shows a number of features resembling those of the Magnoliaceae and the Annonaceae. For example most show spiral arrangement of the perianth segments and numerous stamens and carpels. Thorne regards the Calycanthaceae as very closely related to the Monimiaceae.

Chloranthaceae. 1 *Chloranthus angustifolius* (a) roots, rhizome and aerial stem-bases ($\times\frac{2}{3}$); (b) leafy shoot and inflorescence ($\times\frac{2}{3}$); (c) bisexual flower rear (upper) and front (lower) view ($\times 2$); (d) androecium comprising central anther with two locules and two lateral anthers with one locule ($\times 3$); gynoecium (e) entire and (f) in vertical section ($\times 3$). 2 *Sarcandra glabra*, fruiting shoot ($\times 1$). 3 *Ascarina lanceolata* (a) leafy shoot with male inflorescence ($\times\frac{2}{3}$); (b) bracteate male flower ($\times 6$); (c) anther ($\times 6$); (d) female inflorescence ($\times\frac{2}{3}$); (e) fruits—drupes ($\times 4$); (f) vertical section of fruit ($\times 6$). 4 *Hedyosmum brasiliense* (a) shoot with male inflorescence ($\times\frac{2}{3}$); (b) female flowers ($\times 4$); (c) gynoecium ($\times 4$); (d) male flower ($\times 8$).

**Economic uses.** At least two *Calycanthus* species are grown as ornamental shrubs for their sweetly fragrant summer flowers. Both *Calycanthus floridus* (Carolina allspice) and *C. occidentalis*, native to California, grow to a height of about 3m (10ft) and bear large solitary flowers 5cm–7.5cm (2–3in) across.

In addition to their value as ornamental shrubs two species of *Calycanthus* have other uses. The bark, leaves and roots of *C. fertilis* (eastern United States of America) yield medicinal extracts. The aromatic bark of *C. floridus* is used as spice by American Indians.

Only one species of *Chimonanthus*, *Chimonanthus fragrans* (*C. praecox*), is cultivated widely, and is often known under the name wintersweet. In Japan the flowers are used to make perfumes.                    S.R.C.

# CHLORANTHACEAE

The Chloranthaceae is a small family of herbs, shrubs and trees native to the tropics and southern temperate regions.

**Distribution.** Species in the family occur in tropical America (genus *Hedyosmum*), east Asia and the Pacific, with some of the genus *Ascarina* as far south as New Zealand.

**Diagnostic features.** The leaves are mostly

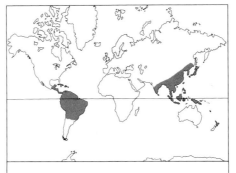

**Number of genera:** 5
**Number of species:** 65
**Distribution:** tropical America, E Asia, Pacific.
**Economic uses:** limited uses as ornamentals, and locally in beverages and medicinally.

aromatic, simple and possess small stipules. They are arranged in opposite pairs, with the opposing petiole bases often meeting and fusing. The flowers are small, devoid of petals, and unisexual or bisexual. The male flowers are arranged in large numbers on a racemose inflorescence. They have no sepals and have either one stamen or three fused

together. Female and bisexual flowers are arranged in small numbers on a short raceme or spike. They possess a three-toothed calyx fused to the unilocular, inferior ovary, which contains only one pendulous ovule, inserted at the top of the locule, and is surmounted by a stigma on a short style. The fruit is a small ovoid or globose drupe with a more or less succulent outer wall. It contains a single seed with a minute embryo surrounded by copious endosperm.

**Classification.** The genera *Chloranthus* and *Sarcandra* have bisexual flowers, the first having three stamens and the latter one stamen. *Ascarina*, *Ascarinopsis* and *Hedyosmum* have unisexual flowers, the former two with bracteate male flowers and naked female flowers, the latter with male flowers with bracts and female flowers enclosed by a cup-like bract.

The family is distinguished from the related pepper family (Piperaceae) by the possession of opposite leaves, united petiole bases, and the inferior ovary, with its single locule and single ovule.

**Economic uses.** At least one species of *Chloranthus* (*Chloranthus glaberi*) is grown as an ornamental shrub. The leaves of *C.*

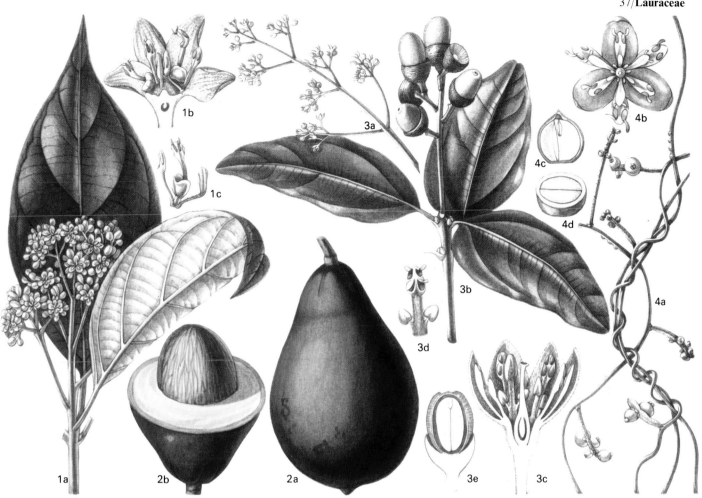

Lauraceae. 1 *Hypodaphnis zenkeri* (a) shoot and inflorescence ( ×⅔) ; (b) half flower ( ×4) ; (c) stamen from second whorl, two from third whorl and gland ( ×4). 2 *Persea gratissima* (avocado) fruit (a) entire and (b) cut away to reveal hard seed ( ×⅔). 3 *Cinnamomum litseifolium* (a) inflorescence ( ×⅔) ; (b) shoot with fruits enclosed in fleshy cupules ( ×⅔) ; (c) half flower showing introrse and extrorse dehiscence of anthers ( ×8) ; (d) stamen with glandular bases and anthers dehiscing by flaps ( ×2) ; (e) half fruit ( ×1). 4 *Cassytha filiformis* (a) habit of this parasitic plant ( ×⅔) ; (b) flower from above ( ×8) ; fruit in vertical (c) and cross (d) section ( ×2).

*officinalis* are used to make a drink in parts of Malaya and Indonesia. The flowers of *C. inconspicuus* are used to flavor tea in various parts of east Asia and an infusion of the flowers and leaves is used to treat coughs. Extracts of leaves from the tropical South American shrub *Hedyosmum brasiliense* are used locally as a diuretic, tonic, aphrodisiac, to induce sweating and to treat stomach complaints.                                                S.R.C.

## LAURACEAE
*Avocado, Bay Laurel and Cinnamon*

The Lauraceae is a family of mostly trees or shrubs, although a few are parasitic climbers without true leaves.

**Distribution.** Although members of the family can be found throughout tropical and subtropical regions of the world, primarily in lowland to montane rain forests, the main centers are in Southeast Asia and tropical America. Genera reaching temperate areas include *Lindera*, *Persea* and *Sassafras*. Several genera occur in the relict laurel forests of the Canary Islands and Madeira, such as *Apollonias*, *Laurus*, *Ocotea* and *Persea*.

**Diagnostic features.** The leaves are alternate

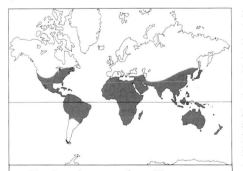

**Number of genera:** about 32
**Number of species:** about 2,500
**Distribution:** tropics and subtropics, centered in SE Asia and Brazil.
**Economic uses:** avocado, cinnamon, camphor, sassafras oil, bay leaves, timber, and some ornamental species.

or opposite, usually leathery and evergreen, and without stipules; along with most other parts of the plants, they contain numerous oil cavities, hence the aromatic nature of some of the products of this family. The inflorescence is usually racemose or cymose, and one group of genera, the tribe Laureae, have the cymes surrounded by large, leafy bracts forming an involucre. The flowers are

regular and either bisexual or unisexual on the same plant. The parts of the flower are in multiples of three, and the petals and sepals, not being well differentiated from each other, form a rather inconspicuous perianth. The stamens, in three or four whorls, are fused to the base of the perianth and form a perigynous zone; some are often reduced to staminodes. The anthers have two or four locules, mostly splitting by valves on the inner face, but often those of the inner third whorl opening towards the outside. The ovary is usually superior, sometimes inferior, and surrounded by, but not fused to, the cup-shaped perigynous zone of the flower; it has a single locule with a single, pendulous, anatropous ovule. The style is simple and the stigma small. The fruit is a berry or drupe-like, often enclosed by the perigynous part of the flower which may become a more or less fleshy cupule. The seed, without endosperm, contains a straight embryo.

**Classification.** The family is divided very unequally into two subfamilies, the small CASSYTHOIDEAE comprising a group of parasitic climbers, and the rest of the genera, the LAUROIDEAE. The latter, which have sometimes been further split into two subfamilies (depending on whether the anthers have two or four locules), are now considered to comprise five tribes. The only tribe with an involucrate inflorescence is the LAUREAE (Litseeae), which include *Laurus* with two species and *Litsea* with over 400 species. Another tribe with the basal part of the fruit surrounded by a fleshy cupule is the CINNAMOMEAE, with important genera such as *Ocotea* and *Cinnamomum*. The tribe PERSEEAE has no fleshy cupule surrounding the fruit, but contains the well-known fruit genus *Persea* (avocado), with over 150 species. *Beilschmiedia* and *Endiandra* are two other large genera in this tribe. The other two tribes have the fruit enclosed in enlarged hardened perigynous tissue: the small tribe HYPODAPHNIDEAE has an inferior ovary, while the CRYPTOCARYEAE, with several spice-producing genera (species of *Cryptocarya* and *Ravensara*) has a superior ovary.

The Lauraceae is a relatively unspecialized family usually placed near the Monimiaceae (from which it differs in the valvate dehiscence of the anthers), and not too far from the Magnoliaceae.

**Economic uses.** *Cinnamomum* provides both cinnamon and camphor. *Sassafras* yields oil of sassafras. *Persea americana* (avocado pear) is an important tropical fruit. Many genera contain valuable timber trees, such as species of *Beilschmiedia*, *Endiandra*, *Ocotea* (green heart) and *Litsea*, the last also being a source of many local medicines. *Laurus* and *Lindera* species are grown as ornamentals. The aromatic leaves of the bay laurel, *Laurus nobilis*, are used as a flavoring in fish and meat dishes. I.B.K.R.

## HERNANDIACEAE

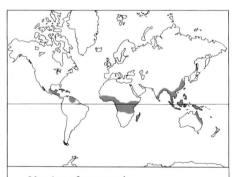

**Number of genera:** 4
**Number of species:** 76
**Distribution:** pantropical, mostly coastal.
**Economic uses:** timber used locally.

The Hernandiaceae is a family of tropical genera of trees, shrubs and some lianas.

**Distribution.** The family is pantropical, most common in coastal areas.

**Diagnostic features.** The leaves are without stipules, large, simple or palmately compound and alternately arranged. The flowers are regular, unisexual or bisexual and arranged in cymose inflorescences. The perianth consists of three to ten sepal-like segments; there are three to five stamens, and one or two outer whorls of staminodes are sometimes present. The ovary is inferior, consists of a single carpel containing one pendulous ovule and surmounted by a single style. The fruit is an achene, sometimes two- or four-winged; the seed, with a leathery testa, has no endosperm.

**Classification.** There are four genera: *Sparattanthelium* (15 species, Central America) with simple leaves and dry, ribbed fruits; *Hernandia* (24 species, Central America, Guiana, West Indies, West Africa, Indomalaysia, Pacific Islands) with male and female flowers on the same plant, peltate leaves and fruits surrounded by the inflated perianth; *Illigera* (30 species, tropical Africa, Madagascar, eastern Asia and western Malaysia); and *Gyrocarpus* (seven species, tropics), which has fruits with two terminal wings and unisexual flowers. The family is most closely related to the Monimiaceae.

**Economic uses.** The soft, light wood of some species of *Hernandia* and *Gyrocarpus* is used locally for crates, boxes and cheap plywood.
S.R.C.

# PIPERALES

## PIPERACEAE
*Pepper and Kava*

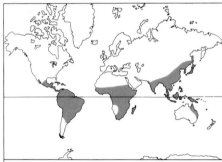

**Number of genera:** about 5
**Number of species:** about 2,000
**Distribution:** pantropical, frequently in rain forests.
**Economic uses:** pepper and kava.

The Piperaceae is a tropical family of small trees, shrubs and woody climbers. Its principal genus, *Piper*, yields the common condiment pepper.

**Distribution.** The family is pantropical, and inhabits mostly rain forests.

**Diagnostic features.** The leaves are characteristically alternate, simple, entire, dotted with glands containing pungent aromatic oil, and have winged petioles which sheathe the conspicuously jointed stem; the stipules when present are usually attached to the petiole. The flowers are tiny, bisexual or unisexual, borne in racemes or spikes which are usually leaf-opposed (in fact terminal, the stem continuing to grow sympodially from a bud in the axil of the last leaf). There are two to several stamens. The ovary is superior with two to four or more fused carpels surrounded by three scale-like bracts and with a single locule which contains a subbasal ovule; there is no perianth. The fruit is a fleshy single-seeded drupe, often sunk into the inflorescence axis or fused with the bracts. The family is very unusual in its vascular anatomy: some of the bundles are scattered within the stem rather as in the monocotyledons although they are closed, not open.

**Classification.** The Piperaceae is treated by some botanists as comprising the single genus *Piper* (about 2,000 species) while other workers recognize several small segregate genera, for example *Ottonia* (70 species, South America), *Pothomorphe* (10 species, tropical), *Sarcorhachis* (four species, Central and tropical South America), which are characterized by differences in the form and position of the inflorescence.

The family most nearly allied to the Piperaceae is the Peperomiaceae, which includes the large genus *Peperomia*; many authorities, indeed, amalgamate the two families. *Peperomia* differs from *Piper* mainly in being herbaceous; it has even more reduced flowers, comprising a single carpel and two anthers subtended by one bract. The Piperaceae is also quite closely related to the Saururaceae; the latter differs in having stem vascular bundles in one ring, stamens and carpels spirally arranged, and (in *Saururus*) free carpels with the ovary and style not completely closed. These features suggest a link, via the Saururaceae, between the Piperaceae and the Magnoliaceae.

**Economic uses.** Useful products of the Piperaceae, though few, are highly appreciated by a large proportion of the world's population. *Piper nigrum* is the source of pepper. Kava, the famous national beverage of the Polynesians is made from the roots of *P. methysticum*. Originally, this narcotic, sedative drink was prepared by children and young women who would chew the peeled roots and eject the pulp into a vessel where it was diluted with water and allowed to ferment. Nowadays, however, the plant is grated mechanically.

*Piper betle* leaves are employed as a masticatory from East Africa to India and Indonesia, in conjunction with the betel nut.
F.K.K.

## SAURURACEAE

The Saururaceae is a small family of temperate and subtropical perennial herbs, commonly called lizard's tails in reference to the appearance of the inflorescences.

**Distribution.** The family is found in eastern Asia and North America.

**Diagnostic features.** The leaves are alternate, simple, and the stipules are partly fused to the petiole. The flowers are regular, bisexual and borne in dense racemes or spikes. Bracts

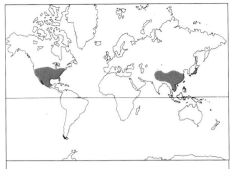

**Number of genera:** 5
**Number of species:** 7
**Distribution:** E Asia and N America.
**Economic uses:** garden ornamentals and folk medicines.

or colored upper leaves often surround the base of the inflorescence. There is no perianth and the flowers consist of three, six or eight stamens which are free or more or less fused to the carpels. The carpels are superior or inferior, three to five in number and sometimes fused together. The styles are free and each free carpel contains two to four ovules, while the locule of fused carpels contains six to ten ovules on each parietal placenta. The fruit may consist of dehiscent capsule-like units when free, or, when fused, a thick capsule opening at the top. Each seed has little endosperm and a small embryo, but a well-defined mealy perisperm.

**Classification.** The five genera are *Houttuynia*, with one species native from the Himalayas to Japan; *Saururus* (from the Greek words for lizard and tail), with two species native to eastern North America and eastern Asia respectively; *Anemopsis*, with a single species from southwest North America; and two Chinese genera, *Gymnotheca* with two species and *Circaeocarpus* with one.

Botanists have related the family to the Piperaceae and Chloranthaceae, but there is also some similarity with the Polygonaceae.

**Economic uses.** Like certain polygonums, *Houttuynia cordata* makes a good ground cover in moist situations, and is perhaps the most commonly cultivated member of the family. It was introduced into Europe about 1820, when seed was sent to the Royal Botanic Gardens at Kew, the plants first flowering there in 1826. The attraction of the species lies partly in the leaves, which are suffused with purple when grown in full sunlight, but green and more luxuriant when grown in shade. The terminal spike of flowers is subtended by a collar of four white bracts, and the whole inflorescence much resembles a single flower; in this respect it is like a number of *Cornus* species. *Houttuynia cordata* is cultivated locally in Vietnam where the leaves are used in salads and for treating eye diseases.

The genus *Saururus* comprises two species of bog plants, *Saururus chinensis* from the Philippines and eastern Asia with creamy cylindrical flower spikes, and *S. cernuus* from eastern North America with fragrant white flower spikes and conspicuously protruding stamens. This genus is occasionally cultivated.

*Anemopsis californica* grows in wet places and is locally known as yerba mansa. It has an aromatic stoloniferous rootstock which was once taken and fashioned into cylindrical necklace beads by Indians, hence the other colloquial name, Apache beads. The same rootstock when infused in water is reputed to ease malarial and dysenteric conditions. The spike of tiny flowers is surrounded by petal-like bracts, making the whole inflorescence resemble a single anemone flower. B.M.

Saururaceae. 1 *Anemopsis californica* (a) shoot with upper leaves and inflorescence surrounded by petal-like involucre of bracts (×⅔); (b) shoot with fruiting head (×⅔); (c) leaf (×⅔); (d) flower which lacks a perianth and the ovary is sunk into the receptacle (×4); (e) cross section of inflorescence (×1½). 2 *Gymnotheca chinensis* (a) flowering shoot with cordate leaves (×⅔); (b) flowers (×2); (c) gynoecium (×4); (d) ovary opened out to show ovules on parietal placentas (×6). 3 *Saururus cernuus* (a) flower and bract (×2); (b) half flower with free carpels (×3). 4 *Houttuynia cordata* (a) flowering shoot (×⅔); (b) flower (×4); (c) fruit dehiscing from apex (×4); (d) fruit opened out to show seeds (×4).

# PEPEROMIACEAE
*Pepper Elders*

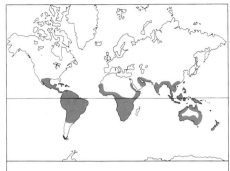

**Number of genera: 4**
**Number of species:** about 1,000
**Distribution:** tropical and subtropical, chiefly America and the W Indies.
**Economic uses:** local uses of *Peperomia* species in medicine and as food; cultivated indoor and greenhouse ornamentals (*Peperomia*).

The Peperomiaceae is a small family of succulent herbs, rarely subshrubs, many being either epiphytic or living in rocky habitats. Many have attractive foliage.

**Distribution.** The family is native to tropical and subtropical regions, especially tropical America and the West Indies.

**Diagnostic features.** The leaves, which are sometimes very large, are without stipules and usually alternate, rarely opposite or in whorls. The flowers are bisexual, very small and usually aggregated in dense spikes. They are subtended by fleshy or succulent bracts. The flowers have no sepals or petals and only two stamens, which are fused in their upper part to form a single anther. The gynoecium consists of a single carpel, containing a single basal ovule. The style is simple (rarely divided). The fruit is a berry with a succulent, thin or dry pericarp and contains a small seed, with a minute embryo surrounded by mealy endosperm.

**Classification.** Three of the genera, *Manekia* (one species), *Piperanthera* (one species), and *Verhuellia* (three species), are native to the West Indies. The tropical and subtropical genus *Peperomia* (about 1,000 species) is native chiefly to tropical America. Many species of *Peperomia* have tuberous rootstocks and many are epiphytes. In *Peperomia* the outer vascular bundles in the stem are scattered in the manner characteristic of many monocotyledons. Most species of *Peperomia* contain translucent cells in the outer cortex and epidermis of the stems and also in the leaves.

The four genera show close relationships with those of the pepper family (Piperaceae) in the possession of spike-like inflorescences, bisexual flowers with no sepals or petals, and an ovary with one locule containing one ovule. However, the Peperomiaceae is distinguished by lacking stipules, by having only two stamens in the flower, and by certain unique anatomical features.

**Economic uses.** *Peperomia* (pepper elders) is the only genus of economic importance. The young leaves and stems of the epiphytic *Peperomia vividispica* are eaten uncooked in various parts of Central and South America. A number of species of *Peperomia* have attractive foliage and are cultivated as ornamentals, eg, *P. caperata* (numerous heart-shaped leaves with deeply-furrowed surfaces and long white flower spikes), *P. obtusifolia* (purple stems, ovate thick fleshy leaves and numerous white flower spikes) and *P. argyreia* (or *P. sandersii*, round to ovate leaves silvery with dark green vein-stripes and red petioles).      S.R.C.

Peperomiaceae. 1 *Peperomia fraseri* (a) shoot with opposite entire leaves and flowers in a terminal spike ($\times\frac{2}{3}$); (b) ovary entire and half section with basal ovule ($\times 6$); (c) anthers ($\times 6$); (d) succulent bract ($\times 6$); (e) part of fruiting head ($\times 2$). 2 *Verhuellia brasiliensis* (a) creeping stem with adventitious roots ($\times\frac{2}{3}$); (b) inflorescence ($\times 3$); (c) flower ($\times 9$). 3 *Peperomia ovalifolia* (a) habit ($\times\frac{2}{3}$); (b) flower with two stamens ($\times 3$); (c) fruit ($\times 3$); (d) vertical section of fruit with minute embryo ($\times 3$). 4 *P. marmorata* (a) shoot with leaves and flower spikes ($\times\frac{2}{3}$); (b) part of spike with flowers ($\times 1\frac{1}{2}$); (c) flower with mushroom-like fleshy bract, two stamens and a single ovary crowned by a dissected stigma ($\times 6$); (d) flower ($\times 6$).

# ARISTOLOCHIALES

## ARISTOLOCHIACEAE
*Aristolochias and Asarums*

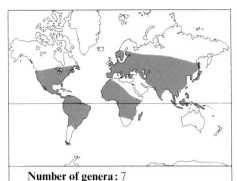

**Number of genera:** 7
**Number of species:** about 625
**Distribution:** tropical and temperate Asia, Africa, Europe, N & S America; usually in forests, often woody lianas.
**Economic uses:** floriculture.

The Aristolochiaceae is a well-defined family of herbs or shrubs, many of them twining lianas, occurring mostly in tropical and warm temperate forest or scrub.

**Distribution.** The family is distributed throughout tropical and temperate Eurasia and America, but excluding Australia.

**Diagnostic features.** The leaves are alternate, often palmately veined, without stipules. The flowers are axillary, solitary or in racemes or cymes, and are bisexual and regular or irregular. The calyx is regularly three- or four-lobed or irregularly one- to three-lobed with three to six teeth, and the corolla is usually absent, three petals (sometimes vestigial) rarely being present. The stamens are 6–40 in number, in one or two rows, free, connate or joined into a stylar column (gynostemium). The ovary is usually inferior, with three to six locules and three to six stigmas. The fruit is usually a many-seeded capsule, and the seeds small, with copious endosperm.

**Classification.** The seven genera are most satisfactorily subdivided into three tribes:
SARUMEAE. Perennial herbs. Flowers regular, solitary. Petals present or absent. Stamens in two rows, free or united at base of filaments. Ovary semi-superior or inferior. Fruit a follicle or capsule. *Saruma* (one species), with well-developed petals (China). *Asarum* (about 100 species), with vestigial or no petals

(mostly temperate Eurasia, North America).
BRAGANTIEAE. Shrubs or subshrubs. Flowers regular, in racemes or cymes. Petals absent. Stamens in one or two rows, free or united with style. Ovary inferior. Fruit a capsule. *Thottea* (10 species), with 12 to about 40 stamens in two rows (Malaysia). *Apama* (12 species), with six stamens in one row (Malaysia, eastern India).
ARISTOLOCHIEAE. Shrubs or perennial herbs. Flowers irregular, solitary or in racemes or cymes. Petals absent. Stamens in one row, united with style. Ovary inferior. Fruit a capsule. *Holostylis* (one species), with slightly bipartite calyx (Brazil, Bolivia, Paraguay, Argentina). *Aristolochia* (about 500 species), with one- to three-lobed, three- to six-toothed calyx (temperate and tropical Eurasia and America). *Euglypha* (one species), with irregularly single-lobed calyx (Brazil, Bolivia, Paraguay, Argentina).

The Aristolochiaceae is difficult to place. Features of the pollen and the presence of cells containing essential oils suggest that the Aristolochiaceae has closest affinities with the Magnoliaceae and related families. Differences are, however, sufficient for it to be included in a distinct order.

Aristolochiaceae. 1 *Aristolochia elegans* (a) twining leafy shoot with alternate leaves and solitary flowers having a colorful calyx ($\times\frac{2}{3}$); (b) receptacle and gynostemium formed from six sessile anthers below and six stigmatic lobes above ($\times 1$); (c) gynostemium ($\times 1\frac{1}{2}$); (d) fruit—a capsule opening from the base ($\times\frac{2}{3}$). 2 *Asarum europaeum* (a) habit showing creeping aerial stems, scale leaves and pairs of foliage leaves and single flower arising at joint of creeping stem ($\times\frac{2}{3}$); (b) flower with perianth removed to show numerous stamens and thick style with lobed stigma ($\times 4$); (c) half flower with regular perianth lobes and an inferior ovary with ovules on axile placentas ($\times 3$).

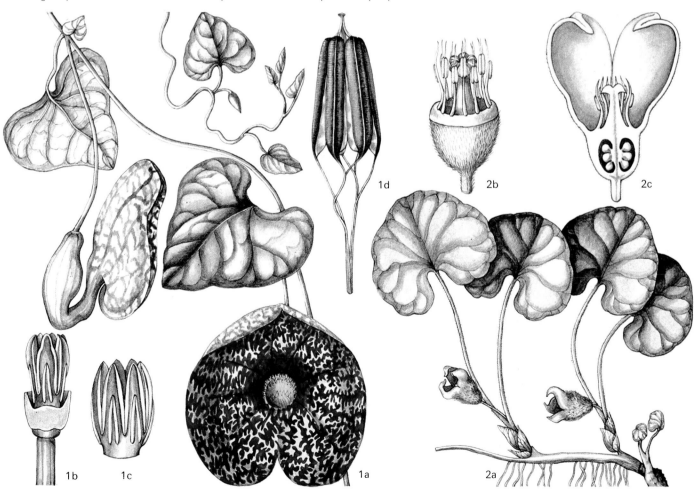

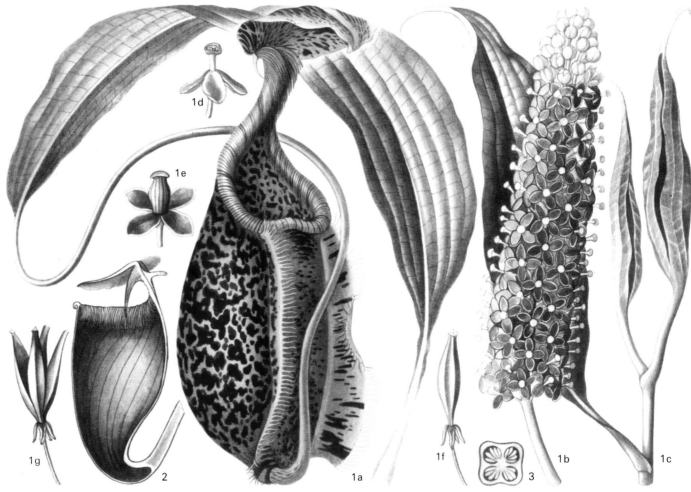

Nepenthaceae. 1 *Nepenthes rafflesiana* (a) leaf, the tip of which is modified to form a pitcher ( ×⅔); (b) inflorescence ( ×⅔); (c) leaves with apical tendrils ( ×⅔); (d) male flower with filaments united into a column ( ×1½); (e) female flower with sessile stigma ( ×1½); (f) fruit—a capsule ( ×⅔); (g) fruit dehiscing by four valves ( ×⅔). 2 *N. bicalcarata* vertical section of pitcher ( ×½). 3 *N. fimbriata* cross section of ovary with four locules and numerous ovules on axile placentas ( ×2⅔).

**Economic uses.** Many *Aristolochia* and *Asarum* species are cultivated for their curious flower shapes and often colorful variegated leaves, such as Dutchman's pipe (*Aristolochia macrophylla*), bird's head (*A. ornithocephala*) and pelican flower (*A. grandiflòra*). The American snake-root (*Aristolochia serpentaria*) and the European birthwort (*A. clematitis*) are reputed to have medicinal properties.                 D.M.M.

## NEPENTHACEAE
### Pitcher Plants

The Nepenthaceae is a family of carnivorous tropical herbs and shrubs, which are well known by their popular name, pitcher plants. Most are woody climbers and many are epiphytes.

**Distribution.** The family has representatives scattered throughout the Old World tropics, with a center of distribution in Borneo, radiating as far east as north Australia and New Guinea and as far west as the Seychelles, Sri Lanka and Madagascar. The family requires very moist conditions and with the exceptions of *Nepenthes distillatoria* (wet savannahs of Sri Lanka) are jungle plants occurring from sea level to over 2,500m (8,000ft) in conditions of high humidity.

**Number of genera:** 1
**Number of species:** about 70
**Distribution:** jungles of Old World tropics.
**Economic uses:** limited local uses.

**Diagnostic features.** The leaves are alternate, without stipules and often without distinct petioles, having a winged basal portion which narrows shortly and then spreads into a ligulate blade. The plants climb by the means of tendrils which are prolongations of the leaf midrib. The end of the tendril generally becomes greatly swollen and hollowed out and develops into a pitcher, with a lid projecting over the mouth, which opens as the pitcher matures. A pitcher develops by

a sheath-like formation in the upper surface of the leaf; the leaf tip takes no part in the development, and the lid grows out below it. The edge of the pitcher is curved inwards, and below the base of the lid on the outside of the pitcher is a spur. At the entrance of the pitcher there are honey glands, below which the interior is slippery. Insects are attracted to the pitcher by its honey or bright color, which may be red or green and is often blotched. Once they enter they are unable to climb out because of the slippery surface, and eventually drown in the water that accumulates in the base of the pitcher. The plant absorbs the products of decay.

Many members of the Nepenthaceae are epiphytic and have climbing stems up to 3cm (1in) in diameter. The small red, yellow or green flowers are regular and unisexual, with males and females on separate plants, and are borne in a spike-like inflorescence. The perianth consists of three or four sepals. In the male flower the filaments of four–24 stamens are united into a column and the anthers crowded into a mass. The female flower has a discoid stigma. The style is short or totally absent. The ovary is superior of few fused carpels and has four locules bearing numerous ovules on central plac-

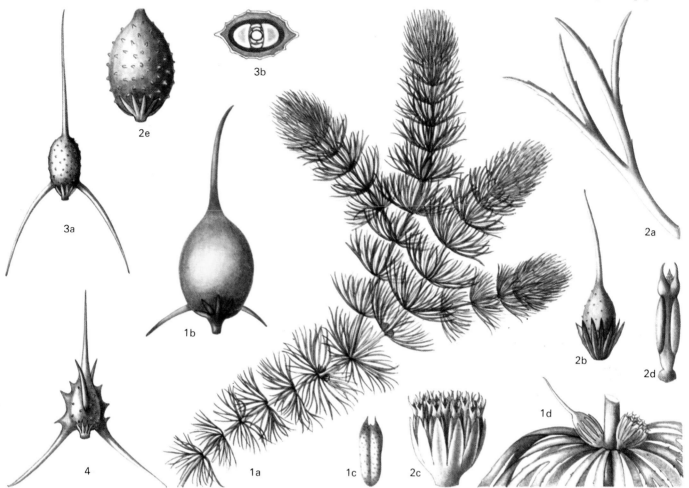

Ceratophyllaceae. 1 *Ceratophyllum demersum* (a) a submerged aquatic herb with whorls of linear leaves and no roots ($\times\frac{2}{3}$); (b) fruit—a one-seeded nut tipped by the persistent, hooked spine-like style and with the marginal spines ($\times4$); (c) anther with spur-like extension of connective ($\times8$); (d) stem node with male (left) and female (right) flowers in the leaf axils ($\times3$). 2 *C. submersum* (a) tip of stem ($\times4$); (b) female flower ($\times10$); (c) male flower ($\times10$); (d) stamen ($\times16$); (e) fruit ($\times4$). 3 *C. oxycanthum* (a) fruit ($\times2\frac{2}{3}$); (b) cross section of fruit ($\times8$). 4 *C. pentacanthum* fruit ($\times2\frac{2}{3}$).

entas. The fruit is a leathery capsule, and the light seeds have hair-like projections on the end. The seeds have a minute embryo and fleshy endosperm.

*Nepenthes ampullaria* has two kinds of leaves, some with tendrils and no pitchers, and others as stalked pitchers arranged in a radical rosette.

Pitchers vary from 5cm (2in) to as much as 30cm (12in) in length, and some in Borneo are large enough to hold 2 litres (3·5 pints) of water. In some species the young pitchers are clothed in tight, rusty, branched hairs.

**Classification.** The family is usually regarded as containing a single genus, *Nepenthes*, but some authorities consider that the Seychelles species *Nepenthes pervillei* represents a separate genus *Anurosperma*.

The Nepenthaceae is related to the Sarraceniaceae and Droseraceae by its insectivorous habit, and on this basis some taxonomists place the three families together in the order Sarraceniales. However, they are usually placed in Aristolochiales, Sarraceniales and Rosales, respectively.

**Economic uses.** The stems of some species (*N. distillatoria* in Sri Lanka, *N. reinwardtiana* in Malaysia) are used locally for basket-making and as a type of cordage.       S.A.H.

# NYMPHAEALES

## CERATOPHYLLACEAE

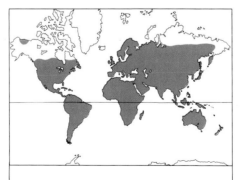

**Number of genera:** 1
**Number of species:** 2 or about 30
**Distribution:** worldwide in fresh water.
**Economic uses:** protects fish fry.

The Ceratophyllaceae is a cosmopolitan family of specialized aquatic herbs entirely without roots, comprising a single genus.

**Distribution.** *Ceratophyllum* is found almost throughout the world in floating masses that can form very rapidly and choke waterways.

**Diagnostic features.** The species of *Ceratophyllum* are entirely submerged herbs that cannot tolerate periods of emergence. Roots are completely lacking, even in the embryo, but sometimes colorless, root-like branches develop which anchor the plant to the substrate. The stems are usually branched but never develop more than one branch at a node. The leaves are whorled with three to ten leaves at a node. Each leaf is rather rigid, often brittle, and one to four times forked; the ultimate leaf-segments bear two rows of minute teeth and are tipped by two bristles.

The flowers are unisexual and solitary in the axil of one leaf in a whorl. Male and female flowers are usually found on alternate nodes. The perianth consists of eight to 12 linear, bract-like segments that are united at the base. The stamens are numerous; the filaments are short or absent but the connectives are prolonged apically into short spurs. The anthers are extrorse. The ovary is superior, solitary, comprising one carpel with a single pendulous ovule.

The fruit is a one-seeded nut, tipped by a persistent, spine-like style and often with additional basal or marginal spines.

**Classification.** *Ceratophyllum* is very variable and at the species level taxonomically

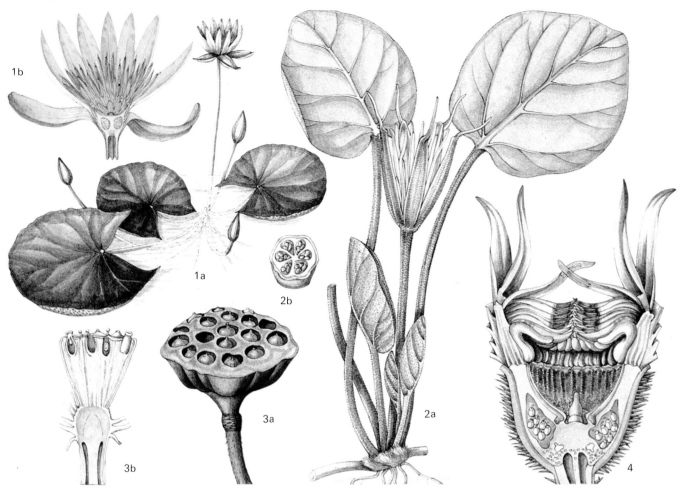

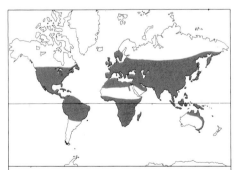

Nymphaeaceae. 1 *Nymphaea micrantha* (a) habit, showing floating leaves and aerial flowers ($\times\frac{1}{3}$); (b) half flower with greenish sepals, many petals grading into numerous stamens and ovary sunk in receptacle ($\times1$). 2 *Barclaya motleyi* (sometimes placed in the Barclayaceae) (a) habit ($\times\frac{2}{3}$); (b) cross section of fruit containing many seeds ($\times\frac{2}{3}$). 3 *Nelumbo nucifera* receptacle containing the fruits is shown (a) entire ($\times\frac{2}{3}$) and (b) in vertical section ($\times\frac{1}{3}$). 4 *Victoria amazonica* vertical section of flower ($\times\frac{2}{3}$).

difficult. There are two well-defined cosmopolitan species, *Ceratophyllum demersum* and *C. submersum*. More than 30 local species have been described and defined mainly on the basis of fruit-spine number and position, but most of them are probably no more than variants of the main species.

The family as a whole is highly specialized and very reduced aquatics. They lack roots, cuticle, stomata and woody or fibrous tissue. The flowers are also reduced and specialized for pollination under water. The pollen is without an aperture, smooth and with a very thin exine (outer wall). The lack of comparable characters makes it difficult to determine the affinities of this family. It is usually considered to be allied to the Nymphaeaceae. The one-seeded fruit, and the seed containing a large embryo with large, fleshy cotyledons, a well-developed plumule but no radicle, endosperm or perisperm suggest a particular relationship with the genus *Nelumbo* (Nymphaeaceae).

**Economic uses.** *Ceratophyllum* usually floats in mats just below the surface of the water. These mats provide a valuable protection for fish fry but also provide protection for bilharzia-carrying snails and malaria- or filaria-carrying mosquito larvae. C.D.C.

# NYMPHAEACEAE
*Water Lilies*

**Number of genera:** 9
**Number of species:** over 90
**Distribution:** cosmopolitan, in freshwater habitats.
**Economic uses:** ornamental aquatics, eg water lilies, lotus, Queen Victoria waterlily; some yield edible seeds and rhizomes.

The Nymphaeaceae is a family of water plants including the water lilies, the sacred lotus (*Nelumbo*) and the spectacular Queen Victoria water lily (*Victoria amazonica*).
**Distribution.** The family is found throughout the world in ponds, streams and lakes.

**Diagnostic features.** All members are perennial aquatics, often with large rhizomes that have massive shoot apices, and peltate or cordate, often floating leaves (and occasionally finely dissected submerged leaves). The flowers are solitary, with three to six green or colored sepals and three to many petals (absent in *Ondinea*), grading in some genera into the numerous stamens. There are five to 35 carpels, united or free, either in a superior or inferior ovary. The fruit is a spongy berry, dehiscing by the swelling of mucilage within. The seeds often have an aril. The flowers are usually pollinated by beetles.

**Classification.** *Nymphaea* has arillate seeds and half-inferior ovaries (50 species, tropical and temperate). *Nuphar* has no arils but a slimy pericarp containing air-bubbles with the same function, and a superior ovary (25 species, north temperate). *Ondinea* has no aril or petals (one species, *Ondinea purpurea*, western Australia). *Victoria* has epigynous flowers and leaves up to 2m (6·5ft) in diameter with upturned rims (two to three species, tropical South America). *Euryale* has large leaves, which are puckered and have no rims (one species, *Euryale ferox*, China and Southeast Asia). *Brasenia* has free carpels

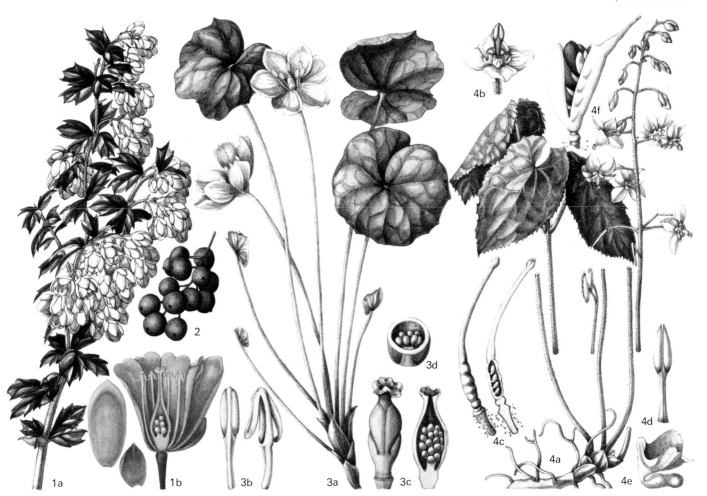

Berberidaceae. 1 *Berberis darwinii* (a) flowering shoot ($\times\frac{2}{3}$); (b) flower dissected to show small petaloid sepals, large petals, numerous stamens and a single carpel ($\times 4$). 2 *Mahonia aquifolium* berries ($\times\frac{2}{3}$). 3 *Jeffersonia dubia* (*Plagiorhegma dubium*) (a) habit ($\times\frac{2}{3}$); (b) stamens ($\times 4$); (c) gynoecium entire (left) and in vertical section (right) ($\times 4$); (d) cross section of ovary ($\times 4$). 4 *Epimedium perralderianum* (a) habit ($\times\frac{2}{3}$); (b) flower ($\times 1$); (c) gynoecium entire (left) and in vertical section (right) ($\times 3$); (d) stamen ($\times 3$); (e) nectariferous petal ($\times 3$); (f) fruit—a capsule ($\times 3$).

(one species, *Brasenia schreberi*, pantropical). *Cabomba* also has free carpels and flower parts in threes like those of monocotyledons (six species, warm Americas). *Victoria* and *Euryale* are sometimes placed in a separate family, the Euryalaceae, and *Cabomba* and *Brasenia* likewise in the Cabombaceae. *Nelumbo* (two species in the Americas and East Asia) is here included in the Nymphaeaceae but differs so much that it is thought by many to be allied to *Podophyllum* (Berberidaceae). *Barclaya*, three to four totally submerged species from the Indomalaysian region, has completely inferior ovaries and is sometimes separated as the Barclayaceae.

The Nymphaeaceae are usually considered allies of the primitive families in the order Ranunculales including the buttercup family Ranunculaceae, but they have many characteristics of monocotyledons.

**Economic uses.** Many species are cultivated, especially *Nymphaea* (water lilies), *Nuphar* (yellow water lilies) *Euryale* and *Victoria*, whilst *Cabomba* is a popular oxygenator for aquaria. The seeds and rhizomes of *Nymphaea* species are sometimes eaten, as are the roasted seeds of *Victoria* and *Nelumbo*. *Euryale* seeds yield arrowroot.     D.J.M.

# RANUNCULALES

## BERBERIDACEAE
*Barberry, Sacred Bamboo and May Apple*

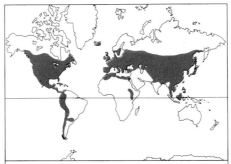

**Number of genera:** 13–16
**Number of species:** 550–600
**Distribution:** chiefly N temperate regions and S American mountains.
**Economic uses:** ornamental shrubs, fruits, dyes and medicines.

The Berberidaceae is a family of shrubs and perennial herbs.

**Distribution.** The herbaceous species are native to north temperate regions in both the Old and New Worlds; the shrubs also extend

to Tierra del Fuego in South America.

**Diagnostic features.** The leaves are normally alternate (almost opposite in *Podophyllum*), basal in some herbaceous species, simple to pinnately or ternately compound (peltate in *Podophyllum* and *Diphylleia*), and usually without stipules: *Mahonia*, *Nandina* and some species of *Berberis* have evergreen persistent leaves. In *Berberis* the leaves of long shoots are often transformed into spines, while the short shoots bear simple leaves, the articulation of blade and petiole indicating their reduction from a compound leaf during the course of evolution, as exemplified in *Mahonia* and more elaborately in *Nandina*. The flowers are borne in panicles (as in *Nandina*) with reduction through racemes to solitary flowers on leafless basal peduncles (as in *Jeffersonia*). They are bisexual and regular, the perianth consisting of several whorls of usually six or four segments variously differentiated, the innermost (petals) being usually distinct from the outer ones (outer and inner sepals) in being petaloid and bearing nectaries or else being reduced to small nectariferous sacs or scales. *Achlys*, however, has no perianth and consequently no nectaries. *Podophyllum*, *Diphylleia* and *Nandina* have perianth segments without nectaries. There are six stamens opposite the innermost perianth segments, except in *Epimedium* with four and *Podophyllum* with up to 18. The anthers open by lengthwise slits in *Nandina* and *Podophyllum*, but by valves hinged at the top in others. The single carpel is superior and has one locule with several or many ovules on the suture or one at its base. The style is short or rarely the stigma is sessile. The fruit is succulent in *Nandina, Berberis, Mahonia* and *Podophyllum* but forms a two-valved capsule in *Epimedium* and *Vancouveria*, and a papery bladder which is wind-dispersed in *Leontice. Caulophyllum* is remarkable in that the enlarging berry-like seeds burst through the wall of the ovary and develop in a completely exposed state; *Gymnospermium* also has exposed seeds.

**Classification.** Although some groups of species come so close together in all but a few diagnostic characters that they could be united generically, ie *Mahonia* with *Berberis*, *Vancouveria* with *Epimedium, Dysosma* with *Podophyllum, Plagiorhegma* with *Jeffersonia, Gymnospermium* with *Leontice*, the genera of the Berberidaceae otherwise stand so apart from one another and manifest such varied combinations of characters that the family is held together by a linkage of these in genera otherwise very different rather than by the possession of diagnostic characters common to them all; however they resemble each other more than they do genera now placed in other families.

A simple classification is that adopted by Hutchinson in which the woody species are put into two distinct families, the Nandinaceae (*Nandina*), and the Berberidaceae (*Berberis* and *Mahonia*), while all the herbaceous perennials are put in the Podophyllaceae, including *Leontice* and its related species, sometimes regarded as a separate family, the Leonticaceae. This ignores the affinity of the woody genera *Berberis* and *Mahonia* with the herbaceous genera *Epimedium* and *Vancouveria* and the great divergence from these of *Podophyllum* and *Diphylleia*. A more natural classification is to put these last two genera in a subfamily PODOPHYLLOIDEAE and the others in the BERBERIDOIDEAE with tribes NANDINEAE, BERBERIDEAE, EPIMEDIAE and ACHLYEAE.

The family is often placed near Ranunculaceae and Papaveraceae without, however, being closely allied to either.

**Economic uses.** Many of the Berberidaceae are prized ornamental plants, for example *Berberis buxifolia, B. darwinii, B. calliantha, B. × stenophylla* and *B. thunbergii, Mahonia aquifolium, M. bealei, M. lomariifolia* and *Nandina domestica* (sacred bamboo). The rhizomes of the may apple, sometimes incorrectly called mandrake (*Podophyllum peltatum* and *P. hexandrum*), yield a resin with drastic purgative and emetic properties which is incorporated into certain types of commercial laxative pills. W.T.S.

# RANUNCULACEAE
*The Buttercup Family*

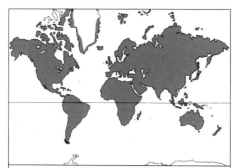

**Number of genera:** about 50
**Number of species:** over 1,800
**Distribution:** centered in temperate and cold regions of the N Hemisphere.
**Economic uses:** many ornamentals (clematis, globe flower, delphinium love-in-a-mist); aconitin (from *Aconitum*) used as a narcotic and analgesic.

The Ranunculaceae is a large family containing a number of well-known wild flowers and garden ornamentals such as buttercups, anemones and hellebores and some very poisonous plants, such as *Aconitum*.

**Distribution.** The family is distributed throughout the world but is centered in temperate and cold regions of the Northern and Southern Hemispheres.

**Diagnostic features.** The plants are mostly herbs, rarely woody climbers, such as *Clematis*. The perennial herbaceous species usually persist by means of a condensed rootstock or rhizome, sympodial in growth. With the death of the shoot after flowering, a basal axillary bud develops to give rise to the following year's flowering shoot. Usually the old roots die and new adventitious ones are formed from the bud; in *Aconitum* and *Ranunculus* these often swell up into storage tubers. The leaves are radical or alternate, although in *Clematis* they are opposite. They are commonly much-divided or palmately lobed, but entire and narrow in *Myosurus* and some species of *Ranunculus*, cordate in *Ranunculus ficaria* and some *Caltha* species, and pinnately lobed in *Zanthorrhiza* and *Clematis*. The aquatic species of *Ranunculus*, belonging to the subgenus *Batrachium*, usually have special submerged, much-dissected leaves with capilliform segments. In *Clematis* the petiole is sensitive to touch and supports the stem as a tendril, in the same manner as species of *Tropaeolum* (garden nasturtium). In *Clematis aphylla* the whole of the leaf becomes a tendril, photosynthesis being carried out by the stem cortex. The petioles are broadened into a sheathing base. Stipules are absent, except in *Thalictrum, Caltha, Trollius* and *Ranunculus*.

The inflorescence is terminal. In *Eranthis* and some species of *Anemone* the flowers are solitary, though more usually they are in racemes or cymes. In *Anemone, Pulsatilla* and *Nigella* there is a protective involucre of leaves beneath the flower, alternate with the calyx segments. The flowers are usually bisexual and regular. The flower has its parts typically arranged in spirals along a more or less elongated receptacle, but often the perianth segments are in whorls. The sepals vary from three to many, though most commonly there are five. They are rarely persistent (*Helleborus*), and often petaloid. Rarely is a true calyx and corolla present, the exception being the genus *Ranunculus* itself. The numerous spirally arranged stamens have extrorse anthers. The carpels range from one (*Actaea*) to many (*Ranunculus*, when they are also spirally arranged). In *Nigella*, the carpels are united into a capsule, otherwise the fruit is an achene or a many-seeded follicle. In *Actaea*, the achene is fleshy and thus a berry. The seeds have a small straight embryo and copious endosperm.

The family shows a wide variation in flower structures, and also a wide variation in pollination methods. The family is insect-pollinated in the main, although some species of *Thalictrum* are wind-pollinated. Many of the annual species are self-pollinated. The remaining insect-pollinated types are visited for either their pollen or their nectar and they can be divided into these two types. The genera *Anemone*, *Pulsatilla* and *Clematis* do not produce nectar and are visited only for their pollen. Nectar flowers with well-developed nectaries are found in *Ranunculus, Aquilegia, Delphinium* and *Helleborus*. In *Anemone* and *Clematis* insects are attracted by brightly colored sepals, in *Ranunculus* by showy petals (with prominent nectar pouches, known as honey-leaves); in *Aconitum* by showy sepals and petals; in some *Thalictrum* species by attractive stamen filaments or anthers. The Ranunculaceae are generally protandrous, the stamens shedding their pollen before the ovaries mature, though the reverse case of protogyny, the ovary maturing before the stamens, also occurs. However, these characteristics may not be strongly developed and some anthers may not have shed their pollen before the ovaries start to mature. These processes favor cross-pollination and outbreeding.

Ranunculaceae. 1 *Trollius europeaus* flowering shoot (×⅔). 2 *Helleborus cyclophyllus* flowering shoot (×⅔). 3 *H. niger* (Christmas rose) half flower showing large sepals, small tubular nectaries (modified petals), numerous stamens and carpels slightly fused at base each containing numerous ovules (×⅔). 4 *Eranthis hyemalis* (winter aconite) (a) habit (×⅔); (b) fruit—a group of follicles (×1⅓); (c) seed (×4). 5 *Aquilegia caerulea* shoot bearing regular flowers with spurred perianth segments (×⅔). 6 *Nigella damescena* fruit—a capsule surrounded by feathery bracts (×⅔). 7 *Aconitum napellus* (a) shoot bearing irregular flowers (×⅔); (b) half flower with large hooded upper sepal (×1⅓).

8

10b

10a

11

9b

12b

9a

12a

13

Seed dispersal is by a variety of agencies. *Clematis* and *Pulsatilla* have styles which lengthen after pollination into long feathery structures adapted for wind distribution. Some species of *Ranunculus* (eg *R. arvensis*) have tubercules or hooked spines on the surface for animal dispersal. *Helleborus* species have an elaiosome or oil-containing swelling on the raphe which attracts ants which then disperse the seeds (myrmecochory).

**Classification.** The family is divided up into two subfamilies and five tribes:

SUBFAMILY HELLEBOROIDEAE

Carpels with more than one ovule and fruits follicular or baccate.

HELLEBOREAE. Flowers regular. *Caltha, Calathodes, Trollius, Helleborus, Eranthis, Coptis, Isopyrum, Nigella, Actaea, Aquilegia, Xanthorrhiza.*

DELPHINIEAE. Flowers irregular. *Aconitum, Delphinium.*

SUBFAMILY RANUNCULOIDEAE

Carpels with a single ovule and fruit almost always of dry achenes.

RANUNCULEAE. Leaves radical or alternate; sepals imbricate; flowers without an involucre of leaves; calyx usually caducous; petals with nectariferous claw, rarely absent; carpels with a single ovule; achenes dry, rarely fleshy. *Trautvetteria, Ranunculus, Hamadryas, Myosurus, Callianthemum, Adonis, Knowltonia, Thalictrum.*

ANEMONEAE. Leaves all radical, or cauline (on the stem), alternate; flowers often with involucre; sepals overlapping, usually petaloid and persistent at anthesis; carpels with a single ovule; achenes dry. *Anemone, Pulsatilla, Hepatica, Barneoudia.*

CLEMATIDEAE. Stems herbaceous or woody and climbing; leaves opposite; sepals valvate, petaloid; petals absent or represented by the outer stamens becoming petaloid; carpels with a single ovule, achenes numerous, often plumed. *Clematis, Clematopsis.*

The Ranunculaceae is now generally regarded as a primitive family, a view put forward by Antoine Laurent de Jussieu as long ago as 1773. It is widely regarded as most probably having evolved little from an ancestral Magnoliales-like stock. The family is related to the Menispermaceae and Lardizabalaceae and the monogeneric Glaucidiaceae, Hydrastidaceae and Circaea-

*Ranunculaceae (continued).* 8 *Anemone coronaria* leaf and flower (×⅔). 9 *Thalictrum minus* (a) leaf and flowers in a terminal inflorescence (×⅔); (b) flower with inconspicuous perianth and numerous large stamens (×4). 10 *Clematis alpina* (a) shoot with opposite leaves with long petioles that aid climbing and axillary flowers (×⅔); (b) fruit—a group of achenes, each with a persistent hairy style (×⅔). 11 *Ranunculus* sp. half flower with free sepals and petals, numerous stamens and carpels, the latter with single ovules (×2). 12 *Myosurus minimus* (a) habit (×⅔); (b) flower with long receptacle bearing numerous carpels (×2⅔). 13 *Ranunculus ficaria* habit showing tuberous roots (×⅔).

steraceae (families discussed below, not in separate entries); and to the Berberidaceae (including *Nandina*). These families are often grouped, as here, into the order Ranunculales. (Engler's name Ranales is not now accepted, and the order is named after the genus *Ranunculus*, the buttercups, not after *rana*, the Latin for frogs.) The Ranunculales differ from the Magnoliales in being mainly herbaceous and in lacking ether oil cells. They also lack the primitive features of unsealed carpels, pollen with a single aperture, beetle pollination, laminar stamens and absence of vessels found in various members of the Magnoliales. Woody members are thought to be secondarily woody as they all have advanced wide wood rays.

There is, however, still much controversy about the actual groups of genera which should be recognized as families in their own right.

The family Glaucidiaceae, with the single genus *Glaucidium*, is often placed in the Ranunculaceae but differs markedly in a number of important respects: the structure of the gynoecium and ovule, the nature of follicle dehiscence and caryology (cell nuclei anatomy and chromosome structure).

The Ranunculaceae is related to the Berberidaceae, particularly with those herbaceous genera of the Berberidaceae that are often segregated as the family Podophyllaceae. *Hydrastis canadensis* from Japan and eastern North America (and now almost extinct through overcollecting of the rhizomes which were used in medicine as a tonic) is somewhat intermediate between the Paeoniaceae, Ranunculaceae and Berberidaceae, and is also often put in its own family, the Hydrastidaceae. It differs from the Paeoniaceae in having palmately lobed leaves, no corolla or disk and stamens polyadelphous (anthers uniting into three or more groups); from the Ranunculaceae in the absence of nectaries and the possession of fleshy carpels; from the Berberidaceae in having apetalous flowers with numerous stamens and carpels; from the Glaucidiaceae on the vernation (arrangement of leaves in bud), vasculature of the pedicel and receptacle, and other characters. The genus *Circaester*, with a single species native from northwest Himalayas to northwestern China, is a very reduced small annual plant. It is included in the Ranunculaceae by some authorities, though some consider it sufficiently different to separate it as the Circaeasteraceae.

The Ranunculaceae and Berberidaceae are also phytochemically related in the possession of the alkaloid berberine.

The order Ranunculales is thought to be closely related to the order Nymphaeales on one side and to the Magnoliales and Illiciales on the other. Some phylogeneticists see relationships with the primitive monocotyledonous order Alismatales, but others have suggested that this relationship is a result of convergent evolution. A close

relationship between some families of the order Ranunculales (Berberidaceae, Glaucidiaceae) and the order Papaverales is widely accepted.

**Economic uses.** Although the family is of great botanical interest it is not of any great economic importance. Many genera have species which provide excellent plants for the garden. Most of the generic names are familiar to nurserymen and gardeners as showy herbaceous border plants (in the case of *Clematis*, climbers): *Trollius*, (globe flower), *Aconitum* (monkshood and aconites), *Helleborus* (Christmas rose, lenten lily and other hellebores). Other genera which include ornamental species (many with cultivated varieties) are *Actaea* (baneberries, herb Christopher), *Anemone, Aquilegia* (columbine), *Ranunculus* (buttercups, spearwort, lesser celandine, crowfoots), *Caltha* (marsh marigold), *Delphinium* (including larkspurs), *Eranthis* (winter aconite), *Hepatica, Nigella* (love-in-a-mist, fennel flower), *Pulsatilla* (pasque flower), *Thalictrum* and *Cimicifuga.*

A number of genera are highly poisonous and have caused deaths. Victorian medical books give lurid details of the symptoms and deaths of gardeners who have inadvertently eaten *Aconitum* tubers, having confused them with Jerusalem artichokes. S.L.J.

# LARDIZABALACEAE
*Akebias and Holboellias*

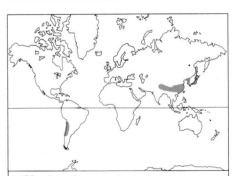

**Number of genera:** 9
**Number of species:** 36
**Distribution:** the Himalayas to Japan; Chile.
**Economic uses:** ornamentals and locally sold fruits.

The Lardizabalaceae is a family of climbing (rarely erect) shrubs, some of which are grown as ornamentals, notably species of the genera *Akebia, Holboellia, Stauntonia, Lardizabala* and *Decaisnea.*

**Distribution.** The family occurs from the Himalayas to China and Japan, and in Chile.

**Diagnostic features.** The leaves are alternate, palmate, rarely pinnate (trifoliolate in *Sargentodoxa*), often with leaflet stalks swollen at the base. The flowers are regular and unisexual, with sexes either on separate plants (dioecious) or the same plant (monoecious), rarely bisexual. They are borne in racemes in the axils of scale leaves situated at

Lardizabalaceae. 1 *Sargentodoxa cuneata* (a) shoot bearing trifoliolate leaf and male inflorescence ($\times\frac{2}{3}$); (b) stamens ($\times 4$); (c) half female flower ($\times 2$); (d) fruits ($\times\frac{2}{3}$); (e) vertical section of fruit containing a single seed ($\times 3$). 2 *Akebia quinata* leafy shoot with separate female (large) and male (small) flowers ($\times\frac{2}{3}$). 3 *Decaisnea fargesii* (a) male inflorescence ($\times\frac{2}{3}$); (b) part of leaf showing swollen bases to leaflet stalks ($\times\frac{2}{3}$); (c) female flower and young fruit ($\times\frac{2}{3}$); (d) androecium from male flower with six partly fused stamens ($\times 2$); (e) gynoecium with vestigial stamens ($\times 2$); (f) vertical section of carpel with numerous ovules ($\times 3$); (g) cross section of carpel ($\times 3$); (h) fruits—follicles ($\times\frac{2}{3}$).

the bases of branches. There are three or six sepals which overlap one another and are often petal-like in appearance. The petals number six and are smaller than the sepals, or may be absent altogether. A double whorl of nectaries often occurs between petals and stamens. The six stamens are free or partly fused together. There are three or more divergent carpels arranged in whorls, each carpel with an oblique stigma. (In *Sargentodoxa* the carpels are numerous and spirally arranged.) The flowers of either sex often contain remnants of the genitalia of the other. Each carpel contains a solitary or numerous ovules inserted in vertical rows on the inside wall (solitary and pendulous in *Sargentodoxa*), and when mature forms a colored, fleshy, many-seeded fruit (a follicle or berry). The seeds, oval or kidney-shaped, contain fleshy endosperm and a small embryo.

**Classification.** *Akebia* (eastern Asia) comprises five deciduous, monoecious species with fragrant female flowers lacking petals, and sausage-like fruits, grayish-violet with a white pulp. The five evergreen, dioecious species of *Holboellia* (Himalayas into China) have free stamens, fleshy sepals and minute petals. *Stauntonia* comprises 15 evergreen, dioecious species (eastern Asia) in which the stamens are united, the sepals are thin and the petals absent. *Lardizabala* (Chile) is dioecious and contains two species with ternate leaves and pedicels with bracts. *Decaisnea* (Himalayas and China) comprises two very similar species of distinctive erect shrubs with pinnate leaves. Other monoecious genera are *Sinofranchetia* (one species, western Chile) and *Parvatia* (three species, eastern Himalayas and China), while *Boquila* (one species, Chile) is dioecious. *Sargentodoxa* has a single dioecious species native to China. This is an interesting genus of scrambling deciduous shrubs often separated as the monotypic family Sargentodoxaceae combining features of the Lardizabalaceae and Schisandraceae.

The Lardizabalaceae, together with the Ranunculaceae, Berberidaceae and Menispermaceae, forms a group with well-established affinities.

**Economic uses.** The edible fruits of *Akebia*, *Lardizabala* and *Stauntonia* species are all sold locally. A number of species are cultivated as ornamentals eg *Akebia quinata* and *Holboellia coriacea*. B.M.

# MENISPERMACEAE
*Curare*

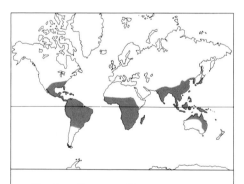

**Number of genera:** 65
**Number of species:** about 350
**Distribution:** mainly tropical rain forest.
**Economic uses:** medicines and drugs (curare); sarsaparilla substitute.

The Menispermaceae is composed of lianas, with a few shrubs and, rarely, trees or herbs.
**Distribution.** Most species grow in tropical rain forest but there are some subtropical and warm temperate species. The genus *Menispermum* (moonseed) is widespread

Menispermaceae. 1 *Tinospora cordifolia* (a) tip of twining stem with flowers in lateral inflorescences ($\times\frac{2}{3}$); (b) male flower with three sepals and petals ($\times 4$); (c) petal and stamen ($\times 6$); (d) female flower ($\times 2$); (e) female flower with calyx removed to show three free carpels each crowned by a sessile stigma ($\times 6$). 2 *Coscinium fenestratum* (a) part of twining stem with lateral inflorescences and fruits ($\times\frac{2}{3}$); (b) male flower viewed from above ($\times 4$); (c) female flower viewed from above ($\times 6$); (d) female flower with perianth removed to show six free carpels crowned by thin stigmas ($\times 10$); (e) vertical section of carpel ($\times 18$); (f) fruits—drupes ($\times\frac{2}{3}$).

from Atlantic North America and Mexico to eastern Asia. *Chondrodendron*, *Disciphania* and *Hyperbaena* occur in South America. The large genus *Cocculus*, also present in Atlantic North America and Mexico, extends across Africa and India into Indomalaysia. It contains gigantic climbers, as well as some small trees, the latter not only in rain forests but also in semidesert scrub and deciduous bushland in Africa, Madagascar and Socotra. *Cissampelos* has a similar distribution. Other African genera are *Jateorhiza* and the large genus *Triclisia* (also in Madagascar and Socotra), and *Tiliacora* which also extends into India, Burma and Sri Lanka. *Tinospora* and *Pycnarrhena* (a large genus) also grow in Indian forests but extend into Thailand, Cambodia, the Philippines, Indomalaysia and northern Australia.

**Diagnostic features.** The leaves are without stipules, alternate and sometimes peltate. The mostly unisexual flowers are very small, often greenish-white; male and female flowers are borne on separate plants. There are usually two rows of three sepals (sometimes more) and two or three of petals; the three to many stamens may be free, united or in bundles. The one to 32 free carpels each

contain two ovules which reduce to one. The stigma is terminal, entire or lobed. The fruit is drupaceous, usually curved, often to horseshoe-shape, and the endocarp has attractive sculpturing.

**Classification.** The family is divided into eight tribes based largely on seed structure:

TRICLISIEAE. Sepals and petals distinguishable, three to many carpels, endocarp straight and without a swelling (condyle) or curved and with a swelling, endosperm absent (eg *Tiliacora* and *Chondrodendron*).

PENIANTHEAE. Sepals and petals distinguishable, three to 12 carpels, endocarp straight and with a swelling, endosperm absent (eg *Penianthus*).

COSCINIEAE. Sepals and petals indistinguishable, three to six carpels, endocarp straight, endosperm little (only genera, *Coscinium*, *Anamirta* and *Arcangelisia*).

FIBRAUREAE. Sepals and petals indistinguishable, usually three carpels, endocarp straight, usually sculptured, endosperm little or absent (eg *Fibraurea*).

TINOSPOREAE. Sepals and petals distinguishable, usually three, rarely four or six carpels, endocarp straight, usually sculptured (eg *Jateorhiza* and *Tinospora*).

ANOMOSPERMEAE. Sepals and petals indistinguishable (none in *Abuta*), three carpels, endocarp usually curved, hardly sculptured, endosperm slight (eg *Anomospermum*).

HYPERBAENEAE. Sepals and petals distinguishable, three carpels, endocarp curved, not sculptured, endosperm absent (eg *Hyperbaena*).

MENISPERMEAE. Sepals and petals distinguishable, petals sometimes absent, six carpels, endocarp curved, ribbed and sculptured, slight endosperm present or absent (eg *Menispermum*, *Cocculus*, *Sinomenium*, *Stephania* and *Cissampelos*).

The Menispermaceae is most closely related to the Lardizabalaceae which differs in its mostly monoecious plants, compound leaves and abundant endosperm.

**Economic uses.** Curare (tubocurarine chloride) obtained mainly from *Chondrodendron tomentosum* is used as a muscle relaxant in surgical operations and in neurological conditions. The drupes of *Anamirta cocculus* ("fish berries") contain a poison (picrotoxin) which is used locally in hunting fish and in treating skin diseases. A tonic and febrifuge is prepared from the roots of *Jateorhiza palmata* (calumba).                H.P.W.

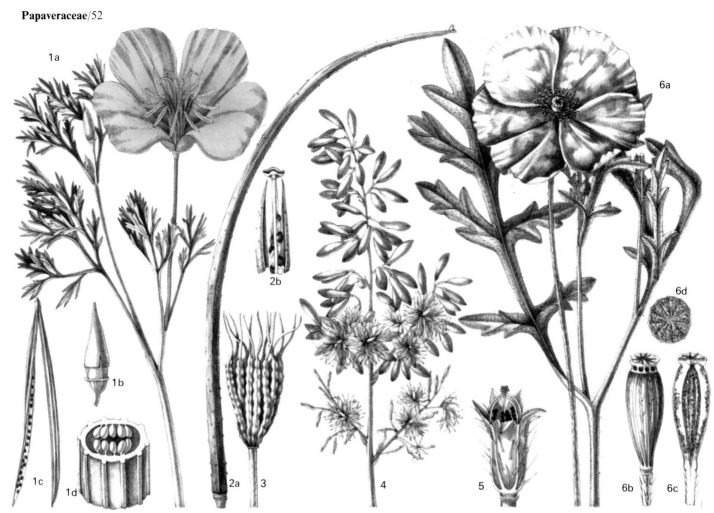

Papaveraceae. 1 *Eschscholzia californica* (a) leafy shoot and flower with four petals ( ×1) ; (b) flower bud with calyx forming cap ( ×1½) ; (c) fruit—a capsule dehiscing by two valves ( ×1) ; (d) cross section of fruit ( ×7). 2 *Glaucium flavum* (a) capsule ( ×⅔) ; (b) tip of opened capsule showing seeds and apical valve ( ×1½). 3 *Platystemon californicus* fruit—a group of follicles ( ×1½). 4 *Macleaya cordata* inflorescence ( ×⅔). 5 *Argemone mexicana* dehisced spiny capsule with seeds exposed ( ×⅔). 6 *Papaver dubium* (a) shoot with dissected leaves and solitary flowers ( ×⅔) ; (b) capsule dehiscing by apical pores ( ×1½) ; (c) vertical section of a capsule ( ×1½) ; (d) cross section of capsule ( ×1½).

# PAPAVERALES

## PAPAVERACEAE
*Poppies*

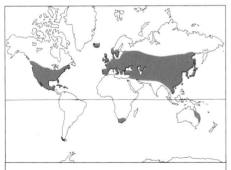

**Number of genera:** 26
**Number of species:** about 250
**Distribution:** temperate.
**Economic uses:** opium and many species cultivated as ornamentals (poppies).

The Papaveraceae is a family of mainly herbaceous annuals or perennials but with some shrubs, all of which produce latex.

**Distribution.** The family is mainly native to the north temperate zone.

**Diagnostic features.** The leaves are alternate, without stipules, entire but often lobed or deeply dissected. The stems, leaves and other parts of the plant contain a well-developed system of secretory canals which produce yellow, milky or watery latex. The flowers are large and conspicuous, either solitary or arranged in cymose or racemose inflorescences. They are regular, bisexual and hypogynous, possessing two free sepals which fall off before the flower opens. There are usually two whorls, each of two free showy petals (absent in *Macleaya*) which are often crumpled in the bud. There are usually several whorls of numerous stamens. The filaments are sometimes petaloid while the anthers have two locules and dehisce longitudinally. The gynoecium consists of two to numerous fused carpels (separate except at the base in *Platystemon*). The ovary is superior and contains usually a single locule with intruding parietal placentas, the number being equal to the number of carpels and each bearing numerous ovules. The stigmas are equal in number to the carpels and are opposite to or alternate with the plac-

entas. The fruit is a capsule, opening by valves or pores (follicular in *Platystemon*) and containing seeds with a small embryo and copious mealy or oily endosperm.

**Classification.** *Papaver* is the largest genus, containing about 100 species. *Meconopsis* (40 species, North America), *Glaucium* (25 species, Europe and Asia) and *Argemone* (10 species, America) have flowers similar to those of *Papaver*, although the leaves of some *Argemone* species are prickly. The flowers of the 10 species of *Eschscholzia* are very variable, while the 60 species of the American genus *Platystemon* are distinctive in that the numerous carpels are united only at the base. The genera *Bocconia* (10 Asian and tropical American species) and *Macleaya* (two east Asian species) are of interest as they possess apetalous flowers which are aggregated into compound racemes. *Sanguinaria* is represented by one North American species. The genus *Pteridophyllum* has flowers with only four stamens, and with *Hypecoum* (Fumariaceae) is sometimes regarded as constituting a separate family (Hypecoaceae) or is given separate family status as Pteridophyllaceae.

The Papaveraceae is sometimes placed

Fumariaceae. 1 *Corydalis lutea* (a) shoot with much divided leaves and irregular flowers in a racemose infloresence ($\times\frac{2}{3}$); (b) half flower showing spurred petal and elongate ovary. ($\times 4$); (c) vertical section of fruit ($\times 2$). 2 *Pteridophyllum racemosum* (a) habit showing fern-like leaves ($\times\frac{2}{3}$); (b) flower—the simplest form in this family ($\times 2$); (c) vertical section of ovary ($\times 4$). 3 *Fumaria muralis* (a) flowering shoot ($\times\frac{2}{3}$); (b) half flower with spurred petal and stamens in two bundles ($\times 3$); (c) vertical section of bladder-like fruit ($\times 6$). 4 *Dicentra spectabilis* (a) leaf and inflorescence ($\times 1$); (b) flower dissected to show varied form of petals and stamens arranged in two bundles ($\times 2$).

with the Cruciferae, Capparidaceae and Resedaceae in an order Rhoeadales. Here it is placed with the irregularly-flowered Fumariaceae into an order Papaverales.

**Economic uses.** Economically the most important species in this family is *Papaver somniferum* (opium poppy) which yields opium. The seeds do not contain opium and are used in baking. They also yield an important drying oil. Likewise the seeds of *Glaucium flavum* and *Argemone mexicana* yield oils which are important in the manufacture of soaps. Many species are cultivated as garden ornamental plants, for example *Dendromecon rigida* (Californian bushy poppy). *Eschscholzia californica* (Californian poppy), *Papaver alpinum* (alpine poppy), *P. nudicaule* (Iceland poppy), *P. orientalis* (oriental poppy) and *Macleaya cordata* (plume poppy). S.R.C.

## FUMARIACEAE
*Dicentras and Fumitory*

The Fumariaceae is a family of annual and perennial herbs, whose best-known genera are *Fumaria* (fumitory), *Corydalis* and *Dicentra*.

**Distribution.** The Fumariaceae occur

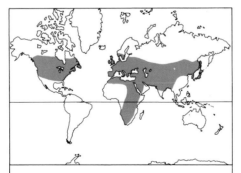

**Number of genera:** 16
**Number of species:** about 400
**Distribution:** mainly temperate.
**Economic uses:** limited use as garden ornamentals (*Corydalis* and *Dicentra*).

throughout north temperate regions. Only a few are found south of the Equator – a few species of *Corydalis* on mountains in East Africa, and the small genera *Phacocapnos*, *Cysticapnos*, *Trigonocapnos* and *Discocapnos* in southern Africa.

**Diagnostic features.** Perennials often have swollen, corm-like stocks. Many species of *Corydalis* have only one developed cotyledon, even though the family clearly

belongs to the dicotyledons when all other characters are considered. A few species are climbers or scramblers with tendril-like leaf axes. All contain alkaloids, though in smaller amounts than the Papaveraceae. The leaves are usually alternate, without stipules, and frequently pinnately or palmately compound. The inflorescence is usually racemose. The flowers are of complex and unusual structure, whose derivation from the simpler papaveraceous type is, however, demonstrated by some of the smaller genera. In the simplest case (*Pteridophyllum*) there are two small sepals, four stamens and an ovary of two united carpels; the two inner petals are slightly larger than the outer two, and a small amount of nectar is secreted at the base of the filaments. A more complex situation is found in *Hypecoum*, in which the two inner petals are prominently three-lobed, with the middle lobe stalked and with an expanded, cup-like apex; the apexes of the middle lobes wrap around the anthers and form a chamber into which the pollen falls. Here again there are four stamens, with nectar secreted at the filament bases. In *Dicentra* there is further elaboration – all four petals are variably fused, particularly towards the apex; the outer petals are spurred at the base, and the apexes of the inner petals are fused around the anthers. The stamens are arranged in two bundles opposite the inner petals; each bundle has a single filament which divides into three parts at the apex; the central division of each bears a complete anther, while the lateral divisions each bear half an anther. This complex structure appears to have evolved from the four-staminate condition by the splitting of the stamens opposite the outer petals. The base of each compound filament is prolonged into the petal spur, and secretes nectar there. In the other genera only one of the outer petals is spurred, producing a very unusual irregular flower; all have similar staminal arrangements to *Dicentra*, and all have a bicarpellate ovary, which is usually many-ovuled, though with one ovule in *Fumaria* and related genera. The fruit is usually a capsule, sometimes swollen and bladder-like; more rarely it is indehiscent, either a one-seeded nutlet, or many-seeded and breaking up into one-seeded indehiscent segments. *Ceratocapnos heterocarpa* has dimorphic fruits. The seed has a small embryo and fleshy endosperm.

**Classification.** The Fumariaceae can be divided into two distinct subfamilies: HYPECOIDEAE (containing the genera *Pteridophyllum* and *Hypecoum*) and FUMARIOIDEAE, which contains the rest of the genera, distributed in two tribes. The largest genus is *Corydalis* (about 320 species); *Fumaria* has about 50 species; all the other genera are small. Many authorities include the family in the Papaveraceae.

**Economic uses.** The family is of little economic importance. A few species of *Corydalis* and *Dicentra* (Dutchman's breeches, bleeding heart) are grown as garden ornamentals. Some species of *Fumaria* are agricultural weeds, such as *Fumaria officinalis* (fumitory).

J.C.

# SARRACENIALES

## SARRACENIACEAE
*Pitcher Plants*

The Sarraceniaceae is a small family containing three genera, all of which are carnivorous pitcher plants.

**Distribution.** They are found on the Atlantic coast of North America (*Sarracenia*), in California (*Darlingtonia*) and tropical South America (*Heliamphora*). All of the species

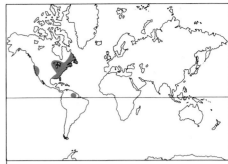

**Number of genera:** 3
**Number of species:** 17
**Distribution:** marshy habitats of Atlantic and Pacific coasts of N America, and north S America.
**Economic uses:** limited use as ornamentals.

are perennial herbs adapted to live in marshy habitats that are devoid of available nutrients.

**Diagnostic features.** The leaves are highly modified organs produced in rosettes arising directly from perennating rhizomes. All or some of the leaves are represented by long, narrow pitchers forming gracefully curved funnels covered by a lid. The insect prey is lured to the mouth of the leaf pitcher by several means: the release of strong-smelling

Sarraceniaceae. 1 *Heliamphora nutans* (a) pitchers (leaves) and flowers on leafless stalks ($\times\frac{2}{3}$); (b) stamens, ovary and style ($\times 2$); (c) section of trilocular ovary with ovules on axile placentas ($\times 3$). 2 *Sarracenia purpurea* (a) pitchers and flower stalk ($\times 2$), (b) flower with green sepals and reddish petals ($\times 1$); (c) style and stigma ($\times 1$); (d) cross section of ovary with ovules on inrolled carpel walls ($\times 2$). 3 *Darlingtonia californica* (a) pitchers and flowers ($\times\frac{2}{3}$); (b) vertical section of gynoecium with stamens attached at the base ($\times 3$).

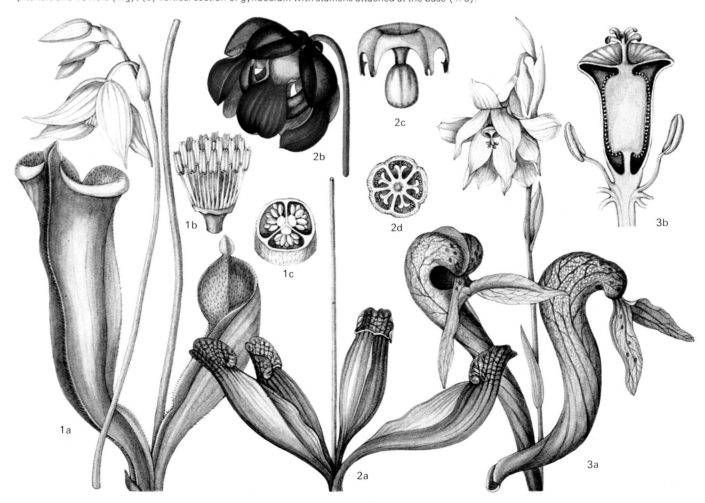

odors, the secretion of nectar from glands at the top and a short distance inside the pitcher, the exhibition of attractive colors (usually as reddish stripes on the pitcher) and bright, window-like perforations (fenestrations) around the necks of the pitchers. Once inside, the insects slip farther down on a "slide zone," then over a series of downwardly projecting hairs and finally into a pool of water in which they eventually drown. The prey is digested by excreted acids and enzymes.

The nodding flowers are borne on leafless stems (scapes) arising from the center of the rosettes; each consists of four, five or six free or overlapping sepals, five simple white or colored petals, many short stamens and an ovary with three to five locules, each locule containing numerous ovules on axile placentas. The fruits are capsules filled with many small seeds.

**Classification.** The 10 species of *Sarracenia* possess solitary flowers; the styles have umbrella-shaped apexes and the pitchers have oval-shaped lids. The monotypic *Darlingtonia* also has solitary flowers but has a style with a five-lobed apex and a recurved pitcher with a fish-shaped appendage; the six species of *Heliamphora* have racemose inflorescences, blunt-ended styles and pitchers with minute, appended lids.

In their insectivorous habit the Sarraceniaceae are allied to two other families, the Nepenthaceae and the Droseraceae; and all three used to be placed together in a single order, Sarraceniales. They are now generally placed in three different orders, the Sarraceniales, Aristolochiales and Rosales.

**Economic uses.** *Darlingtonia californica* and several species and hybrids of *Sarracenia* are cultivated as ornamentals both for indoor and outdoor culture. C.J.H.

# HAMAMELIDAE

## TROCHODENDRALES

### TROCHODENDRACEAE

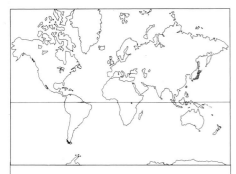

**Number of genera:** 1
**Number of species:** 1
**Distribution:** Japan, Taiwan, Ryu Kyu Islands.
**Economic uses:** birdlime.

The Trochodendraceae is a very small family of Asian forest trees containing a single genus *Trochodendron*.

**Distribution.** The only member of the family, *Trochodendron aralioides*, is native to Japan, Taiwan and the Ryu Kyu Islands, growing at altitudes of up to 3,000m (10,000ft).

**Diagnostic features.** The tree grows to form trunks 5m (16ft) in diameter and has alternate, pseudo-whorled leaves which are evergreen, leathery and without stipules. They have long petioles and are diamond-shaped, with scalloped edges and a glossy surface. The regular or slightly irregular green flowers are bisexual and occur in terminal racemose cymes; they lack sepals and petals. There are many stamens in three or four whorls surrounding a superior ovary consisting of six to numerous slightly fused carpels with free stigmas. Each carpel contains one to several pendulous ovules. The fruit develops into a ring of coalescent, several-seeded follicles, the seeds having an oily endosperm and tiny embryo.

**Classification.** The genus *Euptelea* is frequently placed within this family, although it is sometimes segregated into a family, Eupteleaceae, differing from *Trochodendron* in having carpels free and stipitate and fruits with wings. More debatable is the inclusion of *Tetracentron*, which is also sometimes regarded as constituting a separate family, Tetracentraceae.

The family is probably related to the Winteraceae and Illiciaceae. The wood does not contain vessels, a "primitive" feature it shares with the Winteraceae and other allied families.

**Economic uses.** Birdlime is made locally from the aromatic bark of *T. aralioides*. B.M.

## HAMAMELIDALES

### CERCIDIPHYLLACEAE

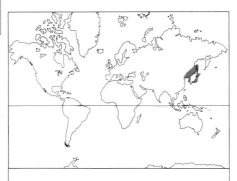

**Number of genera:** 1
**Number of species:** 1
**Distribution:** China and Japan.
**Economic uses:** limited use of wood and as an ornamental.

The Cercidiphyllaceae is a very small family of Asian trees represented by a single genus, *Cercidiphyllum*.

**Distribution.** The family is native to China and Japan.

**Diagnostic features.** *Cercidiphyllum japonicum*, the only species in the genus, bears two kinds of simple deciduous leaves: those on short shoots (spurs) are palmately veined and alternate, while those on the long shoots are pinnately veined and usually opposite. Stipules are fused to the petiole.

The flowers are unisexual and without petals; male and female are borne on separate plants. The male flowers are axillary, solitary or borne in close clusters. They have four small sepals (or bracts) and 15–20 stamens, with slender, long filaments on a conical receptacle. The female flowers are in clusters of two to six, each subtended by a bract, and consisting of only a single carpel, containing many ovules attached in two rows to a parietal placenta in its single locule. The style is thread-like and terminates in two stigmatic ridges. The fruit is a follicle, splitting down the ventral suture to release the many compressed, winged seeds. The seed contains copious endosperm, surrounding the embryo.

**Classification.** The combination of unusual features of this plant, including the dimorphic leaves, short and long shoots, apetalous, unisexual flowers and the nature of the carpels, is not found in other dicotyledonous families. The fossil evidence indicates that this genus may be a primitive relict. Although it has been linked with both the Ranunculaceae and the Magnoliaceae there is no sound evidence that it is closely related to these or to other dicotyledonous families.

**Economic uses.** The tree is valuable for its light, soft, fine-grained wood, which is used for cabinet-making and interior work in houses. It is in limited cultivation as an ornamental. S.R.C.

### PLATANACEAE
*Plane and Buttonwood Trees*

The Platanaceae is a small family of large, attractive trees with scaling bark which are widely cultivated in urban areas particularly the London plane (*Platanus hybrida* (*P. acerifolia, P. hispanica*)).

**Distribution.** Apart from *P. orientalis* (Balkan peninsula and Himalayas) and *P. kerrii* (Indochina), all species are North American.

**Diagnostic features.** The leaves are simple,

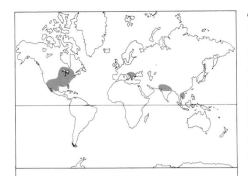

**Number of genera:** 1
**Number of species:** 10
**Distribution:** chiefly N America
**Economic uses:** ornamental town trees
and wood is used for veneers.

palmately three- to nine-lobed (elliptic to
oblong, unlobed, with pinnate, not palmate,
veins in *P. kerrii*). The long petiole is swollen
at the base to form a hood to the axillary
bud. The stipules are large, embracing the
stem, soon falling. All vegetative parts, the
calyx, ovary and fruit bear stellate hairs. The
flowers are unisexual and borne on the same
tree, in one to several crowded globose heads
on separate long peduncles. There are three
to eight small, free, hairy sepals, and as many
larger spathulate petals. The male flowers
have three to eight subsessile stamens, the
connectives fusing above to form a conspicuous peltate scale. Three or four staminodes are sometimes present in male and
female flowers (catkins). The female flowers
have six to nine (sometimes three) free,
superior carpels with tapering styles whose
stigmatic surface is on the inner side. The
ovules are one (rarely two) per carpel,
pendulous. The fruit is a globose head of top-
shaped caryopses or achenes, the styles
persistent and surrounded by a pappus-like
ring of long bristly hairs. The single seed has
little or no endosperm and is wind-dispersed.
**Classification.** The family is perhaps related
to the Hamamelidaceae.
**Economic uses.** The planes (in America
known also as sycamore and buttonwood
trees) are widely grown for ornament. The
hard, fine-textured wood is used for veneer
and other purposes.                F.B.H.

## HAMAMELIDACEAE
*Witch Hazels and Sweet Gums*
The Hamamelidaceae is a medium-sized
family of trees and shrubs including the
witch hazel of pharmacy (*Hamamelis virginiana*), the gum- and timber-producing
sweet gums (*Liquidambar*) and other ornamental species.
**Distribution.** The family shows a very discontinuous distribution in the temperate and
subtropical regions of both hemispheres.

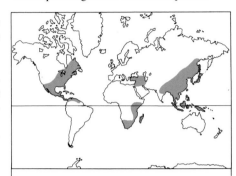

**Number of genera:** 23
**Number of species:** about 100
**Distribution:** very discontinuous, in
subtropical and temperate zones.
**Economic uses:** timber, fragrant gum
(storax) used in perfumery, witch
hazel lotion and ornamentals.

Hamamelidaceae. 1 *Hamamelis mollis* (a) leaf ($\times\frac{2}{3}$); (b) shoot with flowers ($\times\frac{2}{3}$); (c) flower with four hairy sepals, four linear petals, four stamens and bilobed stigma ($\times 3$); (d) vertical section of ovary ($\times 9$); (e) fruit—a woody capsule ($\times 2$). 2 *Fothergilla major* (a) leafy shoot and inflorescence ($\times\frac{2}{3}$); (b) bicarpellate gynoecium ($\times 3$); (c) dehisced fruit ($\times 2$). 3 *Rhodoleia championii* (a) shoot with many-flowered capitula surrounded by numerous bracts giving appearance of single flower ($\times\frac{2}{3}$); (b) five gynoecia on capitulum ($\times 1$); (c) cross section of ovary ($\times 2$); (d) ripe fruits ($\times 1$).

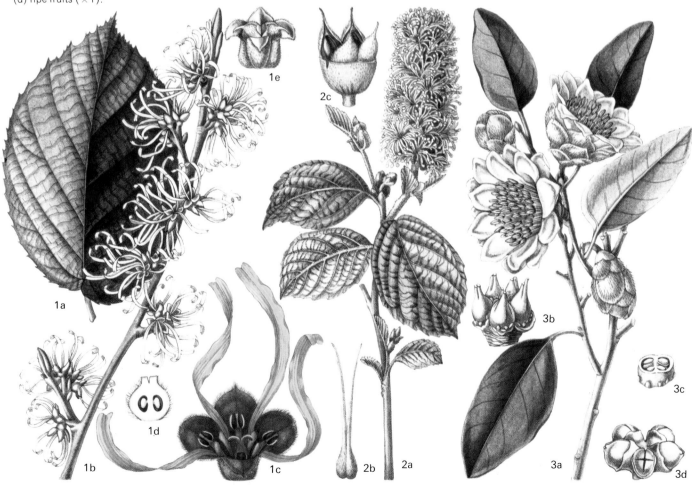

**Diagnostic features.** The Hamamelidaceae are trees and shrubs with generally alternate, simple or palmate leaves with stipules. Stellate hairs are sometimes present. The flowers vary considerably in the different genera, and can be bisexual or unisexual, with different sexes on the same plant (monoecious), or on separate plants (dioecious); they are often in a spike or head, sometimes subtended by colored bracts. The calyx consists of four or five united sepals, the corolla of four or five distinct petals (absent in *Liquidambar*, *Fothergilla* and *Altingia*). The stamens vary from two to 14 in number and the ovary from hypogynous to perigynous and epigynous. The ovary has two locules and two styles; each locule contains one or more ovules. The fruit exocarp is woody and the endocarp somewhat horny. The seeds are straight and with endosperm.

**Classification.** The family is usually divided into five subfamilies (with chief genera):

DISANTHOIDEAE. Flowers separate in two-flowered heads, petals long and narrow, and up to six ovules in each locule: *Disanthus* (endemic to Japan).

HAMAMELIDOIDEAE. Bisexual and female flowers clearly separate from each other (male flowers in the male inflorescence sometimes not so) and one or two ovules in each loculus: *Hamamelis* (eastern Asia and North America), *Trichocladus* (tropical East and eastern South Africa, the only African genus), *Diocoryphe* (endemic to Madagascar), *Corylopsis* (the largest genus, Himalayas to eastern Asia), *Parrotia* (endemic to Iran).

RHODOLEIOIDEAE. Flowers bisexual and borne in a five- to ten-flowered capitulum, surrounded by numerous bracts so as to resemble a single flower: *Rhodoleia* (northern Burma and southern China; Malaysia and Sumatra).

EXBUCKLANDIOIDEAE. Plants polygamomonoecious with uni- and bisexual flowers in capitula; petals in the bisexual flowers are narrow and two to five in number and there are 10–14 stamens; the leaves are palmately nerved from the base; the stipules are broad and closely folded face to face, enclosing the young shoot: *Exbucklandia* (eastern and southeastern Asia).

LIQUIDAMBAROIDEAE. Plants dioecious, although female flowers often have staminodes; the male inflorescence is a terminal raceme of globose stamen clusters, with no perianth; the female inflorescence is a globose head, with a perianth of numerous scales; the ovary has two locules, with the stigmas elongate: *Altingia* (Assam to Southeast Asia, Java, Sumatra), *Liquidambar* (eastern Asia, western Asia and North America).

The genus *Myrothamnus* (tropical Africa and Madagascar), an isolated relict group of undershrubs, is sometimes placed in the Hamamelidaceae, but usually in a segregate family, Myrothamnaceae, in which the plants differ in being always dioecious and having opposite leaves; the flowers have no sepals or petals, the anther filaments are fused and the three or four carpels contain many ovules. Some authorities view the Hamamelidaceae as being intermediate between the Rosales and Amentiferae (an assemblage of catkin-bearers); others have held the view that they are nearly allied to the Saxifragaceae and the small family Cunoniaceae. Current data tends to indicate that the ancestral forms of the Hamamelidaceae gave rise to the Casuarinaceae, Fagales and Urticales.

**Economic uses.** *Liquidambar styraciflua* (the American sweet gum or red gum) gives a useful heavy close-grained heartwood for furniture making and a white sapwood. A fragrant gum (storax) used in perfumery, as an expectorant, inhalant and as a fumigant in treatment of skin diseases is derived from *L. styraciflua* and *L. orientalis* (western Asia). *Liquidambar* species, the largest in the family, are frequently grown for their ornamental autumnal foliage. *Altingia excelsa* (rasamala) yields a heavy timber, and a fragrant gum used in perfumery. *Hamamelis* (witch hazel) provides several species and varieties of ornamental shrubs. *Hamamelis virginiana* yields the widely used astringent and soothing lotion for cuts and bruises. Water diviners favor witch hazel twigs for their dowsers. *Corylopsis*, *Fothergilla*, *Loropetalum* and *Parrotia* are shrubs also cultivated for ornament.                   S.L.J.

# EUCOMMIALES

## EUCOMMIACEAE

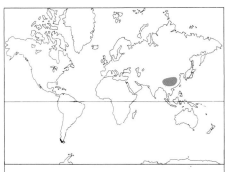

**Number of genera:** 1
**Number of species:** 1
**Distribution:** China
**Economic uses:** limited medicinal use and as rubber source.

The Eucommiaceae comprises a single species of small trees, *Eucommia ulmoides*.

**Distribution.** The tree is native to China, but has yet to be collected in the wild.

**Diagnostic features.** The trees grow to about 9m (30ft) tall with deciduous leaves which are simple, serrate and without stipules. The flowers are unisexual, and male and female are borne on different plants; they lack any perianth. The males are arranged in loose-stalked clusters consisting of the subtending bracts and about 10 elongated anthers which open by slits. The female flowers are solitary in the axils of bracts on the lower parts of the shoot, and comprise a flattened, superior, bicarpellate ovary with two divergent stigmas arising from a notch at the apex. There are two apical, pendulous ovules in the ovary, which matures into a one-seeded samara (like an elm fruit), the wings of which are chaffy and tough but still notched at the apex. The seed hangs from inside the apex of the fruit chamber. The embryo is straight and endosperm plentiful.

Latex is present in all the younger parts of the plant, but not in the wood. When snapped and pulled apart gently a series of fine threads can be drawn out, and this affords a useful means of identification.

**Classification.** The species, although lacking a perianth, is thought to be related to the Ulmaceae.

**Economic uses.** The bark is locally known as "tu-chung" or "tsze-lien" and is well established in the Chinese pharmacopoeia as a tonic and for arthritis. The tree also produces an inferior rubber which it has not been economic to exploit.               B.M.

# LEITNERIALES

## LEITNERIACEAE
*Florida Corkwood*

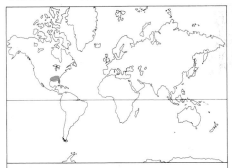

**Number of genera:** 1
**Number of species:** 1
**Distribution:** SE N America.
**Economic uses:** light wood used as floats for fishing nets.

The Leitneriaceae consists of a single deciduous shrub species, *Leitneria floridana*, the Florida corkwood.

**Distribution.** *Leitneria* is native to swampy areas in the southeastern part of the United States of America.

**Diagnostic features.** The leaves are alternate, simple, entire, somewhat leathery, with long petioles and no stipules. The unisexual flowers are borne in catkin-like spikes which appear before the leaves. Male and female are on separate plants. Each male flower is subtended by a bract. It has no perianth, and consists only of three to 12 stamens with free filaments and a rudimentary, sterile ovary. Each female flower is subtended by a bract and possesses a "perianth" of two to six distinct bract-like scales. The ovary is superior and has a single locule containing one ovule attached near the apex. There is a single style, constricted at the base and bearing a single stigma. The fruit is a leathery, rather flattened drupe subtended by the persistent bract. The seed contains a large straight embryo surrounded by thin fleshy endosperm.

**Classification.** The extreme reduction in floral structure suggests that this is an advanced family. However, no close affinities are evident with any other family except possibly the Myricaceae.

**Economic uses.** *Leitneria* is not economically important but the close-grained, soft wood is less dense than cork and is used for floats for fishing nets, hence the common name "corkwood."                            S.R.C.

# MYRICALES

## MYRICACEAE

The Myricaceae is a small family of aromatic trees or shrubs.

**Distribution.** The family is more or less cosmopolitan except for some warm tem-

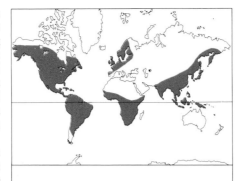

**Number of genera:** 2
**Number of species:** about 35
**Distribution:** almost cosmopolitan.
**Economic uses:** fruits boiled to produce wax.

perate parts of the Old World, and Australia.

**Diagnostic features.** The leaves are alternate, simple or pinnatifid, with or without stipules. The flowers are unisexual and borne on axillary catkin-like spikes, male and female on the same plant. The male flowers usually have two bracteoles, and a disk at the base of the stamens which are usually four, but sometimes two or up to 10 in number. The filaments may be joined at the base. The

female flowers have two to four bracteoles, and no disk; the ovary is superior, comprising two fused carpels which form a single locule with one erect ovule; the style has two short stigmatic branches. The fruit is a small, rough, often waxy drupe with a hard endocarp, and the seed has no endosperm and a straight embryo.

**Classification.** All species of *Myrica* conform to this general pattern. The only other genus, the monotypic *Canacomyrica* from New Caledonia, differs considerably in floral structure: bisexual flowers replace the female flowers of *Myrica*, and they have six stamens which are fused below into a ring surrounding the ovary. Also surrounding the ovary is a six-lobed disk which enlarges in fruit to enclose the ovary completely. *Canacomyrica* also has a certain amount of endosperm in the seeds, and this (along with other, morphological differences) makes for some doubt about its affinities with this otherwise homogeneous family.

The family is thought to be related to the Betulaceae, but is sufficiently distinctive to stand in an order of its own; some authorities relate it to the Juglandaceae.

**Economic uses.** The fruits of some species of *Myrica* are boiled to produce wax. I.B.K.R.

Myricaceae. 1 *Myrica gale* (a) shoot with leaves and catkin-like male and female inflorescences ($\times\frac{2}{3}$); (b) female flower ($\times 10$); (c) fruit ($\times 5$); (d) cross section of fruit ($\times 5$); (e) male flower ($\times 5$). 2 *Canacomyrica monticola* (a) leafy shoot ($\times\frac{2}{3}$); (b) half fruit surrounded by enlarged disk ($\times 10$). 3 *Myrica asplenifolia* (a) flowering shoot ($\times\frac{2}{3}$); (b) male catkin ($\times\frac{2}{3}$); (c) half female flower ($\times 2$). 4 *M. nagi* (a) shoot with fruits ($\times\frac{2}{3}$); (b) female flower ($\times 4$); (c) ovary and styles ($\times 4$); (d) male flower ($\times 4$); (e) fruit cut away to reveal hard endocarp ($\times\frac{2}{3}$).

# FAGALES

## BETULACEAE
*Birches, Alders, Hazels and Hornbeams*

The Betulaceae is a family of trees and shrubs which includes the birches (*Betula*), alders (*Alnus*), hornbeams (*Carpinus*) and hazels (*Corylus*).

**Distribution.** The family belongs predominantly to the north temperate regions, though also occurring on tropical mountains, the Andes of South America and in Argentina.

**Diagnostic features.** The leaves are simple, alternate, deciduous and with stipules. All species have separate male and female flowers which are borne on the same plant, either singly or in cymose groups of three, adhering to their bracts. The groups of male flowers are aggregated into characteristic pendulous cylindrical catkins. The groups of female flowers are borne on a stiff axis, often held erect. The perianth, when present, is of a variable number of scale-like segments. There are two to 12 stamens with two-celled anthers in each male flower, but no traces of vestigial carpels. The female flower, completely lacking vestigial stamens, has a single

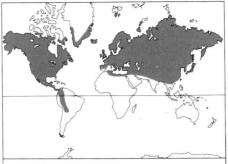

**Number of genera:** 6
**Number of species:** about 170
**Distribution:** N temperate, some species on tropical mountains and in S America.
**Economic uses:** timber, filberts from *Corylus* species and ornamental trees.

inferior ovary of two fused carpels. Pollination is by wind in early spring. The fruit is a single-seeded nut which is often winged for wind-dispersal, maturing in late summer or autumn. The seeds have no endosperm and a straight embryo.

**Classification.** The family has two subfamilies:

BETULOIDEAE. The male flowers are borne in three-flowered groups and have a perianth. The female flowers lack a perianth. The single tribe, BETULEAE, has two genera, *Betula* (about 50 species) and *Alnus* (about 30 species).

CORYLOIDEAE. The male flowers are solitary, the inflorescence therefore a simple spike; perianth absent. Female flowers possess a perianth. The tribe CORYLEAE (male flowers with an average of up to two stamens per flower) has two genera, *Corylus* (about 15 species), and *Ostryopsis* (about two species). The tribe CARPINEAE (male flowers with an average of six stamens per flower) has two genera, *Carpinus* (about 30 species) and *Ostrya* (about 10 species).

The Betulaceae is generally thought to belong with the Fagaceae to the order Fagales, but each family is often placed in its own order. Some workers now go further and split up the Betulaceae, recognizing each of the three tribes as a family in its own right: Betulaceae, Corylaceae and Carpinaceae.

The absence of fossil evidence makes it necessary to determine the evolutionary origin of the family on information from extant forms. On this evidence it is considered to be relatively advanced, with its very small unisexual flowers borne in com-

Betulaceae. 1 *Betula pendula* (a) habit showing typical drooping shoots; (b) leafy shoot with serrate, alternate entire leaves and immature male inflorescences (catkins) ($\times\frac{2}{3}$); (c) pendulous mature male catkins ($\times\frac{2}{3}$); (d) leaves and fruiting catkins ($\times\frac{2}{3}$). 2 Barks of various birch trees (a) *B. pendula*, (b) *B. humilis*; (c) *Betula* sp ($\times\frac{2}{3}$). 3 *Alnus glutinosa* (a) habit; (b) male catkins ($\times\frac{2}{3}$); (c) shoot with (from base) old fruiting cones, immature cones and young male catkins ($\times\frac{2}{3}$).

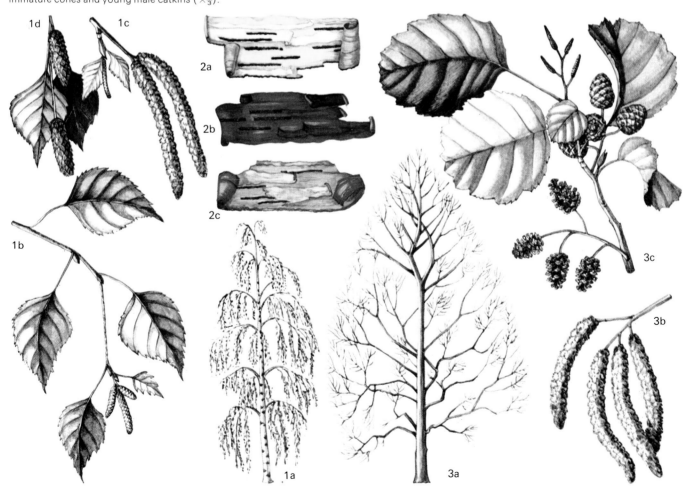

plex inflorescences with inferior ovaries, lack of nectar and scent, and wind pollination. These features together with further evidence from wood anatomy suggest that the Betulaceae is a derived family.

**Economic uses.** The birches provide valuable hardwood timbers. *Betula lutea* and *B. lenta* are important in North America for providing wood for furniture, doors, floors etc, and *B. papyrifera* is used for plywood, boxes, and in turnery. The bark of the latter is also used by the native Indians for making canoes and fancy goods. Branchlets of *Betula* species are used to make the besom brushes used by gardeners. *Alnus rubra* also provides a valuable timber which is a good imitation of mahogany. Both genera provide high-grade charcoal. *Ostrya* has extremely hard wood used for making mallets. Hazelnuts, cobnuts or filberts are produced by species of *Corylus*. Birches, hazels and hornbeams are cultivated as ornamentals.                S.L.J.

# FAGACEAE
*Beeches, Oaks and Sweet Chestnuts*

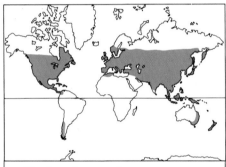

**Number of genera:** 8
**Number of species:** about 1,000
**Distribution:** temperate and tropical forests often as dominants in the former.
**Economic uses:** hardwood timber (*Quercus*, *Fagus*, *Nothofagus*), edible fruits (*Castanea*) and ornamentals.

The Fagaceae is an important temperate to tropical family of hardwood trees and, more rarely, shrubs, which embraces the beeches, oaks, and sweet chestnuts.

**Distribution.** The Fagaceae are prominent and frequently dominant members of the broad-leaved forests which cover, or used to cover, vast areas at middle latitudes in the Northern and, to a much less extent, Southern Hemisphere. In the extensive deciduous and mixed forests of North America and Eurasia beeches (*Fagus*), oaks (*Quercus*) and chestnuts (*Castanea*) figure prominently, while the comparable forests along the southern Andes are dominated by the southern beeches (*Nothofagus*). Evergreen oaks are important members of the forests around the Gulf of Mexico and in southern China and southern Japan, while the "black birch" forests of southern New Zealand are dominated by evergreen southern beeches. In Southeast Asia the

structure of the mixed mountain forest is largely determined by evergreen members of the family, particularly oaks and southern beeches. In total, therefore, the Fagaceae family produces a colossal biomass, possibly exceeded only by the conifers. The Fagaceae have a long fossil record suggesting an origin by at least the middle Cretaceous period, about 90 million years ago.

**Diagnostic features.** The Fagaceae are deciduous or evergreen trees, rarely shrubs, with alternate (rarely whorled) simple, entire to pinnately lobed leaves, and usually deciduous stipules.

The flowers are unisexual and usually arranged in catkins or small spikes (in *Nothofagus* the male flowers may be single) that may comprise only flowers of one sex, as in oaks, or may have female flowers at the base of otherwise male inflorescences, a more ancestral condition found, for example, in chestnuts (*Castanea*). The perianth is bract-like, with four to seven lobes. Male flowers have as many or twice as many stamens as perianth segments, but occasionally have up to 40. The filaments are free, with or without a rudimentary style. Female flowers are in groups of one to three, each group being surrounded by a basal involucre. The ovary is inferior, with three or six styles and locules, and two ovules in each locule. The fruits are single-seeded nuts, in groups of one to three, surrounded or enclosed by an often hardened "cupule"; the seeds lack endosperm.

The pollen and other features, such as a strongly scented inflorescence, suggest that insect pollination is the ancestral condition in the Fagaceae, and this is retained in most members except *Fagus*, *Nothofagus* and the temperate species of oak.

The familiar acorn "cup" of the oak is one example of the great variety of forms shown by the cupule, surrounding one or more fruits, which is a unique feature of the family and the origin of which has caused much controversy. Only with the discovery in 1961 of the remarkable genus *Trigonobalanus*, restricted to north Borneo, the Celebes, north Thailand, Malaya and Sarawak, has it been possible to suggest firmly that the cupule is derived from a three-lobed extension of the pedicel below each flower, which has been variously fused around single flowers, or groups of flowers. It is possible that the cupule gives a link with the pteridosperm ancestors of the flowering plants (angiosperms). The tremendous diversity of scales and spines on the cupules appears to be derived from branched spines.

The fruits of the Fagaceae have a slow and restricted capacity for dispersal and their germination power decreases rapidly with age. These features, together with the great antiquity of the family, have been responsible for its importance in discussions concerning the migration of plants brought about by continental drift in earlier geological epochs, notably evidence from *Nothofagus*.

**Classification.** The Fagaceae is divided into three subfamilies:

FAGOIDEAE. Inflorescence a one- to many-flowered axillary cluster; contains *Fagus* (male inflorescences long-stalked, many-flowered; styles long) and *Nothofagus* (male inflorescences sessile or short-stalked, one-to three-flowered; styles short).

QUERCOIDEAE. Catkin-like inflorescences, flowers usually with six stamens and more or less basifixed anthers 0.5mm–1mm long; contains *Quercus* (female flowers borne singly in the inflorescence; fruit round in transverse section, cupule not lobed) and *Trigonobalanus* (female flowers in clusters of three, sometimes up to seven; fruit three-angled; cupule lobed).

CASTANEOIDEAE. Catkin-like inflorescences, flowers usually with 12 stamens and dorsi-fixed anthers about 0.25mm long; contains *Chrysolepis* (cupule divided into free valves), *Castanea* (cupule valves joined when young; styles six or more; leaves deciduous), *Castanopsis* (cupule valves joined when young; styles three; leaves evergreen) and *Lithocarpus* (cupule without valves).

The family is most closely related to the Betulaceae (birches, hazels and hornbeams).

**Economic uses.** The Fagaceae family is the source of some of the most important hardwood timbers of the world, the most notable being oak (particularly the North American white oaks), beech, chestnut and, to a lesser extent, southern beech. This, together with clearance for agriculture, has resulted in the destruction of large areas dominated by the family. Although their timber is of good quality, the tropical members of *Castanopsis* and *Lithocarpus* have as yet been little exploited. Taken as a whole the timber of these genera exhibits a very wide range of properties and there is a correspondingly wide range of uses, from floorboards and furniture to whisky barrels and, formerly, sailing ships. Commercial cork is derived from the bark of the Mediterranean cork oak (*Quercus suber*), and in southeast Europe and Asia Minor the galls on certain oaks are a source of tannin. Many species of chestnut, but principally the sweet chestnut, *Castanea sativa* of southern Europe, are grown for their large edible nuts, from which are made, for example, purees, stuffings, stews and the famous French delicacy *marrons glacés*. The nuts ("beech-

Fagaceae. 1 *Quercus ilex* leafy shoot with male flowers clustered in pendulous catkins and female flowers at the bases ($\times \frac{2}{3}$). 2 *Fagus orientalis* leafy shoot with slender-stalked globose heads of male flowers ($\times \frac{2}{3}$). 3 *Trigonobalanus veticillata* fruits ($\times 2$). 4 *Quercus robur* leafy shoot showing fruits (nuts) surrounded at base of the familiar cup or cupule ($\times \frac{2}{3}$). 5 *Castanea sativa* young fruit with closed spiny cupule forming at the base of remains of the male catkin ($\times \frac{2}{3}$). 6 *Quercus robur* leafless tree. 7 *Fagus sylvatica* leafless tree. 8 *Nothofagus procera* dehisced four-valved cupule ($\times 2$).

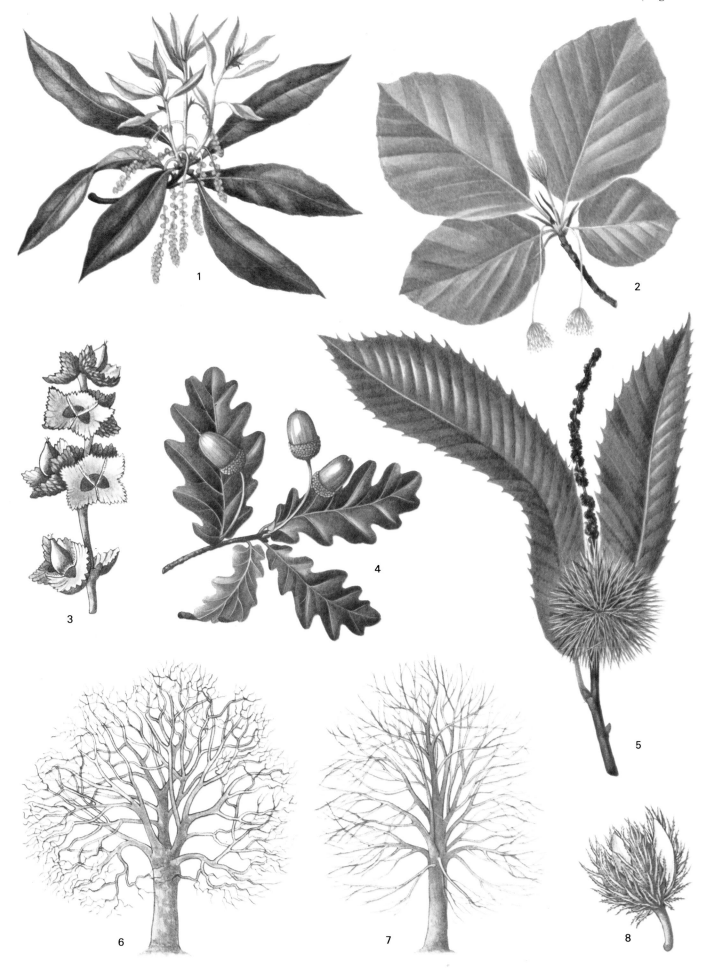

mast") of beech are rich (46%) in oil and in many regions constitute an important food for pigs, as do the fruits (acorns) of oaks.

Because of their form and, in many deciduous species, their rich autumn coloring, many Fagaceae, particularly oaks, chestnuts, beech and, to some extent, southern beech, are grown for ornament in parks and larger gardens. The only American species of *Castanopsis* (*C. crysophylla*) and *Lithocarpus* (*L. densiflorus*) are sometimes grown for ornament in warmer regions. D.M.M.

### BALANOPACEAE

The Balanopaceae (sometimes called Balanopsidaceae) is an isolated family containing a single genus (*Balanops*) of trees and shrubs which are restricted to Queensland in northeastern Australia, New Caledonia and Fiji. The leaves are alternate or pseudo-verticillate (appearing whorled) and without stipules. Male and female flowers occur on separate plants. The male flowers are arranged in catkins and the females are solitary. The males are each subtended by one scale and have two to 12 (usually five or six) stamens, while the females have many scales

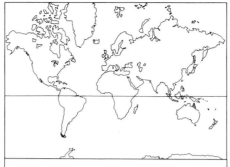

**Number of genera:** 1
**Number of species:** 12
**Distribution:** SW Pacific.
**Economic uses:** none.

and two to three carpels. The fruit is an acorn-like drupe with one or two seeds.

The family is an isolated one, perhaps distantly related to the beeches (Fagaceae). The flowers are so simple as to suggest to some workers that the family may be a very ancient one, representing a stage of "prefloral" evolution. On the other hand, the flowers may be simplified and, therefore highly developed. There are no reported economic uses for the family. D.J.M.

# CASUARINALES

## CASUARINACEAE
*She Oak and River Oak*

The Casuarinaceae is a distinctive family of trees and shrubs adapted to dry habitats, comprising one genus, *Casuarina*.

**Distribution.** The family is distributed in northeast Australia, Malaysia, New Caledonia, Fiji and the Mascarene Islands.

**Diagnostic features.** The Casuarinaceae are mostly tall trees with a characteristic weeping habit caused by their jointed branches with short internodes. The leaves are very peculiar in structure, appearing as whorls of reduced, many-toothed sheaths surrounding the articulations of the jointed stems. The flowers too, are highly reduced, and usually unisexual, with the males and females borne on different parts of the same plant. The male flowers are borne in simple or branched terminal spikes growing towards the tops of the trees, and the whole inflorescence is attached by short green branchlets. The individual flowers tend to be aggregated into groups along the spike giving the appearance of a cup with several stamens protruding out of it. In fact, each flower has only one stamen

Casuarinaceae. 1 *Casuarina suberosa* (a) portion of much-jointed stem with clusters of female flowers on side branches (×⅔); (b) female inflorescence (×3); (c) female flower comprising bract, ovary, short style and bilobed stigma (×9); (d) cluster of cone-like fruits (×⅔); (e) vertical section of seed with persistent woody bracteole behind (×10); (f) tip of shoot bearing male inflorescences (×⅔); (g) cluster of male flowers in inflorescence (×6); (h) male flower comprising a single stamen (×12). 2 *Casuarina* sp portion of ribbed photosynthetic stem with whorl of reduced leaves (×12). 3 *Casuarina sumatrana* habit.

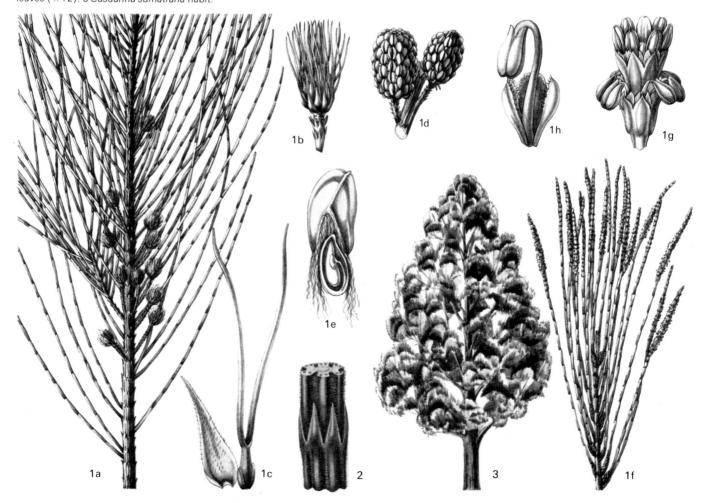

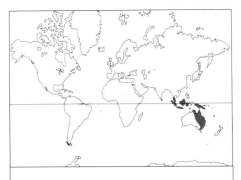

**Number of genera:** 1
**Number of species:** about 65
**Distribution:** SE Asia and SW Pacific
**Economic uses:** hardwood used for furniture-making.

and a perianth of two small lobes each subtended by two more small, leaf-like scales or bracteoles. The females tend to grow on side branches lower down the tree, and are borne in dense, spherical or oval-shaped heads. Each flower is naked, growing out of a leaf-like bract axil, and consists of a tiny ovary of two fused carpels, the posterior locule empty, the anterior locule containing

two ovules. The style is short and divided into two long stigma branches. During development the styles hang out well beyond the bracts to catch wind-borne pollen grains, and after fertilization the ovary develops into a single-seeded samaroid nut. The fruits are enclosed in hard bracteoles which later open to release them, with the result that mature inflorescences resemble pine cones. The seeds have a straight embryo and no endosperm.

Casuarinas are well adapted to very dry habitats in regions of high temperature and low rainfall and are thus remarkably xeromorphic. Besides the obvious reductions to the leaves, already mentioned, there are distinctive modifications to the branches themselves. They are quite slender and more or less circular in outline, but deeply grooved. The green photosynthetic tissue and the respiratory stomata are confined to the grooves, as a protection against low humidity and strong light. The ridges themselves are heavily armored with sclerenchyma so that the branches do not wilt in periods of drought.

**Classification.** At first sight the casuarinas appear to be quite distinct and independent

from all other flowering plant families. The question whether they are a primitive or an advanced family has divided the opinion of taxonomists. Earlier workers tended to think that with their reduced leaves and small, wind-pollinated flowers they were the oldest group of flowering plants and probably related to the gymnosperms (notably the pine family and the genus *Ephedra*). Modern biologists, however, are inclined to the opinion that the family is quite derived and that its peculiar features are the result of extreme specializations in isolated conditions, which have affected all the vegetative and floral parts of the plants. In many of the important morphological features the Casuarinaceae are similar to other petal-less flowering trees of the Hamamelidaceae, and it is now believed that both families have a common ancestry.

**Economic uses.** The wood from several species is extremely hard and valued for furniture manufacture. Red beefwood, *Casuarina equisetifolia*, is the most widespread cultivated species. Other valuable timbers include the Australian native she oak, *C. stricta*, and the cultivated *C. cunninghamiana* (river oak). C.J.H.

# CARYOPHYLLIDAE

## CARYOPHYLLALES

### CACTACEAE
*The Cactus Family*

The Cactaceae is a large family of perennial, xerophytic trees, shrubs or shrublets of distinctive appearance, all more or less succulent and (excepting *Pereskia*) leafless or nearly so. Many species are valued as ornamentals.

**Distribution.** Cacti are mainly plants of semideserts of the warmer parts of North, Central and South America, and are doubtfully native or early naturalized in Africa, Madagascar and Sri Lanka (*Rhipsalis*); *Opuntia* is naturalized in Australia, South Africa, the Mediterranean and elsewhere. The characteristic habitats of cacti experience erratic rainfalls with long drought periods in between, but night dews may be heavy when the temperature falls.

**Diagnostic features.** Most cacti have spines, and the spines, branches and flowers arise from special sunken cushions or areoles which may be regarded as condensed lateral branches; these are either set singly on tubercles or serially along raised ribs. Tufts of short barbed hairs (glochids) may also be present in the areoles. Photosynthesis is undertaken by the young green shoots, but with age these become corky and in the arborescent species develop into a hard, woody, unarmed trunk as in conventional trees. The vascular system forms a hollow cylindrical reticulated skeleton and lacks

**Number of genera:** 87
**Number of species:** over 2,000
**Distribution:** semi-desert regions of North, Central and South America.
**Economic uses:** some garden and house ornamentals with local uses for fruits.

true vessels. The roots are typically superficial and in the larger species widely spreading and adapted for rapid absorption near the soil surface.

The flowers are solitary and sessile (*Pereskia* excepted), bisexual (with rare exceptions) and regular to oblique-limbed. Color range is from red and purple through various shades of orange and yellow to white; blue is lacking. The stamens, petals, sepals and bracts are numerous and spirally arranged, the last three in transitional series without sharp boundaries between them. The ovary is inferior, and consists of two to numerous carpels, forming one locule with numerous ovules on parietal placentas. It is borne on an areole and is commonly covered in hair,

bristles or spines. The style is simple. In *Opuntia* the detached "fruit" (pseudocarp) grows roots and shoots, which form a new plant. The fruit is a berry, which is typically juicy, but may be dry and leathery, splitting open to release the seeds in various ways. The seeds have a straight to curved embryo and little or no endosperm.

**Classification.** The Cactaceae is of especial interest to botanists for its combination of a primitive, unspecialized flower with highly advanced vegetative organs; to the ecologist for its survival under adverse conditions and drought; and to the evolutionist for its parallel life-forms to other unrelated xerophytes, eg *Stapelia*, *Euphorbia* and *Pachypodium*. To the taxonomist it presents great problems, being apparently still in a state of active evolution, and resisting the standard herbarium procedure based on dried specimens. Under pressure from collectors and commercial growers large numbers of "genera" and "species" have been created, more nearly equivalent to subgenera and subspecies or varieties in other plant families. It is here considered to contain 87 genera, while others recognize over 300. There are three subfamilies:

PERESKIOIDEAE. Leaves present, glochids absent, seeds black and without an aril; genera *Maihuenia* and *Pereskia*.

OPUNTIOIDEAE. Leaves and glochids present, seeds covered by pale, bony aril or winged; five genera *Opuntia*, *Pereskiopsis*, *Pterocactus*, *Quiabentia* and *Tacinga*.

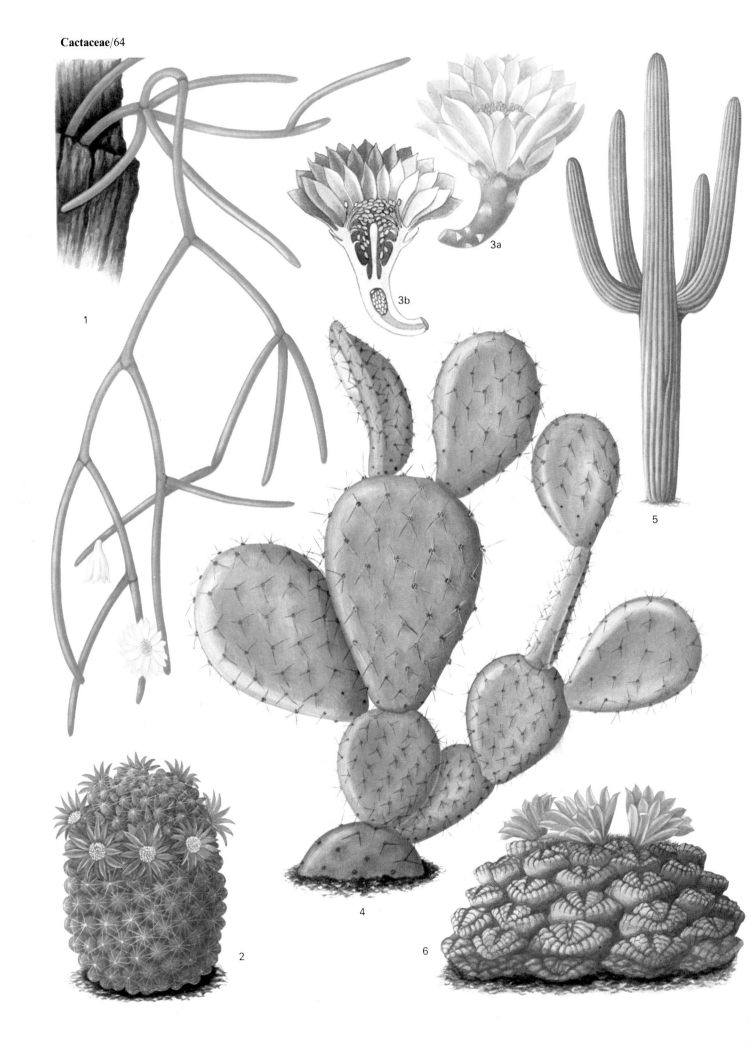

1

3a

3b

2

4

5

6

CACTOIDEAE. Leaves absent or very minute, glochids absent, seeds black or brown, not enveloped by a bony aril; 80 genera in two tribes: CEREEAE, plants predominantly columnar with usually few-ribbed jointed stems, or, if dwarf, flowering from the old areoles, 40 genera including *Cereus, Carnegiea, Echinocereus* and *Echinopsis*; CACTEAE, plants mainly dwarf with usually many ribbed non-jointed stems, flowering from the new areoles, 40 genera including *Rhipsalis, Schlumbergera, Notocactus, Echinocactus, Mammillaria, Lophophora, Ferocactus.*

The Cactaceae has no close relatives and for long it defeated efforts to fit it into existing systems of classification. It is now considered to be most closely related to the Phytolaccaceae, Portulacaceae and Aizoaceae with which it shares embryological and phytochemical peculiarities.

**Economic uses.** Apart from their wide appeal to specialist growers and collectors of the unusual, cacti have few uses. The fleshy fruits of many are collected locally and eaten raw or made into jams or syrups. Some are used for hedging, while those with woody skeletons are used for rustic furniture and trinkets. Opuntias (prickly pears) are grown commercially in parts of Mexico and California for their large juicy fruits.

Probably all the 87 genera are represented in cultivation, collectors being undeterred by difficulties of growing them. The most popular genera among collectors are those that remain dwarf and combine attractive spine colors and rib formations with freedom of flowering: eg *Rebutia, Lobivia, Echinopsis, Mammillaria, Notocactus, Parodia, Neoporteria. Astrophytum* is valued for its very prominent ribs and cottony white tufts, and *Leuchtenbergia* for its extraordinarily long tubercles. More for connoisseurs are the curiously squat, tuberculate, slow-growing species of *Ariocarpus, Pelecyphora* and *Strombocactus. Melocactus,* the "Turk's cap cactus", one of the first cacti to reach Europe, is unique for the large, furry inflorescence terminating the short, stumpy axis.

Even more widely grown are the epi-

Cactaceae. 1 *Rhipsalis megalantha* an epiphytic cactus with many-jointed, spineless stems that are often produced in whorls ($\times\frac{2}{3}$). 2 *Mammillaria zeilmanniana* a dwarf cactus with a solitary stem of spirally arranged tubercles tipped with spine-bearing areoles; flowers arise from the base of the tubercles ($\times\frac{2}{3}$). 3 *Gymnocalycium mihanovichii* (a) entire flower showing the gradual transitional series from bracts to sepals and petals ($\times 2$); (b) vertical section of flower showing inferior ovary and numerous stamens ($\times 2$). 4 *Opuntia engelmannii* showing the characteristic disk-like, many jointed stem bearing numerous glochids—fine readily detached spines ($\times\frac{2}{3}$). 5 *Carnegiea gigantea* showing the characteristic much-branched candelabra habit and ribbed stems ($\times\frac{1}{72}$). 6 *Ariocarpus fissuratus* a dwarf cactus with a many-ribbed, non-jointed stem bearing flowers on new areoles ($\times\frac{2}{3}$).

phytes, so dissimilar in habit and requirements to the foregoing that many a cottage gardener cherishing a 'Christmas cactus' or 'Ackermannii' fails to associate it with cacti at all. The large-flowered epicacti are products of a long line of intergeneric crossings paralleled only in the orchids. These are the only group of cacti where hybridization has been pursued on a grand scale, and the only group grown primarily for flowers. But for the shortness of the flowering season (May to June) and the uninspiring sameness of the green, flat or three-winged stems, these epicacti could become more popular.

The columnar cacti of the tribe Cereeae mostly need to be quite tall before flowering, and are less suitable for small glasshouses.

The name "cactus" is commonly misapplied to a wide range of spiny or fleshy plants quite unrelated to the Cactaceae.

G.D.R.

# AIZOACEAE
*The Mesembryanthemum Family*

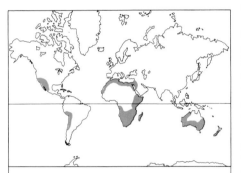

**Number of genera:** 143
**Number of species:** about 2,300
**Distribution:** pantropical, but centered in S Africa.
**Economic uses:** many greenhouse and garden ornamentals (eg *Mesembryanthemum, Lampranthus*) and ornamental curiosity (living stones).

This is a large family of succulent plants with usually showy, daisy-like flowers, often forming brilliant sheets of color when growing in the mass. It contains many popular garden plants. The Aizoaceae is most familiar through the members of the subfamily Ruschioideae (broadly identifiable with *Mesembryanthemum*), which are showy-flowered, highly succulent, tender xerophytes of often curious habit and growth form.

**Distribution.** The family is centered in South Africa, but there are a few pantropical weeds extending to the West Indies, southwestern North America, Florida, South America and Australia.

**Diagnostic features.** The Aizoaceae are annual or perennial herbs or small shrubs, usually more or less succulent, with alternate or opposite, mostly simple, entire leaves with or without stipules. The flowers are solitary or in cymes, and regular, with their parts in whorls, and usually bisexual. The sepals are

four to eight in number (usually five) and are imbricate or rarely valvate, more or less united below. The petals are numerous and of staminodial origin; occasionally they are absent. The stamens are perigynous, usually numerous, sometimes with connate filaments. The ovary is superior or inferior, with one up to 20 (usually five) stigmas and locules and usually numerous ovules. The fruit is a dry capsule, rarely a berry or nut. The seed has a large curved embryo surrounding a mealy endosperm.

Many of the features of the Aizoaceae are the result of adaptations to extremely dry climates (xeromorphy) and typical members are able to survive long periods of extreme insolation and drought, eg in the desert regions of South Africa. The leaves are more or less succulent, and in addition some plants have succulent roots or caudices. Often the plant is reduced to a single annual pair of opposite leaves, which may be so condensed as to approach a sphere, with minimal surface in relation to volume, enabling the plant to resist desiccation.

The internal tissues also show modifications. Large, watery cells rich in sugars called pentosans are characteristic of succulents. In *Muiria* these cells may be 1mm in diameter and can retain their moisture for weeks when separated and exposed to dry air. The possession of two different leaf forms (heterophylly) is common (*Mitrophyllum, Monilaria*), the leaf pair formed at the start of the dormant season being more united and compact than that formed when in full growth, and acting as a protective sheath to the stem apex. Other genera are partly subterranean, with only the clear 'window' in each leaf tip exposed above soil. A type of optical system exists whereby a layer of apical tissue rich in calcium oxalate crystals acts as a filter to intense sunlight before it reaches the thin chlorophyllous layer below (*Fenestraria, Frithia, Ophthalmophyllum*).

Other so-called mimicry plants show a striking similarity to their background rocks and are difficult to detect when not in flower. These are the pebble plants or living stones (*Lithops*); each species is associated with one particular type of rock formation and occurs nowhere else. *Titanopsis*, with a white incrustation to the leaves, is confined to quartz outcrops. It is suggested that this is a rare case of protective coloration in plants akin to that found in the animal kingdom.

The phenomenon of crassulacean acid metabolism occurs in members of the Aizoaceae, having evolved independently in a number of different families of succulent plants.

The mostly showy, many-petalled, diurnal flowers have a superficial resemblance to the flower heads of the Compositae. They are insect-pollinated and most require full sun before they will expand. Several have set hours for opening and closing. *Carpobrotus*

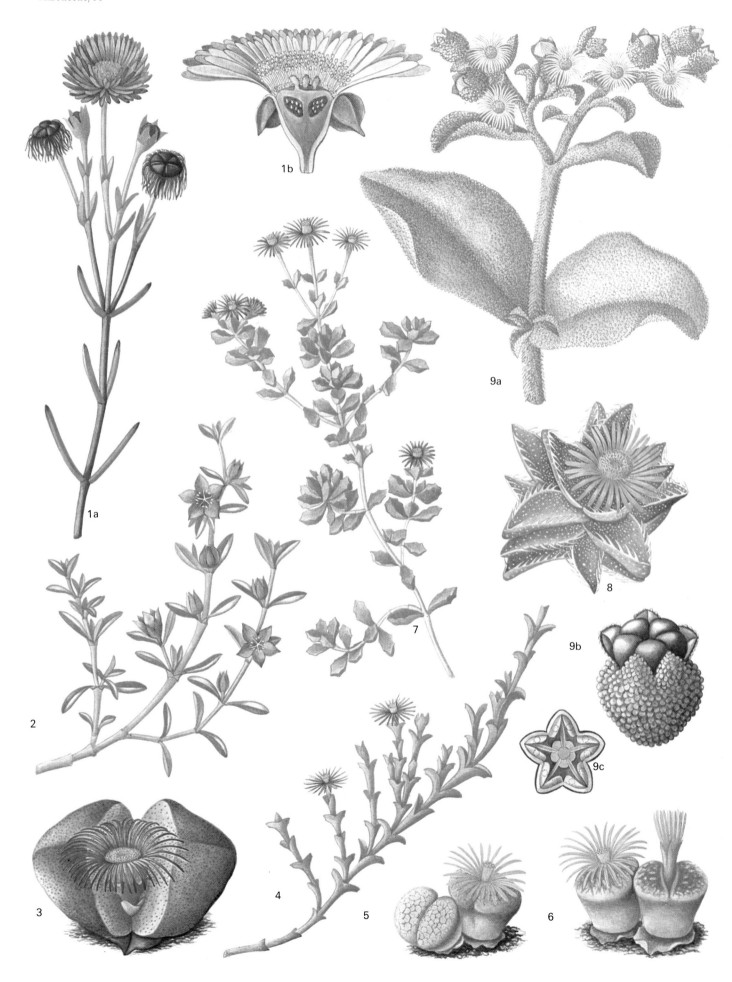

produces an edible berry, the hottentot fig, but the remainder form dry, dehiscent capsules operated by a hygroscopic mechanism which expands the valves in response to moistening, closing them again on drying out. In desert conditions this ensures germination during the brief rainy periods.

*Conicosia* and certain related genera have three different methods of seed dispersal. The capsule first opens hygroscopically and some seeds are washed out by the impact of raindrops. It remains open when dry and the remaining loose seeds are shaken out as from a pepper-pot over a longer period. Finally the whole fruit breaks up into segments which are dispersed by the wind, each wing-like lamella containing up to two seeds trapped in two pocket-like folds. These rank among the most specialized of all angiosperm fruit structures.

**Classification.** The Aizoaceae is divided into five subfamilies, four of which are based on the large *Mesembryanthemum* complex. What Linnaeus treated as a single large genus *Mesembryanthemum* is nowadays split up on the basis of fruit structure into 125 genera, and accounts for some 2,000 of the species of the family. Some authors treat the *Mesembryanthemum* complex as a separate family Mesembryanthemaceae, whereby the Aizoaceae is restricted to a small group of rather insignificant and small-flowered weedy genera.

The Aizoaceae belongs to the Caryophyllales and shares with most families therein the possession of betalains in place of anthocyanin as a floral pigment. Its affinities lie with the Phytolaccaceae.

**Economic uses.** Shrubby members of the Ruschioideae (*Lampranthus, Oscularia, Drosanthemum, Erepsia,* etc) are half-hardy and grown for summer bedding especially in southern Europe where they flower profusely. Only one species, *Ruschia uncinata,* verges on complete hardiness, although *Carpobrotus* survives most winters in coastal areas and is much planted (and naturalized) as a sandbinder. Hybrids of the annual *Dorotheanthus* enjoy great popularity, and

Aizoaceae. 1 *Lampranthus* sp (a) shoot with opposite succulent leaves and terminal, solitary flowers ( ×⅔); (b) half flower with free sepals, several series of petals, numerous stamens and gynoecium with separate styles and numerous ovules ( × 2). 2 *Sesuvium portulacastrum* jointed, succulent stem with opposite leaves and tubular flowers ( ×⅔). 3 *Pleiospilos bolusii* a plant comprising two large succulent leaves with flowers produced between ( ×⅔). 4 *Ruschia uncinata* flowering shoot ( ×⅔). 5 *Lithops pseudotruncatella* and 6 *L. lesliei* pebble-like plants (living stones) of two succulent leaves with flowers arising from the fissure ( ×⅔). 7 *Oscularia deltoides* flowering shoot ( ×⅔). 8 *Faucaria tigrina* with a dense rosette of spiny leaves ( ×⅔). 9 *Mesembryanthemum crystallinum* (a) the ice plants, so-called for the glistening papillae that cover the whole plant ( ×⅔); (b) dehiscing capsule ( × 2); (c) capsule from above ( × 1⅓).

have supplanted the original ice plant *Mesembryanthemum crystallinum,* which has glossy papillae-like water droplets covering the foliage, in popularity in gardens.

G.D.R.

# CARYOPHYLLACEAE
*The Carnation Family*

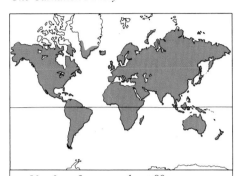

**Number of genera:** about 80
**Number of species:** about 2,000
**Distribution:** temperate regions, centered in the Mediterranean area.
**Economic uses:** many popular garden ornamentals, notably pinks and carnations; others are widespread weeds.

The Caryophyllaceae is a large family of mainly temperate herbaceous plants, of somewhat doubtful relationships. It includes the popular carnations and pinks, and some well-known wild flowers and weeds such as campions and chickweed.

**Distribution.** The family is found in all temperate parts of the world and sparingly on mountains in the tropics; several species of *Stellaria* (chickweed) and *Cerastium* (mouse-ear chickweed) have become almost cosmopolitan weeds. The center of distribution is, however, in the Mediterranean region and adjoining parts of Europe and Asia. Representation in the temperate Southern Hemisphere is small in terms of genera and species. All the larger genera (*Silene, Dianthus, Arenaria,* etc) are found in the Northern Hemisphere, with a strong concentration in the Mediterranean region.

**Diagnostic features.** In spite of its large size, the Caryophyllaceae is a relatively uniform and easily recognized family. Plants are usually herbaceous, either annual or perennial and dying back to the crown. A few species are somewhat shrubby with persistent woody stocks. The leaves are almost invariably opposite; the stem nodes are swollen, and the bases of each leaf pair often join around them to form a perfoliate base. The leaves themselves are always simple and entire. Stipules are usually absent; when present (as in the subfamily Paronychioideae) they are usually scarious.

The inflorescences are of cymose construction, although varied in detail; the most complex is the dichasial panicle, in which each of the two bracteoles of the terminal flower subtends an inflorescence branch itself bearing a terminal flower and two

bracteoles, which repeats the structure. Suppression of individual flowers can lead to raceme-like monochasia, and, ultimately, to a single-flowered inflorescence (as in the carnation).

The flowers are regular and usually bisexual. The calyx consists of four or five free sepals (subfamily Alsinoideae) or of united sepals with a four- or five-lobed apex (subfamily Silenoideae); a number of subtending bracts are present at the base of the calyx in some genera, notably *Dianthus.* The corolla consists of four or five petals which are free from each other (petals absent in part of subfamily Paronychioideae, and some other species) and often sharply differentiated into limb and claw; two small outgrowths (ligules) may be present at the junction of the two parts on the inner surface. The apexes of the petals are often notched or deeply cut, producing more or less bilobed or even fringed petals. The stamens are typically twice as many as the petals, but may be reduced to as many, or even fewer (eg *Stellaria media*). They are usually free from each other and attached directly to the receptacle, but in some apetalous members of subfamily Paronychioideae they are attached to the sepals, rendering the flower perigynous. The ovary is superior, of two to five united carpels; usually it has a single locule with free central placentation, but in a few species of *Silene* and *Lychnis* it is septate at the base. The styles are free, as many as there are carpels. The ovules are usually numerous, but may be reduced to one, when the placentation is basal. The fruit is most frequently a capsule, dehiscing by means of teeth at the apex; these may be as many as, or twice as many as, the number of carpels. More rarely, in single-ovuled genera, the fruit is an achene. Seeds are usually numerous, with the embryo curved around the food-reserve material, which is usually perisperm (tissue derived from the diploid nucleus) rather than endosperm (derived from the triploid nucleus of the fertilized embryo sac). In *Silene* and *Lychnis* the petals, stamens and ovary are separated from the calyx by a shortly extended internode (anthophore).

**Classification.** The family is usually divided into three subfamilies:

ALSINOIDEAE. Stipules absent; sepals free from each other. This subfamily contains many well-known and widely distributed genera, such as *Arenaria, Minuartia, Honkenya, Stellaria, Cerastium, Sagina, Colobanthus* and *Lyallia.*

SILENOIDEAE. Stipules absent; sepals connate. This subfamily is of about the same size as the Alsinoideae, and contains equally well-known genera: *Silene, Melandrium, Dianthus, Gypsophila, Agrostemma, Lychnis* etc. Both of these subfamilies are well distinguished and easily recognized, though the recognition of genera within them is taxonomically difficult and controversial.

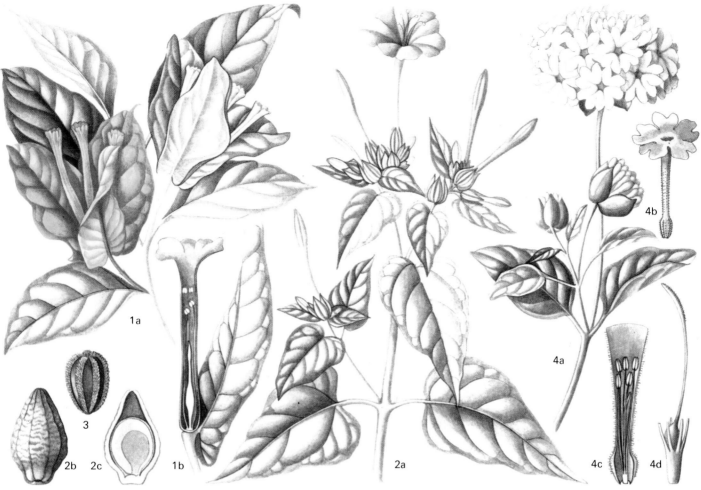

Nyctaginaceae. 1 *Bougainvillea spectabilis* (a) leafy shoot with flowers subtended by conspicuous bracts ( ×⅔) ; (b) bract and half flower showing petaloid tubular calyx and no petals ( ×2). 2 *Mirabilis jalapa* (a) leafy shoot with each flower enclosed by a calyx-like involucre of bracts ( ×⅔) ; (b) indehiscent fruit ( ×3⅓) ; (c) vertical section of fruit showing single seed ( ×3⅓). 3 *Pisonia aculeata* glandular fruit that is dispersed by birds ( ×1). 4 *Abronia fragrans* (a) shoot bearing dense clusters of flowers ( ×⅔) ; (b) flower ( ×1) ; (c) section of base of perianth-tube showing stamens and style ( ×2) ; (d) gynoecium with elongate, hairy stigma ( ×4).

PARONYCHIOIDEAE. Stipules present, usually scarious; sepals free or connate. This sub-family is more variable than the other two, and consists essentially of two groups of genera: (a) Those in which a corolla is present, the flower being hypogynous and the fruit is usually a several- to many-seeded capsule. This is a group consisting of the tribes SPERGULEAE and POLYCARPEAE, containing, among others, the genera *Spergula*, *Spergularia* and *Polycarpon*. These plants are very similar to members of the Alsinoideae, differing only in the possession of

Caryophyllaceae. 1 *Stellaria graminea* shoot with opposite leaves, swollen nodes and cymose inflorescence ( ×⅔). 2 *Telephium imperati* (a) shoot with flowers and fruits ( ×⅔) ; (b) flower with five sepals, five petals, five stamens and three styles ( ×4). 3 *Dianthus deltoides* habit ( ×⅔) ; (b) half flower with fused sepals, deeply notched petals, stamens twice as many as petals and superior ovary crowned by two styles ( ×3) ; (c) cross section of ovary ( ×8) ; (d) vertical section of ovary ( ×8) ; (e) fruit ( ×3). 4 *Arenaria purpurascens* habit ( ×⅔) 5 *Silene dioica* inflorescence bearing flowers with deeply cleft limb and white claw to each petal ( ×⅔). 6 *Herniaria ciliolata* (a) shoot with small axillary inflorescences ( ×⅔) ; (b) flower with clawed petals ( ×12).

stipules. (b) Those in which the corolla is absent, the flower being perigynous and the fruit usually single-seeded and indehiscent. This group contains the tribe PARONYCHIEAE and one or two others, and is sometimes recognized as a separate family, the Illecebraceae. Its best-known genera are *Paronychia*, *Herniaria* and *Corrigiola*. Its status is unsettled, but it appears to be only distantly related to the Sperguleae.

The family is generally grouped with a number of others (Chenopodiaceae, Aizoaceae, Phytolaccaceae, etc) in a group known as the Centrospermae (Caryophyllales). This disposition was based on morphological structure, in particular that of the embryo, which is usually strongly curved. The discovery that most of the Centrospermae (but not the Caryophyllaceae) contain a group of chemical pigments known as betalains, instead of the more usual anthocyanins, has led to a separation of the Caryophyllaceae from the rest of the group. In spite of this difference, however, the Caryophyllaceae still shows a great deal of similarity to the betalain-containing families, even though this may not imply as close a relationship as was once thought.

**Economic uses.** The Caryophyllaceae provides a large number of widely cultivated garden ornamentals. The most important single species is the carnation (*Dianthus caryophyllus*), which is now found in numerous garden forms (cultivars), and forms a specialized crop for the cut-flower market. Other widely cultivated genera are *Dianthus* (pinks), of which many species are cultivated, including alpines, *Silene* (catchfly, campion), *Gypsophila* (baby's breath), *Agrostemma* (corn cockle), *Lychnis* and *Saponaria* (soapworts).

Several species, notably *Stellaria media*, are widespread annual weeds of fields, gardens,and other disturbed places. *Agrostemma githago* was at one time a troublesome weed of grain fields, but, with the use of selective herbicides, this species is no longer a considerable problem.

*Spergula arvensis* var *sativa* is occasionally used as a fodder plant in dry, sandy areas.                J.C.

# NYCTAGINACEAE
*Bougainvilleas*
The Nyctaginaceae is a family of chiefly tropical herbs, shrubs and trees.

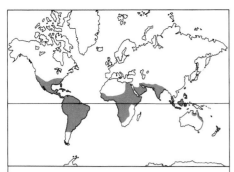

**Number of genera:** 30
**Number of species:** about 290
**Distribution:** pantropical, particularly America.
**Economic uses:** *Bougainvillea* and *Mirabilis* cultivated as ornamentals; *Pisonia* used as vegetables.

**Distribution.** The family is found throughout the tropics, particularly in America.

**Diagnostic features.** The leaves are alternate or opposite, simple and without stipules. The inflorescence is usually cymose and the flowers are bisexual or unisexual, and are sometimes surrounded by colored bracts which resemble a calyx. The perianth is usually petal-like, tubular, and often the lower part persists into fruiting time. There are no petals and the one to many stamens, usually five, alternate with the five perianth lobes, the filaments being free or fused together at the base, or sometimes branched above. The ovary is superior, comprising a single carpel with a single basal erect ovule; it is surmounted by a long style. The fruit is an indehiscent achene, sometimes enclosed by the persistent base of the calyx which may serve to assist fruit dispersal. The seeds have endosperm, perisperm and a straight or curved embryo.

**Classification.** *Mirabilis*, including *Oxybaphus*, comprises about 60 American species. The base of the flower is surrounded by a five-part involucre of bracts which greatly resembles a calyx. The derivation of the involucre is from a much-condensed three-flowered cyme, with only one of the flowers developed. Evidence for this is seen in species such as *Mirabilis coccinea* where the involucre encloses more than one flower. In *Mirabilis* the involucre has taken on the role of a parachute-like fruit dispersal structure. The flowers of *Mirabilis jalapa* open in the evening, whence one of the common names, four o'clock. Another common name, marvel of Peru, relates to the polychromic flowers, which are white, yellow or red.

In *Bougainvillea*, a genus of 18 South American species, the three decorative and colored "sepals" are in fact bracts which subtend groups of three inconspicuous tubular flowers.

*Pisonia* comprises 50 tropical and subtropical species, and the fruits are glandular so that they adhere to animals and are dispersed.

Other genera in the family include the monotypic southern North American and Mexican *Nyctaginia*; the 40 tropical and subtropical species of herbs in the genus *Boerhavia*; 35 North American species of the genus *Abronia*; and *Neea*, comprising some 80 tropical American species.

The Nyctaginaceae belongs to a group of families often referred to as the Centrospermae (Caryophyllales).

**Economic uses.** Bougainvilleas are often grown as defensive and decorative hedges in warmer climates, and as greenhouse plants farther north. The two most commonly grown species are *Bougainvillea glabra* and *B. spectabilis*. From these, and from *B. peruviana* and *B. × buttiana*, many cultivars have been produced. *Mirabilis jalapa* and *M. coccinea* are amongst many species of *Mirabilis* cultivated for their ornamental value as garden plants. The tuberous roots of *M. jalapa* are the source of a purgative drug used as a substitute for jalap. The leaves of the brown cabbage tree (*Pisonia grandis*) and the lettuce tree (*P. alba*) can be used as a vegetable. Decoctions of the leaves of *P. aculeata* and of the fruits of *P. capitata* are used medicinally to treat a range of complaints. **B.M.**

# AMARANTHACEAE
*Cockscombs and Celosias*

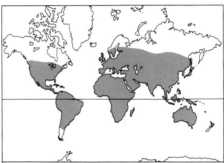

**Number of genera:** 65
**Number of species:** 900
**Distribution:** cosmopolitan, with tropical members centered in Africa and America.
**Economic uses:** widely cultivated as garden ornamentals and a few used as pot herbs and vegetables.

The Amaranthaceae is a large family of herbs and shrubs containing several well-known species of horticultural importance such as the cockscombs (*Celosia* species) and love lies bleeding (*Amaranthus caudatus*).

**Distribution.** The family of about 65 genera and 900 species grows in tropical, subtropical and temperate regions, the tropical members occurring mainly in Africa and America.

**Diagnostic features.** Most members are herbs or shrubs, rarely climbers, with entire, opposite or alternate leaves without stipules. The flowers may be solitary or in axillary dichasial cymes arranged in spike-like or head-like, usually bracteate, inflorescences. They are bisexual, rarely unisexual, nor-mally regular with four to five perianth segments which are sometimes united; stamens are one to five, free or often united at the base in a tube from which petaloid appendixes arise in some genera between the stamens; the ovary is superior, comprising two to three fused carpels that are free from or united with the perianth; it has a single locule containing one to numerous ovules. The perianth is often dry, membranaceous and colorless. Sometimes the lateral flowers are sterile and develop spines, wings or hairs which serve as a dispersal mechanism, as in *Froelichia*. The flowers are subtended by well-developed dry, chaffy scales (bracteoles) and in the mass are often very showy. The fruit may be a berry, pyxidium or nut; the seeds usually have a shiny testa and the embryo is curved.

**Classification.** The family may be divided into two subfamilies, each containing two tribes, as follows:

AMARANTHOIDEAE. Stamens are four-locular and the ovary has one to many ovules; tribe CELOSIEAE, main genera *Celosia* and *Deeringia*; tribe AMARANTHEAE, main genera *Amaranthus*, *Ptilotus*, *Achyranthes*.

GOMPHRENOIDEAE. Stamens are bilocular and the ovary has a single ovule; tribe BRAYULINEAE containing the two small genera *Brayulinea* and *Tidestromia*; tribe GOMPHRENEAE including *Froelichia*, *Pfaffia*, *Alternanthera*, *Gomphrena* and *Iresine*.

The Amaranthaceae belongs to the group of families known as the Centrospermae, and in common with most other members of the group contains betalain pigments instead of the customary anthocyanins found in most other angiosperm families. It is closely related to the Chenopodiaceae but differs in having a scarious perianth and the stamens frequently united in a ring.

**Economic uses.** The family contains many weedy species but several are widely cultivated as garden ornamentals and a few are used as pot herbs or as vegetables. The edible seeds of some species of *Amaranthus* were widely used, especially in Central and South America, and a few are still grown today. Celosias are often grown as pot plants or as tender bedding plants. *Celosia cristata* is the cockscomb, a tropical herbaceous annual. *Alternanthera* species from the New World tropics are grown for their ornamental leaves. The fleshy leaves of *A. sessilis* are eaten in various tropical countries. Also grown for their colorful, usually scarlet leaves as house or tender bedding plants are the iresines, especially *Iresine herbstii* and *I. linderii*, both from South America. Some species are reputed to have medicinal properties. The Australian *Ptilotus* (*Trichinium*) *manglesii*, with globose heads of feathery pink and white flowers, is sometimes cultivated as a tender bedding plant or under glass. *Gomphrena globosa*, a tropical annual with white, red or purple heads, is grown as an "everlasting." **V.H.H.**

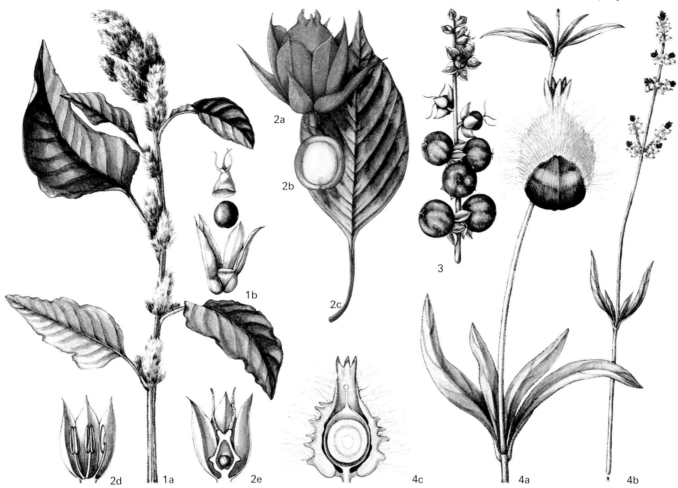

Amaranthaceae. 1 *Amaranthus retroflexus* (a) leafy shoot with flowers in axillary tassels ($\times\frac{2}{3}$); (b) fruit dehiscing by lid to disperse globular seed ($\times6$). 2 *A. caudatus* (a) flower with reddish subtending bracteoles and reddish perianth ($\times14$); (b) seed ($\times14$); (c) leaf ($\times\frac{2}{3}$); (d) vertical section of male flower ($\times12$); (e) vertical section of female flower ($\times4$). 3 *Deeringia amaranthoides* inflorescence with fruits at the base ($\times3$). 4 *Froelichia gracilis* (a) shoot with large lateral, hairy, sterile flowers ($\times\frac{2}{3}$); (b) inflorescence ($\times\frac{2}{3}$); (c) vertical section of sterile flower ($\times\frac{2}{3}$).

# PHYTOLACCACEAE
*Pokeweed and Bloodberry*

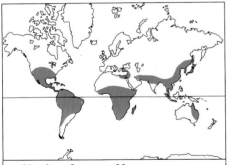

**Number of genera:** 22
**Number of species:** about 125
**Distribution:** chiefly tropical and subtropical, America and W Indies
**Economic uses:** many medicinal uses, yield red dyes and are used as ornamentals and potherbs.

The Phytolaccaceae is a family of trees, shrubs, woody climbers and herbs, including some medicinal plants and ornamentals.

**Distribution.** Most members of the family are native to tropical America and the West Indies, but some are found in Central and temperate South America, the eastern Mediterranean area, in tropical and South Africa, Madagascar, the Indian subcontinent. Malaysia, China, Japan and Australasia.

**Diagnostic features.** The leaves are alternate, simple and entire, typically without stipules or with minute stipules. The flowers are small, mostly in terminal or axillary racemes or cymes. They are regular, bisexual (rarely irregular), rarely unisexual, then with both sexes on one plant (monoecious) or on different plants (dioecious). The perianth is almost always of one whorl of four or five usually free and persistent segments, but sometimes disciform; petals are sometimes present. The stamens are usually hypogynous, sometimes united at the base, as many as the perianth lobes or more numerous, often as a result of branching. The ovary is usually superior, sometimes raised on a gynophore, rarely more or less inferior; it comprises one to many separate or united carpels, each carpel typically with a single basal or axillary ovule. Styles may be as many as the carpels, short or more or less filiform, or absent. The fruit is a fleshy berry, dry nut or, rarely, a loculicidal capsule; the seed has mealy perisperm and the curved embryo characteristic of the order.

**Classification.** The family is divided into four subfamilies. The subfamily PHYTO-LACCOIDEAE (perianth single; one ovule per carpel, apotropous) has four sections; PHYTOLACCEAE (*Anisomeria, Ercilla, Phytolacca*); BARBEUIEAE (*Barbeuia*); GYROSTEMONEAE (*Gyrostemon, Codonocarpus, Didymotheca*); RIVINEAE (*Hilleria, Ledenbergia, Monococcus, Petiveria, Rivina, Seguieria, Trichostigma*); and AGDESTIDEAE (*Agdestis*). Subfamily STEGNOSPERMA-TOIDEAE (perianth double; ovules epitropous, fruit capsular, with red aril; glands with oxalate deposits) contains one genus, *Stegnosperma*. Subfamily MICROTEOIDEAE (perianth single; gynoecium with one locule but two to four stigmas and a single basal ovule; fruit dry) includes *Microtea* and *Lophiocarpus* (sometimes united with *Microtea*). Subfamily ACHATOCARPOIDEAE (flowers unisexual, plants dioecious; ovary with one locule of two fused carpels with two stigmas, containing one ovule; fruit a berry) includes *Achatocarpus* and *Phaulothamnus*).

In recent years, the woody climber *Barbeuia madagascariensis* has been made by some authorities the only species of the family Barbeuiaceae. Another monotypic family (Agdestidaceae) has been proposed for *Agdestis clematidea*. The family Stegnospermataceae has also been proposed for the

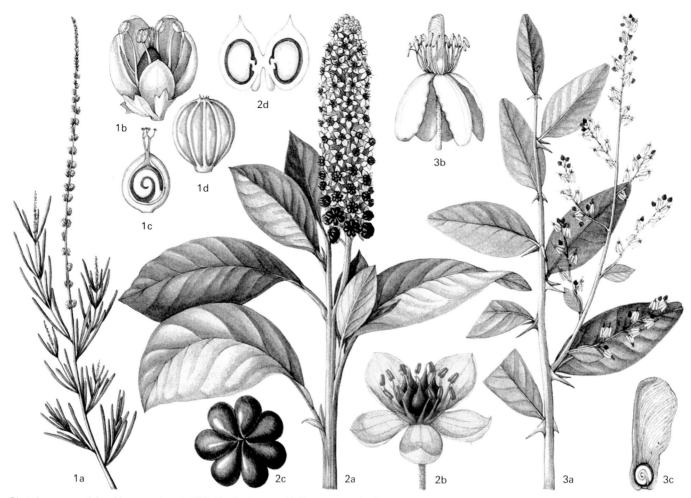

Phytolaccaceae. 1 *Lophiocarpus burchellii* (a) leafy shoot and inflorescences ($\times\frac{2}{3}$); (b) flower with single whorl of five perianth segments ($\times10$); (c) vertical section of ovary showing curved basal ovule and forked styles ($\times20$); (d) fruit—a nut ($\times10$). 2 *Phytolacca clavigera* (a) shoot with axillary leaf-opposed racemose inflorescence bearing flowers and fruits ($\times\frac{2}{3}$); (b) flower with numerous stamens with swollen bases and seven free carpels ($\times6$); (c) fleshy fruit ($\times3$); (d) vertical section of fruit ($\times4$). 3 *Seguieria coriacea* (a) shoot with pairs of thorny stipules and axillary inflorescence ($\times\frac{2}{3}$); (b) flower ($\times4$); (c) vertical section of winged fruit with curved embryo ($\times1$).

three species of the genus *Stegnosperma*. Section Gyrostemoneae (five genera) has also been recognized as a family, the Gyrostemonaceae, as has Achatocarpoideae (Achatocarpaceae), with two genera. Finally, the whole section Rivineae (single carpellate ovary, styloid oxalate cells in the leaves, and unusual wood anatomy) may be separated as the family Petiveriaceae.

With its curved embryo, perisperm and betalain pigments, the Phytolaccaceae is properly included among the Centrospermae (Caryophyllales). The family has a close affinity with the Nyctaginaceae and the subfamily Microteoideae is held to show a transition to the family Chenopodiaceae.

**Economic uses.** *Rivina humilis*, the blood-berry or rouge plant, is grown in greenhouses and a red dye is extracted from the berries. *Petiveria alliacea*, remarkable for its smell of onions, is used medicinally in South America. *Trichostigma peruvianum* is often grown as a hothouse plant. *Phytolacca americana* (*P. decandra*), the pokeweed or red ink plant, is cultivated as an ornamental shrub and like some other *Phytolacca* species yields edible leaves, sometimes used as potherbs, and, from the berries, a red dye.

The medicinal uses are many and varied, from treatment of rabies, insect bites, lung diseases and tumors by species of *Phytolacca*, mainly in root preparations, to the treatment of syphilis by *Agdestis*, again in root preparations.                    F.B.H.

# CHENOPODIACEAE
*Sugar Beet, Beetroot and Spinach*

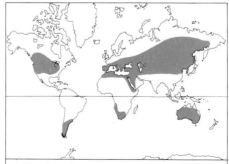

**Number of genera:** about 100
**Number of species:** about 1,500
**Distribution:** temperate and subtropical, principally in saline habitats.
**Economic uses:** sugar beet, beetroot, leaf beets and spinach beet.

The Chenopodiaceae is a large family of perennial herbs (and rarely shrubs or trees) which are halophytic, that is adapted to live in soils containing an unusually high percentage of inorganic salts. Thus they are dominant components of salt marshes, and since saline soils are often associated with arid conditions, many species exhibit xerophytic adaptations.

**Distribution.** The family is widely distributed in temperate and subtropical saline habitats, particularly around the Mediterranean, Caspian and Red seas, the salt-rich steppes of central and eastern Asia, the edge of the Sahara desert, the alkaline prairies of the United States of America, the Karroo of South Africa, Australia and the Pampas of Argentina. They also grow as weeds in salt-rich soils around human habitations.

**Diagnostic features.** Typically, members of the family have deep penetrating roots to obtain any available water supply, and small, mealy-textured or hair-covered, lobed or spiny alternate leaves without stipules. Some genera, eg *Salicornia*, have fleshy jointed stems with no leaves at all, which gives the plants a curious cactus-like appearance. The inconspicuous flowers, arranged into spike-like or cymose inflorescences,

Chenopodiaceae. 1 *Kochia scoparia* (a) leafy shoot with inconspicuous flowers ($\times\frac{2}{3}$); (b) bisexual flower with five perianth-segments and conspicuous stamens ($\times 8$); (c) female flower ($\times 8$); (d) gynoecium of female flower with two styles ($\times 8$); (e) vertical section of fruit ($\times 8$); (f) section of seed with circular embryo surrounding endosperm ($\times 16$). 2 *Salicornia europaea* (a) habit showing fleshy leafless stem ($\times\frac{2}{3}$); (b) part of flowering shoot ($\times 4$); (c) flower sunk in stem ($\times 8$). 3 *Atriplex triangularis* (a) flowering shoot ($\times\frac{2}{3}$); (b) fruiting stem ($\times\frac{2}{3}$); (c) enlarged bracteoles enclosing the fruit ($\times 4$); (d) fruit ($\times 4$). 4 *Salsola kali* (a) fruit with one segment of persistent bracteole removed ($\times 4$); (b) fruit ($\times 6$); (c) seed ($\times 6$).

are small, usually regular, bisexual or unisexual and wind-pollinated. The sepals and petals are very similar to one another in appearance and usually consist of five, three or two brown or greenish lobes. There are usually the same number of anthers as perianth segments, sometimes fewer, arranged on top of the ovary or on a disk. The ovary is superior (semi-inferior in *Beta*) and consists of three fused carpels and a single locule containing a single basal ovule. There is a terminal style with two (rarely one or three) stigmas. The fruit is a small round nut or achene.

**Classification.** The 100 genera can be divided up into two main tribes on the basis of embryo shape.

CYCLOBEAE. Embryo ring-shaped, horse-shoe-shaped or semicircular, wholly or partially enclosing the endosperm; important members of this group include the temperate and subtropical genera *Chenopodium*, *Kochia* and *Atriplex*, the salt marsh herbs of *Salicornia*, *Halocnemum* and *Arthrocnemum* and the well-known food-plant genera *Beta* and *Spinacia* (spinach).

SPIROLOBEAE. Embryo spirally twisted, and endosperm divided into two halves by the embryo or entirely wanting; important members of this group include the coastal salt marsh and steppe genera *Salsola*, *Anabasis*, *Halimione*, *Haloxylon* and *Suaeda*.

The family belongs to the assemblage of families commonly known as the Centrospermae (Caryophyllales).

**Economic uses.** *Beta vulgaris* is really the only species of any major agricultural importance; it includes amongst its many different varieties the sugar and fodder beets. Today, sugar beet serves as a source of sugar in almost every country in Europe, in the USSR, the United States of America, Turkey, Iran, parts of Africa, Korea, Japan and parts of Australia. In South America, sugar beets are cultivated in Argentina, Chile and Uruguay, but the major source of sucrose there is still sugar cane. Other varieties of *B. vulgaris* include the deep-red beetroot and mangel-wurzels, the large leaved spinach beet or perpetual spinach and seakale beet or Swiss chard.

Other cultivated plants of the family include quinoa, *Chenopodium quinoa*, which is grown for its edible leaves and seeds to provide a staple diet for Andean Peruvians, *C. anthelminticum* which is used as a vermifuge and *C. amaranticolor*, with its decorative green and violet-red colored foliage, is used as a border plant in gardens.

C.J.H.

## DIDIEREACEAE

The Didiereaceae is a curious family of chiefly columnar cactus-like plants.

**Distribution.** The four genera of the family are confined to semi-desert areas in Madagascar.

**Diagnostic features.** The leaves are alternate, simple, entire and without stipules, and in some species wither and fall off to expose the spiny stems. The flowers are unisexual and borne on different plants (bisexual and female in *Decaryia*), the males consisting of two opposite petal-like sepals and four overlapping petals surrounding the eight to ten stamens which are shortly united at the base. The female flowers have a perianth like the males, and a superior ovary of three fused carpels comprising three locules, only one of which is fertile, containing a single basal ovule. The style is single and usually expanded into three or four irregular stigmatic lobes. The triangular dry fruits do not dehisce and the seed has a folded embryo,

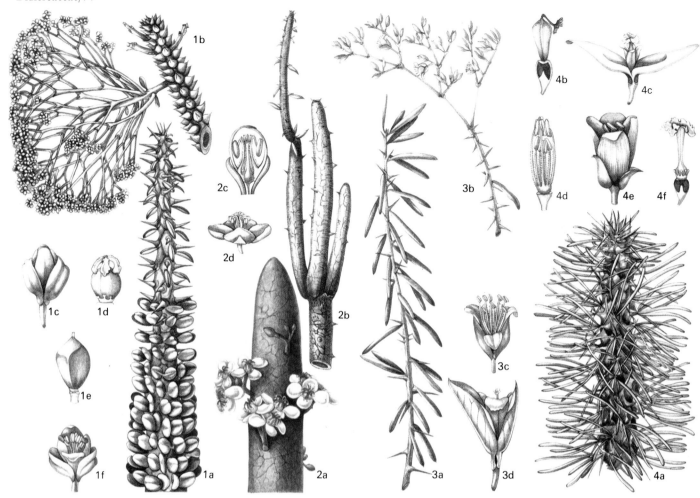

Didiereaceae. 1 *Alluaudia procera* (a) upper portion of flowering shoot ($\times\frac{1}{3}$); (b) shoot bearing inflorescence of male flowers ($\times\frac{1}{3}$); (c) female flower ($\times4$); (d) gynoecium ($\times7$); (e) fruit ($\times7$); (f) male flower ($\times2$). 2 *A. dumosa* (a) upper portion of flowering shoot ($\times\frac{2}{3}$); (b) spiny branches ($\times\frac{1}{3}$); (c) vertical section of male flower bud ($\times3\frac{1}{2}$); (d) male flower ($\times1\frac{1}{2}$). 3 *Alluaudiopsis fiherenensis* (a) spiny shoot ($\times\frac{1}{3}$); (b) shoot with male flowers ($\times\frac{1}{3}$); (c) male flower ($\times1\frac{2}{3}$); (d) female flower ($\times1\frac{1}{3}$). 4 *Didierea madagascariensis* (a) upper portion of branch ($\times\frac{1}{9}$); (b) fruit ($\times3$), (c) female flower opened out ($\times2$); (d) androecium ($\times2$); (e) male flower ($\times2$); (f) gynoecium ($\times3\frac{1}{2}$).

arborescent euphorbias. *Didierea madagascariensis* has erect branches 4m–6m (13ft–20ft) tall, while *D. trollii* is smaller with branches which spread horizontally. The internal structure of the stem is divided into a series of chambers by transverse diaphragms of pith which confer light weight but rigidity to the member. The single species of *Decaryia* is characterized by its spreading branches and thorny zig-zag twigs.

The Didiereaceae is one of the group of families known as the Centrospermae (Caryophyllales) containing betalain pigments rather than anthocyanins. It has earlier been placed in the Euphorbiales.

**Economic uses.** The plants occasionally appear in succulent collections.          B.M.

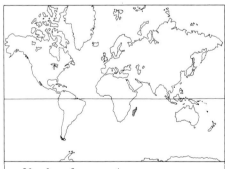

**Number of genera:** 4
**Number of species:** 11
**Distribution:** dry parts of Madagascar.
**Economic uses:** rarely cultivated.

fleshy cotyledons and little or no endosperm.
**Classification.** There are four genera in the family: *Alluaudia* (six species), *Alluaudiopsis* (two species), *Decaryia* (one species) and *Didierea* (two species). The strangest species is perhaps *Alluaudia procera*. It looks like a bent and thorny telegraph pole up to 15m (50ft) high, with flowers produced in incongruous apical clusters.

The two species of *Didierea* are a little more conventional, reminiscent of certain

# PORTULACACEAE
*Lewisias and Portulacas*

The Portulacaceae is a medium-sized family of annual or perennial more or less succulent herbs or subshrubs, several of which are valued as ornamentals.

**Distribution.** The family is cosmopolitan, but is especially well represented in South Africa and America.

**Diagnostic features.** The leaves are more or

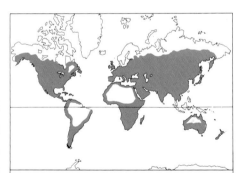

**Number of genera:** 19
**Number of species:** about 500
**Distribution:** cosmopolitan, but centered in S Africa and America.
**Economic uses:** several ornamentals and a potherb.

less fleshy, alternate or opposite, entire and bear stipules which may form hairs or membranous scales that may envelop the whole stem and leaves (*Anacampseros papyracea*). The flowers are rather small, except in a few species cherished by the gardener (*Portulaca grandiflora*, *Lewisia tweedyi*). They are regular, bisexual and typically composed of two green sepals, five (sometimes four or six) free petals and stamens and

Portulacaceae. 1 *Lewisia cotyledon* (a) habit showing basal rosette of leaves and inflorescence ( ×⅔) ; (b) vertical section of ovary with basal ovules ( ×3). 2 *Portulaca grandiflora* (a) flowering shoot showing hairy stipules ( ×⅔) ; (b) cross section of unilocular ovary ( ×4) ; (c) half flower showing overlapping petals ( ×4). 3 *Montia fontana* (a) habit ( ×⅔), (b) fruit—a capsule dehiscing by three valves ( ×6) ; (c) flower ( ×6). 4 *Claytonia perfoliata* (a) habit showing flower stalks erect before and curved downwards after pollination, and erect when bearing fruit ( ×⅔) ; (b) mature capsule with one of two persistent sepals removed ( ×6) ; (c) flower with petals partly removed ( ×12).

a more or less superior ovary of three (sometimes four or five) fused carpels containing a single locule with two to numerous ovules on a central basal placenta. The style is usually divided. Some authorities consider the two green sepals to be bracts and the corolla to be a perianth. Nectar is present and pollination is by insects. *Anacampseros papyracea* only rarely opens a flower and is almost wholly cleistogamous. The fruit is a capsule dehiscing by two or three valves or a lid, and the seeds have a large embryo coiled around the endosperm.

**Classification.** The most recent classification splits the family into seven tribes and 19 genera. Some genera are ill-defined and can be distinguished from one another only by microscopic characters of the seed. The Portulacaceae is placed in the Centrospermae (Caryophyllales), sharing with other families of that order the possession of betalains in place of anthocyanin in the flowers. Morphologically it comes closest to the Caryophyllaceae and Basellaceae, and is linked to the Aizoaceae by the many-stamened genera *Portulaca* and *Lewisia*.

**Economic uses.** *Portulaca oleracea*, the common purslane, has been cultivated since classical times as a salad and pot herb, and is still so used in some countries. The plant long known as *P. grandiflora*, although now thought by some to be a cultivar of hybrid origin, is much in demand as a summer bedding annual, and has been bred for a wide range of colors as well as double flowers. Species of *Lewisia* are collectors' pieces for the alpine garden or cold house. The starchy rootstocks of *Lewisia rediviva* are eaten by the American Indians. Several tender genera are popular in collections of succulent plants ; for example the African *Anacampseros* with over 50 miniature rosette species, the shrubby *Ceraria* and *Portulacaria*, and the leafy caudiciform American and African *Talinum*. Several species of *Claytonia* are grown in rock gardens.                    G.D.R.

# BASELLACEAE
## Madeira Vine

The Basellaceae is a small family of climbing vines, mostly from tropical America.

**Distribution.** Apart from America, species are also found in tropical Africa, Madagascar, southern Asia, New Guinea and some Pacific islands.

**Diagnostic features.** The leaves are mostly simple, broadly-ovate and fleshy, without stipules and arranged alternately on the long climbing stem of the vine. The rootstock is often tuberous. The flowers are small, regular, bisexual or unisexual, supported by a pair of bracts and arranged in spikes or in racemes in the leaf axils. The floral organs are usually arranged perigynously. The perianth has five segments which may be partially fused at their bases and in some species may be colored. There are five stamens, each opposite a perianth segment and adnate, or partially fused, to its base. The ovary is superior and comprises three fused carpels which form a single locule containing one basal ovule. There is usually one style and three stigmas. The fruit is a drupe and usually enclosed within the persistent fleshy perianth. The seed usually has copious endosperm and a twisted or ring-like embryo.

**Classification.** The four genera can be grouped into two tribes:

BASELLEAE. Filaments erect and straight in bud; *Basella*, or Malabar nightshade, with two principal species, *Basella rubra* and *B. alba*, bearing red and white flowers respectively; *Tournonia*; *Ullucus* (ulluco).

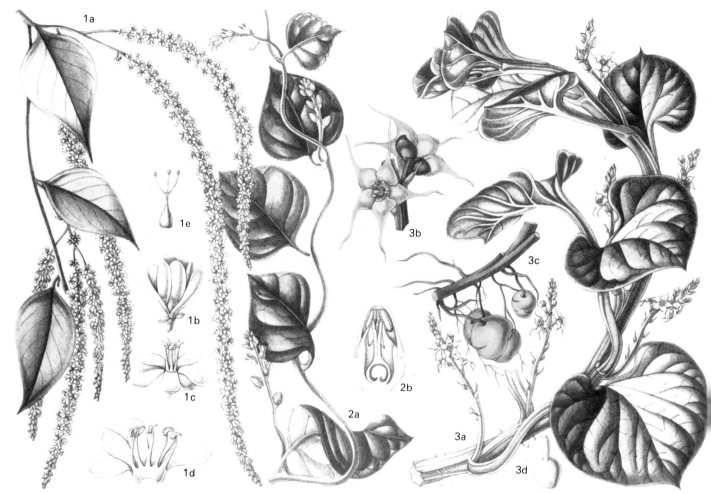

Basellaceae. 1 *Anredera cordifolia* (a) twining stem bearing alternate broadly ovate simple leaves and flowers in panicles ($\times\frac{2}{3}$); (b) flower partly open showing pair of bracteoles below the flower ($\times4$); (c) fully open flower with five perianth segments and five stamens ($\times4$); (d) perianth opened out to show stamens attached to its base ($\times6$); (e) gynoecium with terminal style deeply-divided into three stigmas ($\times3$). 2 *Basella alba* (a) twining stem ($\times\frac{2}{3}$); (b) half flower bud showing ovary with single locule containing one ovule ($\times6$). 3 *Ullucus tuberosus* (a) twining stem ($\times\frac{2}{3}$); (b) flowers with brightly colored bracteoles and five greenish perianth segments ($\times2$); (c) tuberous roots ($\times\frac{2}{3}$); (d) gynoecium ($\times6$).

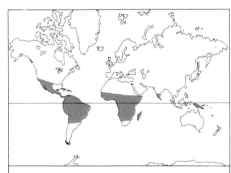

**Number of genera:** 4
**Number of species:** 22
**Distribution:** mainly tropical America.
**Economic uses:** limited uses as vegetables and ornamentals.

ANREDEREAE. Filaments curved outwards in bud; *Anredera* (*Boussingaultia*), Madeira vine.

The family is related to the Portulacaceae and the other families in the group known as the Centrospermae (Caryophyllales).

**Economic uses.** The leaves of *Basella rubra* and *B. alba* are sometimes eaten like spinach, and ulluco produces small tubers which are eaten as a substitute for potatoes. *Anredera*

*baselloides* is grown for its decorative value for covering porches, arbors etc. It is fast growing and mainly propagated by means of the small bulbils produced in the leaf-axils.

B.N.B.

# BATALES

## BATIDACEAE
*Saltwort*

The Batidaceae is a family consisting of only a single species, the saltwort (*Batis maritima*).

**Distribution.** The saltwort is found in the coastal regions of the West Indies and Brazil, as well as in the Pacific from California to the Galapagos Islands and Hawaii.

**Diagnostic features.** Saltwort is a straggly shrub with rather fleshy leaves which are opposite, simple, narrow, and without stipules. The flowers are unisexual, and male and female are borne on separate plants. In both types, the flowers are very small with very short stalks and are arranged in cone-like spikes. The male flowers arise in the axils of bracts and consist of a two-lipped calyx and four petals (sometimes described as stami-

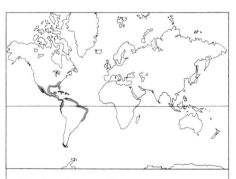

**Number of genera:** 1
**Number of species:** 1
**Distribution:** coastal regions of the W Indies, Brazil, California, Central America and Hawaii.
**Economic uses:** minor use in salads.

nodes), which are fused at the base and alternate in position with four free stamens. The spike bearing the female flowers is rather fleshy and the bracts are smaller than those of the male-flowered spikes. There is no indication of either calyx or petals in the female flowers, which essentially consist of only an ovary divided into four locules, each of which contains a single ovule inserted at

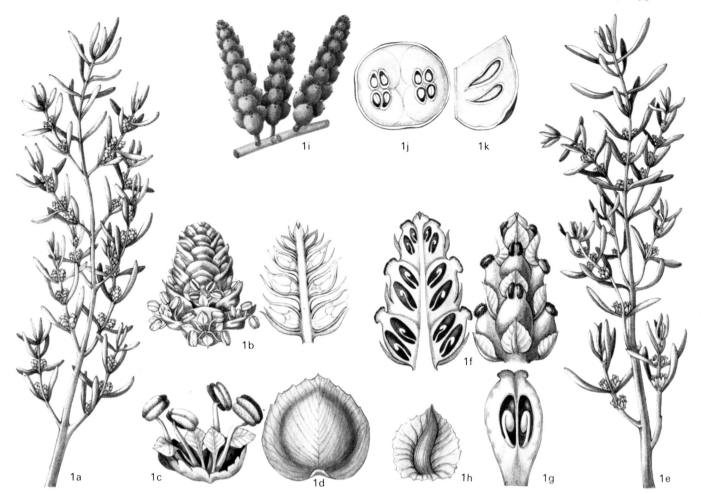

Batidaceae. 1 *Batis maritima* (a) male flowering shoot ($\times\frac{2}{3}$); (b) male inflorescence entire (left) and vertical section (right) ($\times 7$); (c) male flower with two-lipped calyx, four petals and four stamens ($\times 20$); (d) bract from male inflorescence ($\times 14$); (e) female flowering shoot ($\times\frac{2}{3}$); (f) female inflorescence entire (right) and in vertical section (left) ($\times 10$); (g) half female flower with sessile stigma and single basal ovules in each chamber ($\times 14$); (h) bract from female inflorescence ($\times 14$); (i) ripe fruit ($\times 3$); (j) cross section of fruiting spike showing two individual fruits each with four seeds ($\times 12$); (k) vertical section of fruit ($\times 20$).

the base. There is no distinct style and the flattened stigma sits directly on top of the ovary. As the fruits of the four up to 12 flowers in each spike ripen into berries, they often become fused into a fleshy structure. The seed contains a large straight embryo but no endosperm.

**Classification.** The relationships of the Batidaceae are uncertain, but it seems to have some affinities with a number of families in a group known as the Centrospermae (Caryophyllales), and in particular with the Amaranthaceae and the Chenopodiaceae, which also contain species to be found in coastal regions. There is also evidence to suggest that it ought to be separated into an order of its own, as followed here.

**Economic uses.** The leaves of saltwort are edible and are sometimes eaten raw in salads.
S.R.C.

# POLYGONALES

## POLYGONACEAE
*Buckwheats, Rhubarb and Sorrels*

The Polygonaceae is a very large cosmopolitan family of herbs, some shrubs and a few trees, with a number of cultivated ornamentals and plants with edible seeds (buckwheat), stalks (rhubarb), leaves (sorrel) or berries (*Coccoloba*).

**Distribution.** Most genera inhabit the temperate northern regions. A few are tropical or subtropical, notably *Antigonon* (Mexico and Central America), *Coccoloba* (tropical America and Jamaica) and *Muehlenbeckia* (Australasia and South America).

**Diagnostic features.** The leaves are usually alternate and simple, with a characteristic ochrea or membranous sheath uniting the stipules. The small white, greenish or pinkish flowers are solitary or grouped in racemose inflorescences and are usually bisexual; occasionally the sexes are separate. There are three to six sepals that often become enlarged and membranous in the fruit. Petals are absent. The six to nine stamens have bilocular anthers opening lengthwise. The ovary is superior and composed of two to four carpels united into one locule containing a single basal ovule. There are two to four usually free styles. The fruit is a triangular nut; the seeds have abundant endosperm and a curved or straight embryo.

**Classification.** Three major groups of genera

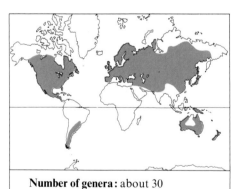

**Number of genera:** about 30
**Number of species:** about 750
**Distribution:** cosmopolitan, but chiefly in northern temperate regions.
**Economic uses:** ornamentals and crop plants (buckwheat, rhubarb).

may be distinguished. The first comprises plants which are tropical, or half-hardy in temperate zones. *Antigonon* has three or four species of flamboyant climbers. *Brunnichia* is a genus of two deciduous climbers closely related to *Polygonum*. The large genus *Coccoloba* comprises some 125 species of tropical trees, shrubs and climbers. *Muehlenbeckia* includes about 15 species of climbing or woody, hardy, half-hardy and greenhouse plants related to *Coccoloba*. Notable in the second group of desert-loving hardy and half-hardy shrubs and subshrubs are *Atraphaxis*, the shrubby buckwheats from semidesert regions of southeast Europe and central Asia; *Calligonum*, 100 species of broom-like xerophytic shrubs from central Asia; *Padopterus*, whose single species *Podopterus mexicanus* is an attractive spiny pink-flowered shrub, cultivated in dry greenhouse conditions; and *Eriogonum*, a genus of over 100 species of annual or perennial herbs and subshrubs mostly from the dry, warm regions of western North America: most species are densely woolly and have clusters of small white flowers. The third group of genera, whose typical habitat is in temperate regions, includes edible species, vigorous ornamental herbs and shrubs and weeds. *Fagopyrum* has 15 perennial and annual herbaceous species often with succulent stems, native to temperate regions of Eurasia. The 50 species of *Rheum*, strong, large-leaved herbs, come from Siberia, the Himalayas and eastern Asia. *Polygonum* comprises 150 vigorous and often invasive species of mainly perennial and hardy herbs

with a few woody climbers. The often pink or white flowers are pollinated by insects. The greenish, occasionally reddish or yellowish flowers of *Rumex* (150 northern temperate species) are wind-pollinated.

The Polygonaceae is most closely related to the Plumbaginaceae; both families have affinities with the Caryophyllaceae.

**Economic uses.** Cultivated ornamentals include *Antigonon leptopus* (coral vine or rosa de montaña), *Muehlenbeckia axillaris*, a small creeping rock garden shrub, *Atraphaxis frutescens*, grown for its pink and white flowers, the rock garden species of *Eriogonum*, grown for their gray and white foliage, the waterside *Rheum palmatum* and *Rumex hydrolapathum*, and many fast-growing border, ground cover and rock garden species of *Polygonum*. The purple berries of the West Indian seaside grape *Coccoloba uvifera* are eaten, as are the leaves of the common sorrel (*Rumex acetosa*), used as a salad and potherb, and the stalks of the common rhubarb *Rheum rhaponticum*. *Fagopyrum esculentum* (common buckwheat) is widely cultivated for its seeds and as a manure and cover crop and similar but less extensive uses are made for *F. tataricum*

(tartary buckwheat) although the seeds are not eaten by Man.                    T.J.W.

# PLUMBAGINALES

## PLUMBAGINACEAE
*Sea Lavender and Thrift*

The Plumbaginaceae is a medium-sized family of annual or perennial herbs and shrubs or climbers, many of which are cultivated as garden ornamentals.

**Distribution.** The family is cosmopolitan but especially frequent in dry or saline habitats such as sea coasts and salt steppes.

**Diagnostic features.** The leaves are either arranged in a basal rosette or alternately on the aerial branched stems. They are simple, glandular and without stipules. The inflorescence is cymose or racemose (eg *Limonium*), spicate (eg *Acantholimon*) or in dense, capitulate clusters (eg *Armeria*). The bracts are scarious and sometimes form an involucre. The flowers are bisexual and regular, with their parts in fives. The five persistent sepals are fused to form a five-toothed or five-lobed tube which is often

Polygonaceae. 1 *Rumex hymenosepalus* (a) leafy shoot with flowers and winged fruit ($\times\frac{2}{3}$); (b) mature fruits showing persistent perianth ($\times\frac{2}{3}$); (c) cross section of fruit ($\times 1$); (d) flower ($\times 2$). 2 *Oxyria digyna* (a) habit ($\times\frac{2}{3}$); (b) winged fruit ($\times 4$); (c) cross section of fruit ($\times 4$). 3 *Polygonum amplexicaule* (a) flowering spike showing sheathing stipules or ochreas clasping the stem above the leaf bases ($\times\frac{2}{3}$); (b) perianth opened out to show eight stamens ($\times 2$); (c) vertical section of ovary ($\times 4$). 4 *Coccoloba platyclada* (a) flowering shoot ($\times\frac{2}{3}$); (b) flower buds and young fruits ($\times 4$); (c) mature fruit ($\times 4$); (d) seed ($\times 4$); (e) cross section of seed ($\times 4$); (f) flower viewed from above ($\times 7$).

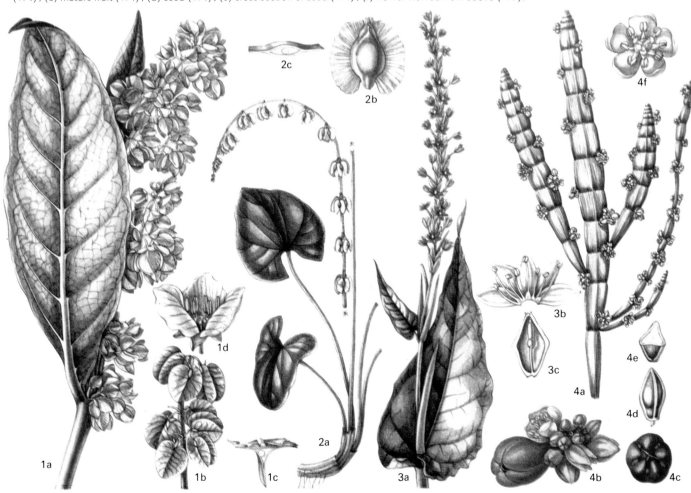

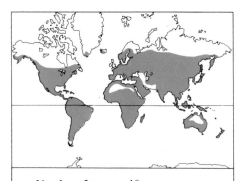

**Number of genera:** 10
**Number of species:** about 560
**Distribution:** cosmopolitan, particularly in dry or saline habitats.
**Economic uses:** garden ornamentals (thrift, *Plumbago*) and medicinal uses.

membranous, ribbed and colored. The five petals are free, connate only at the base or fused into a long tube. The five stamens, which are arranged opposite the petals, are free or inserted at the base of the corolla. The anthers are bilocular and split open longitudinally. The ovary is superior, of five fused carpels and a single locule. It contains a single basal anatropous ovule and is sur-

mounted by five styles or five sessile stigmas. The fruit is usually enclosed by the calyx and is normally indehiscent. The seed contains a straight embryo surrounded by mealy endosperm.

**Classification.** The chief genera are *Limonium* (about 300 species), *Acantholimon* (150 species), *Armeria* (80 species), *Plumbago* (12 species), *Limoniastrum* (10 species), and *Ceratostigma* (eight species). *Plumbago* is similar to *Ceratostigma* in that it has leafy stems, but differs in having free stamens and a glandular calyx as opposed to the epipetalous condition and non-glandular calyx in species of the latter. *Limonium*, *Armeria* and *Acantholimon* usually have radical leaves. Species of *Armeria* and *Acantholimon* generally possess tighter flower heads than those of *Limonium* but can be distinguished from each other on various floral and vegetative features.

The family is related to the Polygonaceae, and both families have affinities with the order Caryophyllales. Relationships have also been suggested with the Primulaceae and Linaceae.

**Economic uses.** A number of species yield extracts that are used medicinally. For

example those from the subshrub *Plumbago europaea* and the herbaceous climber *P. scandens* are used to treat dental ailments while extracts from the leaves and roots of the tropical *P. zeylanica* are used to treat skin diseases. Extracts from the roots of the European *Limonium vulgare* are used to treat bronchial haemorrhages.

Many members of the family are grown in gardens. For example, many species of *Armeria* (sea pink or thrift) are grown in borders and as rock plants. They are tufted perennials with moss-like leaves and globose white or pink flowering heads on wiry stems.

Many species of *Limonium* (sea lavender) are grown in borders and beds and for cut flowers which may be dried and used as everlastings. *Acantholimon glumaceum* and *A. venustum* are grown in rock gardens, producing loose spikes of small rose-colored flowers. Two climbing species, *Plumbago auriculata* (pale blue flowers) and *P. rosea* (red flowers) are grown in warm glasshouses; *P. europaea*, a perennial herb with violet flowers and *P. micrantha*, a white-flowered annual, are hardy in cool temperate conditions. *Ceratostigma willmottianum* is a popular ornamental garden shrub.     S.R.C.

Plumbaginaceae. 1 *Limonium imbricatum* (a) habit showing part of tap root, rosette of leaves and flowers in branched panicles ($\times\frac{1}{3}$); (b) part of inflorescence ($\times 2$). 2 *L. tunetanum* (a) half flower showing stamens inserted at base of corolla tube ($\times 8$); (b) cross section of ovary with a single ovule ($\times 40$). 3 *L. thouini* vertical section of fruit ($\times\frac{2}{3}$). 4 *Aegialitis annulata* indehiscent fruit with persistent calyx ($\times 1$). 5 *Armeria pseudarmeria* habit showing radical leaves and flowers in dense capitulate clusters ($\times\frac{2}{3}$). 6 *A. maritima* (a) half flower with lobed petals, epipetalous stamens and gynoecium with simple hairy styles and a single basal ovule ($\times 4$). 7 *Plumbago auriculata* shoot bearing simple leaves and inflorescences ($\times\frac{2}{3}$).

Dilleniaceae. 1 *Hibbertia tetrandra* (a) flowering shoot ( $\times \frac{2}{3}$ ) ; (b) half flower showing lobed petals and free carpels with one basal ovule ( $\times 3$ ). 2 *Dillenia indica* (a) gynoecium showing ovoid ovary and free styles and stigmas ( $\times \frac{2}{3}$ ) ; (b) cross section of ovary showing numerous partly united carpels ( $\times 1$ ). 3 *Dillenia suffruticosa* (a) shoot with wing-like stipules on leaf-stalks ( $\times \frac{2}{3}$ ) ; (b) cross section of ovary with united carpels ( $\times 1$ ) ; (c) gynoecium ( $\times \frac{2}{3}$ ). 4 *Tetracera masuiana* (a) flowering shoot ( $\times \frac{2}{3}$ ) ; (b) vertical section of gynoecium with free carpels ( $\times 6$ ).

# DILLENIIDAE

## DILLENIALES

### DILLENIACEAE
*Dillenias and Hibbertias*
This large family of tropical trees, shrubs and climbers contains a number of handsome species, some with large white or yellow flowers.

**Distribution.** The family is almost pantropical, though only one genus, *Tetracera*, has members in Africa. Among the major genera with 40 or more species, *Dillenia* (including *Wormia*) is found in Asia, New Guinea, Australia, Fiji and Madagascar, and *Hibbertia* is native to Australasia and Madagascar.

**Diagnostic features.** Most plants in the family have deciduous alternate leaves with prominent lateral veins (resembling those of the loquat); stipules may be present or absent. The white or yellow flowers are usually regular and bisexual, borne solitary or in cymes. There are five persistent

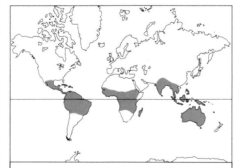

**Number of genera:** 18
**Number of species:** 530
**Distribution:** almost pantropical, but centered in Asia and Australasia.
**Economic uses:** *Dillenia* and *Hibbertia* are cultivated as ornamentals.

overlapping sepals, five free overlapping deciduous petals (often crumpled in the bud) and numerous stamens often persisting on the fruit. The numerous carpels are free or slightly fused (rarely completely fused), each

carpel containing one or more erect ovules. The styles are separate and divergent, each with a single stigma, and fruits may be dehiscent or berry-like. There are one to few seeds with an aril which is convolute or fringe-like; the seed coat is crusty, the endosperm fleshy and the embryo very small.

**Classification.** The Dilleniaceae is here allied with the Crossosomataceae and Paeoniaceae within the order Dilleniales, and although each of the families is sharply set off from the others, they have a unique combination of features in common. This group is related to the Magnoliales, and allied orders.

**Economic uses.** Some species are decorative climbers or trees, the rough leaves of others are used in rural economies as a sandpaper, and the timber of *Dillenia* species is used in general construction and boat building. The flowers of *Dillenia* are like "gigantic buttercups" to quote E. J. H. Corner; in *Dillenia obovata* they are about 15cm (6in) in diameter and yellow, and in *D. indica* about 20cm (8in) across and white. Such flowers

Paeoniaceae. 1 *Paeonia peregrina* (a) shoot with upper leaves and solitary terminal flower ( ×⅔) ; (b) young fruit comprising three follicles ( ×⅔) ; (c) dehisced fruit ( ×⅔) ; (d) vertical section of seed with copious endosperm and a small embryo ( ×2) ; (e) young leafy shoots ( ×½). 2 *P. wittmanniana* (a) cross section of carpel ( ×⅔) ; (b) vertical section of carpel ( ×⅔) ; (c) young fruit ( ×⅔). 3 *P. mascula* fruit of five follicles ( ×⅔). 4 *P. emodi* fruit ( ×⅔). 5 *P. tenuifolia* flower ( ×⅔).

last only one day and are borne on the leafless tree in *D. obovata*, but when new leaves are emerging in *D. ovata*.

A group of *Dillenia* species, with dehiscent fruits which open on the tree, has sometimes been separated as the genus *Wormia*, but shares the colloquial name simpoh. Of these *D. excelsa*, a 25m (80ft) tree with red fruit buds, yellow flowers to 10cm (4in) across with purple anthers is one of the most noble.

The Australian *Hibbertia scandens* is often encountered as a glasshouse climber or outdoors in warmer climates. *Tetracera sarmentosa* is also an evergreen climber often grown in the semishade.                B.M.

## PAEONIACEAE
*Peonies*

The family Paeoniaceae comprises one genus (*Paeonia*) of 33 species of perennial, rhizomatous herbs or shrubs, including many popular ornamentals, cultivated for their showy flowers and attractive foliage.

**Distribution.** The family is native to north temperate regions, chiefly south Europe, China and northwestern America.

**Diagnostic features.** The leaves are composed of several leaflets which may be lobed. They are alternate and without stipules. The

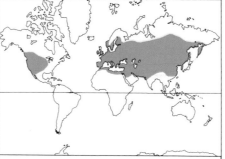

**Number of genera: 1**
**Number of species: 33**
**Distribution:** north temperate, chiefly southern and central Europe, China and NW America.
**Economic uses:** ornamental herbs and shrubs (peony) and some local medicinal uses.

large conspicuous flowers are regular and bisexual, rather globular in appearance. Leafy bracts may be present, which by reduction of the blade and expansion of the base pass into five green, persistent sepals. The five to ten petals are large, with a gradual transformation into successively narrower forms. The many anthers are centrifugally arranged and bear anthers with two locules

opening lengthwise by slits which face outward. The two to five carpels are free and borne on a fleshy nectar disk. They are themselves fleshy, each surmounted by a thick stigma and containing numerous ovules in a double row. The fruit consists of two to five large, leathery follicles containing several large red seeds which turn shiny black on maturity. The seeds have arils, copious endosperm and a small embryo.

Most species of peony, such as *Paeonia officinalis*, possess large flowers up to 14cm (5.5in) across. There is substantial variation in the degree of subdivision of the leaves, and this can be of diagnostic value in separating the various species and subspecies.

**Classification.** At one time *Paeonia* was placed in the Ranunculaceae, where it was anomalous in possessing persistent sepals, petals with a sepalar rather than staminodial derivation, a perigynous disk and seeds with arils. Corner in 1946 first pointed out that the Paeoniaceae was much closer to Dilleniaceae, and recent research supports his view.

**Economic uses.** Many species are grown as garden flowers, making attractive border plants. There are a number of varieties of *P. officinalis* (native to southern Europe) including 'Alba-plena' (double white-flowered), 'Rosea-plena' (double pink-flowered) and 'Rubra-plena' (double crimson-red-flowered).

A species native to Siberia, *P. lactiflora*, produces scented, pure white single flowers but is rarely cultivated as such, having been superseded by a range of forms often with double flowers, such as 'Albert Crousse' (double pink, carmine at the center), 'Bower of Roses' (double rose-crimson) and 'Bowl of Beauty' (semi-double pink with prominent gold/yellow stamens). Other popular cultivated species are the southern European *P. arietina* with hairy stems and single pink flowers and *P. mlokosewitschii* (Caucasus) with single yellow cup-shaped flowers.

A number of shrubby species are also popular ornamentals. *P. lutea* (China) is a deciduous wide-spreading shrub with deeply segmented pale-green leaves and fragrant yellow single flowers. 'Chromatella' and 'L'Esperance' are hybrids between this species and *P. suffruticosa* with double or semi-double yellow flowers. *P. suffruticosa* (*P. moutan*), moutan or tree peony, also native to China, grows to 2.5m (8.0ft) with large bluish-pink flowers up to 18cm (7in) across.                                S.R.C.

## CROSSOSOMATACEAE

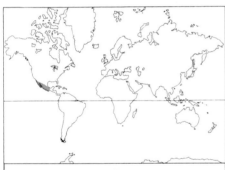

**Number of genera:** 1
**Number of species:** 4
**Distribution:** southern N America.
**Economic uses:** limited use as ornamentals.

The Crossosomataceae is a small family of New World shrubs with conspicuous white or purplish flowers.

**Distribution.** The family comprises the single genus *Crossosoma* of four species, restricted to dry habitats in the southwest United States of America and Mexico.

**Diagnostic features.** The shrubs are glabrous and sometimes spiny. The leaves are simple, alternate, without stipules and more or less glaucous, sometimes deciduous or retained whilst dead (ie marcescent, like the leaves of seedling beech trees). The flowers are superficially very like those of the Dilleniaceae. They are solitary, regular and bisexual, with five sepals, five petals and numerous stamens. The ovary is superior, with three to six carpels, each with a short style, capitate stigma and numerous ovules. The fruit is a stalked dehiscent follicle of many seeds,

which have conspicuous fringed arils, thin endosperm and a slightly curved embryo.

**Classification.** The absence of stipules and the presence of other characters have suggested to some that *Crossosoma* is allied to the subfamily Spiraeoideae of the Rosaceae, but the family is usually allied to Dilleniaceae particularly on the basis of seed characters, and has been included in it by some workers.

**Economic uses.** Some species are cultivated for ornament in sunny situations in Europe.
                                D.J.M.

# THEALES

## THEACEAE
*Tea, Camellias and Franklinia*

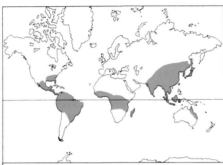

**Number of genera:** about 29
**Number of species:** about 1,100
**Distribution:** tropical and subtropical, centered in America and Asia.
**Economic uses:** tea (*Camellia sinensis*), tea-seed oil (*C. sasanqua*), timber and ornamentals (camellias).

The Theaceae, formerly called the Ternstroemiaceae by some authorities, is a medium-sized family of trees and shrubs, rarely scrambling vines. The former type genus *Thea*, now merged in the older name *Camellia*, includes the economically important tea plant, *Camellia sinensis*.

**Distribution.** The family is mainly restricted to tropical and subtropical regions, and centered chiefly in America and Asia.

**Diagnostic features.** The leaves are alternate (rarely opposite), often evergreen and leathery, and usually without stipules. The flowers are regular, usually bisexual, generally solitary but occasionally in a branched inflorescence, and are often very showy; there are four to seven sepals and petals while there may be four, eight or, more generally, numerous stamens which may be free or in bundles or united into a tube. The ovary is superior (rarely inferior or semi-inferior) and has three to five (rarely two or eight to 25) fused carpels and a corresponding number of locules, each of which contains two to numerous ovules (rarely one) on axile placentas. There are as many styles as locules, free or united (rarely a single style). The fruit is a capsule, berry or achene with the sepals persistent at the base. The

seed usually has little or no endosperm and a straight or curved embryo which may be folded or spirally twisted.

**Classification.** As described here the family comprises about 29 genera and almost 1,100 species. However, many authorities have split it into a number of small families. In its narrowest sense (16 genera and 500 species) it comprises two tribes:

CAMELLIEAE. Anthers versatile and the fruit usually a loculicidal capsule; chief genera *Camellia*, *Gordonia*, *Stewartia*.

TERNSTROEMIEAE. Anthers basifixed and the fruit a berry or achene; chief genera *Ternstroemia*, *Adinandra*, *Eurya* (in *Annesleya* and *Visnea* the ovary is semi-inferior, while in *Symplococarpon* it is inferior).

The small families now generally accepted as being included within the Theaceae are:

ASTEROPEIACEAE (one genus, seven species, Madagascar). Stamens nine to 15, ovary of three (rarely two) carpels, ovules two to numerous in each locule, style single.

BONNETIACEAE (three genera, 22 species, tropical Asia and America). Stamens numerous, ovary of three to five carpels, ovules numerous in each locule, styles three to five.

PELLICIERACEAE (one genus, one species, tropical America). Stamens five, ovary of two carpels, one ovule in each locule, style single.

PENTAPHYLACACEAE (one genus, two species, Southeast Asia). Stamens five, ovary of five carpels, two ovules in each locule, style single.

TETRAMERISTICACEAE (one genus, three species, Malaysia). Stamens four, ovary of four carpels, one ovule in each locule, style single.

More debatable is the inclusion of the following families:

MEDUSAGYNACEAE (one genus, one species, Seychelles). Leaves opposite or whorled, stamens numerous, ovary of 17–25 carpels, two ovules in each locule, styles 17–25.

STACHYURACEAE (one genus, 10 species, eastern Asia). Leaves with stipules, stamens eight, ovary of four carpels, numerous ovules in each locule, style single.

CARYOCARACEAE (two genera, 25 species, tropical America). Leaves opposite or alternate with stipules, stamens numerous, ovary of four or eight to 20 carpels, ovules solitary in each locule, styles four to 20.

SYMPLOCACEAE (two genera, 500 species, tropics and subtropics, except Africa). Corolla of five or ten petals, stamens five, 10, 15 or more numerous, ovary inferior or semi-inferior, of two to five carpels, ovules two to four in each locule, style single. (The position of this family is the most debatable and several systems place it within the Ebenales.)

Within the Theales, the Theaceae is most closely related to the Dipterocarpaceae, Guttiferae and Marcgraviaceae, while the order Theales appears to be the stock for a number of other orders including the Capparales, Ebenales, Ericales, Malvales, Primulales and Violales.

**Economic uses.** *Haploclathra paniculata* yields a handsome red wood, mura piranga. The seeds of *Camellia sasanqua* yield tea-seed oil. Among the various ornaments is *Franklinia altamaha*, originally distributed over a very small area near Forth Barrington in Georgia in the southern United States of America, but now cultivated in North America and Europe. The best-known ornamentals are cultivars of *Camellia japonica*, which have long been grown in the Far East for their beautiful, often scented, flowers. They are now popular garden plants in many temperate regions, but the flowers are sensitive to frost damage.

The tea plant, *C. sinensis*, is native to Southeast Asia and has long been cultivated in China, probably at first for medicinal use, but subsequently as a beverage plant. The distinctive flavor of the stimulant beverage is due to the constituents caffeine, polyphenols and essential oils, the proportions of which vary according to the age of the leaves and the method by which they are processed after picking. Tea is used by about half of the world's population. The chief producers are India and Sri Lanka, with East Africa, Japan, Indonesia and Russia next.     B.S.F.

# OCHNACEAE
*African Oak and Ochnas*

The Ochnaceae is a large tropical family of trees, shrubs and some herbs which includes a number of hothouse ornamentals.

**Distribution.** The family is pantropical with the greatest concentration of genera and species in tropical South America.

**Diagnostic features.** The family has alternate, simple leaves with stipules (*Godoya splendida* is an exception with large pinnate leaves). Regular, bisexual flowers are borne in panicles, racemes or false umbels. The calyx is free or united at the base. The corolla has five, rarely 10–12, contorted petals. The five or ten or numerous stamens are hypogynously arranged or on an elongated axis. The ovary is superior, of two to five (rarely 10–15) carpels which are often free below, but have a common style. There are one or two or numerous erect (rarely pendulous) ovules in each carpel. The axis of some genera swells and becomes fleshy under the fruit, which is usually a cluster of drupes, or sometimes a berry or capsule. The seed may or may not contain endosperm and the embryo is usually straight.

**Classification.** The major genera are distin-

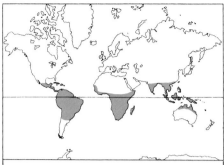

**Number of genera:** about 40
**Number of species:** about 600
**Distribution:** tropical, chiefly S America.
**Economic uses:** timber (African oaks), hothouse and tropical garden ornamentals (*Ochna* and *Cespedesia*).

guished mainly on floral characters. *Ochna* (85 species of deciduous trees) has an indefinite number of stamens with filaments as long or longer than the anthers and a three- to 15-carpelled ovary with numerous ovules in each carpel. The genus has a characteristic colored calyx which does not enlarge in fruit as in some members of the

Theaceae. 1 *Camellia rosiflora* flowering shoot ($\times \frac{2}{3}$). 2 *C. japonica* cv 'Kimberley' half flower with semi-inferior ovary and numerous stamens, united at their bases ($\times \frac{2}{3}$). 3 *C. salicifolia* (a) gynoecium with three-lobed stigma ($\times 4$); (b) cross section of trilocular ovary ($\times 6$); (c) stamens with fused filaments ($\times 3$); (d) stamen ($\times 4$); (e) fruit—a capsule ($\times 2$). 4 *Symplocarpon hintonii* (a) half flower showing inferior ovary ($\times 6$); (b) dry indehiscent fruit ($\times 1$). 5 *Eurya macartneyi* (a) leafy shoot with female flowers ($\times \frac{2}{3}$); (b) female flower ($\times 6$); (c) male flower ($\times 6$); (d) epipetalous stamens with lengthwise dehiscence ($\times 12$). 6 *E. japonica* fruit—a berry.

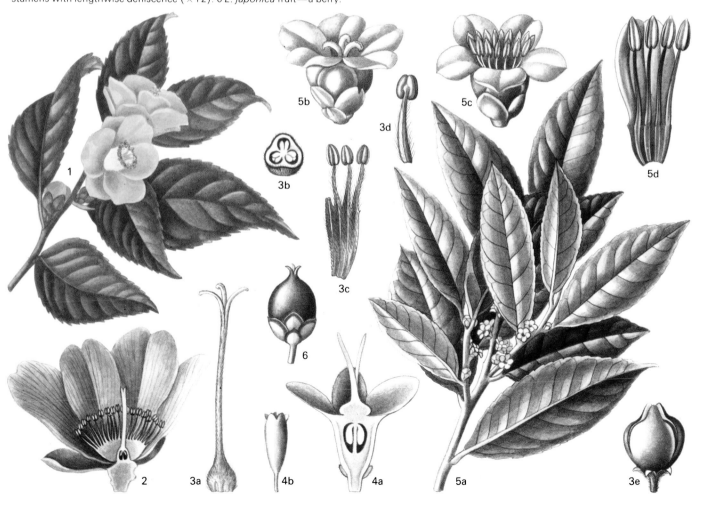

family and is a brilliant red or purple. The petals are greenish yellow. Each carpel forms a drupe, the receptacle becoming fleshy beneath it. *Ouratea* (300 mostly South American evergreen tree and shrub species) has only 10 stamens, with filaments shorter than the anthers, and a five-lobed ovary. Most species have yellow flowers, and fruit consisting of five or fewer drupes borne sessile on an enlarged disk. *Euthemis* (two Asian species) has five stamens and five staminodes; the ovary comprises four or five locules with only one or two ovules in each locule. *Godoya*, a genus of trees, is represented by five species from Peru and Columbia, with pure white fragrant flowers. *Sauvagesia* comprises 30 species of chiefly South American herbs and shrubs, with ciliate stipules and white, pink or purple flowers with convolute petals and five fertile stamens surrounded by numerous staminodes.

The Ochnaceae is most closely related to the Dipterocarpaceae.

**Economic uses.** The family includes few plants of economic value. The only genus with commercial timber is *Lophira*, whose two species, *Lophira alata* and *L. lanceolata*, are both commonly called the African oak,

meni oil tree or red iron tree. Plants cultivated in tropical gardens and in hothouses in temperate zones include *Ochna kibbiensis* and *O. flava*, *Sauvagesia erecta* (the West Indian iron shrub), *Cespedesia bonplandii* and *C. discolor*. S.A.H.

## DIPTEROCARPACEAE

The Dipterocarpaceae is a family of small to very lofty trees which dominate the lowland rain forests of Asia and which are among the grandest in the tropics and a major source of hardwood.

**Distribution.** Concentrated in Malaya, Indonesia, Borneo and Palawan, the family is distributed throughout tropical Asia and Indomalaysia, with two genera in tropical Africa. Many species are gregarious; in the monsoon region of India and Burma there are vast forests comprised almost entirely of a single species, such as *Shorea robusta* or *Dipterocarpus tuberculatus*, while the moist evergreen forests of Malaysia are dominated by members of this family.

**Diagnostic features.** Dipterocarps tend to share a characteristic shape: the trunk is buttressed at the base and then rises, often to a great height, smooth and unbranched, before reaching the open, cauliflower-

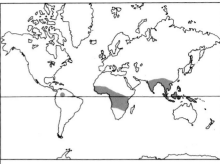

**Number of genera:** about 15
**Number of species:** about 580
**Distribution:** centered in tropical rain forests of Malaysia.
**Economic uses:** the world's main source of hardwood timber.

shaped crown. Except in African species, all parts of the plant contain special resin canals and most species exude an oily aromatic resin called dammar from wounds.

The leaves are simple, entire and alternate, and subtended by stipules, although these often fall early on. The flowers are regular, bisexual, usually borne in racemes, and are adapted for insect pollination, being large, showy and often scented. The calyx and

Ochnaceae. 1 *Luxemburgia ciliosa* (a) leaf ($\times\frac{2}{3}$); (b) flower ($\times 1$); (c) fruit ($\times 1$); (d) cross section of fruit ($\times 2$). 2 *Ochna atropurpurea* (a) flowering shoot ($\times\frac{2}{3}$); (b) stamen with apical pores on anthers ($\times 8$); (c) fruit—a cluster of drupes on a fleshy receptacle, with a persistent colorful calyx ($\times 1$); (d) gynoecium ($\times 6$). 3 *Ouratea intermedia* (a) leafy shoot with terminal inflorescence ($\times\frac{2}{3}$); (b) flower with cluster of ten stamens ($\times 2$); (c) flower with perianth removed to show stamens (with filaments much shorter than anthers) and gynoecium ($\times 4$).

corolla each consist of five parts which may be free or connate at the base. The stamens vary in number from five to many, and the anthers are distinctive in having a sterile tip. The ovary is superior of three carpels; it has three locules, each with two pendulous ovules. Only one ovule develops, the fruit being a single-seeded nut enclosed in the persistent winged and membranous calyx. The seed has no endosperm.

**Classification.** The Dipterocarpaceae is divided into two subfamilies: MONOTOIDEAE (all African species) and DIPTERO-CARPOIDEAE (all Asian species). However, a new species of an undescribed genus and subfamily has recently been discovered in the highlands of South America.

The Dipterocarpaceae is related to members of the Theales, especially the Ochnaceae, Guttiferae and Theaceae.

**Economic uses.** Many closely related species of several genera (principally *Dipterocarpus, Hopea, Shorea* and *Vatica*) grow together in mixed dipterocarp rain forests. These forests are the world's main source of hardwood timber. However, they are doomed largely to disappear by the end of the century unless present conservation and replanting programs are invigorated and extended. Dip-

terocarp timber can be grouped for sale into a few grades, which immeasurably assists marketing. Clear cylindrical boles of 20m–30m (65ft–98ft) length and girths of 2m–4m (6.5ft–13ft) are common. The wood is light in weight and pale in color and therefore in great demand in modern conditions. Much is processed into plywood either locally or in Korea and Japan. The principal market is North America. Minor products are the resin (damar) used for special varnishes; and a substitute for cocoa butter from Bornean *Shorea* species.

# GUTTIFERAE

*Mangosteen and Mammey Apple*

The Guttiferae is a large cosmopolitan family of trees and shrubs many of which produce timber, drugs, dyes and fruits.

**Distribution.** The family is almost worldwide in distribution, although only *Hypericum* (widespread) and *Triadenum* (Asia and North America) occur outside the tropics.

**Diagnostic features.** Most members have simple, usually opposite, entire leaves without stipules. The flowers may be bisexual or unisexual and borne on separate plants, and single or grouped in a cymose or thyrsoid inflorescence. The perianth most often has

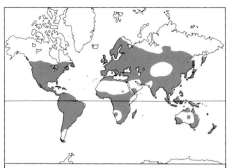

**Number of genera:** about 40
**Number of species:** about 1,000
**Distribution:** cosmopolitan, but centered in the tropics.
**Economic uses:** timber, fruits (mangosteen, mammey apple), drugs and gums, and cosmetics from essential oils.

whorls in five or four parts, usually of distinct sepals and petals, the latter being free and either imbricate or contorted when in bud. The androecium consists basically of two whorls, each of five stamen-bundles with filaments free almost to the base; the outer antisepalous whorl is usually sterile and may be absent, whereas the inner whorl is always fertile (except in female flowers), modified by

Dipterocarpaceae. 1 *Shorea ovalis,* tree in leaf. 2 *Monoporandra elegans* (a) leafy twig with fruit and inflorescence ( ×⅔); (b) flower dissected to show sepal (left), petals and gynoecium in vertical section (center), and stamens fused at the base and anthers with long connectives (right) ( ×6); (c) cross section of ovary ( ×6). The fruits are often enclosed in winged extensions of the calyx; shown here are 3 *Dipterocarpus incanus* ( ×⅔), 4 *Doona ovalifolia* ( ×1½) and 5 *Monentes tomentellus* ( ×⅔). 6 *Dipterocarpus oblongifolia* vertical section of single-seeded fruit ( ×1).

fusion or reduction to form masses of united anthers or more or less numerous apparently free stamens. (In *Hypericum gentianoides* each stamen bundle is often reduced to a single stamen.) The ovary is superior and consists of usually five to three, rarely two or up to 13, united carpels and one to many locules, each containing one to many ovules on axile or parietal placentas. There are as many stigmas as placentas, and the styles may be free, united or totally lacking. The fruit is usually a capsule (usually dehiscing lengthwise along the compartment walls) but may be berry- or drupe-like. The seeds, sometimes winged or with an aril, lack endosperm and usually have a straight embryo with cotyledons either well-developed or reduced and replaced in function by a swollen hypocotyl.

Glandular secretions are produced in canals or lacunae within stems, leaves and floral parts. They consist of essential oils and fats, anthocyanins and resins.

**Classification.** The subdivision of the Guttiferae into five subfamilies is based primarily on sex-distribution, androecium, ovary, fruit and seeds. Hypericoideae has often been treated as a separate family

Hypericaceae; but this segregation does not appear to be warranted unless each subfamily is similarly treated, for they are all quite distinct.

KIELMEYEROIDEAE. Flowers usually bisexual, stamens "free," styles elongate and wholly united; the many seeds without an aril, the embryo with thin free cotyledons; includes *Kielmeyera, Mahurea, Caraipa.*

HYPERICOIDEAE. Flowers bisexual, the outer stamen whorl sterile or absent, and styles elongate, free or more or less united; the fruit is normally either a capsule or a berry, rarely a drupe; the five to many seeds have no aril; the embryo has thin free cotyledons; includes *Hypericum, Cratoxylum, Vismia.*

CALOPHYLLOIDEAE. Flowers unisexual, bisexual or polygamous (male, female and bisexual on same plant); the outer stamen whorl is absent, the styles elongate and free; the fruit is capsular or usually drupaceous; the one to four seeds have no aril; the embryo is enlarged, usually with united cotyledons (including *Mesua, Calophyllum, Endodesmia*).

MORONOBEOIDEAE. Flowers bisexual, the outer androecium whorl sterile, the styles elongate, more or less united; the fruit is

berry-like, the many, or rarely solitary, seeds have no aril, and the embryo is undifferentiated; includes *Pentadesma, Montrouziera, Symphonia.*

CLUSIOIDEAE. The two tribes in Engler's treatment of this subfamily appear to be quite distinct and each may deserve subfamilial status. CLUSIEAE has flowers unisexual or rarely polygamous; the outer androecium whorl fertile or sterile; styles short and free to completely united or absent; the fruit is capsular (rarely berry-like); the five to many seeds have an aril; the embryo has an enlarged hypocotyl and vestigial or no cotyledons; includes *Clusia, Tovomita,* possibly also *Decaphalangium, Allanblackia.* GARCINIEAE has unisexual or rarely bisexual flowers; the outer androecium whorl sterile or absent; styles very short and united or usually absent; the fruit is drupaceous; the one to five (sometimes up to 13) seeds have no aril; the embryo has an enlarged hypocotyl and no cotyledons; includes *Garcinia, Mammea.*

The Guttiferae is related to the Bonnetiaceae (here included within the Theaceae) and through them to the Dilleniaceae, differing from these families in having

Guttiferae. 1 *Symphonia globulifera* (a) leafy shoot and terminal inflorescence ( $\times\frac{2}{3}$ ); (b) androecium, comprising tube of fused stamens, surrounding a five-lobed stigma ( $\times$ 3); (c) fruit—a berry ( $\times$ 1$\frac{1}{3}$ ). 2 *Hypericum calycinum* (a) shoot with decussate leaves and terminal solitary flower ( $\times\frac{2}{3}$ ); (b) half flower with stamens in bundles and numerous ovules on axile placentas ( $\times$ 1); (c) fruit—a capsule ( $\times$ 1). 3 *H. frondosum* cross section of ovary showing single locule with ovules on three parietal placentas ( $\times$ 3). 4 *Calophyllum inophyllum* (a) shoot with inflorescences in axils of terminal leaves ( $\times\frac{2}{3}$ ); (b) gynoecium showing long, curved style ( $\times$ 4); (c) stamen ( $\times$ 4).

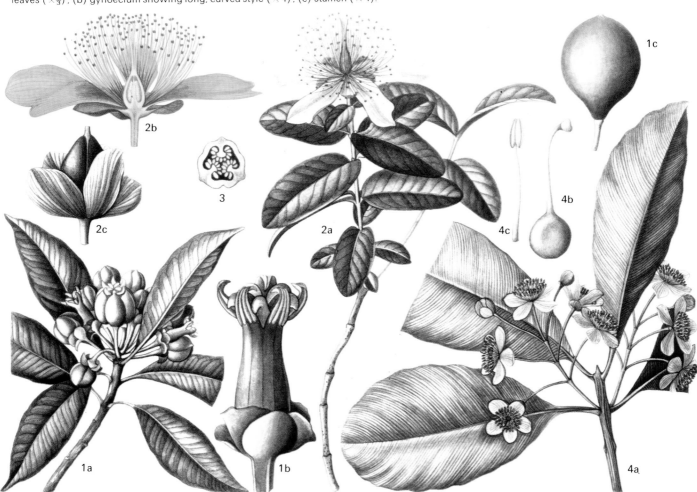

glandular secretions, leaves usually opposite and petals usually contorted in bud. It may also be related to the Myrtaceae.

**Economic uses.** The Guttiferae have been used as a source of: hard and/or durable wood (species of *Cratoxylum*, *Mesua*, *Calophyllum*, *Montrouziera*, *Platonia*); easily worked wood (species of *Harungana*, *Calophyllum*); drugs or dyes from bark (species of *Vismia*, *Psorospermum*, *Harungana*, *Calophyllum*); gums, pigments and resins from stems (including species of *Garcinia*, gamboge, and *Clusia*, healing gums); drugs from leaves (species of *Hypericum*, *Harungana madagascariensis*); drugs and cosmetics from flowers (*Mesua ferrea*); edible fruits (species of *Garcinia* including *Garcinia mangostana*, mangosteen, and of *Mammea* including *Mammea americana*, mammey apple; *Platonia insignis*); fats and oils from seeds (species of *Calophyllum*, *Pentadesma*, *Allanblackia*, *Garcinia*, *Mammea*).

N.K.B.R.

# ELATINACEAE

The Elatinaceae is a widespread family of herbs and shrub-like plants found in aquatic and semi-aquatic or dry habitats.

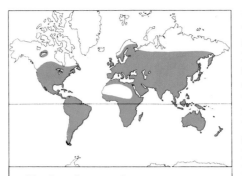

**Number of genera:** 2
**Number of species:** about 33
**Distribution:** cosmopolitan, centered in temperate and subtropical regions.
**Economic uses:** weeds in ricefields and irrigation ditches.

**Distribution.** The genus *Elatine* is almost cosmopolitan with the majority of species in temperate regions. *Bergia* is found throughout warmer regions of the world with a few species in temperate zones.

**Diagnostic features.** The species of Elatinaceae are annual or perennial herbs or occasionally somewhat woody shrub-like plants. The leaves are opposite and decussate in all species except *Elatine alsinastrum* which has leaves in whorls. A pair of small stipules is present at the base of each leaf; the leaf blade is simple and linear to ovate in shape. The flowers are bisexual, regular, either solitary in leaf axils or in cymes, and usually inconspicuous. The sepals are two to five in number and are free or united at the base. The petals are free and as many as the sepals. The ovary is superior, with two to five locules, and numerous ovules on axile placentas. The fruit dehisces septicidally (along a wall dividing compartments). The seed wall bears a characteristic complex net-like pattern. The seeds have a curved or straight embryo and very little or no endosperm.

**Classification.** The genus *Bergia* is characterized by having five free, acute sepals with a distinct vein or midrib and flowers in dense clusters. Of the 20 species about half are aquatic or semiaquatic while the other species grow in dry regions. The species of *Bergia* are robust and conspicuous while those of *Elatine* are small and inconspicuous. *Elatine* has two to four sepals which are united at the base, obtuse at the apex and without a distinct vein. Most species are

Elatinaceae. 1 *Bergia ammannioides* (a) habit (×⅔); (b) inflorescences in axils of leaves (×1⅓); (c) flower viewed from above with three each of sepals, petals and stamens and a four-lobed ovary (×8); (d) dehiscing capsule with seeds exposed (×20); (e) cross section of fruit (×20). 2 *B. capensis* (a) part of creeping stem with adventitious roots (×⅔); (b) fruit (×20). 3 *Elatine hydropiper* (a) habit showing long adventitious roots (×⅔); (b) solitary flower in leaf axil (×4); (c) flower with four sepals and petals, eight stamens and globose, superior ovary (×8); (d) dehiscing fruit (×8); (e) curved seed (×20).

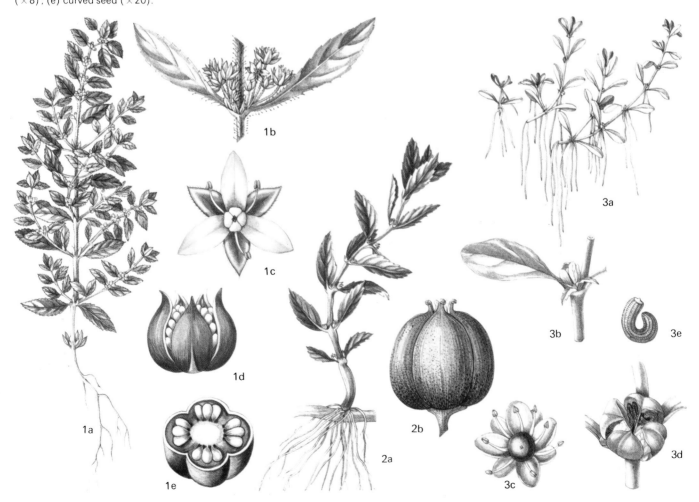

adapted to fluctuating water levels and are found in shallow water that seasonally dries out. They are particularly common in ricefields and periodically drained fishponds.

The Elatinaceae has been little studied and its relationship to other families is not clear. It shows similarities with the Guttiferae, Frankeniaceae, Lythraceae and Haloragaceae.

**Economic uses.** *Elatine* is usually considered beneficial as it effectively consolidates mud. However, *Elatine* and *Bergia* are frequently found as weeds in ricefields and irrigation ditches.                                    C.D.C.

## QUIINACEAE

The Quiinaceae is a family of little-known trees and large shrubs native to the tropical forests of South America and the West Indies.

The leaves are opposite or in whorls, simple or pinnately divided, with a smooth satiny texture. Stipules are present, in one to four pairs at the nodes. The flowers are small, bisexual or sometimes unisexual, in racemes or panicles, with four or five sepals and four or five or eight petals; there are 15–20 or many (160–170) stamens; the ovary is superior, has three, or seven to 11, locules,

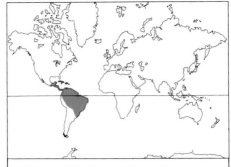

**Number of genera:** 4
**Number of species:** 52
**Distribution:** tropical S America.
**Economic uses:** none, but potentially useful as timber.

with the same number of long curving styles each with an oblique stigma; each locule contains two ovules. The fruit is a single berry with one to four seeds, or three separate carpels each with one seed; the seed has no endosperm and the seed coat is often velvety.

*Quiina* is the largest genus, with 35 species. The other genera are *Touroulia* (four species), *Lacunaria* (11 species) and *Froesia* (two

species). There are some similarities between the Quiinaceae and Guttiferae and Ochnaceae, but the relationships between the families are not well defined.

The heartwood of some species could prove to be a useful timber but at present is of no economic importance.          S.W.J.

## MARCGRAVIACEAE

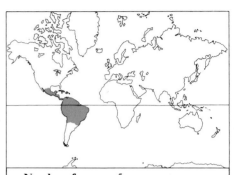

**Number of genera:** 5
**Number of species:** about 125
**Distribution:** tropical America.
**Economic uses:** none.

The Marcgraviaceae is a family of tropical climbers with a number of interesting

Marcgraviaceae. 1 *Marcgravia umbellata* (a) climbing stem with juvenile foliage and adventitious roots ($\times\frac{2}{3}$); (b) nectar cup ($\times\frac{2}{3}$); (c) flower with petals united into deciduous cap ($\times 2$). 2 *M. exauriculata* (a) tip of shoot with adult leaves and inflorescence ($\times\frac{2}{3}$); (b) young flower ($\times 1\frac{1}{3}$). 3 *Norantea peduncularis* flower and nectar cup ($\times\frac{2}{3}$). 4 *Ruyschia clusiifolia* (a) flower ($\times 2$); (b) stamen ($\times 4$); (c) fruit ($\times 1\frac{1}{3}$). 5 *Souroubea* sp cross section of ovary ($\times 3$). 6 *Marcgravia nepenthoides* (a) shoot with adult foliage and inflorescence with infertile flowers bearing conspicuous nectar cups ($\times\frac{2}{3}$); (b) half flower ($\times 1\frac{1}{3}$); (c) gynoecium ($\times 2$); (d) cross section of ovary ($\times 2$). 7 *M. affinis* half flower ($\times 1\frac{1}{3}$).

features including highly modified nectaries.
**Distribution.** The family is restricted to tropical America.

**Diagnostic features.** Members of the family are climbing shrubs, often epiphytic. The leaves are simple, often leathery, and without stipules. In some genera (including *Marcgravia* and *Norantea*) the climbing shoots bear different leaves from the mature shoots (dimorphic foliage). The flowers are bisexual and borne in pendulous racemes or racemose umbels, with the bracts of sterile flowers modified into variously shaped pitchers which secrete nectar. The flowers have four or five imbricate sepals, and five free, or fused petals forming a cap which falls off when the flower opens (*Marcgravia*). There are three to many stamens which are free or variously fused. The ovary is superior and has at first one, but later two to many, locules with multiple ovules by ingrowth of the parietal placentas. The stigma has a short style and radiates out in five lobes. The fruits are globose, fleshy and thick, often indehiscent, and contain many small seeds lacking endosperm. Each seed has a somewhat curved embryo, a large radicle and small cotyledons.

**Classification.** *Marcgravia*, the type-genus, commemorates Georg Marcgraaf (b.1610), an early writer on Brazilian natural history. The 55 or so species in this interesting genus are climbing epiphytes with dimorphic foliage, the climbing shoots having small, two-ranked, rounded, sessile, often reddish leaves pressed against the surface up which the shoot is climbing. Such leaves are regarded as juvenile. The stem, meanwhile, puts out adventitious roots onto this surface. With increasing age or environmental factors not yet fully understood, the mature shoot develops, the stem becoming pendulous, and the leaves change gradually to a stalked, green, leathery, lanceolate condition. Should the mature shoot be required to climb once more, the leaves switch back to the juvenile form: the process is reversible.

The mature shoot is terminated by dense racemes of green flowers with stalked pitcher-like nectaries borne at the very tip. In species such as *Marcgravia rectiflora* the inflorescence is erect, the flowers erect but nectaries hanging down; in others the inflorescence hangs down. The pollination of some species involves the visits of hummingbirds, bananaquits and todies, while others are visited by lizards and bees, and some are self-pollinated before the flower opens. Only in *Marcgravia* are there four not five sepals and the petals form a fused cap over the flower.

*Ruyschia* comprises 10 species, with petals slightly fused at the base, five stamens, an ovary with two locules, and trilobed solid nectaries which are globose or spoon-shaped. The numerous fragrant flowers are borne in elongated racemes.

In *Norantea* (about 35 species) the stamens are often many, the nectaries spoon- or pitcher-like, the ovary with three to five locules, and the flowers borne in umbel-like clusters. *Souroubea* (about 25 species) has three to five stamens, three-lobed bracts and an ovary with five locules. *Caracasia* (two species) has three stamens and free petals and filaments.

The Marcgraviaceae is thought to be related to the Theaceae.
**Economic uses.** There are no known uses for this family. B.M.

# MALVALES

## SCYTOPETALACEAE

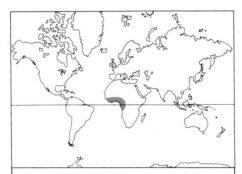

**Number of genera:** 5
**Number of species:** 20
**Distribution:** tropical W Africa.
**Economic uses:** wood used locally in house building.

The Scytopetalaceae is a small family of tropical trees and shrubs.
**Distribution.** The family is native to tropical West Africa.
**Diagnostic features.** The leaves are without stipules, alternate, often in two ranks, simple, sometimes toothed and asymmetrical at the base.

The flowers are bisexual, regular, often with long pedicels, borne either on the old woody stems or terminally (in panicles) or in the axils of leaves (in racemes) on young stems. The calyx has three or four sepals, fused into a cup-shaped structure, which is either entire or toothed. The petals are three to 16, either free or joined at the base, valvate. The petals are sometimes thick and they either become reflexed as the anthers mature, or do not separate and fall off the flower as a cap. The flower has a disk on which the numerous stamens are inserted in three to six series; the filaments are sometimes joined below; anthers dehisce by means of longitudinal slits or by apical pores. The ovary is superior, with three to eight locules and two or more pendulous ovules in each locule borne on axile placentas. The ovary is surmounted by a simple style with a small stigma.

The fruit is normally a woody capsule, but in a few cases is drupe-like. The seeds vary in number from one to eight and are sometimes covered with mucilaginous hairs. The seed contains a narrow embryo surrounded by copious endosperm which may have a rough or mottled appearance.

**Classification.** The five genera, *Oubanguia*, *Scytopetalum*, *Rhaptopetalum*, *Brazzeia* and *Pierrina*, can be classified on features of inflorescence and flowers. For example, *Oubanguia* has long, lax panicles and has stamen filaments joined at the base, in contrast to the other genera which have flowers either in fascicles (close clusters) on old woody branches (*Brazzeia*) or in racemes in the axils of leaves, as in *Scytopetalum* (petals several to many and stamen filaments unequal in length) and *Rhaptopetalum* (free petals and stamen filaments all the same length).

Various affinities have been suggested for this family, notably with Tiliaceae, Sterculiaceae, Malvaceae or Olacaceae. The features that Scytopetalaceae shares with all these families are the woody habit, the alternate, simple leaves, the bisexual flowers with usually numerous stamens, sometimes partly fused, and a several-loculed ovary.
**Economic uses.** There are no known economic uses for this family apart from *Scytopetalum tieghemii*, a medium-sized tree reaching about 25m (80ft) in height, whose wood is resistant to decay and is used as poles for house-building in Ghana and Sierra Leone. S.R.C.

## ELAEOCARPACEAE
*Crinodendrons and Aristotelias*

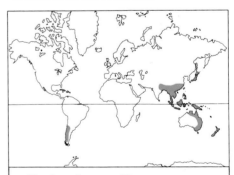

**Number of genera:** 12
**Number of species:** about 350
**Distribution:** tropics and subtropics.
**Economic uses:** limited as ornamentals and local uses of edible fruits.

The Elaeocarpaceae is a smallish family of tropical and subtropical trees and shrubs, some of which are cultivated as ornamentals.
**Distribution.** Members of the family are found in eastern Asia, Indomalaysia, Australasia, the Pacific area, South America and the West Indies.
**Diagnostic features.** The leaves are alternate or opposite, and have stipules. The flowers are regular, bisexual, borne in racemes, panicles or cymes, and have four or five sepals, free or partly united. The petals are usually free, either four or five in number or absent altogether; they are often

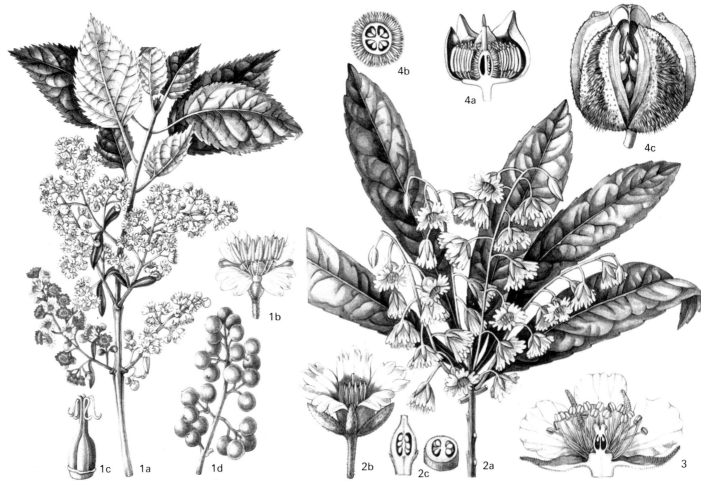

Elaeocarpaceae. 1 *Aristotelia racemosa* (a) shoot with axillary racemes of flowers ( ×$\frac{2}{3}$) ; (b) male flower with three-lobed petals and numerous stamens ( ×3) ; (c) gynoecium with free curled styles ( ×6) ; (d) fruits ( ×$\frac{2}{3}$). 2 *Elaeocarpus dentatus* (a) shoot with inflorescence ( ×$\frac{2}{3}$) ; (b) bisexual flower ( ×2) ; (c) bilocular ovary in vertical (left) and cross (right) section ( ×6). 3 *Muntingia calabura* half flower ( ×2). 4 *Sloanea jamaicensis* (a) half flower with numerous stamens having short filaments and large anthers ( ×1$\frac{1}{2}$) ; (b) cross section of ovary with four locules and numerous ovules on axile placentas ( ×3) ; (c) dehiscing fruit ( ×$\frac{1}{3}$).

fringed or lacerated at their tips. There are numerous free stamens, and the anthers have two locules which release pollen through two apical pores. The ovary is superior and contains two to many locules (rarely one only), and each locule contains two to many pendulous ovules. The style is simple and sometimes lobed at the tip. The fruit is a capsule or drupe; the seeds contain straight embryos and abundant endosperm.

**Classification.** The largest genus, *Elaeocarpus*, comprises about 200 species from eastern Asia, Indomalaysia, Australasia and the Pacific area. The other large genus in the family is *Sloanea*, with about 120 species of tropical Asian and American trees. It differs from *Elaeocarpus* in not having a succulent fruit, but a hard capsular one covered with rigid bristles. The best known of the small genera are *Aristotelia*, with five species from Australasia and South America, and *Crinodendron*, with two species from temperate South America.

The family is related to the Tiliaceae and to the Combretaceae and Rhizophoraceae. Also associated with the family are the Sarcolaenaceae (Chlaenaceae) and the Sphaerosepalaceae (Rhopalocarpaceae).

**Economic uses.** Several species of *Elaeocarpus*, *Crinodendron* and *Aristotelia* are cultivated. *Elaeocarpus reticulatus* (*E. cyaneus*), a native of Australia with cream flowers and blue drupes, and *E. dentatus* from New Zealand, with straw-colored flowers, are both known in cultivation in Europe. Better known, however, are the two ornamental evergreen species in the genus *Crinodendron*. The pendulous, urn-like, crimson flowers of *Crinodendron hookerianum*, about 2.5cm (1in) long on crimson stalks to 7cm (3in) long, are complemented by dark green narrow leaves to 10cm (4in) long. It grows to 9m (30ft) tall and is a spectacle in flower. The leaves of *C. patagua* are smaller and more oval, and the flowers white and bell-shaped. Both species are frost-tender, and require moist acid soils. The Chilean *Aristotelia chilensis* (*A. macqui*) grows to about 3m (10ft) and has insignificant green-white flowers, and black pea-sized fruits.

*Aristotelia chilensis* produces maqui berries which are said to have medicinal properties and are made into a local wine in Chile. The fruits and seeds of several *Elaeocarpus* species are eaten, eg *E. calomala* (Philippines), *E. dentatus* (New Zealand), *E.*

*serratus* (Ceylon olives). The West Indian *Sloanea berteriana* (motillo) and *S. woollsii* (gray or yellow carabeen) yield a heavy and a light timber respectively. The New Zealand *Aristotelia racemosa* and Australian *Elaeocarpus grandis* produce wood used for cabinet-making. B.M.

# TILIACEAE
*Limes, Lindens, Basswoods and Jute*

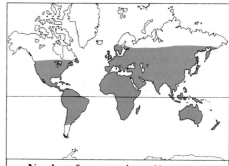

**Number of genera:** about 41
**Number of species:** about 400
**Distribution:** tropical with a few temperate species.
**Economic uses:** jute, timber and ornamental trees.

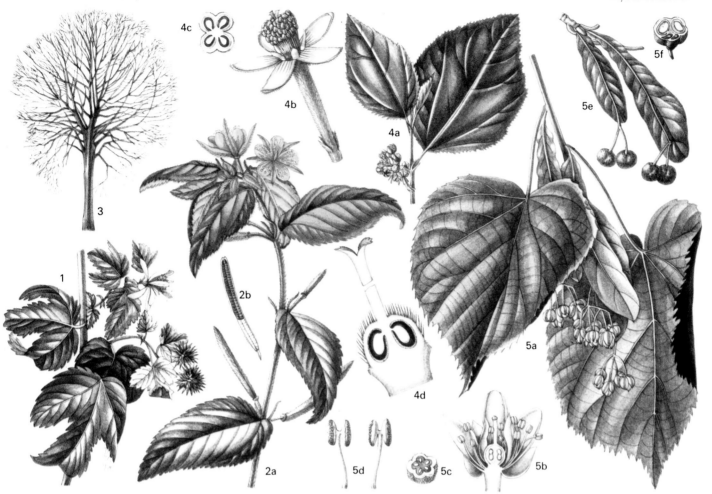

Tiliaceae. 1 *Triumfetta subpalmata* leafy shoot with flowers and fruits ( ×⅔). 2 *Corchorus bullatus* (a) leafy shoot with narrow stipules, flowers and fruits ( ×⅔) ; (b) fruit (a capsule) with wall removed to show seeds ( ×⅔). 3 *Tilia platyphyllos* leafless tree. 4 *Grewia parvifolia* (a) flowering shoot ( ×⅔) ; (b) flower with sepals fused into a tube and small free petals ( ×5) ; (c) cross section of ovary with four locules ( ×10) ; (d) vertical section of gynoecium ( ×15). 5 *Tilia petiolaris* (a) flowering shoot ( ×⅔) ; (b) half flower ( ×4) ; (c) cross section of ovary ( ×6) ; (d) anthers ( ×12) ; (e) fruits with persistent bract ( ×⅔) ; (f) cross section of fruit ( × 1½).

The Tiliaceae is a medium-sized family of tropical and temperate trees and shrubs. It includes the lindens (limes) and the economically important jute.

**Distribution.** This family is widely distributed throughout the tropical regions of the world, especially South America, Africa and Southeast Asia. The largest genus, *Grewia*, occurs in tropical Africa, Asia and Australasia. *Triumfetta* is an important genus of the New World tropics. *Corchorus*, the only genus which contains herbaceous species, is found in Africa and Asia, and *Sparmannia* is a tropical African genus which extends into the temperate region of the Cape of Good Hope. The genus *Tilia* also extends into the temperate regions of Europe and North America.

**Diagnostic features.** The leaves are alternate in two ranks, both tending to lie towards the upper side of the horizontally spreading shoots which bear them. They are usually deciduous, simple and asymmetrical and have branched hairs; their bases bear small stipules. The bark is often fibrous and mucilaginous.

The flowers are borne in complex cymes in the leaf-axils, and are normally bisexual, regular, small and green, yellow or white. They often secrete scent and nectar. There are five sepals which are either free and valvate or united, and usually five free petals, with glandular hairs at their bases; in a few species the corolla may be absent. The many stamens are arranged in groups, or fascicles, of five to ten which are inserted at the bases of the petals or on an androphore. The ovary is superior with two to ten or many locules each containing one to many ovules. The style is simple with a capitate or lobed stigma.

The several-seeded fruits take various forms, from the globose, indehiscent, nut-like fruits of *Tilia* to the spheroidal, dehiscent capsules of *Corchorus* and *Sparmannia*. The seeds contain endosperm and well-differentiated, straight embryos.

**Classification.** The family is related to the Malvaceae from which it differs in having anthers with two locules (not one), and to the Bombacaceae and Vitaceae.

**Economic uses.** The genus *Tilia* contains several woodland trees which yield valuable timber. These are *Tilia cordata*, *T. × platyphyllos* and *T. vulgaris*, the European limes or lindens, *T. americana*, the American basswood, and *T. japonica*, the Japanese linden. The wood of *T. cordata* is particularly good for making furniture and musical instruments since it is easily cut and carved. Bees produce an excellent honey from lime flowers. The decorative leaves and perfumed flowers of these species have also made them popular as ornamental trees for the public squares, parks and streets of many European towns. *Sparmannia africana*, the rumslind tree, is grown in South Africa for its beautiful clusters of white flowers.

Jute is obtained from the bast fibers of *Corchorus capsularis*, grown chiefly in the Ganges-Brahmaputra delta, and to a lesser extent from *C. olitorius* grown in Africa. The leaves of *C. olitorius* are used for food in eastern Mediterranean countries. The bark of many tropical trees and shrubs in this family, such as species of *Grewia*, *Triumfetta* and *Clappertonia*, is also used as a source of fibers for rope-making.                    B.N.B.

# STERCULIACEAE
## Cocoa and Kola

The chiefly tropical family Sterculiaceae consists of soft-wooded trees and shrubs, and a few herbaceous and climbing species.

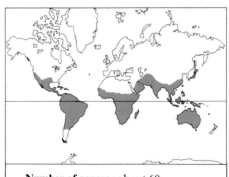

Sterculiaceae. 1 *Dombeya burgessiae* (a) inflorescence and leaf with stipules ($\times\frac{2}{3}$); (b) two outer sterile stamens (staminodes) and three fertile stamens ($\times 1\frac{1}{2}$); (c) gynoecium with five-lobed stigma ($\times 1\frac{1}{2}$); (d) cross section of ovary with five locules and ovules on axile placentas ($\times 4$). 2 *Melochia depressa* habit ($\times\frac{2}{3}$). 3 *Theobroma cacao* (a) unripe fruit cut away to show seeds—the cocoa beans of commerce ($\times\frac{2}{3}$); (b) flowers and young fruits which form on old wood ($\times\frac{2}{3}$). 4 *Cola acuminata* (a) flowering shoot ($\times\frac{2}{3}$); (b) male flower ($\times\frac{2}{3}$); (c) stamens united into a column ($\times 2$); (d) gynoecium ($\times 2$); (e) cross section of ovary ($\times 2$). 5 *Sterculia rupestris* tree in leaf.

**Number of genera:** about 60
**Number of species:** about 700
**Distribution:** pantropical.
**Economic uses:** cocoa and cola, with a number of ornamental shrubs.

It includes the economically important genera *Cola* and *Theobroma* (cocoa).

**Distribution.** The family is pantropical, extending into subtropical regions.

**Diagnostic features.** The leaves are alternate, with stipules, and simple or partly divided into lobes. Many species bear star-shaped hairs on their parts. The flowers are regular and borne in complex cymes, and are bisexual or unisexual on the same plant.

There are three to five sepals more or less united, and five petals which are free or fused together by a staminal tube. The petals are often small, and occasionally absent altogether. An outer whorl of stamens is often reduced to staminodes or quite lacking, while the inner whorl bears the anthers with two locules each. The ovary is superior and with usually two to 12 carpels, rarely more, or rarely reduced to one. Each locule contains two or more ovules on axile placentas, and the style is simple or divided into lobes, or rarely free to the base. The fruits may be dry or rarely berry-like, and often dehiscent. The seeds contain fleshy, little or no endosperm and the embryo is curved or straight.

**Classification.** *Sterculia*, the largest genus, comprising about 300 species, derives its name from the Roman god of privies, Sterculius (Latin, *stercus* = dung), on account of the smell of the flowers and leaves of certain species. *Cola* is another large genus of about 125 African species. The 30 or 50 species of *Theobroma* are native to America. *Dombeya*, comprising some 350 species from Africa, Madagascar and the Mascarene Islands, commemorates J. Dombey

(1742–94), a French botanist who accompanied Ruíz and Pavón on their South American expedition from Spain. Other major genera include *Pterospermum* and *Reevesia*, with 40 and 15 Asian species respectively, *Firmiana* (15 species from Africa and Asia) and *Brachychiton* (11 species from Australia). The family is closely related to the Tiliaceae, Malvaceae and Bombacaceae.

**Economic uses.** The two economically important products from this family are cocoa (*Theobroma cacao*) and kola (*Cola nitida* and *C. acuminata*).

There are a number of ornamental genera in the family, the best-known being the attractive shrubs of the two small genera *Fremontodendron* from California and Mexico and *Abroma* from Asia and Australia. There are several attractive *Dombeya* species and hybrids: large, rounded shrubs with mallow-like flowers borne in pink or white erect or pendulous heads. *Pterospermum acerifolium* from India has remarkable large, erect, brown-felted flower buds which open lily-like to reveal a boss of stamens about 12cm (5in) long. The foliage is bold, tough and felted bronze when young, gray

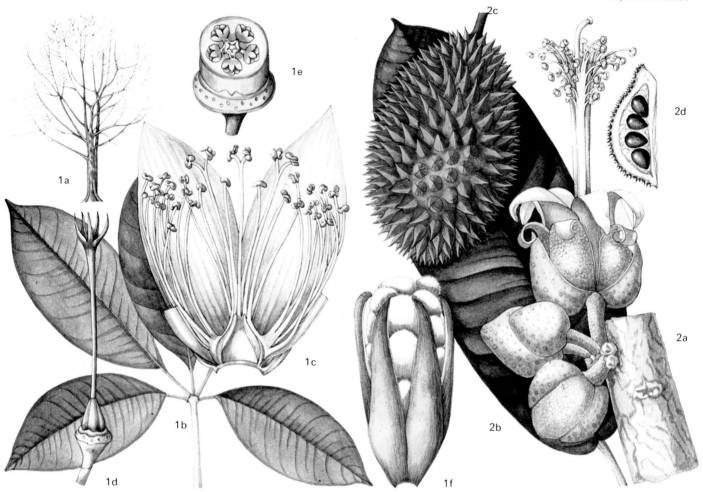

Bombacaceae. 1 *Bombax ceiba* (a) habit; (b) digitate leaf ($\times\frac{2}{3}$); (c) flower dissected to show petals and numerous bundles of stamens ($\times\frac{2}{3}$); (d) gynoecium with superior ovary, simple style and divided stigma ($\times\frac{2}{3}$); (e) cross section of ovary ($\times 3$); (f) fruit with seeds embedded in hairs ($\times\frac{2}{3}$); 2 *Durio zibethinus* (a) flowers arising from old wood showing the pair of sepal-like appendages subtending the whorls of sepals and petals, and the stamens united into a tube surrounding the style ($\times\frac{2}{3}$); (b) leaf ($\times\frac{2}{3}$); (c) spined fruit (the evil smelling durian) ($\times\frac{1}{2}$); (d) vertical section of part of fruit showing seeds ($\times\frac{1}{4}$).

beneath when mature, making it a splendid specimen tree. *Reevesia thyrsoidea*, native to Hong Kong, is an evergreen shrub with scented white heads of 45–50 flowers borne at the shoot tips. *Firmiana* and *Bracychiton* species are also cultivated.　　B.M.

## BOMBACACEAE
### Baobab, Durian and Balsa Trees

The Bombacaceae is a small family of tropical flowering trees, whose members include the well-known baobab, balsa, durian and kapok trees.

**Distribution.** Most species are found in South America, above all in Brazil. A few occur in Southeast Asia, and some unusual ones in Africa and Madagascar. They live mostly in dense rain forest in South America, and in open savanna and weedy habitats in Africa.

**Diagnostic features.** Several of the trees (eg *Adansonia digitata* – baobab, *Cavanillesia platanifolia* – Colombian ciupo, and *Chorisia* species) which have adapted to dry places have a peculiar appearance, with small apical crowns and unusually thick bottle-shaped, egg-shaped or barrel-shaped trunks – an adaptation for water storage. Many

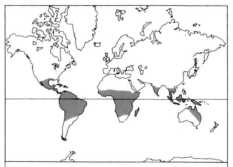

**Number of genera:** 20
**Number of species:** about 180
**Distribution:** tropics especially rain forests of S America.
**Economic uses:** baobab used locally in Africa, silk cotton tree yields kapok, and durian fruits from *Durio zibethinus*.

species are deciduous and their entire, palmate or digitate leaves and stipules are shed at the end of each rainy season. During this leafless period the flowers open. They are massive for most genera of the family, and even when small they are showy, with white or brightly colored flowers. They are always

bisexual and frequently emerge from the branches and trunks and even near the base in some tropical forest genera such as *Durio*. The calyx and corolla respectively consist of five separate sepals and petals, which are sometimes fused into a tube. The whole flower is subtended by another whorl of sepal-like appendages – the epicalyx. There are five to numerous stamens, free or joined into a tube. The ovary is superior with two to five fused carpels and many locules, with two to numerous erect ovules in each locule. The style is simple, lobed or capitate. The fruit is a capsule containing smooth seeds often with long cotton-like fibers, as in *Ceiba*, giving the kapok of commerce. The seeds have little or no endosperm.

Several species of the Old World genus *Adansonia* and their allies are pollinated by ants, and have ant colonies living within the spines of the branches, the extrafloral nectaries of the leaves, calyx and flower stalks providing a valuable food source for the ants, which protect the flowers from predators.

**Classification.** Important genera of the family include *Adansonia*, the kapok or silk cotton trees (*Bombax* and *Ceiba*), the

Malvaceae. 1 *Malva sylvestris* (a) flowering shoot ( $\times\frac{2}{3}$ ); (b) gynoecium ( $\times 4$ ); (c) androecium and base of corolla ( $\times 4$ ); (d) fruit and persistent calyx viewed from above ( $\times 1\frac{1}{2}$ ). 2 *Malope trifida* (a) flowering shoot ( $\times\frac{2}{3}$ ); (b) young fruit with remains of styles and stigmas removed ( $\times 2$ ); (c) vertical section of young fruit ( $\times 2$ ); (d) ripe fruit enclosed in calyx and epicalyx ( $\times\frac{2}{3}$ ); (e) flower ( $\times 1$ ). 3 *Hibiscus schizopetalus* (a) leafy shoot bearing flower and fruit—a capsule ( $\times\frac{2}{3}$ ); (b) vertical section of lower part of flower showing ovary containing ovules on axile placentas ( $\times 1$ ); (c) cross section of ovary showing five locules ( $\times 2$ ).

Southeast Asian durians (*Durio*) and the tropical South American trees of *Eriotheca, Chorisia, Cavanillesia, Ochroma* and *Matisia*.

The Bombacaceae is very closely related to the Malvaceae and has occasionally been placed in it, but, because it forms a group of closely related trees with smooth instead of rough pollen grains, it is kept distinct.

**Economic uses.** Perhaps the most important species commercially is *Ochroma pyramidale* or balsa. Other light timbers used for matchstick manufacture, boxes and veneers are obtained from various species of *Bombax*. The baobab tree plays an important role in African life. The fruit fibers of *Ceiba pentandra* and *Bombax ceiba* (*B. malabaricum*) constitute the valuable filling material kapok. The edible arils of the evil-smelling *Durio zibethinus* are the much-sought-after durian fruit of Malaya.                C.J.H.

## MALVACEAE
### Cotton, Mallows and Hollyhocks
The Malvaceae is a cosmopolitan family of herbs, shrubs and trees. Its most important members are cotton, okra and China jute. It is also the mallow family, producing many popular garden plants, including the holly-

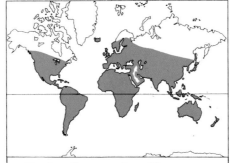

**Number of genera:** about 80
**Number of species:** over 1,000
**Distribution:** cosmopolitan, but centered in S America.
**Economic uses:** fiber crops (cotton, China jute, aramina), fruits (okra) and ornamentals (mallows, hollyhock).

hock (*Althaea*), mallows (*Malva*) and *Hibiscus*.

**Distribution.** Members of this family occur over most of the world, except the very cold regions, and are particularly abundant in tropical South America. *Hibiscus* is the largest genus with some 300 species widely distributed, although most are tropical; *Hibiscus trionum* and *H. roseus* are the only

European species. *Abutilon* is basically tropical and *Lavatera* Mediterranean. *Althaea* is cosmopolitan in temperate and warm regions, *Althaea rosea* being the garden hollyhock native to the eastern Mediterranean. *Malva* itself comes from north temperate regions.

**Diagnostic features.** The leaves are alternate, with stipules; often stellate hairs are present. The flowers are bisexual and regular, with parts usually in fives. The calyx is composed of five sometimes joined sepals, and is often subtended by an epicalyx. This has been interpreted both as fused bracteoles and as stipules. The corolla consists of five free petals, usually convolute. The numerous stamens are monadelphous, that is united below into a tube basally joined to the corolla. Division of the filaments has resulted in the anthers being unilocular. The ovary is superior and composed of five or more fused carpels with axile placentation. The style is branched. The fruit is dry, capsular or schizocarpic, except in *Malvaviscus* where it is a berry. The seeds are often covered in fine hairs, which may be tufted, as in *Gossypium*. They have little or no endosperm and a straight or curved embryo.

**Classification.** The classification of the Malvaceae poses many problems, and opinions differ as to delimitation of genera and tribes. Most modern treatments recognize five tribes:

MALOPEAE. Carpels in two or more superposed or spirally arranged whorls; *Malope*, *Palaua* and *Kitaibelia*.

HIBISCEAE. Fruit a capsule, the carpels persistent; *Hibiscus* and *Gossypium*.

MALVEAE. Fruit a schizocarp and stigmas decurrent on the style-branches; *Malva*, *Malvastrum*, *Lavatera*, *Althaea*, *Sidalcea*, and *Hoheria*.

ABUTILEAE. Fruit a schizocarp and stigmas apical or nearly so; *Abutilon*, *Sphaeralcea*, *Modiola*, and *Sida*.

URENEAE. Style-branches and stigmas twice as many as carpels; *Malvaviscus*, *Pavonia* and *Urena*.

The Malopeae with the carpels in two or more superposed or spirally arranged whorls is the most primitive tribe. All other members have carpels in a single whorl. Hibisceae, Malveae and Abutileae have equal numbers of style-branches or stigmas and carpels, but the tribe Ureneae, possibly the most recently evolved, has style-branches twice the number of carpels.

The Malvaceae is related to the Tiliaceae, but differs in the possession of one-celled anthers and monadelphous stamens. It probably represents a climax group derived from ancestral woody Tiliaceous stock.

**Economic uses.** *Gossypium* provides cotton and is by far the most important genus of the whole family. The young fruits of *Hibiscus esculentus* are known as okra, a common vegetable in warm climates. Several species yield tough fibers, notably *Abutilon avicennae* (China jute) and *Urena lobata* (aramina). *Abutilon*, *Althaea*, *Hibiscus*, *Lavatera*, *Malope*, *Malva*, *Malvastrum*, *Pavonia*, *Sida* and *Sidalcea* species are among those grown as ornamentals.                      S.L.J.

## SPHAEROSEPALACEAE

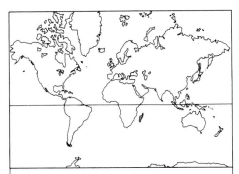

**Number of genera:** 2
**Number of species:** 14
**Distribution:** Madagascar.
**Economic uses:** none.

The Sphaerosepalaceae (Rhopalocarpaceae) is a small family of trees and shrubs native to Madagascar. The leaves are simple and possess stipules. The flowers are regular, bisexual, in terminal and axillary panicles or cymes, with four or six unequal sepals, and four (rarely three) petals. Sepals and petals are imbricate. The stamens are numerous with slender filaments which are more or less joined at the base: the anthers have two widely separated locules. A single style rises between the two lobes of the ovary which is superior and comprises two (rarely three) united carpels partly sunk in a large cup-like disk. There are two locules each with about three erect, anatropous, basal ovules. The one- or two-seeded fruit is round or didymous, covered in short, stout spines. The seeds have a more or less ruminate endosperm and a minute embryo.

*Rhopalocarpus* has 13 species, *Dialyceras*, the only other genus, one. It has been suggested that the family is related to Thymelaeaceae, resembling particularly the tropical Asian *Gonystylus* in its leaves and flowers. *Gonystylus* differs, however, in having no stipules, five valvate sepals and petals, and a disk outside rather than inside the stamens; it also differs in many features of the fruit. Others have allied the family to the Malvaceae or to the Sarcolaenaceae, which is also restricted to Madagascar. Sometimes the family is submerged in the Cochlospermaceae or allied to the Guttiferae or Flacourtiaceae. No economic uses are known.                      D.J.M.

## SARCOLAENACEAE

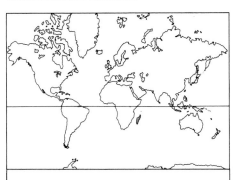

**Number of genera:** 8
**Number of species:** 39
**Distribution:** eastern Madagascar.
**Economic uses:** wood used locally in house-building.

The Sarcolaenaceae (Chlaenaceae) is a small family of beautiful trees mostly restricted to the rain forest of eastern Madagascar; the forests of the western slopes of the high plateaus were dominated by them until they were destroyed by persistent burning and replaced by *Heteropogon* grassland. The trees have simple alternate leaves with stipules, regular, often showy flowers with three to five sepals, five or six large petals, sometimes a disk, five to ten or many stamens and a superior ovary of one to five united carpels containing one to five locules with few to several anatropous basal, apical or axile ovules in each locule. The fruit is a many-seeded capsule or single-seeded and indehiscent, often surrounded by woody bracts. The seeds have a straight embryo and fleshy or horny endosperm.

The largest genera are *Sarcolaena* (10 species), *Leptolaena* (12 species) and *Schizolaena* (7 species). The family is usually allied with the Malvaceae with which it shares mucilage cells in the pith, though the petiole structure suggests some closer relationship with the Dipterocarpaceae, from which it differs in having no resin canals; the pollen is unlike that of any of these families or their allies, however, and although it is included here doubtfully in the Malvales, the Sarcolaenaceae is possibly more allied to the Theaceae in the order Theales.

Apart from local use of the wood of *Leptolaena pauciflora* in house-building, no uses are recorded.                      D.J.M.

# URTICALES

## ULMACEAE
*Elms and Hackberries*

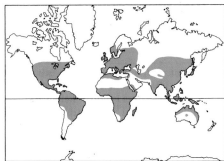

**Number of genera:** 16
**Number of species:** about 2,000
**Distribution:** two groups, north temperate, and tropical and subtropical.
**Economic uses:** major source of superior hardwood timber and valued ornamental trees.

The Ulmaceae is a family of tropical and temperate trees and shrubs whose best-known members are the elms.

**Distribution.** The two tribes into which the Ulmaceae are customarily divided have different patterns of distribution. The Ulmeae are mainly a north temperate group, one genus, *Ulmus*, occurring in all three northern continents, while the other genera are less wide-ranging, such as *Hemiptelea* and *Pteroceltis* in China, and *Planera* in North America. *Holoptelea* has a disjunct distribution in peninsular India and West Africa. The Celtideae are mainly tropical or subtropical, with *Celtis* also represented in temperate regions. *Trema* occurs throughout the tropics; three of the smaller genera are restricted to tropical Asia and the Malaysian regions, three to tropical and subtropical America, and one to Africa.

**Diagnostic features.** All members of the family are trees or shrubs. The leaves are usually alternate and simple, with stipules that are shed as the leaves unfold. The flowers may be either bisexual or unisexual; they are typically green and inconspicuous and borne in clusters. There is a calyx at the base of which the stamens arise, usually one stamen opposite each calyx-lobe; petals are absent. The ovary is superior (rudimentary in the male flowers) of two carpels usually forming one locule (rarely two) with a single pendulous ovule. There are two styles with the stigmatic surface on inner face. The fruit is a nut, samara or drupe, containing a single seed with a straight embryo and little or no endosperm. The anatomy of the wood is variable. Several genera include species in which the vessels are arranged in wavy concentric bands, the so-called ulmiform pattern. Chemically, the family is distinguished more by the absence of diagnostic substances than by their presence. Leucoanthocyanins are, however, usually present.

**Classification.** The Ulmaceae is divided into two tribes: the ULMEAE with pinnately veined leaves and the flat seeds without endosperm, and the CELTIDEAE which includes species with both pinnately veined leaves and, more typically, leaves having three main veins

diverging from the base; the seeds are usually rounded and with at least some endosperm. The principal genera of the Ulmeae are *Holoptelea*, *Planera*, *Ulmus* and *Zelkova*, and of the Celtideae, *Aphananthe*, *Celtis*, *Gironniera* and *Trema*.

The Ulmaceae are similar in basic floral structure to the large families Moraceae and Urticaceae. The former differs in the presence of milky latex, the latter in its generally herbaceous habit. Also possibly related is the family Eucommiaceae.

**Economic uses.** The economic significance of the family is mainly in its timber. Most species of *Ulmus* produce a superior timber with a distinctive patterned grain. It is much used for furniture-making and for posts. The timber is resistant to decay under waterlogged conditions and is used for underwater piles. Some species of *Aphananthe*, *Celtis*, *Holoptelea*, *Trema* and *Zelkova* also produce good timber. The leaves of some *Celtis* species and formerly of *Ulmus* have been used as a forage. Coarse fibers from the inner bark of species of *Celtis*, *Trema* and *Ulmus* are in local use. The only edible fruits of any consequence are those borne by *Celtis* (hackberries). The principal

medicinal product is the mucilaginous inner bark of *Ulmus rubra*, the slippery elm.

R.H.R.

# MORACEAE
## Fig, Hemp and Mulberries

The Moraceae is an economically important and scientifically interesting family of mainly tropical and subtropical trees and shrubs.

**Distribution.** The family is widely distributed in the tropics, subtropics and some temperate regions of both hemispheres.

**Diagnostic features.** All are trees or shrubs except *Cannabis*, *Dorstenia* and *Humulus*. The character which distinguishes them most clearly from related families is the presence of milky sap containing latex. Latex production begins in the embryo and continues throughout the plant's life, yet despite its apparent importance its function is still unknown. The leaves are alternate, spiral or opposite, entire or lobed, the margins entire or toothed. Stipules are present. Throughout the family the flowers are unisexual, borne on the same or separate plants, small, with four perianth segments (five in *Cannabis* and *Humulus*), and are arranged in heads on flattened or hollowed

Ulmaceae. 1 *Ulmus campestris* (a) habit; (b) leafy shoot ($\times \frac{2}{3}$); (c) flowering shoot ($\times \frac{2}{3}$); (d) flower with calyx, no petals, five stamens and two styles ($\times 8$); (e) anther posterior (right) and anterior (left) view ($\times 24$); (f) gynoecium crowned by two styles with stigmas on inner faces ($\times 12$); (g) winged fruit ($\times 1$). 2 *Trema orientalis* (a) leafy shoot with flowers ($\times \frac{2}{3}$); (b) fruit ($\times 6$). 3 *Celtis integrifolia* (a) male flower with five stamens and hairy vestige of ovary ($\times 6$); (b) bisexual flower with five stamens and gynoecium crowned by two styles forked at apices ($\times 4$); (c) vertical section of ovary with single pendulous ovule ($\times 4$).

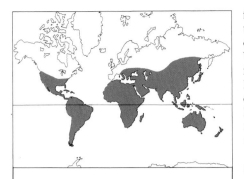

**Number of genera:** 75
**Number of species:** about 3,000.
**Distribution:** centered in tropics and subtropics but some temperate.
**Economic uses:** important fruits (figs, mulberries, breadfruit), fiber crops (hemp), narcotic drugs (cannabis) and hops.

receptacles or in catkins. A solitary flower is found in only one genus, *Chlorophora*. In *Cannabis* and *Humulus* female flowers are surrounded by large, persistent bracts. The male flower has four stamens, rarely one or two, and five in *Cannabis* and *Humulus*. The female flower has two carpels of which one is usually abortive, the style remaining. The ovary contains a single pendulous ovule. Within the Moraceae can be seen a transition from a superior ovary to an inferior ovary. For example, *Artocarpus* has a superior ovary whereas *Castilla* has an inferior ovary. Intermediate states can clearly be seen in related genera. The same trend is repeated in other parts of the family.

The infructescences are remarkably variable and the ripe fruit is often fleshy and edible. The fleshy parts are produced not by the ovary but by the swollen receptacle in which the seeds are embedded, eg *Ficus, Artocarpus, Morus*. Sometimes the fruit is an achene, covered by persistent perianth segments, eg *Humulus, Cannabis*. The seed is with or without endosperm.

**Classification.** The most important scheme of classification is based on inflorescence types, and divides the family into six tribes. The Moraceae has been included in the Urticaceae, but the latter can readily be distinguished from the Moraceae by the single style and stigma, basal ovule and clear sap (two styles, apical ovule and milky sap in Moraceae). However, the boundary between the two families is not entirely clear, because there are some transitional genera which do not produce latex but still possess latex ducts. *Cannabis* and *Humulus* are often considered to be a distinct family (Cannabaceae), differing most notably from the Moraceae by their five floral parts instead of four.

**Economic uses.** Most important are the fig (*Ficus*), mulberry (*Morus*) and breadfruit or jackfruit (*Artocarpus*). Other useful fruits come from species of *Brosimum* (breadnut) and *Pseudolmedia* (bastard breadnut). *Cannabis sativa* provides hemp fiber and narcotic drugs. Hops (*Humulus lupulus*) are cultivated for their use in brewing, giving the characteristic bitter taste to beer. *Castilloa elastica* (Panama rubber) and some species of *Ficus* were formerly used as latex sources in rubber manufacture. Useful timbers include osage orange (*Maclura*), iroko-wood (*Chlorophora excelsa*) or fustic (*Chlorophora tinctoria*).

S.W.J.

## URTICACEAE
*Stinging Nettles and Ramie Fiber*

The Urticaceae is a medium-sized family which includes *Urtica* and related genera, the

Moraceae. 1 *Ficus religiosa* (a) shoot with drip tip leaf and fruits ($\times\frac{2}{3}$); (b) fruit from below ($\times\frac{2}{3}$); (c) stalked and sessile female flowers ($\times 5$); (d) sterile female or gall flower ($\times 5$); (e) male flower ($\times 5$); 2 *Morus nigra* (a) shoot with fruit, the mulberry—an aggregation of achenes and the fleshy perianth ($\times\frac{1}{3}$); (b) female inflorescence ($\times 1\frac{1}{2}$); (c) female flower ($\times 3$). 3 *Ficus benghalensis*—the banyan tree. 4 *Ficus carica*—the fig, a multiple fruit of numerous achenes and the swollen receptacle ($\times\frac{1}{2}$). 5 *Artocarpus communis* (a) multiple fruit comprising numerous achenes, swollen perianth and swollen receptacle ($\times 1$); (b) female inflorescence ($\times\frac{1}{2}$); (c) leaf ($\times\frac{1}{4}$).

stinging nettles of north temperate regions.

**Distribution.** The family is found in most tropical and temperate regions, although it is relatively poorly represented in Australia.

**Diagnostic features.** Most genera are herbs or small shrubs, although one of the tribes contains primarily trees. The leaves may be alternate or opposite, and have stipules; the epidermis is usually marked with prominent cystoliths. The inflorescence is basically cymose and is often condensed into heads. The flowers are greenish, regular and unisexual (male and female being borne on the same individuals), although rarely bisexual, and usually have four or five sepal-like lobes. The male flowers have four or five stamens, usually bent downwards and inwards in bud and exploding when ripe. The female flowers possess a superior ovary which has one locule containing a single erect basal ovule. There is a single style. The fruit is usually dry, an achene, with the seed rich in oily endosperm, and with a straight embryo.

**Classification.** The family is divided into six tribes, based mainly on floral differences. The URTICEAE, however, is distinguished by its characteristic stinging hairs. This tribe contains *Urtica* (50 species) and the wide-

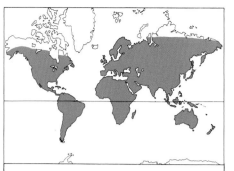

**Number of genera:** about 45
**Number of species:** over 1,000
**Distribution:** most tropical and temperate regions
**Economic uses:** ramie fiber and several troublesome weeds.

spread tropical and subtropical genus *Urera* (35 species) with its persistent, fleshy perianth making a berry-like fruit. *Laportea*, containing the stinging trees of Australia, also belongs here. The largest tribe is the PROCRIDEAE; the perianth of the female flowers is three-lobed, and the stigma has the characteristic appearance of a paint brush. *Pilea*, the widespread tropical genus contain-

ing the "artillery plant", so-named because of the puffs of pollen released when the anthers explode, *Elatostema* and the tropical Asian and Polynesian *Pellionia* belong here. The small tribe FORSKOHLEEAE, containing *Forskohlea* with six species from the Canary Islands to western India, has very reduced male flowers consisting of a single stamen. The two tribes BOEHMERIEAE and PARIETARIEAE are distinguished from each other by the presence of an involucre of often united bracts in the latter, with its principal genus *Parietaria* of 30 species possessing bisexual flowers (an unusual feature for the family). The former of these two tribes contains the tropical and northern subtropical genus *Boehmeria*, with 100 species, and *Maoutia*, from Indomalaya and Polynesia, unusual in its 15 species having no perianth in the female flowers, The sixth tribe, the CONOCEPHALEAE, includes *Cecropia*, trees with light wood used for making floats, and *Poikilospermum*, 20 species from the eastern Himalayas to southern China extending into Malaysia. They differ from the other genera in the family in having straight stamens, and because of this and some other similarities, used to be placed in a neighboring family, the

Urticaceae.1 *Pilea microphylla* (a) bud of male flower (×10); (b) male flower (×12); (c) perianth of female flower (×24); (d) gynoecium (×24); (e) vertical section of ovary (×24). 2 *Parietaria judaica* (a) bisexual flower (×4); (b) perianth opened out to show single style and hairy capitate stigma (×6). 3 *Forskohlea angustifolia* male flower (×6). 4 *Myrianthus serratus* (a) part of shoot with inflorescences in leaf axil (×⅔); (b) vertical section of ovary (×8); (c) dehiscing stamen (×18); (d) fruits (×⅔). 5 *Urtica magellanica* (a) flowering shoot (×⅔); (b) bud of male flower (×10); (c) male flower (×12); (d) vertical section of male flower with cup-like vestige of ovary (×12); (e) female flower (×12).

Moraceae. This affinity is still generally recognized, but the presence of latex in most genera of the Moraceae is a useful diagnostic feature, along with details of the fruit structure. The Ulmaceae is also closely related.

**Economic uses.** As well as being a well-known and noxious weed, *Urtica dioica,* the common stinging nettle, contains some small amount of silky bast-fiber. Fiber from other members of this family is used in the textile industry, notably *Boehmeria nivea* (ramie fiber or China grass) from Southeast Asia.
I.B.K.R.

# LECYTHIDALES

## LECYTHIDACEAE
*Brazil Nuts and Cannonball Tree*
The Lecythidaceae is a family of tropical trees, the best known of which is *Bertholletia excelsa,* which gives Brazil nuts.
**Distribution.** The family is centered in the wet regions of tropical South America, with some genera in Africa and Asia.
**Diagnostic features.** Tree size ranges from very small to very large. The leaves are

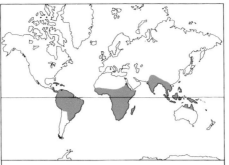

**Number of genera:** about 20
**Number of species:** about 450
**Distribution:** tropical, centered in S America.
**Economic uses:** nuts (Brazil nuts, sapucaia nuts) and useful timber.

spirally arranged in clusters at the tips of the twigs, each leaf large, simple, usually without gland-dots. Stipules are very rarely present. The flowers are bisexual, and borne in spikes (up to 1m (3ft) long, as in *Couroupita guianensis*), terminally, on side shoots, or on the older parts of the stems. They are generally large and showy, in shades of red, pink, yellow or white, and have an attractive,

fluffy appearance due to numerous stamens. They seldom last long (in *Barringtonia* remaining open for only one night before stamens and petals fall at dawn). The flowers are regular or irregular, with four to six calyx segments and four to six petals (absent in *Foetidia*) free or united into a ribbed tube. The stamens also are joined to each other and to the petals at their base in one or more rings. The ovary is inferior of two to six fused carpels, totally fused with the top of the receptacle or partially sunk into it. There are two to six or more locules, each with one to many ovules, a long simple style and a lobed or disk-shaped stigma. Pollination is frequently by bats, attracted by the sweet-smelling flowers. The usually large fruits have fleshy outer layers, hard and woody inner layers, and are indehiscent, with a lid through which the seeds leave the fruit. The seed is large, woody and lacks endosperm.
**Classification.** *Barringtonia* is the most important Old World genus, a well-known tree throughout Southeast Asia and Africa; others include *Careya, Combretodendron, Crateranthus, Foetidia, Napoleona* and *Planchonia.* However, the majority of the Lecythidaceae and those of the greatest

Lecythidaceae. 1 *Gustavia pterocarpa* (a) flowering shoot (×⅔); (b) flower with stamens and petals removed to show simple style surmounted by a lobed stigma (×⅔). 2 *Napoleona imperialis* (a) petal-less flower showing several whorls of stamens and outer ring of staminodes forming a corolla-like saucer-shaped body (×⅔); (b) half flower with calyx and staminodes removed, showing fertile outer whorls of stamens within which are outgrowths (disk) of the receptacle and short flat-topped style (×1⅓); (c) berry-like fruit with apical lid (×⅔). 3 *Barringtonia racemosa* (a) spike bearing flowers with many stamens (×⅔); (b) vertical section of flower base showing ovary fused into the receptacle (×2); (c) fruit (×⅔).

Violaceae. 1 *Anchietea salutans* shoot with alternate leaves and fruit ( ×⅔). 2 *Rinorea moagalensis* half flower showing stamen filaments fused at the base and the anthers with a membraneous extension to the connective ( ×10). 3 *Viola hederacea* (a) habit ( ×⅔); (b) vertical section of flower showing irregular petals and anthers in a close ring around the ovary ( ×4). 4 *Corynostylis hybnanthus* (a) shoot with leaf and inflorescence ( ×⅔); (b) cross section of ovary with one locule containing numerous ovules on parietal placentas ( ×4). 5 *Hybanthus* (*Ionidium*) *enneaspermum* var *latifolium* dehiscing fruit—a capsule ( ×4). 6 *Hymenanthera obovata* leafy shoot with fruit ( ×⅔).

economic importance are natives of South America. These include *Bertholletia, Couroupita, Grias, Gustavia* and *Lecythis*. Some classification schemes assign the African and Asian genera to a separate family, the Barringtoniaceae. Hutchinson combined the African genera *Crateranthus* and *Napoleona* with an American genus, *Asteranthus*, in another separate family, Napoleonaceae.

By tradition, the family has generally been associated with, or even included in, the Myrtaceae, the two families sharing many characters, particularly of the flowers. However, there are important differences of development and anatomy. Closer relatives might be found in the order Malvales, for, besides similarities of the flowers, the mucilage canals which are so characteristic of that order are also found in some genera of the Lecythidaceae.

**Economic uses.** *Bertholletia excelsa* is the source of Brazil nuts or para nuts. Sapucaia or paradise nuts, from several species of *Lecythis*, such as *Lecythis zabucajo*, are reputedly superior to Brazil nuts. The fruits are woody fibrous capsules resembling pots with a terminal lid. After the nuts are shed the empty pots are used as traps for wild monkeys (hence the common name monkeypots). *Couroupita guianensis* (cannonball tree), a native of tropical South America, is grown as an ornamental for its large 10cm (4in) waxy, sweet-smelling red and yellow flowers which are followed by spectacular spherical fruits 15cm–20cm (6in–8in) in diameter, the "cannonballs" borne on the trunks. Useful timber comes from *Lecythis grandiflora* (wadadura wood), *Careya* (tummy wood) from Malaya and India and from *Bertholletia excelsa*.                     S.W.J.

# VIOLALES

## VIOLACEAE
*Violets and Pansies*
The Violaceae is a medium-sized family of perennial (rarely annual) herbs or shrubs, including the violets and pansies.

**Distribution.** The family is cosmopolitan, but more typical of the temperate regions; in the tropics it tends to be restricted to higher mountainous areas.

**Diagnostic features.** Most species have simple alternate leaves, with small stipules. Only *Hybanthus* and *Ionidium* have species

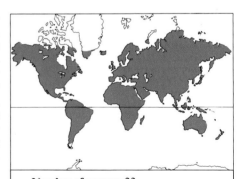

**Number of genera:** 22
**Number of species:** about 900
**Distribution:** cosmopolitan, chiefly temperate.
**Economic uses:** valued garden ornamentals (violets and pansies) and uses in perfumery and flavoring.

with opposite leaves. The flowers are in racemes, or solitary in the leaf-axils. They are regular (except in *Viola*), bisexual, and have five sepals and five petals. There are five stamens which usually have their filaments fused at the base to form a ring around the ovary. The ovary is superior, consisting of three fused carpels, with one locule contain-

Flacourtiaceae. 1 *Oncoba spinosa* (a) flowering shoot armed with axillary spines ($\times\frac{2}{3}$); (b) cross section of fruit ($\times\frac{2}{3}$); (c) entire fruit with remains of flower below ($\times\frac{2}{3}$). 2 *Idesia polycarpa* (a) pendulous inflorescence ($\times\frac{2}{3}$); (b) cross section of unilocular ovary with ovules on five large parietal placentas ($\times 2$); (c) vertical section of female flower showing branched style ($\times 2$); (d) male flower with numerous short stamens ($\times 2$); (e) ripe fruit ($\times\frac{2}{3}$). 3 *Azara microphylla* (a) flowering shoot ($\times\frac{2}{3}$); (b) vertical section of flower with single style and large stamens ($\times 9$); (c) part of inflorescence ($\times 3$); (d) cross section of unilocular ovary ($\times 18$).

ing numerous ovules arranged on a parietal placenta. The style and stigma are usually simple. The fruit is a capsule which usually dehisces, often explosively, along lateral lines into three or five valves. The seeds are usually spherical, and winged in some tropical climbing species. They contain a straight embryo and a fleshy endosperm.

All species of *Viola* have unequal petals with the lowermost pair, often the largest, forming a prominent spur. Their colored petals and scent attract pollinating insects which are said to be guided into the spur by linear markings, or honey guides, on the petals. Nectar is secreted into the spur from the bases of the two lowermost stamens. To reach it the insect's body must touch the stigma, so pollinating it with the pollen it may be carrying. The insect also touches the spurs on the anthers which shower pollen onto its back which it then carries to the stigma of another flower. Many species of *Viola* have small, insignificant, cleistogamous flowers which never open and are self-pollinating.

**Classification.** The chief genera are *Viola* (about 400 species, mainly north temperate herbs but some small shrubs), *Rinorea*

(*Alsodeia*, about 340 species of tropical and warm temperate shrubs), *Hymenanthera* (seven species of shrubs from Australia, New Zealand and Norfolk Island), *Paypayrola* (seven species of tropical South American trees), *Hybanthus* (about 150 species of herbs from tropics and subtropics), *Anchietea* (eight species of shrubs and climbers from tropical South America), *Leonia* (six species of tropical South American shrubs) and *Corynostylis* (four species of tropical American shrubs).

The Violaceae has its closest affinities with the Flacourtiaceae but some links with the Resedaceae are probable.

**Economic uses.** The Violaceae has little commercial significance except for *Viola*, which contains *Viola odorata*, grown in the south of France for essential oils used in manufacture of perfumes, flavorings, toiletries and the very sweet, violet-colored liqueur *parfait amour*; the flowers are also preserved in sugar as crystallized violets used largely for decoration. Many species of *Viola* are grown as ornamentals (pansies and violets). The garden pansies are a hybrid group (*Viola × wittrockiana*).

Some species are of limited medicinal

importance: the root of *Hybanthus ipecacuanha* is used as a substitute for true ipecacuanha as an emetic; the roots of *Anchietea salutaris* are used as an emetic and to treat sore throats and lymphatic tuberculosis; and the roots of *Corynostylis hybanthus* are used as an emetic.          B.N.B.

# FLACOURTIACEAE
*Chaulmoogra Oil and West Indian Boxwood*

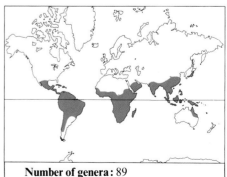

**Number of genera:** 89
**Number of species:** about 1,250
**Distribution:** tropical and subtropical, with some in temperate regions.
**Economic uses:** a few ornamentals; chaulmoogra oil and timber.

The Flacourtiaceae is a large family, chiefly of trees and shrubs, and of limited economic importance.

**Distribution.** The family is widespread in the tropics and subtropics, with some species in temperate regions.

**Diagnostic features.** The leaves are simple, alternate, opposite or in whorls, toothed or entire. The branches are sometimes spiny, and flowers are regular, bisexual or unisexual with the sexes on separate plants. They are solitary or borne in axillary or terminal inflorescences which may be racemose, corymbose, paniculate, in bundles. There are two up to 16 sepals, free, overlapping or contorted, sometimes intergrading with the petals in appearance. The usually overlapping or contorted petals, sometimes united, are normally rather small and the same number as the sepals, rarely more and then not arranged in relation to the sepals, or rarely absent altogether. The petals are usually hypogynous or more or less perigynous, and each may or may not have a scale-like outgrowth inside the base.

The stamens are frequently numerous or sometimes the same number as and opposite the petals. The filaments may be free, fused in bundles and alternating with glands, or rarely fused into a tube. The anthers usually open by slits, rarely by pores. The ovary position varies from superior through semi-inferior to inferior, and comprises two to ten united carpels with a single locule containing several ovules on one or more parietal placentas. The styles may be fused or free, rarely much branched or lacking, and are the same in number as placentas, the stigmas likewise. The fruits are capsules, berries, drupes, dry indehiscent and winged, horny or prickly, and containing seeds sometimes with arils or silky hairs and usually with much endosperm. There is a straight or curved embryo and the cotyledons are broad.

**Classification.** On the basis of floral characters 12 tribes have been recognized. Chief genera are: *Erythrospermum* (six species, Madagascar, Sri Lanka, Burma, China, Malaysia, Polynesia); *Hydnocarpus* (40 species, Indomalaysia); *Scolopia* (45 species, tropical and South Africa, Asia and Australia); *Homalium* (200 species, tropical and subtropical); *Flacourtia* (15 species tropical and South Africa, Mascarene Islands, Southeast Asia, Malaysia and Fiji); *Azara* (10 species, southern Bolivia and Brazil to Chile and Argentina); *Xylosoma* (100 species warm regions); *Casearia* (160 species tropical).

In evolutionary terms, the Flacourtiaceae lies between the Dilleniaceae and Tiliaceae, and is the most primitive family within the Violales, although this does not mean that it is the ancestor of other families in the group, rather that it is the least advanced of a series of parallel evolutionary lines.

**Economic uses.** The family contains few plants of economic importance. Notable ornamentals are *Oncoba spinosa*, *Berberidopsis corallina*, *Carrierea calycina*, *Idesia polycarpa* and *Azara* species. *Hydnocarpus wightiana* (southwestern India) and *Taraktogenos kurzii* (Burma) have seeds which yield chaulmoogra oil of use in the treatment of leprosy. The few timber trees include *Gossypiospermum praecox*, West Indian boxwood. B.M.

# LACISTEMATACEAE

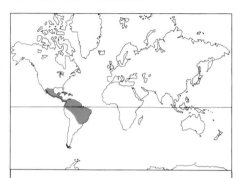

**Number of genera:** 2
**Number of species:** 27
**Distribution:** tropical America, W Indies.
**Economic uses:** none.

The Lacistemataceae is a small family of tropical shrubs.

**Distribution.** The family is native to the West Indies and tropical America.

**Diagnostic features.** The leaves are simple and alternate, with two small deciduous stipules or without stipules. The flowers are very small and inconspicuous, without petals. They are unisexual or bisexual, closely clustered into axillary spikes or racemes. Each flower is subtended by a concave overlapping bract and there are also two bracteoles at the base of each flower. Sepals are four to six in number and unequal in size, or absent. The axis is expanded into a fleshy concave disk on which is inserted a single stamen. The two anthers are separate, sometimes borne on a short stalk. The ovary is superior, of two or three fused carpels forming a single locule with two or three parietal placentas each bearing one or two pendulous ovules. There are two or three stigmas borne on a short style. The fruit is a capsule bearing one to three seeds each of which has a straight embryo surrounded by copious fleshy endosperm.

**Classification.** The two genera in this family can be distinguished on the basis of both vegetative and floral characters. *Lacistema* (20 species) has entire leaves, conspicuous bracts and spike-like inflorescences, while *Lozania* (seven species) has dentate leaves, small bracts and racemose inflorescences.

The family shows marked similarities with a number of other families in the order Violales, especially the Flacourtiaceae.

**Economic uses.** There are no known economic uses for this family. S.R.C.

# PASSIFLORACEAE
*Passion Flowers and Granadillas*

The Passifloraceae is a medium-sized family of vines, trees, shrubs and herbs, some species producing showy flowers and edible berries.

**Distribution.** The family is native to the tropics and subtropics. *Passiflora* itself is widely distributed, with 400–500 species in the warmer parts of America, a few species in Asia and Australasia, and one in Madagascar; it is widely cultivated throughout the world. *Adenia* has about 80 species in tropical Africa and Asia. Thirteen genera are found only in tropical and subtropical Africa.

**Diagnostic features.** The plants are trees, shrubs, herbs or frequently climbers with axillary coiled tendrils, corresponding to sterile pedicels. The leaves are alternate, entire or lobed, and have small, often deciduous stipules. The flowers are regular and usually bisexual; when they are unisexual they are usually borne on the same plant, rarely (in *Adenia* only) on separate plants. The single-flowered peduncle is jointed and furnished with three bracts. The five sepals are distinct or fused at the base. The petals are usually five, occasionally absent, and like the sepals may be distinct or basally connate. The receptacle is often terminated by one or more rows of petal-like or stamen-like filaments forming a corona. The stamens are usually equal in number to the petals and often opposite them, but frequently arising from a gynophore or androphore. The ovary is superior, with one locule containing three to five parietal placentas and numerous ovules. The styles are free or united. The three to five stigmas are capitate or discoid. The fruit is an indehiscent berry or capsule. The seeds are surrounded by a pulpy aril, and contain fleshy endosperm and a large, straight embryo.

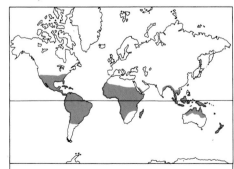

**Number of genera:** 20
**Number of species:** about 600
**Distribution:** tropical and subtropical, chiefly in America and Africa.
**Economic uses:** several species of *Passiflora* cultivated for edible fruit (passion fruit, granadilla) and as ornamentals.

**Classification.** *Passiflora* is the only genus of importance. The genera *Smeathmannia*, *Soyauxia* and *Barteria*, erect shrubs lacking tendrils from Africa, are considered the most primitive members of the family. They relate the Passifloraceae to the Flacourtiaceae. Both families have ovaries with parietal placentation, but the Passifloraceae is more advanced in possessing a perigynous calyx and corolla, a corona and tendrils for climbing. The family is also considered closely related to the Cucurbitaceae and Loasaceae. Recent works regard all these families as very closely related.

**Economic uses.** Between 50 and 60 species of *Passiflora* have edible fruits, but few are commercially cultivated. *Passiflora quadrangularis*, (the giant granadilla) is cultivated throughout the tropics for its juicy edible fruits up to 25cm (10in) long. In Hawaii, *P. edulis* var *flaviocarpa*, the yellow passion fruit, forms the basis of the entire passion fruit juice industry. In Australia, India and Sri Lanka it is *P. edulis*, passion fruit, or purple granadilla, that is cultivated. In Australia both species are grown, but the former is used solely as a stock on which the latter is grafted. The purple granadilla is used in beverages, candy and sweets. *Passiflora maliformis* (West Indies) is the sweet calabash which has grape-like juice. Other species are used locally in the South American tropics.

About twenty species of *Passiflora* are cultivated for their attractive and unusual flowers. The family has no other species of economic or domestic importance.　　S.L.J.

## TURNERACEAE

The Turneraceae is a family of mostly shrubs or small trees and a few herbs with no significant economic importance.

**Distribution.** The family is confined to tropical and subtropical parts of the New World, Africa, Madagascar and the Mascarene Islands.

**Diagnostic features.** The leaves are alternate, usually without stipules, often with glandular teeth on the margin. The flowers, which are usually solitary in the leaf axils, are bisexual and regular. The five regularly overlapping sepals usually have a swelling on the inner surface; the petals and stamens also each number five; all are united below into a cup which surrounds the ovary. The ovary is superior and formed of three united carpels

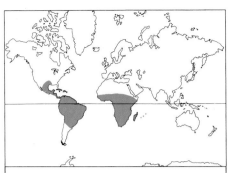

**Number of genera:** 8
**Number of species:** about 100
**Distribution:** tropical and subtropical regions, mainly in America.
**Economic uses:** limited medicinal and flavoring uses.

with a single locule and three parietal placentas bearing three to many ovules; it is surmounted by three styles. The fruit is a capsule splitting into three parts; the seeds have an aril and contain copious endosperm.

Most species have two types of flower, some with short, and some with long styles. Many species have nectaries on parts of the plant other than in the flowers. Self-

Passifloraceae. 1 *Passiflora caerulea* (a) twining stem with coiled tendrils, solitary flower with conspicuous filamentous corona and five-lobed leaves subtended by leafy bracts (×⅔); (b) vertical section of flower with, from the base upwards, subtending bracts, hollowed-out receptacle bearing spurred sepals, petals and filaments (the latter forming the corona) and a central stalk (androphore) with at the apex the ovary bearing long styles with capitate stigmas and at the base downward curving stamens (×1½); (c) fruit (×⅔); (d) cross section of fruit containing numerous seeds (×1½); (e) seed (×6).

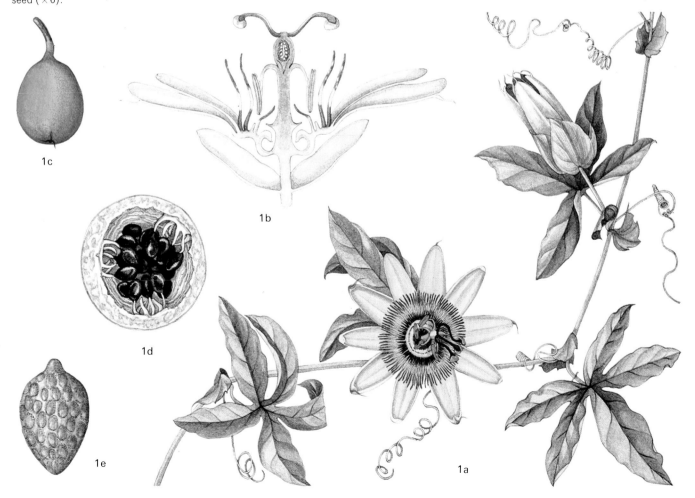

1c

1d

1e

1b

1a

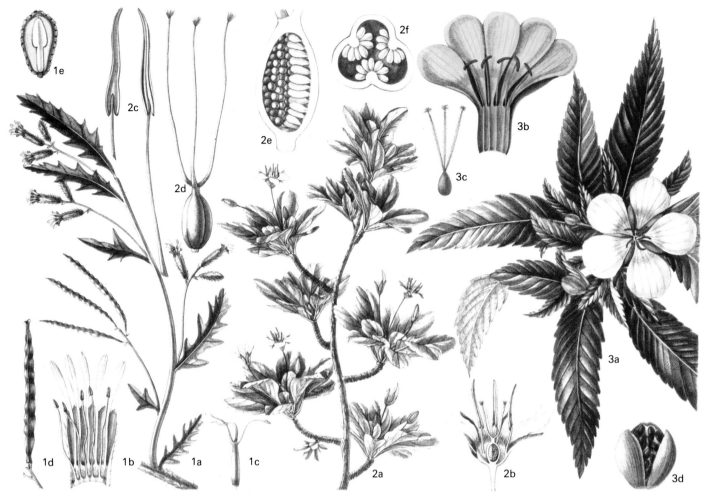

Turneraceae. 1 *Wormskioldia heterophylla* (a) shoot with alternate leaves, flowers in racemes and elongate fruits ($\times\frac{2}{3}$); (b) perianth opened out ($\times 2$); (c) gynoecium ($\times 2$); (d) fruit—a capsule ($\times 1$); (e) vertical section of seed ($\times 4$). 2 *Turnera berneriana* (a) shoot with leaves and solitary flowers ($\times\frac{2}{3}$); (b) half flower ($\times 3$); (c) bilocular anthers ($\times 8$); (d) gynoecium showing fringed stigmas ($\times 8$); (e) vertical section of ovary ($\times 15$); (f) cross section ovary with ovules on three parietal placentas ($\times 15$). 3 *Turnera angustifolia* (a) flowering shoot ($\times\frac{2}{3}$); (b) perianth opened out ($\times\frac{2}{3}$); (c) gynoecium ($\times 1$); (d) fruit dehiscing by three valves ($\times 1\frac{1}{2}$).

fertilization occurs in the absence of insect visits, by the corolla withering and pressing anthers and stigmas together.

**Classification.** The eight genera may be distinguished on a number of characters. Flowers are pendulous and the seeds have an aril covered in long thread-like hairs in *Mathurina*. A corona is present within the calyx of *Erblichia* and *Piriqueta*, the former genus having leaves with minute stipules and the latter lacking them altogether. Of the genera that lack a corona *Hyalocalyx* has sepals united only halfway, *Turnera* has solitary flowers with variously lobed stigmas, *Loweia* has solitary flowers and leaves with resin glands and stellate hairs, *Streptopetalum* is a herb with flowers in a raceme, petals inserted in the throat of the calyx tube and short fruits, and *Wormskioldia* is as *Streptopetalum* except that the petals are inserted in the calyx tube and fruits are elongate.

The affinities of the family lie with the Passifloraceae and neighboring families.

**Economic uses.** The leaves of a few species of *Turnera* have medicinal uses locally. They are also used in Mexico as a tea substitute and to flavor wines, etc.                     I.B.K.R.

# MALESHERBIACEAE

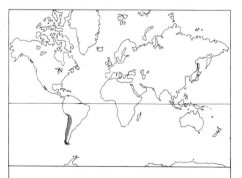

**Number of genera:** 1 or 2
**Number of species:** 27
**Distribution:** western S America.
**Economic uses:** none.

The Malesherbiaceae is a small family of herbs and undershrubs.

**Distribution.** The family is native to western South America.

**Diagnostic features.** The leaves are alternate, simple, without stipules, and frequently covered with a dense mat of hairs. The flowers are bisexual, regular and arranged in racemes, cymes or panicles. The calyx is five-lobed and with a long tube; the five petals are valvate, as are the sepals.

The ovary is superior, borne on a hairy stalk from which also arise the five stamens (androgynophore). The anthers have two locules which open lengthwise. The ovary is formed of three fused carpels forming one locule with three or four parietal placentas bearing numerous ovules, and is surmounted by three or four slender, thread-like styles separated at the base. The fruit is a capsule enclosed by the persistent calyx. The seed contains a straight embryo surrounded by fleshy endosperm.

**Classification.** Opinions vary as to whether all the species belong to a single genus (*Malesherbia*) or whether some can be ascribed to another genus (*Gynopleura*). The basis for such a division lies in floral characters: species belonging to *Malesherbia* have flowers in racemes with petals smaller than the calyx lobes, while those belonging to *Gynopleura* have flowers in panicles or dense clusters with petals larger than the calyx lobes.

This family is closely related to the passion flower family (Passifloraceae) and to the Turneraceae, but differs from the former in having more widely separated styles and from the latter in having a persistent receptacle. A further distinction is that in the Malesherbiaceae the seeds do not possess an aril.

**Economic use.** The family is of no economic importance. S.R.C.

## FOUQUIERIACEAE
*Ocotillo*

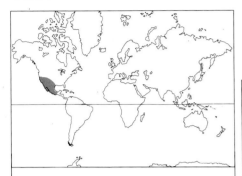

**Number of genera:** 2
**Number of species:** 8
**Distribution:** Mexico and SW USA.
**Economic uses:** limited uses as hedging and as a source of wax.

The flowering, spiny trees or shrubs that make up this family are a familiar sight in the dry southwest of North America.

**Distribution.** The family occurs only in Mexico and bordering states of the United States of America.

**Diagnostic features.** The leaves are small, succulent, are borne in groups or singly and are without stipules. The showy flowers are yellow (*Idria*) or red (*Fouquieria*). Borne in panicles or spikes, they are regular, bisexual, with five imbricate sepals, five hypogynous petals fused into a tube at the base, and 10 or more hypogynous stamens in one or more series with free or slightly fused filaments. The ovary is superior comprising three fused carpels and has a single locule with axile placentation below, but with septiform parietal placentation above dividing it incompletely into three locules. The style is branched (*Fouquieria*) or entire (*Idria*). The fruit is a capsule with oblong, winged and compressed seeds containing thin endosperm and a straight embryo with thick flat cotyledons.

**Classification.** The Fouquieriaceae comprises two genera, the monotypic *Idria* (stem usually branched), and the seven species of *Fouquieria* (usually unbranched stem).

*Fouquieria splendens* (ocotillo) a characteristic plant of parts of the Mojave and Colorado deserts, flowers between March and July. The leaves are deciduous on the few branched or cane-like stems to 7m (23ft) tall, being only briefly replaced after the spring and late summer rains. *Idria columnaris* has a weird habit with a thick trunk at the base, conical and often hollowed out, with a few long erect branches which may

become top heavy and fall over sideways to assume fantastic shapes.

The family is of disputed affinity, being sometimes placed in the Violales, as here, regarded as a peripheral member of the Centrospermae (Caryophyllales) or as closely allied to the Polemoniaceae.

**Economic uses.** With its curious spiny habit, *F. splendens* is sometimes used for hedging or boundary fences. Local use is made of wax obtained from the bark or stem of some *Fouquieria* species. B.M.

## CARICACEAE
*Papaw*

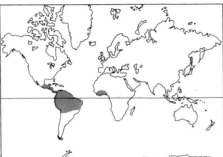

**Number of genera:** 4
**Number of species:** 30
**Distribution:** tropical America and tropical W Africa, centered in S America.
**Economic uses:** *Carica papaya* grown for fruit (papaw, papaya) and latex containing enzyme papain; other fruits eaten locally.

The Caricaceae is a tropical family of small trees, famous for the edible fruit of *Carica papaya*, grown throughout the tropics and widely known as the papaw, pawpaw, papaya or melon tree.

**Distribution.** The family is almost confined to the New World, where it is most abundant in South America, but with representatives also in Central America. *Cylicomorpha*, with two species, is native to parts of tropical West Africa.

**Diagnostic features.** The plants are small, sparsely branched trees, with soft wood; all parts contain milky latex in articulated laticifers which may be pungent. The leaves are alternate, crowded at branch tips, digitately lobed or foliolate and without stipules. The flowers are unisexual, rarely bisexual, the trees being usually dioecious, rarely monoecious or polygamous. The flowers are regular, with the parts in fives; the sepals are more or less free; the corolla is contorted or valvate with the corolla tube long in male, short in female flowers. The stamens are attached to the petals and the anthers open inward. The ovary is superior, of five fused carpels, has one to five locules, with many anatropous ovules on parietal placentas. The style is short and crowned by five stigmas. The fruit is a berry. The numerous seeds each have a gelatinous

envelope, oily endosperm and a straight embryo.

The stem of *Carica papaya* is very unusual in that there is little development of secondary xylem. The wood is formed largely from the phloem, which gives the soft, large-pithed trunk much of its rigidity.

**Classification.** There are four genera. *Jacaratia* (six species, tropical America and Africa) has sepals opposite the petals, and digitate leaves. In the spiny *Cylicomorpha* (two species, Africa) the sepals alternate with the petals and the leaves are mostly entire. *Jarilla* (one species, Mexico) has fruits with five protruding basal horns. The fruit of *Carica* (21 species) is without horns.

The Caricaceae is related most closely to the Passifloraceae, with which it has sometimes been united, and perhaps to the Cucurbitaceae.

**Economic uses.** The family's economic importance resides largely in *Carica papaya*, which has a large, bland juicy fruit and is extensively cultivated throughout the tropics. This species has nocturnal, sweet-scented, moth-pollinated flowers. As with many other cultivated plants, its origin is unknown and it may have arisen from hybridization. The fresh fruit is eaten with lemon or lime juice or in fruit salad; it may be tinned, crystallized or made into jam, ice cream, jellies, pies or pickles; when unripe, it is used like marrow or apple sauce. The green fruit produces latex which contains the proteolytic enzyme papain that breaks down proteins, and leaves are sometimes wrapped round meat to soften it. Commercial grade papain will digest 35 times its own weight of lean meat and is an important article of commerce for medicinal and industrial usages (for example, in canned meat and leather tanning). A few other species of *Carica* are cultivated in Latin America for their edible pericarp or the sweet, juicy seed envelopes, including *C. chrysophila* (higicho) and *C. pentagona* (babaco) both from Colombia and Ecuador, and *C. candicans* from Peru.

The mountain papaw is *C. cundinamarcensis* of the Andes: it has smaller fruits than the papaw and is successfully grown at altitudes in the tropics where *C. papaya* would fail. The fruits of *Jarilla caudata* and *Jacaratia mexicana* are eaten locally. T.C.W.

## BIXACEAE
*Annatto*

The Bixaceae is a family consisting of a single tropical genus (*Bixa*) of shrubs and small trees.

**Distribution.** The family is native to tropical America and the West Indies and one species, *Bixa orellana*, is naturalized in tropical West Africa.

**Diagnostic features.** The leaves are alternate, simple, entire, palmately nerved and with stipules. They have long petioles and are

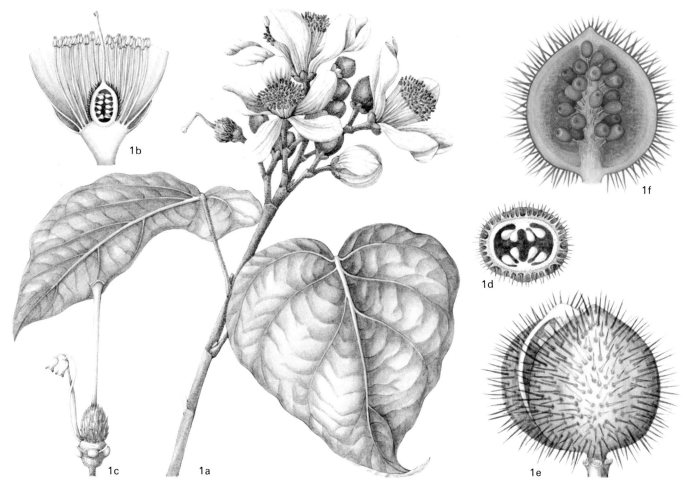

Bixaceae. 1 *Bixa orellana* (a) shoot with alternate leaves and flowers in a terminal panicle (×1); (b) half base of flower showing numerous stamens and ovary with numerous ovules on parietal placentas (×2); (c) stamens dehiscing by short slits and ovary crowned by long style and bilobed stigma (×3); (d) cross section of ovary showing ovules on parietal placentas (×8); (e) capsule dehiscing by two valves (×1); (f) vertical section of fruit with numerous red seeds (×1).

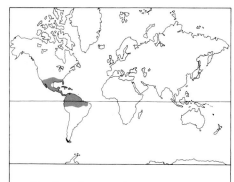

**Number of genera:** 1
**Number of species:** 1 or 4
**Distribution:** S America and W Indies.
**Economic uses:** *Bixa orellana* cultivated for annatto dye and as an ornamental.

covered with tufted hairs. Both leaves and stems contain reddish sap. The flowers are regular, bisexual and borne in showy, paniculate inflorescences. They have five imbricate, deciduous sepals. There are five large petals, imbricate and twisted in the bud. The stamens are numerous with free filaments and with two-celled, horseshoe-shaped anthers which dehisce by short slits at

the apex. The ovary is superior, of two to four fused carpels with one locule and two parietal placentas bearing numerous ovules. It is surmounted by a long slender style which terminates in a two-lobed stigma. The fruit dehisces along the two valves of the capsule and is often covered with hairs or stout prickles. It contains many obovoid seeds, with a bright red, rather fleshy testa. The seeds contain a large embryo surrounded by copious starchy or granular endosperm.

**Classification.** The genus *Bixa* consists of one or four species, depending on taxonomic interpretation. The Bixaceae is perhaps related to the Cochlospermaceae, which also has alternate leaves with stipules, and sometimes orange or reddish sap. Both families have showy flowers, but the Cochlospermaceae differs mainly in having palmately lobed, not entire, leaves, three- to five-valved (not two-valved) fruits and oily endosperm within the seeds. The Bixaceae may also have affinities with the more primitive Dilleniaceae. The horseshoe-shaped anthers are also to be found in the Thymelaeaceae.

**Economic uses.** Economically the family is important for the species *Bixa orellana*

(annatto) which is often cultivated as a quick-growing ornamental shrub in warm countries.

The testa is the source of a reddish-yellow dye, used for coloring foodstuffs such as cheese and butter. On a plantation scale, where the trees are grown at spaces of about 5m (16ft), mature specimens yield 80kg–100kg of seed per hectare (some 4cwt–5cwt per acre). The brown or dark red ovoid fruits are borne in large clusters at the ends of branches. Harvesting takes place when the fruits are nearly ripe. When the capsule splits open on drying the seeds are either dried and sold as annatto seed or paste.

S.R.C.

# COCHLOSPERMACEAE

*Rose Imperial*

This small family of tropical trees and shrubs contains a number of species with attractive large yellow flowers.

**Distribution.** Members of the family are native to the tropics, often in drier parts, in America, western central Africa, Indomalaysia and northern and western Australia.

**Diagnostic features.** The Cochlospermaceae are mostly trees and shrubs which may

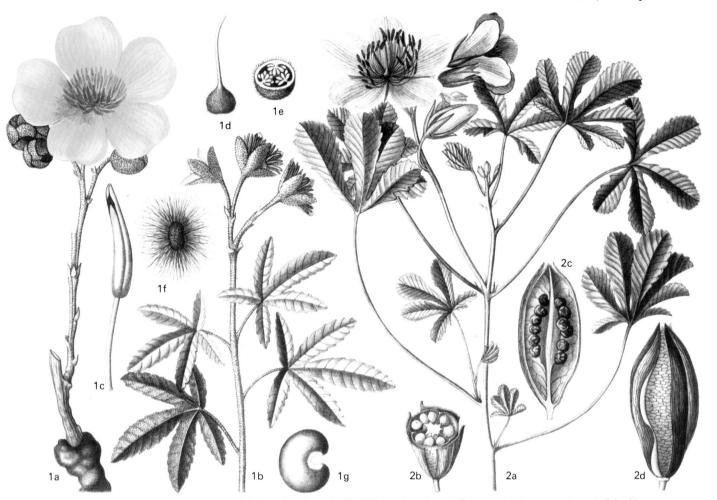

Cochlospermaceae. 1 *Cochlospermum tinctorium* (a) flowering shoot ($\times\frac{2}{3}$); (b) shoot bearing old flowers and alternate, palmately-lobed leaves ($\times\frac{2}{3}$); (c) stamen with anther opening by apical, pore-like slits ($\times 6$); (d) gynoecium ($\times 2$); (e) cross section of unilocular ovary ($\times 3$); (f) seed covered with woolly hairs ($\times 2$); (g) seed with hairs removed ($\times 4$). 2 *Amoreuxia schiedeana* (a) flowering shoot ($\times\frac{2}{3}$); (b) half fruit showing three locules and valves formed by separation of inner and outer layers of the fruit ($\times\frac{2}{3}$); (c) vertical section of fruit ($\times\frac{2}{3}$); (d) entire fruit ($\times\frac{2}{3}$).

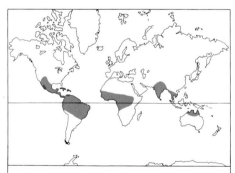

**Number of genera:** 2
**Number of species:** about 38
**Distribution:** tropical, often in drier regions.
**Economic uses:** tropical ornamental and limited uses as kapok, for cordage and as a medicinal gum.

contain colored sap. Some species of *Cochlospermum* have stout tuberous underground stems. The leaves are alternate, palmately lobed and with stipules. The flowers are regular or slightly irregular, bisexual, often showy and borne in racemes or panicles; there are five free, imbricate, deciduous sepals, five imbricate or contorted petals,

and numerous free stamens. The anthers open at the tip by pore-like slits. The ovary is superior, with three to five fused carpels containing a single locule with parietal placentas in *Cochlospermum*, and three locules with axile placentas in *Amoreuxia*. The ovules are numerous. The style is simple, with a minute stigma. The fruit is a three- to five-valved capsule, containing seeds which may be hairy or not, and coiled or straight in shape. The seeds contain oily endosperm; the cotyledons are broad, and the embryo is the same shape as the seed.

**Classification.** The family comprises two genera, *Cochlospermum* (about 30 species) and *Amoreuxia* (eight species, southern United States of America to Peru). The ovary in *Cochlospermum* has a single locule (although there may be more at the apex or base) and the seeds are covered in woolly hairs, while in *Amoreuxia* the ovary is completely three-locular, and the seeds have no hairs or are slightly hairy.

The Cochlospermaceae was once included in Bixaceae, from which it differs in its palmate leaves, three- to five-valved, not two-valved, fruit and oily endosperm.

**Economic uses.** The large, often copious flowers of species of *Cochlospermum* make

them useful, if sometimes gaunt, ornamentals in dry tropical gardens. The Central American species *Cochlospermum vitifolium*, rose imperial, is cultivated in many parts of the tropics, especially in drier places. It grows to about 10m (33ft) with a slender, open-crowned habit. The plants are leafless when the large yellow flowers appear in January to April at the shoot tips. Flowering may occur sporadically through the year in areas far from tropical America. The five- to seven-lobed leaves appear in June. A double-flowered form does not set fruit, but its spectacular 14cm (5.5in) diameter blossoms are larger than the single flowers. The buttercup tree or yellow cotton tree, *C. religiosum*, is a native of India, and has flowers like the rose imperial but often in greater numbers. In addition, in Queensland a red-flowered form of *C. gillivraei* is reported as being showy, as is *C. heteroneurum* from northern Australia. The former yields a useful kapok, as do some other species. The bark fiber of *C. vitifolium* is used locally to make cordage. The tissues of *C. religiosum* contain a fragrant colorless sap; in older trees an amber-colored gum (karaya or kutira gum) with medicinal value may exude.                                        B.M.

Cistaceae. 1 *Fumana procumbens* (a) habit showing simple opposite leaves and solitary flowers ($\times\frac{2}{3}$); (b) flower with petals removed showing outer whorl of articulated sterile stamens and inner whorl of fertile stamens surrounding the elongated style with curved base and discoid stigma ($\times 6$); (c) dehiscing capsule showing three valves containing seeds ($\times 2\frac{2}{3}$). 2 *Tuberaria guttata* habit ($\times\frac{2}{3}$). 3 *Cistus ladanifer* var *maculatus* (a) habit ($\times\frac{1}{2}$); (b) dehiscing capsule with ten valves ($\times 2$); (c) cross section of capsule showing seed ($\times 2\frac{2}{3}$). 4 *C. symphytifolius* cross section of ovary showing projecting placentas bearing numerous ovules ($\times 4$). 5 *Lechea mexicana* half flower with ovary containing two ascending ovules ($\times 8$).

# CISTACEAE

## Rockroses

The Cistaceae is a medium-sized family of shrubs and subshrubs (occasionally herbs) which are characteristic of dry, sunny habitats. They bear large showy, short-lived flowers and many species are grown as ornamentals (rockroses).

**Distribution.** The family occurs in Old World temperate regions, especially the Mediterranean, with some in North and South America.

**Diagnostic features.** The leaves are opposite (rarely alternate: *Halimium*, *Hudsonia*, *Crocanthemum*), simple, usually with stipules, and often bear ethereal oil glands or glandular hairs. The flowers are regular, bisexual and are either solitary or borne in cymes. There are five sepals (two of which are often small and then sometimes regarded as bracteoles) and five, three (*Lechea*) or no petals, which are overlapping (contorted in *Hudsonia*) and often crumpled in the bud. The stamens are numerous, hypogynous and have free filaments, which in some species are sensitive to touch. In *Tuberaria* and *Fumana* the outer stamens are sterile. The ovary is superior, of five to ten or three fused carpels,

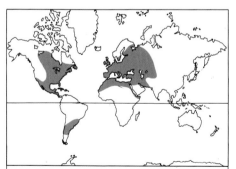

**Number of genera:** 8
**Number of species:** about 165
**Distribution:** chiefly north temperate (especially the Mediterranean region).
**Economic uses:** ornamentals (notably *Cistus*, rockroses) and fragrant ladanum resin.

with one locule having two or numerous ovules on each parietal placenta. There are three to five styles, which may be free or united. The fruit is a loculicidal capsule with five or ten valves (three in *Helianthemum*) containing numerous seeds with endosperm and a curved embryo.

**Classification.** *Cistus* includes about 20 species of evergreen shrubs bearing large

white or pink flowers with crumpled petals. They are often abundant in Mediterranean shrub communities. *Halimium* (nine species) comprises smaller shrubs, all but one with yellow flowers, found mainly in the Iberian Peninsula in similar habitats to *Cistus*. *Tuberaria* (11 species) includes small annual and perennial herbs with yellow flowers; the widespread Mediterranean annual *Tuberaria guttata* extends north in exposed coastal heath to Wales and Ireland. *Fumana* (10 species) and *Helianthemum* (about 70 species) are typically dwarf shrubs with yellow flowers, often growing gregariously in dry, base-rich grassland or on open rocky ground in the Mediterranean region and the neighboring mountains. *Helianthemum nummularium* occurs widely north to Britain and Scandinavia; *H. canum* and *H. oelandicum* form a polymorphic group with a broad but disjunct distribution in the mountains of central and southern Europe with outlying sites reaching to the British Isles, Öland in the Baltic, and Russia. *Crocanthemum* includes about 25 species of dwarf shrubs, with both conspicuous yellow flowers and minute cleistogamous flowers, distributed from the Atlantic coast of the United

Tamaricaceae. 1 *Tamarix aphylla* (a) shoot with minute leaves and dense raceme of flowers ($\times\frac{2}{3}$); (b) part of inflorescence ($\times 4$); (c) stamens and gynoecium ($\times 6$); (d) four-lobed gynoecium ($\times 8$); (e) vertical section of ovary with basal ovules ($\times 10$). 2 *Reaumuria linifolia* (a) leafy shoot with solitary flowers ($\times\frac{2}{3}$); (b) gynoecium ($\times 2$); (c) petal ($\times 1\frac{1}{2}$). 3 *Tamarix africana* habit ($\times\frac{1}{120}$). 4 *Myricaria germanica* (a) flowering shoot ($\times\frac{2}{3}$); (b) flower ($\times 2$); (c) stamens united at the base ($\times 4$); (d) fruit—a capsule ($\times 2$); (e) dehiscing capsule showing cluster of hairy seeds ($\times 2$); (f) seed which is hairy at the apex only ($\times 4$).

States of America to Chile. *Lechea* has about 17 species and *Hudsonia* three species in eastern North America; all are slender subshrubs or herbs with tiny flowers.

Within the Violales, the Cistaceae is most closely allied to the Bixaceae. It also bears a superficial resemblance to the Papaveraceae.

**Economic uses.** Several species and hybrids of *Cistus* and *Halimium*, and variously colored cultivars and hybrids of *Helianthemum nummularium*, are often cultivated and make attractive garden plants. The leaves of several species of *Cistus*, especially *Cistus ladanifer* and *C. incanus* subspecies *creticus*, produce the aromatic resin ladanum, formerly used in medicine. *Cistus salviifolius* has been used in Greece as a substitute for tea. M.C.F.P.

## TAMARICACEAE
### Tamarisks

The Tamaricaceae is a family of small heath-like shrubs or small trees.

**Distribution.** The family is chiefly temperate and subtropical, growing in maritime or sandy places in Norway and from the Mediterranean, North Africa and southeastern Europe through central Asia to India

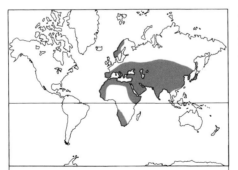

**Number of genera:** 4
**Number of species:** about 120
**Distribution:** mainly temperate and subtropical in maritime and sandy habitats.
**Economic uses:** some ornamental species and others yielding dyes, medicines and tannins from galls.

and China, and in southwestern Africa.

**Diagnostic features.** The plants are halophytes or xerophytes with slender branches bearing alternate, small, tapering or scale-like leaves without stipules. The flowers are minute, regular, bisexual, without bracts and either solitary (as in *Reaumuria*) or in dense spikes or racemes (as in *Tam-*

*arix*). The sepals and petals are both four or five in number and free. The petals and the five to ten or numerous stamens are inserted on a fleshy, nectar-secreting disk. The stamens are free or slightly fused at the base, with bilocular anthers which dehisce longitudinally. The ovary is superior of two, four or five fused carpels and has a single locule, with few to numerous ovules inserted on parietal or basal placentas. There are as many placentas as there are carpels. The styles are usually free and in some species they are absent and the stigmas are sessile (eg *Myricaria*). The fruit is a capsule containing seeds with or without endosperm.

The seeds are sometimes winged but in most species they are covered with long hairs either all over or in a tuft at the apex. The embryo is straight with flat cotyledons.

**Classification.** The four genera are disposed into two tribes, REAUMURIEAE and TAMARISCEAE. The former contains *Reaumuria* (20 species) and *Hololachna* (two species), both of which have solitary flowers and endospermic seeds. The species of *Reaumuria* are halophytic shrubs or undershrubs native to the eastern Mediterranean and central Asia. The solitary flowers are terminal with showy

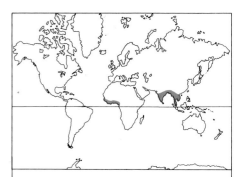

Ancistrocladaceae. 1 *Ancistrocladus vahlii* (a) flowering shoot showing hooked tip to twig, simple, alternate leaves and flowers in a loose inflorescence ($\times\frac{2}{3}$); (b) shoot with fruits enclosed in enlarged wing-like sepals ($\times\frac{2}{3}$); (c) flower with petals, stamens and one sepal removed to show thickened style surmounted by three stigmas ($\times 8$); (d) corolla opened out showing stamens with short fleshy filaments ($\times 8$); (e) stamen with basifixed anthers ($\times 12$); (f) half fruit containing single seed ($\times 1$). 2 *A. heyneanus* (a) flowering shoot ($\times\frac{2}{3}$); (b) flower ($\times 2$); (c) flower opened out ($\times 2$); (d) stamen ($\times 6$); (e) fruit (a nut) surrounded by the persistent calyx ($\times\frac{2}{3}$).

petals, each bearing two longitudinal appendages on the inside. The stamens are numerous and either free or more or less fused into five bundles opposite the five petals. The ovary bears five styles and contains few seeds. The two species of *Hololachna* are native to Central Asia and bear axillary solitary flowers whose petals are devoid of appendages. There are only five to ten stamens, free or shortly connate at the base. The ovary has two to four styles and contains few seeds.

The Tamarisceae consists of *Tamarix* (90 species) and *Myricaria* (10 species), both characterized by their numerous flowers in a spike-like or racemose inflorescence and numerous non-endospermic seeds.

*Tamarix* species are halophytic shrubs and small trees native to Western Europe, the Mediterranean, North Africa, northeast China and India. The inflorescence is borne either on the woody lateral branches or on the terminal young shoots.

*Myricaria* consists of 10 species of undershrubs distributed in Europe, China and Central Asia. The inflorescence is a long terminal raceme and the flowers possess 10 stamens, fused or connate at the base, and

the ovary is crowned by three sessile stigmas.

The Tamaricaceae is probably most closely related to Frankeniaceae. Both share the heath-like shrubby habit, leaves without stipules, regular flowers, sepals and petals, superior ovary with one locule and parietal or basal placentation and the capsular fruit containing seeds with a straight embryo.

**Economic uses.** The twigs of the shrub *T. mannifera* (from Egypt to Afghanistan) yield the white sweet gummy substance manna as a result of puncture by the insect *Coccus maniparus*. Insect galls on species of *Tamarix* (*T. articulata* and *T. gallica*) are a source of tannin, dyes and medicinal extracts. The wood of *T. articulata* is used in North Africa for house construction.

*Tamarix gallica* and *T. africana*, which produce profuse slender branches, are often grown as ornamental shrubs for their rather feathery appearance and their catkin-like inflorescences. *T. pentandra*, a somewhat taller shrub (3.5m–4.5m, 12ft–15ft), produces long spikes of rose-pink flowers ('Rubra' has deep red flowers) and is sometimes grown as a hedge or windbreak, as is *T. gallica* in Mediterranean coastal resorts.                                    S.R.C.

# ANCISTROCLADACEAE

| | |
|---|---|
| **Number of genera:** 1 | |
| **Number of species:** 20 | |
| **Distribution:** tropics. | |
| **Economic uses:** limited local uses. | |

This tropical family of lianas contains the single genus *Ancistrocladus*.

**Distribution.** The 20 species grow in tropical Africa, Sri Lanka, Burma, the eastern Himalayas and southern China, extending to western Malaysia.

**Diagnostic features.** Most are lianas (rarely shrubs), sympodially branched, each branch ending in a coiled hook. The leaves are

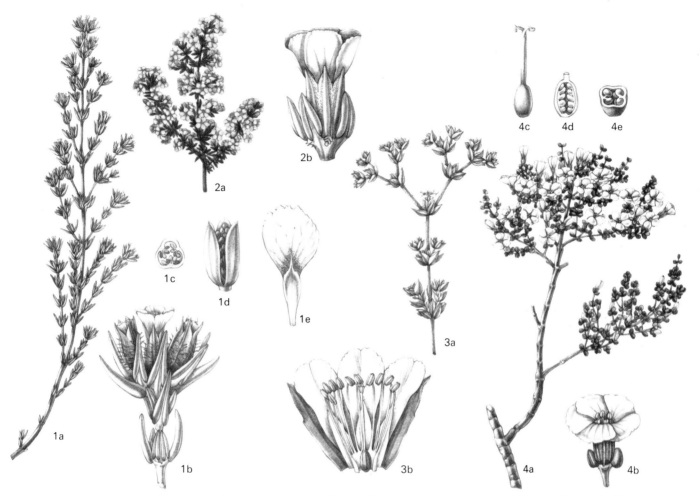

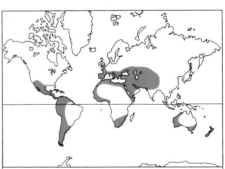

Frankeniaceae. 1 *Frankenia boissieri* (a) flowering shoot showing small heather-like opposite leaves and cymose clusters of flowers ( $\times \frac{2}{3}$ ) ; (b) inflorescence showing folded leaves and flowers with hairy, tubular calyx and regular perianth with free petals ( $\times 3$ ) ; (c) cross section of ovary with single locule and ovules on three parietal placentas ( $\times 8$ ) ; (d) dehiscing fruit—a capsule ( $\times 6$ ) ; (e) petal showing scale on the claw ( $\times 6$ ). 2 *F. laevis* (a) flowering shoot ( $\times \frac{2}{3}$ ) ; (b) flower ( $\times 6$ ). 3 *Hypericopsis persica* (a) flowering shoot ( $\times \frac{2}{3}$ ) ; (b) half flower ( $\times 4$ ). 4 *Beatsonia portulacifolia* (a) flowering shoot ( $\times \frac{2}{3}$ ) ; (b) flower ( $\times 3$ ) ; (c) gynoecium ( $\times 6$ ) ; (d) vertical section of ovary ( $\times 6$ ) ; (e) cross section of ovary ( $\times 6$ ).

simple, alternate and entire, with very small stipules which soon drop; sometimes the stipules are absent. The flowers are small, regular and bisexual with articulated pedicels, arranged in dichotomous cymes with branches curved back. There are five overlapping sepals with a short calyx tube fused to the base of the ovary; in fruit the sepals become unequally enlarged and wing-like. There are five petals which are more or less fleshy, contorted, slightly joined or cohering (joined but apparently free). There are 10 stamens in a single series, rarely five, with short fleshy filaments joined beneath and anthers with two locules dehiscing lengthwise. The ovary comprises three fused carpels and is semi-inferior. It has one locule, containing a single erect, basal ovule. The styles are three, free or joined and with three stigmas. The fruit is dry and woody, surrounded by the wing-like, enlarged and persistent (accrescent) calyx lobes. The seeds have a markedly ruminate endosperm and the cotyledons are deeply folded.

**Classification.** The affinities of the Ancistrocladaceae are very uncertain. It has previously been included in the Dipterocarpaceae and relationships have been suggested with the Violaceae and Ochnaceae.

**Economic uses.** The roots of *Ancistrocladus extensus* are reportedly boiled and used to counter dysentery in Malaya and the young leaves used for flavoring in Thailand.

V.H.H.

## FRANKENIACEAE

Most members of this small family are salt-tolerant herbs (halophytes) of a characteristic appearance with smallish, somewhat heath-like leaves.

**Distribution.** The relatively large type-genus *Frankenia* (about 80 species) is distributed throughout the warm temperate and subtropical regions, although richest in the Mediterranean region and others with a similar climate. Because of their salt-loving physiology members of this family are well represented in arid and maritime environments. Two small genera, *Anthobryum* (four species) and *Niederleinia* (three species), are restricted to South America, while the monotypic *Hypericopsis* occurs in southern Iran.

**Diagnostic features.** The leaves are decussate, simple and entire, without stipules, and often have inrolled margins as an

**Number of genera:** 4
**Number of species:** about 90
**Distribution:** warm temperate and subtropical.
**Economic uses:** limited use as ornamental oddities and local use of *Frankenia* species for fish poison.

adaptation for the desiccating conditions in which they usually grow. The flowers are regular and bisexual, except in the Patagonian *Niederleinia*, where they are unisexual and borne either on the same plant or on separate plants. They are usually arranged in terminal or axillary cymes, although sometimes solitary. The calyx is tubular with four to seven short lobes. The petals, equal in

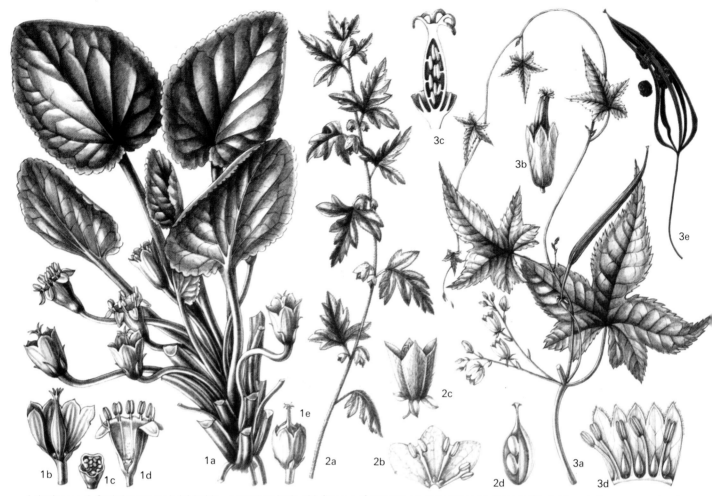

Achariaceae. 1 *Guthriea capensis* (a) habit—a stemless herb with flowers of separate sexes on the same plant ($\times \frac{2}{3}$); (b) female flower opened out showing ovary surmounted by lobed stigma ($\times \frac{2}{3}$); (c) cross section of ovary showing single locule and parietal placentas ($\times 1$); (d) male flower opened out ($\times \frac{2}{3}$); (e) fruit (a capsule) surmounted by the persistent corolla tube ($\times \frac{2}{3}$). 2 *Acharia tragodes* (a) habit—a woody dwarf shrub ($\times \frac{2}{3}$); (b) male flower opened out ($\times 2$); (c) female flower ($\times 2$); (d) capsule with one valve removed ($\times 1$). 3 *Ceratiosicyos ecklonii* (a) habit ($\times \frac{2}{3}$); (b) young female flower ($\times 2$); (c) vertical section of gynoecium ($\times 2$); (d) male flower opened to show fleshy staminodes ($\times 2$); (e) dehiscing capsule ($\times \frac{2}{3}$).

number to the calyx lobes, are free (although with converging claws in *Anthobryum*) and have a long claw in the calyx tube but expanded and spreading outward above. Each petal has a scale at the base of this expanded limb, which is continued down the sides of the claw. There are normally six stamens (about 24 in *Hypericopsis*) in two often unequal whorls, sometimes united at the base of the ovary, sometimes free; the bilocular anthers are turned outward and dehisce by longitudinal slits. The ovary is superior, of two to four united carpels, with a single locule containing several parietal placentas. The few or many ovules are anatropous, ascending or with a long, recurved stalk. The style is simple and has a usually two- or three-lobed stigma. The fruit is a capsule which dehisces lengthwise and remains in the persistent calyx, and the usually numerous endospermic seeds have a straight embryo; *Niederleinia* has a single placenta with one seed.

**Classification.** The family has strong affinities with the Tamaricaceae, one difference being the presence of opposite (decussate) leaves in members of the Frankeniaceae, not alternate as in the Tamaricaceae.

**Economic importance.** The family is of little economic importance; some species are occasionally seen as ornamental oddities. The largest member of the family is sometimes separated as a distinct genus *Beatsonia*, and is a relatively spectacular shrub some 60cm (2ft) high, native to St. Helena. Local exploitation of *Frankenia* species includes the use of *Frankenia ericifolia* in the Macaronesian Islands as a fish poison, and *F. berteroana*, a small shrub in Chile, is burnt and the ash used as a source of salt.

I.B.K.R.

# ACHARIACEAE

The Achariaceae is a very small family of three genera and three species of small shrubs and stemless or slender-stemmed climbing herbs.

**Distribution.** Members of this family are exclusive to South Africa.

**Diagnostic features.** The leaves are alternate or radical, simple, serrated or palmately lobed and without stipules. In *Guthriea* they arise from the crown of the root.

The flowers are borne either singly or in a racemose inflorescence and are regular and unisexual, with both sexes borne on the same

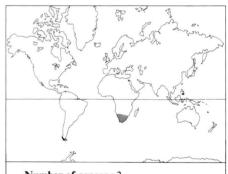

**Number of genera:** 3
**Number of species:** 3
**Distribution:** S Africa.
**Economic uses:** none.

plant. Both male and female flowers have three to five sepals which are fused at the base. The three to five petals are united into a tube-like structure opening out at the top into as many lobes as there are petals. The stamens equal the petals in number with filaments fused to the corolla tube at different levels according to species. The portion of the stamen connecting filament to anther is broadly expanded. In the female flower the ovary is composed of three to five

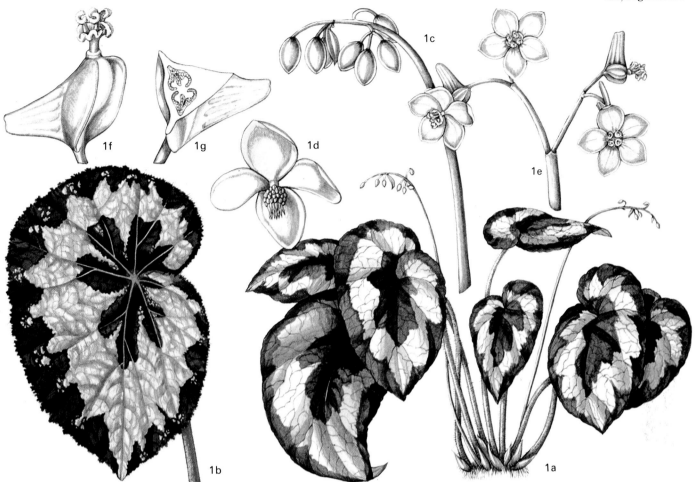

Begoniaceae. 1 *Begonia rex* (a) habit showing leaves with stipules and axillary inflorescences ($\times\frac{1}{3}$); (b) young leaf with one side larger than the other—a characteristic feature ($\times$1); (c) male flower buds ($\times$1); (d) male flower showing four perianth segments and cluster of stamens each with an elongated connective ($\times$1); (e) female flowers showing five perianth segments ($\times\frac{2}{3}$); (f) young winged fruit with persistent styles that are fused at the base and bear twisted, papillose stigmatic surfaces ($\times$2); (g) cross section of young winged fruit with two chambers and numerous seeds on branched, axile placentas ($\times$2).

fused carpels and is superior, with one locule containing few to numerous ovules on three to five parietal placentas. There is a single style surmounted by a two-lobed stigma. There is no indication of a rudimentary gynoecium in the male flower or of rudimentary stamens or staminodes in the female flower. The fruit is a globular capsule which on ripening splits into three to five valves releasing the few to numerous seeds. The seeds have an aril and contain a small embryo surrounded by copious endosperm.

**Classification.** The three genera, *Acharia*, *Ceratiosicyos* and *Guthriea*, have only one species each. They can readily be distinguished from each other by their habit, the persistence of the petals (which may either fall off early or persist as the fruit ripens), the position of the stamens and the shape of the fruit. Both *Guthriea* (a stemless herb) and *Acharia* (a woody shrublet) have stamens inserted on the corolla tube which is persistent, while *Ceratiosicyos* (a slender-stemmed herb) has stamens inserted at the base of the corolla tube which falls off shortly after the anthers have shed their pollen.

This family has a number of features in common with the passion flower family (Passifloraceae) to which it is considered to be related. The presence of both sepals and petals, stamens equal in number to the petals, a unilocular superior ovary with three to five parietal placentas and numerous ovules developing into endospermic seeds are features of both families, which in turn are thought also to show relationships with the gourd family (Cucurbitaceae).

**Economic uses.** There are no known economic uses for any of the three species of Achariaceae.

S.R.C.

# BEGONIACEAE
## Begonias

The Begoniaceae is a family of perennial herbs and shrubs, the great majority being included in the genus *Begonia*, which includes many popular ornamentals grown for their foliage and flowers.

**Distribution.** *Begonia* itself is widespread in the tropics and subtropics. Two of the four satellite genera are South American: *Semibegoniella* (two species, Ecuador) and *Begoniella* (five species, Colombia). The other two genera are Pacific: the monotypic

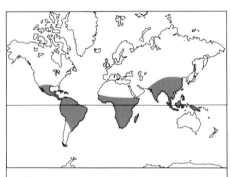

**Number of genera:** 5
**Number of species:** over 900
**Distribution:** tropics and subtropics.
**Economic uses:** ornamentals species (begonias), and limited local uses.

*Hillebrandia* in Hawaii and the 13 species of *Symbegonia* in New Guinea.

**Diagnostic features.** Most species have a characteristic succulent, often jointed, stem, with large membranous stipules. Many also have a thick rhizome or tubers, and some climb with the aid of aerial roots; a few have woody stems and are fibrous-rooted. The leaves, which are usually alternate and arranged in two ranks, are asymmetrical and have large, membranous, deciduous stipules; small axillary bulbs sometimes develop as a form of vegetative reproduction. The cymose, or more rarely racemose, inflorescence bears separate male and female flowers which differ even in the structure of the perianth; the segments are usually free, although united in some of the satellite genera, and they may be regular or irregular. In male flowers there are usually four petal-like segments in two opposite pairs at right angles to one another, the lower, larger ones representing the calyx and the upper the corolla; both pairs are valvate in bud. The female flowers usually have two to five imbricate petal-like segments, although *Hillebrandia* has 10. The stamens in the male flowers are numerous (four in *Begoniella*), free or joined at their bases, and have two-celled anthers often with an elongated connective and opening by longitudinal slits, rarely by pores. The ovary in female flowers is inferior (partly so in *Hillebrandia*), most often winged, usually comprising two or three fused carpels and locules with simple or branched axile placentas bearing numerous anatropous ovules. The styles may be united or joined at their bases, and the stigmas are often twisted and papillose. The fruit is a tough or fleshy, usually winged, loculicidal capsule, and the seeds have a straight embryo and no endosperm.

**Classification.** The most recent studies of *Symbegonia* and *Begoniella* suggest that these should be included within *Begonia*.

The family is a homogeneous assemblage of no obvious affinities. It is usually placed in the Violales but differs from the majority of families in this order on account of the general flower structure (eg ovary inferior). It is probably most closely related to the Datiscaceae.

**Economic uses.** *Begonia* species provide many very popular ornamentals. Two groups of species and hybrids are particularly popular, namely those related to *Begonia rex*, rhizomatous pot plants with hairy, warty and decorated leaves, and those related to *B. semperflorens*, fibrous-rooted bedding plants which are totally hairless and bear numerous flowers. Leaves of *B. tuberosa* in the Moluccas are cooked locally as a vegetable, while others have medicinal uses.                                    I.B.K.R.

# LOASACEAE
*Blazing Star and Rock Nettle*
The Loasaceae is a family comprising mostly herbs with rough, often stinging hairs.
**Distribution.** The family is native to the American tropics and subtropics, with the two species of *Kissenia* native to southern Arabia and southwest Africa.
**Diagnostic features.** The plants are herbs or rarely shrubs, sometimes climbers. They are generally covered with rough hairs, which may be barbed (as in *Petalonyx*, the sandpaper plant of southwestern North America)

and frequently are nettle-like, often delivering a severe sting. The leaves are alternate or opposite, entire or variously divided, without stipules. The flowers are bisexual, regular and either solitary or borne in cymes or heads. The calyx is formed of five, or four to seven, sepals joined at their bases into a tube fused with the ovary. The corolla is formed of four or five petals which are free or rarely fused, often concave. There are usually numerous stamens, sometimes two or five, the inner often staminodes united into a large colorful nectary (as in *Blumenbachia, Loasa*). The ovary is inferior of one or normally three to five fused carpels and has one to three locules containing one to many ovules borne on parietal placentas. The style is simple. The fruit is a capsule. The seeds, with or without endosperm, have a straight embryo.

**Number of genera:** 15
**Number of species:** 250
**Distribution:** American tropics and subtropics, Arabia and SW Africa.
**Economic uses:** garden ornamentals.

**Classification.** The Loasaceae is usually divided into three subfamilies: MENTZELI-OIDEAE, with three to five (rarely six) carpels each with numerous ovules, ten to many stamens, and usually no staminodes; LOAS-OIDEAE, with three to five carpels each with numerous ovules, 12 to many stamens and episepalous staminodes; and GRONOVI-OIDEAE, with one apical ovule and five stamens, some of which are occasionally staminodes.

The Loasaceae is most closely related to the Begoniaceae.
**Economic uses.** Among genera grown as annuals or perennials for their showy flowers are *Loasa, Mentzelia* (blazing star), *Eucnide* (rock nettle), the usually climbing *Blumenbachia* and the twining *Grammatocarpus volubilis*.                            D.M.M.

# DATISCACEAE
*Datiscas*
The Datiscaceae is a small family of tropical and subtropical trees and herbs of little economic value.
**Distribution.** *Datisca glomerata* occurs in dry southwestern North America, *D. cannabina* from the Mediterranean to central Asia, *Octomeles* chiefly in the East Indies, and *Tetrameles* in Indochina and Malaysia.

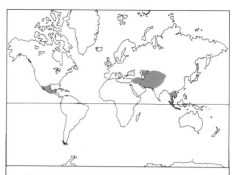

**Number of genera:** 3
**Number of species:** 4
**Distribution:** tropics and subtropics, centered in SW America, W Asia, Indochina, Malaysia and E Indies.
**Economic uses:** some ornamentals.

**Diagnostic features.** The family comprises trees (*Octomeles, Tetrameles*) or perennial herbs (*Datisca*). The leaves are alternate, either pinnate (*Datisca*) or simple, and without stipules. The flowers of *Datisca* are either unisexual with the sexes on different plants, or bisexual and crowded in clusters on long leafy branches. The calyx has three to nine unequal lobes; there are no petals. The stamens are eight to many, with short distinct filaments and large bilocular anthers which dehisce longitudinally. The bisexual flowers often bear staminodes. In the female and bisexual flowers the calyx tube is adnate to the ovary which is inferior formed of three to five fused carpels and a single locule with three to five parietal placentas bearing numerous ovules. There are three to five free thread-like styles which are deeply forked at the apex. The fruit is a membranous capsule bearing seeds with a straight embryo and little or no endosperm.

In *Octomeles* and *Tetrameles* the flowers are unisexual and borne on different plants; they have four to eight calyx lobes and six to eight petals in the male flowers (*Octomeles*) or no petals (*Tetrameles*). There are four to eight stamens with long filaments and bilocular anthers dehiscing longitudinally. The ovary is inferior with one locule and four to eight parietal placentas bearing numerous ovules and either eight (*Octomeles*) or four (*Tetrameles*) styles. The fruit is similar to that of *Datisca*.

**Classification.** Depending on interpretation the family comprises either three genera and four species or one genus (*Datisca*) and two species. It has been suggested that *Octomeles* and *Tetrameles* are related to the Lythraceae, and *Datisca* to the Haloragaceae. There is some dispute about the relationship of *Datisca* to the other two genera, which are sometimes placed in a separate family (Tetramelaceae).

**Economic uses.** The *Datisca* species are sometimes cultivated as ornamentals. The leaves, roots and stems of *D. cannabina* yield a yellow dye once used for dyeing silk.
                                              S.R.C.

# CUCURBITACEAE
*The Gourd or Pumpkin Family*

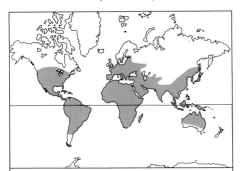

**Number of genera:** about 90
**Number of species:** about 700
**Distribution:** centered in the tropics, with some semidesert species.
**Economic uses:** major food plants, including cucumbers, marrows and squashes, gourds, melons, chayote, pumpkins, courgette and gherkin and many other uses.

The Cucurbitaceae is a medium-sized and botanically highly specialized family of mainly climbing plants. It is of major importance to Man as a source of food.

**Distribution.** The Cucurbitaceae is well represented in the moist and moderately dry tropics of both Old and New Worlds, particularly in rain forest areas of South America and wood-, grass- and bushland areas of Africa. Some species occur in semidesert or even desert vegetation. Cucurbits are poorly represented in Australasia and all temperate regions.

**Diagnostic features.** Members of the family are typically climbing plants with palmately veined leaves, spiraling tendrils, inferior ovaries and unisexual flowers with yellowish petals, as in the familiar cucumber, melon, watermelon, squashes and gourds. The few non-climbing species are probably derivative. Most are perennial herbs with a swollen tuberous rootstock which can be subterranean or wholly or partly superficial. Some are annual, a few are softly woody lianas. The rootstock is formed by swelling of the hypocotyl, which begins early in the seedling stage. The stems of most are characterized by bicollateral vascular bundles with internal as well as external phloem. The leaves are alternate, simple and palmately lobed or palmately compound with three or more leaflets. In most species a solitary, branched or unbranched tendril arises at the side of each leaf base. The tendril tip curls round any suitable nearby object, such as a plant stem; the rest of the tendril then coils in a spring-like manner, drawing the stem in close to its support. The flowers are nearly always unisexual, on the same or different plants. The sepals and petals (usually five of each) are borne at the top of a cup- or tube-like expanded receptacle (hypanthium); the petals are often more or less united at the base. The stamens are distinc-

tive for the union, in various degrees, of filaments and anthers, the reduction of the number of pollen sacs in each anther from four to two, and for the elaboration and convolution of the pollen sacs themselves. Commonly there are three stamens, two double with four pollen sacs each, one single with two pollen sacs, inserted on the lower part of the hypanthium. In other cases, the filaments are more or less completely fused into a single central column. In the female flowers, the ovary is inferior, usually with a single locule with one to three placentas attached to the walls. Ovules and seeds vary from one to many in each fruit; the ovules are anatropous and the seeds, which are without endosperm, usually large. The fruits may be berries, firm-walled berries (such as the melon) known as pepos, fleshy or dry capsules, or leathery and indehiscent.

**Classification.** There are two subfamilies: the Cucurbitoideae with eight tribes, and the single-tribed Zanonioideae.

### SUBFAMILY CUCURBITOIDEAE

Plants with one style, tendrils unbranched or two- to seven-branched from the lower part, spiralling only above the point of branching; the pollen is various and the seeds unwinged.

JOLIFFIEAE. Ovules usually horizontal, hypanthium short; petals fimbriate or with basal scales; pollen reticulate. The tribe includes *Telfairia* (tropical Africa); *Momordica* (Old World tropics).

BENINCASEAE. Ovules horizontal; hypanthium in female flowers short; fruit usually smooth and indehiscent; pollen sacs convoluted and pollen usually reticulate. The tribe includes *Acanthosicyos* (south tropical Africa, with *Acanthosicyos horridus* a spiny shrub of the Namib desert dunes); *Ecballium* (Mediterranean); *Benincasa* (tropical Asia); *Bryonia* (Eurasia); *Coccinia* (Old World tropics); *Citrullus* (Old World tropics and subtropics).

MELOTHRIEAE. Ovules horizontal; hypanthium bell-shaped or cylindrical and alike in both sexes; pollen sacs straight or almost so, and pollen reticulate. Included are the genera *Dendrosicyos* (Socotra, with *Dendrosicyos socotranus* a small succulent-stemmed tree); *Trochomeria* (tropical Africa); *Corallocarpus* (Old World tropics); *Cucumeropsis* (tropical Africa); *Cucumis* (Africa and Asia); *Ibervillea* (southern North America); *Kedrostis* (Old World tropics); *Seyrigia* (Madagascar, leafless succulent lianas); *Zehneria* (Old World tropics); *Gurania* (New World tropics, lianas with a red or orange hypanthium and sepals, pollinated by hummingbirds).

SCHIZOPEPONEAE. Ovules pendulous in a three-locular ovary; three free stamens; fruit dehiscing explosively into three valves; pollen reticuloid. The tribe includes *Schizopepon* (eastern Asia).

CYCLANTHEREAE. One to many ovules, erect or ascending in a unilocular ovary with one to three placentas; stamen filaments united into

a central column; fruit often spiny, usually dehiscent, often explosively; pollen punctate, not spiny. The tribe includes *Apatzingania* (Mexico, fruits geocarpic); *Cyclanthera* (New World); *Elateriopsis* (New World tropics); *Marah* (southwestern United States); *Echinocystis* (North America).

SICYOEAE. Ovule single, pendulous in a unilocular ovary; filaments united into a central column; fruit one-seeded, indehiscent, usually hard or leathery; pollen spiny. Members of the tribe include *Polakowskia* (Central America); *Sechium* (Central America); *Sicyos* (New World, Pacific and Australia).

TRICHOSANTHEAE. Ovules horizontal; hypanthium long and tubular in both sexes; petals fimbriate or entire; fruit fleshy or dry and dehiscent by three valves; pollen striate, smooth or knobbly, not spiny. Members include *Hodgsonia* (tropical Asia); *Peponium* (Africa and Madagascar), *Trichosanthes* (tropical Asia).

CUCURBITEAE. Ovules horizontal or erect; fruit fleshy, indehiscent, and one- to many-seeded; pollen large, spiny, with numerous pores. Included are *Calycophysum* (tropical South America, pollinated by bats); *Cucurbita* (New World, pollinated by specialized solitary bees of the genera *Peponapis* and *Xenoglossa*); *Sicana* (tropical New World).

### SUBFAMILY ZANONIOIDEAE

This subfamily comprises a single tribe, ZANONIEAE, with two or three styles, tendrils two-branched from near the apex, spiraling above and below the point of branching; ovules pendulous; pollen small, striate and uniform; seeds often winged. It includes *Fevillea* (tropical South America); *Alsomitra* (tropical Asia, with *Alsomitra macrocarpa*, a liana with large fruits and large, beautifully winged seeds); *Gerrardanthus* (tropical Africa); *Xerosicyos* (Madagascar, leaf-succulents); *Cyclantheropsis* (tropical Africa and Madagascar, fruit a one-seeded samara); *Zanonia* (Indomalaysia).

The relationships of the Cucurbitaceae are completely obscure, a fact recognized by their frequent classification in a single-family order, the Cucurbitales. They are unrelated to the bulk of families amongst which they were once placed, in or near the Campanulales. They resemble only superficially the Passifloraceae, near which they have been placed by some authors, while the Caricaceae, Loasaceae, Begoniaceae and Achariaceae, though similar in some features, differ widely in others. The Cucurbitaceae are highly specialized in habit, floral structure and biochemistry. Of the two subfamilies, the Zanonioideae is in some ways the less specialized.

**Economic uses.** Major food crops are produced in tropical, subtropical and temperate regions by *Cucurbita* species (pumpkins, squashes, gourds, vegetable marrows, courgettes, vegetable spaghetti), *Cucumis* species (*Cucumis melo*, melon, cantaloupe, honey-

dew, and *C. sativus*, cucumber) and *Citrullus lanatus* (watermelon; citron). Other important food sources are *Cucumis anguria* (West Indian gherkin), *Lagenaria siceraria* (calabash, bottle gourd), *Benincasa hispida* (wax gourd), *Sechium edule* (chayote), *Luffa cylindrica* (loofah) and *L. acutangula*, *Trichosanthes cucumerina* var *anguina* (snake gourd), *Momordica charantia* (bitter melon; balsam apple), *Sicana odorifera* (cassa-

banana), *Cyclanthera pedata* (achocha), *Hodgsonia heteroclita* (lard fruit), *Telfairia occidentalis* (oyster nuts, the seeds yielding an edible oil), *Cucumeropsis mannii* (egussi), and *Praecitrullus fistulosus* (dilpasand, tinda). *Luffa cylindrica* is the source of loofah sponges (dried skeleton of the fruit), while dry fruits of *Lagenaria siceraria* have been used as containers since ancient times. This species is one of the earliest of man's cultivated plants and the only one with an archaeologically documentated prehistory in both Old and New Worlds. Fruits of wild *Citrullus lanatus* (tsamma), *Acanthosicyos naudinianus* and *A. horridus* (narras) are important sources of food and water in the desert areas of southern Africa.

Bitter substances, known as cucurbitacins, are widespread in the family. Many of the edible species occur in both bitter (inedible) and non-bitter (edible) variants.

As ornamentals, Cucurbitaceae are of minor importance; *Cucurbita pepo* produces the ornamental gourds and species of *Momordica*, *Kedrostis*, *Corallocarpus*, *Ibervillea*, *Seyrigia*, *Gerrardanthus*, *Xerosicyos* and *Cyclantheropsis* are sometimes cultivated by succulent enthusiasts.                          C.J.

# SALICALES

## SALICACEAE
*Aspens, Poplars and Willows*
The Salicaceae is a family of mostly north temperate trees and shrubs containing the aspens, the poplars and the willows.
**Distribution.** The family is common throughout the north temperate zone. Two of the genera, *Salix*, the willows, and *Populus*, the aspens and poplars, are well known, but the other two contain very few species restricted to northeast Asia. There are a few Southern Hemisphere and tropical species.

Most willows are shrubs or small trees; few are forest trees and few are ever ecological dominants. Most are scrub or marginal species, being particularly common in wet places and on mountains. Most poplars, on the other hand, are tall trees and some can dominate the landscape in northern areas.
**Diagnostic features.** The leaves are simple, usually alternate, have stipules and are almost invariably deciduous. The flowers are unisexual, male and female flowers borne on separate plants.

Cucurbitaceae. 1 *Gurania speciosa* female flowers ( ×⅔). 2 *Curcurbita moschata* (a) male flower ( ×⅔); (b) cross section of ovary ( ×⅔); (c) female flower with petals and sepals removed ( ×⅔). 3 *Sechium edule* (a) female flower with discoid stigma ( ×1⅓); (b) stamens partly joined in a single column ( ×2); (c) vertical section of ovary with single pendulous ovule ( ×2). 4 *Kedrostis courtallensis* male flower opened out to show two double and one single epipetalous stamens ( ×4). 5 *Trichosanthes tricuspidata* leaf, tendril and female flower ( ×⅔). 6 *Gynostemma pentaphyllum* (a) female flower ( ×6); (b) young fruit with remains of styles ( ×8); (c) leafy shoot with tendrils and inflorescence ( ×⅔). 7 *Zanonia indica* (a) winged seed ( ×¼); (b) fruits ( ×⅔). 8 *Echinocystis lobata* fruit ( ×⅔). 9 *Coccinea grandis* leaves, tendrils, female flowers and fruit ( ×⅔).

Salicaceae. 1 *Populus sieboldii* (a) leafy shoot and pendulous fruiting catkins ( ×⅔); (b) young female catkin ( ×⅔); (c) female flower with cup-like disk ( ×6); (d) ovary ( ×6); (e) stigmas ( ×6); (f) shoot with young male catkin ( ×⅔); (g) male flower ( ×6); (h) mature male catkins together with remains of one from the previous year ( ×⅔). 2 *P. nigra* 'Italica' (Lombardy poplar) habit. 3 *Salix caprea* (a) leaves ( ×⅔); (b) young female catkins ( ×⅔); (c) female flower and bract ( ×6); (d) vertical section of female flower ( ×6); (e) cross section of ovary ( ×8); (f) mature female catkins ( ×⅔); (g) male catkin; (h) male flower ( ×6).

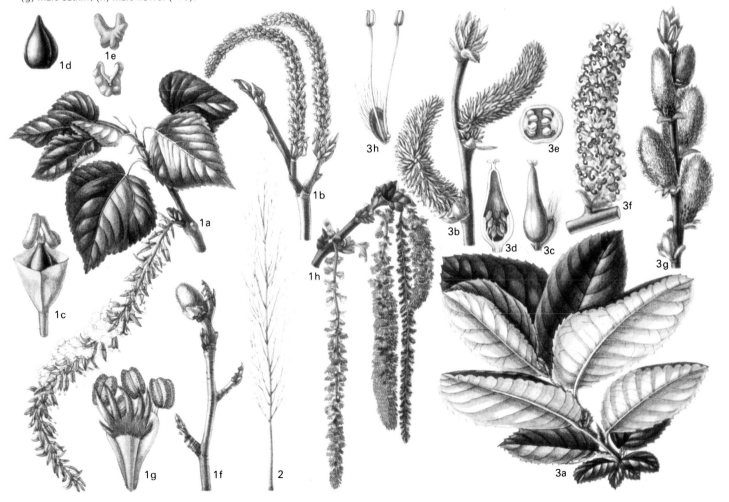

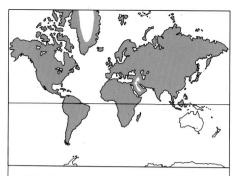

**Number of genera:** 4
**Number of species:** about 350
**Distribution:** temperate with few in tropics and S Hemisphere.
**Economic uses:** wood used for pulp, matches, and boxes, osiers for basket making and some are popular ornamentals.

The flowers are borne in catkins which usually appear before or at the same time as the leaves, in early spring. The individual flowers occur in the axil of a small bract, and lack petals or sepals. The male flowers usually have two up to 30 free or united stamens. The female flower has a bicarpellate, unilocular, superior ovary with numerous anatropous ovules on parietal or basal placentas. The style may be short or long and is often divided.

The method of pollination is markedly different in the two main genera. In *Populus* the catkins are pendulous and scentless; they dangle in the breeze and pollination is effected by wind. There are no nectaries but there is a disk- or cup-shaped gland of unknown function at the base of each flower. In *Salix* the catkins are rigid and furnished with one or two small knob-like glands at the base of each flower. These glands secrete a sweetly-scented nectar and are very attractive to insects, particularly bees and moths, which effect pollination. Willows can be an important source of food for hive bees in early spring, and are also popular with lepidopterists hunting for early moths. Of the other two small genera, *Chosenia* is wind- and *Toisusu* insect-pollinated.

The fruits of all the species are small capsules bearing numerous seeds, each furnished with a tuft of hairs to aid in dispersal by wind. Many species shed the whole catkin once the seed is ripe, often forming cottony drifts under the trees in late spring. The seeds have no endosperm and a straight embryo.

Hybrids are very common in both the poplars and willows, and there is a hybrid in Japan between a species of *Chosenia* and one of *Toisusu*. A great many artificial hybrids have also been synthesized by plant breeders, and, since these hybrids are often subsequently propagated vegetatively, many of them are known as one sex only.

**Classification** The Salicaceae is a taxonomically isolated family. Some authorities suggest that it is a florally reduced derivative of the order Violales, and others propose a closer relationship with the Flacourtiaceae and Tamaricaceae.

**Economic uses.** Although the timber of willows and poplars is not of high quality it is put to a great many uses, and the rapid growth of the plants is an asset. They are important natural resources in some countries and have been extensively planted in others, for example poplars in France. The wood is used principally for pulp, matches and boxes. Willows, particularly the weeping willow, are popular ornamentals, and their supple twigs are used in basket-making. A specialized use of the willow is in the making of cricket bats. Willow bark is used for tanning and some medicines.

C.A.S.

Capparaceae. 1 *Capparis spinosa* (a) leafy shoot with spiny recurved stipules and large, solitary, axillary flowers ($\times\frac{2}{3}$); fruit (b) entire and (c) in cross section ($\times\frac{2}{3}$). 2 *Dipterygium glaucum* winged seed ($\times 6$). 3 *Cleome hirta* (a) leafy shoot with flowers and capsular fruits ($\times\frac{2}{3}$); (b) half flower with toothed sepals, two petals, six curved stamens and ovary with numerous ovules ($\times 2\frac{1}{2}$). 4 *Podandrogyne brachycarpa* fruit—a capsule ($\times 1$). 5 *Buhsia trinervia* dry inflated fruit ($\times 1$).

# CAPPARALES

## CAPPARACEAE
*Capers*

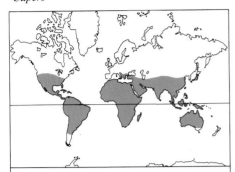

**Number of genera:** 40–50
**Number of species:** about 700
**Distribution:** tropics (particularly Africa) and subtropics.
**Economic uses:** condiments (capers) and garden flowers.

The Capparaceae is a medium-sized family related to the Cruciferae, containing herbs, trees and shrubs and some lianas. Few of its members are of horticultural or economic importance; the best-known are the capers.

**Distribution.** The family is found in the warmer parts of the world, mainly in the tropics and subtropics of both hemispheres and in the Mediterranean. It is well represented by about 15 genera in Africa where it forms a conspicuous element of the flora of dry regions.

**Diagnostic features.** The leaves are alternate, rarely opposite, and simple or palmate or digitate. They have two to seven leaflets and minute or spiny stipules which are persistent or caducous. The inflorescences are terminal or axillary, and may be racemose or corymbose, or the flowers may be solitary or fascicled and often showy. The flowers are bisexual or rarely unisexual (the sexes then on separate plants). They are usually irregular, with four to eight sepals which are free or variously joined, in some cases forming a hood which dehisces and falls off at flowering time. There are four to sixteen petals (sometimes absent), which may be equal, or the two posterior ones larger. The stamens range from four to many: when there are six stamens, four of them may be considered as derived from the splitting of the middle pair during development; higher numbers are similarly derived, and then many of the filaments lack anthers. Often there is a development of the axis between the petals and the stamens producing an internode called an androgynophore, bearing the androecium and gynoecium. The ovary may be sessile, but is more usually borne on an internodal development between the androecium and gynoecium called a gynophore; it is superior, of two fused carpels, and has one locule with parietal placentas, or may be divided by false septa into two or more locules; the ovules are few to numerous. There is a single style and the stigma is bilobed or capitate. The fruit is a capsule dehiscing by valves, sometimes with the appearance of a siliqua, or a round to cylindrical berry, rarely a single-seeded indehiscent nut. The seeds are without endosperm and the embryo is variously folded. A curious feature is the development of the axis in the form of swellings, disks or tubular structures, sometimes inside, sometimes outside the corolla. Pollination is by insects, and possibly by bats in some South American species.

**Classification.** It is generally agreed that the Capparaceae is allied to the Cruciferae and may have evolved from a common ancestor. Relationships with the Papaveraceae and Resedaceae have also been suggested. The Capparaceae contains many woody members, and the anatomy of the wood shows several advanced features.

**Economic uses.** Capers of commerce are derived from the flower buds of *Capparis* species. Several species are cultivated as garden plants, especially *Cleome spinosa*, the spider flower, a strongly scented annual with white or pink flowers. Species of *Capparis, Gynandropsis* and *Polanisia* are also grown occasionally as garden plants.

V.H.H.

## TOVARIACEAE

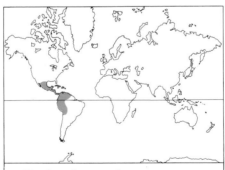

**Number of genera:** 1
**Number of species:** 2
**Distribution:** tropical America and the Caribbean.
**Economic uses:** none.

The Tovariaceae is a small family consisting of one genus (*Tovaria*) and two species of tropical shrubs or annual herbs.

**Distribution.** The family is native to tropical America and the Caribbean area.

**Diagnostic features.** The two species are herbs or shrubs with pungent smelling parts, and alternate, trifoliolate leaves without stipules. The flowers are regular, bisexual, hypogynous in lax terminal racemes, and consist of eight narrow sepals with overlapping edges, eight petals and eight free stamens with hairy filaments, dilated near the base. The ovary is superior and has six to eight fused carpels, with six to eight locules containing numerous ovules borne on axile placentas; the locules are formed by membranous dividing walls. The style is short with a lobed stigma. The fruits are berries, mucilaginous when young, but with a membranous outer coat. They are about 1cm (0.4in) in diameter. The many small, shiny seeds have curved embryos and a sparse endosperm.

**Classification.** Both species are green-barked shrubs which sometimes behave like annuals. In *Tovaria pendula*, which occurs in Peru, Bolivia and into Venezuela, the flowers and fruits are greenish, the anthers brown or yellow. The other species, *T. diffusa*, grows in dense, wet thickets in the mountains of Central America and the West Indies, and has pale green or yellow flowers and very long sparsely-flowered racemes.

The Tovariaceae is related to the Capparaceae, but has fruits reminiscent of certain members of the Phytolaccaceae.

**Economic uses.** No economic uses have been recorded for this family. B.M.

## CRUCIFERAE
*The Mustard Family*

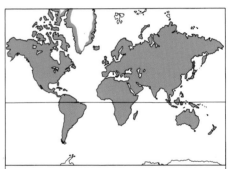

**Number of genera:** about 380
**Number of species:** about 3,000
**Distribution:** cosmopolitan, mostly temperate, with centers in the Mediterranean region, SW and central Asia.
**Economic uses:** many important vegetable, fodder and oilseed crops and popular garden flowers.

The Cruciferae is a large natural family of major economic importance, containing a wide array of crop plants grown as salads, vegetables, for oilseed, animal feed and condiments, and several well-known garden ornamental plants such as the wallflower, honesty and aubretia.

**Distribution.** Members of the family are found in most parts of the world but are mainly concentrated in the north temperate region and more especially in the countries surrounding the Mediterranean basin and in southwestern and central Asia, where more genera occur than anywhere else in the world. The family is only sparingly represented in the Southern Hemisphere, and there are very few species in tropical regions. In the Mediterranean area 113 genera occur of which 21 (17%) are endemic, and 625 species of which 284 (45%) are endemic. The Irano-Turanian region has 147 genera of which 62 (42%) are endemic and 874 species of which 524 (60%) are endemic, while in the Saharo-Sindian region there are 65 genera, 19 (30%) being endemic and 180 species, 62 (34%) of which are endemic. Two of the tribes (see below) are confined to South Africa – the Chamireae containing a single species, *Chamira circaeoides*, and the Heliophileae comprising mainly the genus *Heliophila* with about 70 species mostly confined to the winter rainfall area around the Cape. Another tribe, the Pringleeae, contains as its sole member the species *Pringlea antiscorbutica*, the Kerguelen Island cabbage, found only on the remote islands of Kerguelen and Crozet in the Southern Hemisphere.

**Diagnostic features.** The Cruciferae are mostly annual to perennial herbs, rarely small shrubs such as *Alyssum spinosum* or tall shrubs reaching up to 2m (6.5ft) in height (eg *Heliophila glauca* from South Africa), and very rarely "climbers" such as *H. scandens* which attains a height of 3m (10ft). An unusual aquatic species is *Subularia aquatica,* a usually submerged annual with long narrow leaves, circular in cross-section. The curious hummocks of *Xerodraba pycnophylloides* from the Andes of Argentina are known as vegetable sheep. The stems become spiny in *Vella* and some *Alyssum* species. A very curious habit is shown by *Anastatica hierochuntica*, rose of Jericho, which becomes detached from the soil when the seeds begin to ripen in the dry season; the leaves fall off and the branches inroll so that the whole plant becomes an intricate ball which is blown about, containing the enclosed seed pods, until it reaches a wet area suitable for germination.

The leaves are usually alternate and without stipules. The hairs of the indumentum vary from simple to forked, many-branched, star-like or peltate, features which are useful for identification of genera and species.

The inflorescence is usually a raceme or corymb, usually without bracts or bracteoles. The basic floral structure is highly characteristic and constant: four sepals, four cruciform petals, six stamens, (four long and two short) and an ovary with two parietal placentas. There are, however, some exceptions to this plan. The flowers are usually bisexual, regular and hypogynous. There are four sepals; sometimes the inner ones are convexly swollen at the base and contain nectar secreted by the nectaries at the base of the stamens. The petals are four, arranged in the form of a cross (cruciform, hence the name of the family), rarely absent as in some *Lepidium* and *Coronopus* species, free, often clawed, imbricate or contorted. In a few genera such as *Teesdalia* and *Iberis* (candytuft) the outer petals are radiate and larger than the inner. The stamens are typically six, and tetradynamous, that is one outer pair with short filaments, and two inner pairs, one

posterior, one anterior, with long filaments. There may be only four stamens in some species of *Cardamine* and up to 16 in *Megacarpaea*. The filaments are sometimes winged or with tooth-like appendages. The shape and disposition of the nectaries at the base of the stamens is variable and widely used in the classification of the family. The nectaries appear like swellings or little cushions. The ovary is superior, of two carpels and is syncarpous, with two parietal placentas, usually with two locules through the formation of a membranous false septum or replum by the union of outgrowths from the placentas; sometimes the ovary is transversely plurilocular. The stigma is capitate to bilobed.

As characteristic as the flower is the fruit, which is basically a bilocular capsule with a false septum (replum), usually dehiscent, opening by two valves from below. When it is at least three times as long as wide it is called a siliqua, and when less than three times as long as wide, a silicula. The fruit may sometimes be indehiscent, breaking into single-seeded portions; rarely it is transversely articulate with dehiscent and indehiscent segments, sometimes breaking at maturity into single-seeded portions (lomentum). The fruits range from linear-oblong to ovate to spherical; they may be winged or not and stalked or not; the seeds may be in one or two rows.

The range of variation shown in fruit types is vast and fruit characters are relied upon very extensively in the classification of the family at tribal, generic and specific level. Examples of unusual or anomalous fruits are *Cakile* which has siliquas which divide into two single-seeded joints, the lower sterile and forming a thick stalk, the upper indehiscent, globose and single-seeded; *Lunaria* (honesty) in which the silicula is flattened laterally to give a very broad septum; or the silicula may be compressed anteriorposteriorly to give a very narrow septum as in *Capsella* (shepherd's purse). *Geococcus pusillus* from Australia buries its fruits by sharply reflexing its pedicels at fruiting time, thus forcing the fruit into the soil. Similarly *Morisia hypogea* from Sardinia and Corsica is a stemless species whose peduncles bend downwards after flowering and bury the closed pod in the ground. The seeds are nonendospermous and the testa often contains mucilaginous cells of various types which swell up when wetted and produce a halo of mucilage. The ovules are campylotropous, the embryo being curved with the radicle in one half of the seed and the cotyledons in the other.

Great taxonomic importance is attached to the shape of the embryo and the position of the radicle relative to the cotyledons. The main types recognized are: (1) notorhizal with the radicle incumbent, ie lying on the back of one cotyledon, the cotyledons not being folded on themselves; (2) pleurorhizal,

with the radicle accumbent, against the edges of the cotyledons; (3) orthoplocous, with the cotyledons conduplicate; (4) spirolobous, as in (1) but with the cotyledons once folded; (5) diplecolobous, as in (4) but with the cotyledons folded twice or more.

**Classification.** Various attempts have been made to produce a natural subdivision of the family into tribes, using fruit characters, embryo features, nectary glands, distribution of myrosine cells in the embryos and other such features, but the most widely used is that proposed by O. E. Schulz in a posthumous classification published in 1936. This classification was based on a wide range of features. Various modifications have been suggested, notably by Janchen, and it has to be admitted that several of the tribes are far from being satisfactory. Only two of the tribes, the Brassiceae and the Lepideae, can be regarded as natural, apart from the monotypic ones (Pringleae and Chamireae) which are confined to South Africa. The tribes usually recognized are:

THELYPODIEAE. *Stanleya, Macropodium*
PRINGLEEAE. *Pringlea.*
SISYMBRIEAE. *Sisymbrium, Braya, Alliaria, Arabidopsis.*
HESPERIDEAE. *Hesperis, Cheiranthus, Matthiola, Anastatica.*
ARABIDEAE. *Arabis, Aubrieta, Barbarea, Cardamine, Armoracia, Nasturtium, Isatis, Rorippa.*
ALYSSEAE. *Alyssum, Lunaria, Lesquerella, Draba, Berteroa.*
LEPIDIEAE. *Lepidium, Cochlearia, Camelina, Capsella, Iberis, Biscutella, Thlaspi.*
BRASSICEAE. *Brassica, Raphanus, Sinapis, Diplotaxis, Crambe, Rapistrum, Cakile, Morisia, Eruca, Moricandia.*
CHAMIREAE. *Chamira.*
SCHIZOPETALEAE. *Schizopetalum.*
STENOPETALEAE. *Stenopetalum.*
HELIOPHILEAE. *Heliophila.*
CREMOLOBEAE. *Cremolobus, Hexaptera.*

The closest ally to the Cruciferae is generally accepted to be the Capparaceae with which it shows close similarity in the androecium, gynoecium and other features. The genus *Cleome* in the Capparaceae is

Cruciferae. 1 *Iberis pinnata* leafy shoot and inflorescence with flowers having outer petals longer than the inner ($\times \frac{2}{3}$). 2 *Heliophila coronopifolia* shoot with leaves, flowers and fruit ($\times \frac{2}{3}$). 3 *Moricandia arvensis* (a) shoot with sessile leaves, flowers and fruit ($\times \frac{2}{3}$); (b) half flower showing stamens with long and short filaments ($\times 3$). 4 *Biscutella didyma* var *leiocarpa* (a) shoot with leaves, flowers and fruit ($\times \frac{2}{3}$); (b) fruit – a silicula ($\times 4$). 5 *Crambe cordifolia* spherical fruit ($\times 4$). 6 *Isatis tinctoria* fruit—a siliqua ($\times 2$). 7 *Lunaria annua* fruit—a flattened silicula ($\times 1$). 8 *Capsella bursapastoris* dehiscing fruit – a silicula with a narrow septum ($\times 6$). 9 *Berteroa incana* fruit – a silicula dehiscing from the apex ($\times 4$). 10 *Thlaspi arvense* cross section of bilocular ovary showing false septum ($\times 12$). 11 *Cheiranthus cheiri* fruit – a siliqua ($\times 1$).

especially close to members of the Cruciferae. Chemically the two families also show similarities. On the other hand there is no reason to regard the Cruciferae as having been derived from the Capparaceae, as has been suggested in the past, and the two families are better regarded as both having evolved from a common stock.

**Economic uses.** While the Cruciferae contains a considerable number and diversity of crop plants, it is not comparable with, say, the Leguminosae or Gramineae, and although the crop species are mainly grown as food plants they do not form a substantial part of staple diets. Many cruciferous species are used as condiments or garnishes, such as mustard and cress, and many are collected from the wild rather than cultivated. Many cruciferous crops have been cultivated since ancient times, such as *Brassica oleracea*; the ancestral cabbage was cultivated about 8,000 years ago in coastal areas of northern Europe whence it was introduced into the Mediterranean and eastern Europe. The first selection of sprouting broccoli was probably made in Greece and Italy in the pre-Christian era.

All important cruciferous crops are propagated from seed; only minor crops such as watercress, horseradish and seakale are vegetatively propagated.

The seed crops can be divided into oils and mustard condiments; forage and fodder crops; and vegetables and salads for human consumption. Cruciferous oil seeds now rank fifth in importance behind soybeans, cotton seed, groundnut and sunflower seed. The main crops are derived from *Brassica campestris* (*B. rapa* (oilseed rape)) but *B. juncea* is important in Asia and *B. napus* (oilseed rape, colza) is cultivated in temperate Europe and Asia. Mustard is obtained from the ground seed of *Brassica juncea*, *B. nigra* and *Sinapis alba*.

Animal feeds are supplied by cruciferous crops in the form of silage, seed meal left over after oil extraction, forage crops grazed in the field, and stored root fodder, used for winter feeds. The characteristic glucosinolates produced by Cruciferae affect the use of many species economically. Glucosinolates are the precursors of the mustard oils which are responsible for the pungency of most crucifers. Although desirable in the case of some crops such as mustard, radish and horseradish, they may also be responsible for toxic manifestations when used as animal feed or in human nutrition. Because of this the production of silage from crucifers is limited. Seed meal is obtained from species such as *Brassica napus*, especially races with a low glucosinolate content. Forage and fodder crops of crucifers are restricted mainly to countries such as Britain, the Netherlands and New Zealand which specialize in intensive small-scale farming of ruminants. The range of species used as forage crops includes *Brassica*

*oleracea* (kale, cabbage), *B. campestris, B. napus* (rapes) and *Raphanus sativus* (fodder radish); fodder crops include those with swollen stems or root storage organs such as *B. oleracea* (kohlrabi), *B. campestris* (turnip) and *B. napus* (swede).

Considerable proportions of the vegetable crop acreage in Europe and Asia are formed by cruciferous species – some European countries have up to 30% of their vegetable acreages devoted to crucifers compared with 6% in the United States of America. There are some curious geographical differences in what is cultivated, reflecting more national taste rather than the geographical origins of the crops. Thus the Brussels sprout is very much a British crop, with the British production equalling that of the rest of Europe and ten times that of the United States. Likewise cauliflowers are mainly a European crop. The most important species are *Brassica oleracea*, cultivars of which produce kales, Brussels sprouts, kohlrabi, cabbage, broccoli, calabrese and cauliflower, and *B. campestris* which produces turnip, Chinese cabbage, etc.

Ornamental genera include wallflower (*Cheiranthus*), honesty (*Lunaria*), candytuft (*Iberis*), sweet alysson (*Lobularia maritima*), golden alyssum (*Alyssum* spp.), stocks (*Matthiola*), rocket (*Hesperis*), rock cress (*Arabis*), *Draba, Aethionema Erysimum* and *Aubrieta*.                                  V.H.H.

# RESEDACEAE
*Mignonette*

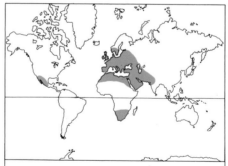

**Number of genera:** 6
**Number of species:** about 75
**Distribution:** centered in the Mediterranean region.
**Economic uses:** minor uses include sources of a yellow dye, perfume oil and an ornamental (*Reseda odorata*).

The Resedaceae is a small family of herbs and shrubs mostly of dry places, and contains some ornamentals.

**Distribution.** Centered on the Mediterranean region, the family extends into parts of northern Europe, and eastward to Central Asia and India. *Oligomeris* (nine species) is widely distributed, with outliers in South Africa, the Canary Islands and a single species in the southwestern United States and Mexico. *Caylusea* (three species) spreads from the Cape Verde Islands across northern

Africa to India. *Reseda* (60 species), by far the largest genus, is restricted to Europe and the Mediterranean region to Central Asia.

**Diagnostic features.** The leaves are alternate and entire or divided; glandular stipules are present. The flowers are irregular, usually bisexual, arranged in a bracteate raceme or spike. There are two to eight sepals which are usually free, sometimes unequal. The petals (occasionally absent) also number two to eight but are not always equal in number to the sepals; they are mostly broadly clawed, with a scale-like appendage at the base and usually a more or less deeply cut lamina. The stamens and ovary are often on a short androgynophore, an upward extension of the receptacle, and there is usually an irregular disk outside the three to 45 stamens. The ovary is superior, of two to seven more or less fused carpels, which are open above, and one locule in which the ovules are usually on parietal placentas. The fruit is usually an indehiscent open capsule, rarely of separate, spreading carpels; *Ochradenus* has a berry. The seeds are kidney-shaped, with a caruncle, curved embryo and very little endosperm.

**Classification.** The family is divisible into three tribes, based mainly on the placentation. Thus *Sesamoides*, the only genus in the ASTROCARPEAE, has one, rarely two pendulous ovules in the center of the abaxial wall of the almost free carpels. The CAYLUSEEAE, again with a single genus (*Caylusea*), have 10–18 erect ovules on a united basal placenta. The rest of the genera belong to the RESEDEAE, with numerous, pendulous ovules on parietal marginal placentas.

The family is generally considered to be allied to the Cruciferae and Capparaceae.

**Economic uses.** *Reseda odorata* (mignonette) provides a perfume oil and is grown for ornament and *R. luteola* yields a reddish yellow dye.                                  I.B.K.R.

# MORINGACEAE

The Moringaceae is a family of small, quick-growing deciduous trees. They are gummy, pale-barked and their stems are often thickened with myrosin cells.

**Distribution.** The family has a distribution from the Mediterranean and North Africa to the Arabian peninsula and India, and is also represented in southwestern Africa and Madagascar.

**Diagnostic features.** The very graceful leaves are two or three times pinnate and alternate; stipules may be present or replaced by stipitate glands. The numerous flowers are irregular, bisexual, sweet-scented, cream or red, and produced in axillary panicles. The petals are unequal, five in number and slightly longer than the sepals, which have five short, unequal, spreading, reflexed lobes. Five stamens alternate with five staminodes, and all are joined at the base into a cupular disk. The anthers have one locule which

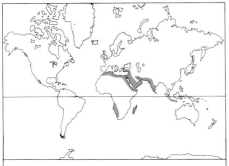

**Number of genera:** 1
**Number of species:** 12
**Distribution:** mainly Mediterranean to India and N Africa.
**Economic uses:** source of ben oil and some cultivated as ornamentals.

**Classification.** The family is represented by only one genus, *Moringa*, with 12 species including *Moringa oleifera*, the ben oil or horseradish tree. It forms a link between the Capparaceae and Leguminosae.

**Economic uses.** The seed of *M. oleifera* is the source of ben oil, formerly used only as a lubricant by watch-makers and others, but now used as a salad oil and in soap manufacture. The young swollen roots are eaten as a vegetable, the similarity of which to horseradish accounts for the common name attributed to the species. All species grow rapidly from seed and are used as boundary markers; some are grown as ornamentals.                    S.A.H.

# ERICALES

## CLETHRACEAE
*Lily-of-the-valley Tree*

The Clethraceae is a family of tropical and subtropical evergreen or deciduous shrubs, represented by one genus, *Clethra*, with about 120 species. "Clethra" is the Greek name for alder, and is applied to this genus on account of the resemblance of some

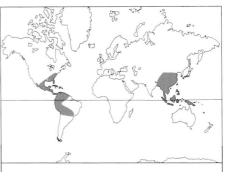

**Number of genera:** 1
**Number of species:** about 120
**Distribution:** tropical and subtropical America and Asia; Madeira.
**Economic uses:** some ornamentals.

species of *Clethra* to those of alder (*Alnus*).

**Distribution.** The family is found in tropical and subtropical Asia and America, and also Madeira.

**Diagnostic features.** The majority of the members of the family, excluding *Clethra arborea*, attain only large shrub-size and bear alternate, simple leaves without stipules. The flowers are bisexual, regular, white and borne in racemes or panicles, without

dehisces by slits. The ovary is superior, of three fused carpels, usually becoming unilocular with parietal placentas bearing numerous biseriate pendulous anatropous ovules. There are two to four slender styles. The fruit is an elongated pod-like capsule with three valves, and contains many black, rounded, winged or wingless seeds without endosperm and a straight embryo.

Resedaceae. 1 *Randonia africana* (a) shoot bearing flowers and fruit ($\times\frac{2}{3}$); (b) flower showing petals with incised margins ($\times 4$); (c) vertical section of flower showing numerous stamens and superior ovary with ovules on axile placentas ($\times 4$); (d) fruit—a capsule ($\times 4$). 2 *Sesamoides canescens* (a) leafy shoot, inflorescences and fruits ($\times\frac{2}{3}$); (b) flower with equal green sepals and both incised and linear petals ($\times 10$); (c) dehiscing fruit ($\times 8$). 3 *Reseda villosa* (a) shoot bearing fruits ($\times\frac{2}{3}$); (b) tip of inflorescence ($\times\frac{2}{3}$); (c) vertical section of flower showing sessile stigma on top of ovary ($\times 3$); (d) flower showing small petals ($\times 2$); (e) dehisced fruit with apical opening ($\times 3$).

bracteoles. There are five sepals, five free petals and two whorls each of five stamens; the anthers are bent outwards in bud and open by pores. The ovary is superior, of three fused carpels and contains three locules and numerous ovules on axile placentas in each locule. The style is three-lobed. The fruit is a loculicidal capsule with many, often winged seeds that have fleshy endosperm and a cylindrical embryo.

**Classification.** The family is closely related to the Ericaceae and Cyrillaceae.

**Economic uses.** The best-known ornamental is *C. arborea*, the lily-of-the-valley tree of Madeira, which grows larger than other members of the family and can form a multi-stemmed tree about 8m (25ft) high. In late spring and early summer, the tree is a cascade of fragrant white bell-like flowers, borne in terminal drooping panicles up to 15cm (6in) long. The flowers are similar to those of *Erica*, and borne on delicately spikey stalks, giving the general appearance of a large specimen of the lily of the valley. The leaves are similar to those of *Rhododendron*, about 5cm–10cm (2in–4in) long and half as wide, with serrated edges and woolly beneath. The young shoots are covered with fine hairs and are rust-colored.

Several other species including *C. arborea, C. alnifolia* (sweet pepper bush), *C. acuminata* (white alder), *C. monostachya* and *C. tomentosa*, are also known in cultivation as ornamental, fragrant-flowered shrubs.

S.A.H.

# GRUBBIACEAE

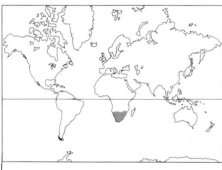

**Number of genera:** 2
**Number of species:** 5
**Distribution:** S Africa.
**Economic uses:** none.

The Grubbiaceae is a family of two genera and five species of small heath-like shrubs.

**Distribution.** The family is native to South Africa.

**Diagnostic features.** The leaves are opposite and in four rows, simple, narrow and without stipules.

The flowers are bisexual, small and sessile. They are arranged in small cone-like inflorescences in the axils of the leaves. The structure of the flower is variously interpreted but is probably as follows: two bract-like sepals and two vestigial sepals,

petals absent or in a 2+2 arrangement surrounding eight stamens. The stamens are in two whorls, sometimes with the filaments laterally compressed. The anthers have two locules and dehisce laterally, by reflexion of the pollen sac wall. The flower has an annular, hairy disk. The ovary is inferior, composed of two fused carpels, and has two locules, each with a single pendulous, anatropous ovule. The style is simple, terminating in a simple bifid stigma. The fruit is a drupe. In some cases the ovaries of several adjacent flowers become coherent or connate, in which case the fruit is a drupe-like or achene-like syncarp. There is only one seed per carpel in the syncarp and it possesses a thin testa and fleshy or oily endosperm surrounding a central linear embryo.

**Classification.** According to some systems this family is composed of only a single genus, *Grubbia*, with five species, but one, *Grubbia strictus*, is considered sufficiently distinct to warrant a separate generic status as *Strobilocarpus*.

It has been suggested that the Grubbiceae could be related either to the Olacaceae or to the Santalaceae. It shares with both these the features of woody habit, simple leaves, small perianth, nectar-secreting disk and endospermic seeds. However, there is perhaps a stronger affinity with the Empetraceae, which is another family of heath-like shrubs, with linear leaves and no stipules, bracteate flowers, solitary ovule in each locule of the ovary, and an endospermic seed with a straight embryo.

**Economic uses.** No economic uses for this family have been recorded.

S.R.C.

# CYRILLACEAE
*Leatherwood and Buckwheat Trees*

The Cyrillaceae is a small family of three genera of deciduous or evergreen shrubs or small trees.

**Distribution.** Two of the three genera, *Cyrilla* and *Cliftonia*, each containing one species, are native to southeast North America. The 12 species of *Purdiaea* are found mostly in Cuba, but extend into Central and South America.

**Diagnostic features.** The leaves are alternate, simple, and without stipules. The flowers are regular, bisexual and arranged in racemose inflorescences. The five sepals are separate or fused at the base, persistent and often enlarged in the fruit. The five petals are equal, free or fused at the base, and may be imbricate.

The stamens are normally ten, in two whorls of five, but the inner whorl may be replaced by sterile staminodes or in some species may be absent altogether. They have free filaments and are inserted on the receptacle, surrounding a superior ovary formed of two to five fused carpels. The ovary has two to four locules with one to numerous pendulous ovules, borne on axile placentas. The ovary is surmounted by a

**Number of genera:** 3
**Number of species:** 14
**Distribution:** SE N America, Central and S America.
**Economic uses:** a few ornamentals.

short style (absent in some species) and there are one to three narrow stigmas. The fruit is a capsule or drupe and has two to four wings in those species where the calyx is enlarged. The seeds have a small straight embryo, enclosed in fleshy endosperm.

**Classification.** The genera can be separated on the presence or absence of an enlarged calyx at fruiting, and on details of floral structure such as stamen number and nature of the style. Thus *Cliftonia* is characterized by possessing 10 stamens, deciduous calyx (not enlarged after flowering) and very short style divided into three stigmas. *Purdiaea* also possesses 10 stamens, but the calyx becomes enlarged, enveloping the fruit, and the style is slender and undivided. On the other hand *Cyrilla* has only five stamens and a thick, short style ending in a two- or three-lobed stigma.

The family is often regarded as belonging to the Ericales. However, it is distinguished from the rest of the order by the presence of racemose inflorescences, the slight degree of fusion of the petal bases, and by the winged capsular or drupe-like fruit and is sometimes placed in another order, the Celastrales.

**Economic uses.** There is no economically important species in this family but two shrubs, the leatherwood (*Cyrilla racemosa*) and the buckwheat tree (*Cliftonia monophylla*) possess attractive white flowers and reddish-tinted autumnal foliage, and are grown as ornamentals in gardens.

S.R.C.

# ERICACEAE
*The Heath Family*

The Ericaceae is a large family, mainly of shrubs, containing many well-known genera such as *Rhododendron, Erica* (heath), *Calluna* (heather), *Vaccinium* (blueberries, cranberries etc) and *Gaultheria* (wintergreen).

**Distribution.** Considered overall, the family is found in almost all parts of the world. It is, however, absent from most of Australia where it is largely replaced by the related family Epacridaceae. The distributions of some of the genera are of more interest than that of the family as a whole. The two largest

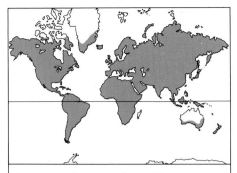

**Number of genera:** about 100
**Number of species:** about 3,000
**Distribution:** cosmopolitan with large concentrations in the Himalayas, New Guinea and southern Africa.
**Economic uses:** many important horticultural species (rhododendrons, azaleas, heaths and heathers), some edible berries (cranberries and bilberries) and occasional use for timber.

genera are *Rhododendron*, with about 1,200 described species, and *Erica* with over 500 species. Both of these genera show remarkable concentrations of species in relatively small areas. *Rhododendron* has a massing of more than 700 of its species in the area where China, Tibet, Burma and Assam all meet – the area in which the great rivers of eastern Asia (the Brahmaputra, Irrawaddy, Salween, Mekong and Yangtse) break through the Himalayan chain. The genus also has a secondary center of almost 300 species in New Guinea. The rest of the species are found in the Himalayas and Japan, with a small number occurring in Europe, southern Asia and the United States of America. *Erica* has an even more remarkable concentration of species in southern Africa, where about 450 occur, many of them restricted to the Cape Province. The rest of the species are found in Africa, in the Mediterranean region, and in southern and western Europe. *Gaultheria* also has an interesting distribution, though not one which involves marked concentrations of species, in that it rings the Pacific Ocean almost completely. Such distributions present challenges to plant geographers and students of evolution, and no generally acceptable explanation of them has so far been proposed.

**Diagnostic features.** As is to be expected with a large family, great morphological variability renders any general description subject to a large number of exceptions. Almost all members of the family are found in acidic habitats, and all that have been examined are dependent to some extent on a fungal mycorrhiza for successful growth. This dependence of the plants on the mycorrhiza varies in extent; most species have developed leaves and photosynthesize normally. But one subfamily, the Monotropoideae (often regarded as a separate family, Monotropaceae) is totally dependent on its symbiont; in this group chlorophyll is absent, the leaves are reduced to small yellowish scales, and the plants live saprophytically among rotting leaf litter, from which, with the help of the fungus, they absorb all their nutrients.

Most of the family are shrubs or climbers, though herbaceous forms are found in the subfamily Monotropoideae. The leaves are always simple and without stipules, usually alternate, and often evergreen. The leaves of several genera show adaptations to dry conditions: such leaves are needle-like or folded, presenting a reduced surface area to the environment, and occur in *Erica*, *Calluna* and *Cassiope*, among others.

The inflorescences of the Ericaceae are extremely variable, ranging from umbel-like

Cyrillaceae. 1 *Purdiaea nutans* (a) leafy shoot and terminal inflorescence ($\times \frac{2}{3}$); (b) flower partly closed showing unequal sepals ($\times 1\frac{1}{3}$); (c) flower fully open showing ten stamens and slender, undivided style ($\times 1\frac{1}{3}$); (d) young fruit with persistent calyx ($\times 1\frac{1}{3}$); (e) vertical section of gynoecium showing each locule with a single pendulous ovule ($\times 14$); (f) stamen ($\times 2\frac{2}{3}$). 2 *Cyrilla racemosa* (a) shoot with axillary inflorescences ($\times \frac{2}{3}$); (b) flower ($\times 4$); (c) cross section of ovary ($\times 14$); (d) half flower ($\times 6$); (e) gynoecium ($\times 6$); (f) fruit ($\times 6$). 3 *Cliftonia monophylla* (a) shoot with terminal inflorescences ($\times \frac{2}{3}$); (b) winged fruit ($\times 3$).

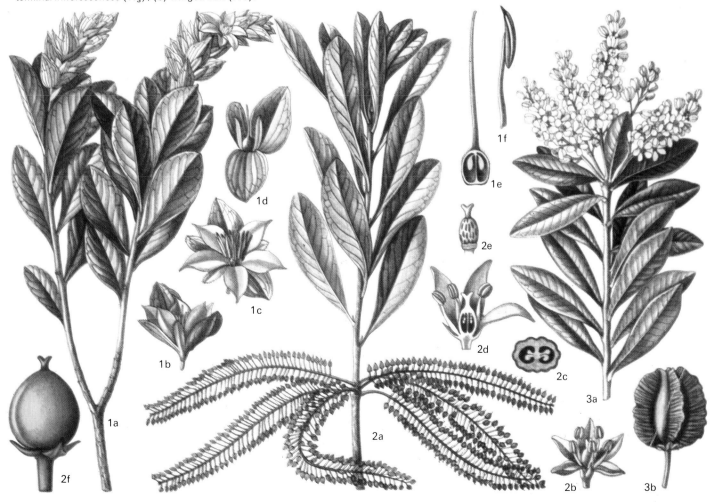

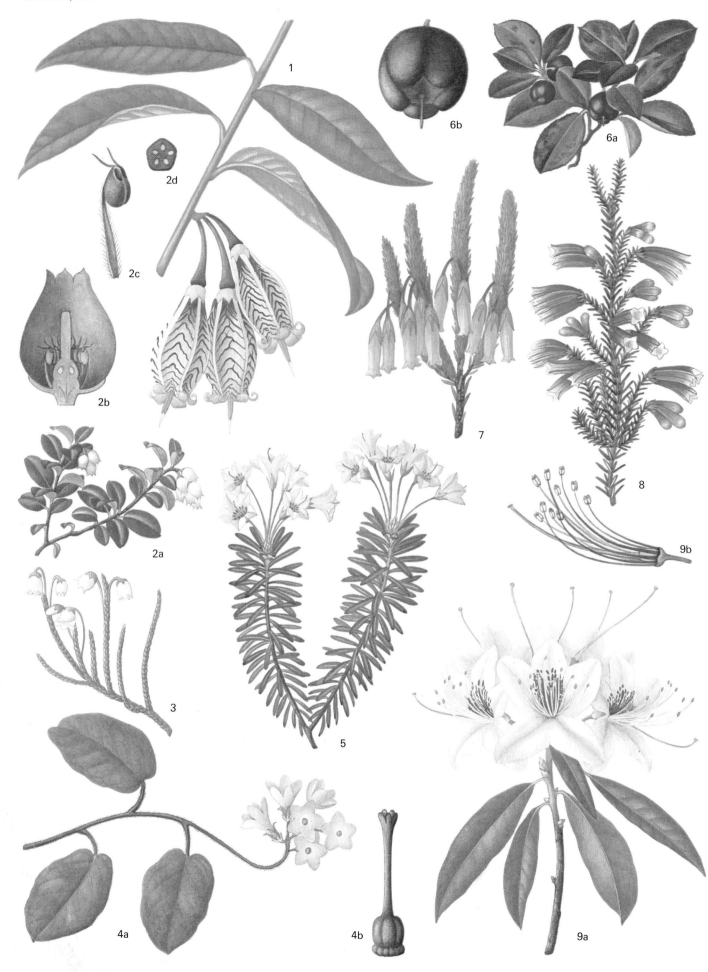

racemes to clusters or to single flowers. The flowers are usually regular and bisexual. The calyx consists of four or five sepals which are fused at the base; in many species of *Rhododendron* the calyx is reduced to an undulate rim. The corolla consists of four or five petals (rarely more, as in some species of *Rhododendron*) which are usually fused to form a tube at the base (the petals are completely free in *Ledum*, *Leiophyllum* and a few other genera). The stamens are usually twice as many as the petals (ie eight or 10) but are reduced to four or five in some genera and species; they are usually attached directly to the floral receptacle. In *Kalmia* the anthers fit into pockets in the corolla, from which they are released to shed their pollen. The anthers are inverted during growth, and open by means of pores at the apparent apex (which is morphologically the base), and are often ornamented with outgrowths (awns). The pollen is shed mostly in tetrad form, though single grains are found in some genera. The ovary is formed from four or five united carpels (sometimes more in *Rhododendron*) and has one to five (sometimes up to 10) locules with many ovules on usually axile placentas (parietal placentation occurs in the Monotropoideae). It is usually superior, but inferior ovaries are found in many genera of the subfamily Vaccinioideae. The style is simple and the stigma usually capitate. The fruit is usually a capsule opening loculicidally or septicidally, or, in many of the genera with inferior ovaries, a berry. The seeds have fleshy endosperm and a straight embryo.

**Classification.** The overall classification of the family and some of its allies has been a matter of dispute for many years. Recently, however, Stevens has studied the problem again, and has produced a workable classification. He gives convincing reasons for the inclusion within the Ericaceae of the traditionally recognized families Pyrolaceae and Monotropaceae (as subfamilies Pyroloideae and Monotropoideae). In this work, however, the Pyrolaceae is being considered as a separate family. An outline of Stevens' classification (omitting Pyroloideae) is given below, with five subfamilies:

Ericaceae. 1 *Agapetes macrantha* part of leafy shoot with axillary inflorescence ($\times\frac{2}{3}$). 2 *Arctostaphylos uva-ursi* (a) leafy shoot with terminal inflorescences ($\times\frac{2}{3}$); (b) half flower ($\times 4$); (c) stamen with broad hairy filament and anthers crowned by recurved arms and opening by terminal pores ($\times 10$); (d) cross section of ovary ($\times 4$). 3 *Cassiope selaginoides* stem covered with small clasping leaves ($\times\frac{2}{3}$). 4 *Epigaea repens* (a) leafy stem and inflorescence ($\times 1$); (b) gynoecium with lobed ovary and stigma ($\times 4$). 5 *Phyllothamnus erectus* flowering shoot ($\times\frac{2}{3}$). 6 *Gaultheria* sp (a) leafy shoot and berries ($\times\frac{2}{3}$); (b) berry ($\times 2\frac{2}{3}$). 7 *Erica vallis-aranearum* flowering shoot ($\times\frac{2}{3}$). 8 *E. versicolor* var *costata* flowering shoot. 9 *Rhododendron yunnanense* (a) flowering shoot ($\times\frac{2}{3}$); (b) androecium and gynoecium ($\times 1\frac{1}{3}$).

RHODODENDROIDEAE. Shrubs; inflorescences usually terminating the main branches; corolla deciduous; pollen often mixed with a sticky substance (viscin) that causes it to cohere in masses or strings; ovary superior. Divided into seven tribes with 19 genera, of which *Rhododendron*, *Andromeda*, *Kalmia*, *Ledum* and *Daboecia* are the most familiar.
ERICOIDEAE. Shrubs or shrublets; inflorescences not terminating the main branches; corolla persistent in fruit; viscin absent; ovary superior. About 20 genera, most of them found in southern Africa, all heath-like in appearance. *Erica* (heath) and *Calluna* (heather or ling) are the best known.
VACCINIOIDEAE. Shrubs or climbers; inflorescences not usually terminating the main branches, viscin absent; ovary superior or inferior; fruit often a berry. Divided into five tribes with about 50 genera, many of which are found in Andean parts of South America. *Agapetes*, *Arbutus*, *Enkianthus*, *Gaultheria*, *Cassiope*, *Lyonia* and *Vaccinium* are well known.
WITTSTEINIOIDEAE. Small shrublet; stamens epipetalous; anthers opening by slits; ovary inferior. One genus (*Wittsteinia*) from Australia, which has sometimes been placed in the Epacridaceae, and whose relationships are still uncertain.
MONOTROPOIDEAE. A variable number of genera is recognized in this subfamily which is in need of considerable study. Attempts to cultivate these plants have usually failed, and many details of their structure and relationships are in doubt. Some authorities place members of this subfamily within its own family Monotropaceae. Chief genus *Monotropa*.

The Ericaceae belongs to a group formerly known as the Bicornes (now Ericales), and is closely related to the Clethraceae, Epacridaceae, Empetraceae and Diapensiaceae.

**Economic uses.** Many of the Ericaceae are ornamental shrubs, and are widely cultivated in gardens. Possibly the most important is *Rhododendron*, with about 700 of its species in cultivation. Of these, most are Sino-Himalayan species which are hardy in north temperate regions. A wide variety of types is available, ranging from creeping shrublets to moderately sized trees, with shades of white, pink, red, mauve and yellow represented in the flowers. Most are evergreen, but some (azaleas in horticultural terminology) are deciduous. Very many hybrids have been raised by specialists, and plants suitable for various habitats and garden situations can be obtained very easily. A recent development has been the introduction of many New Guinea species into horticulture; they are, however, not generally hardy and cannot be grown in Northern Europe and the United States of America.

*Erica* is also an important genus in horticulture. Towards the end of the 19th century there was a fashion for the exotic

Cape heaths (South African species), which were grown in cool glasshouses. More recently, however, this fashion has been supplanted by one for the hardy heaths, ie those native to Europe and southwestern Asia. By means of careful selection and propagation, a wide range of small shrublets, with high weed-smothering potential and with the ability to provide color throughout the year, has been developed.

Other genera important in amenity horticulture are *Menziesia*, *Ledum*, *Cladothamnus*, *Elliottia*, *Kalmia*, *Phyllodoce*, *Daboecia*, *Calluna*, *Arbutus*, *Arctostaphylos* (bearberry), *Enkianthus*, *Cassiope*, *Pieris*, *Leucothoe*, *Zenobia*, *Gaultheria*, *Vaccinium*, *Cavendishia*, *Macleania*, *Oxydendrum* and *Pernettya*.

Some species of *Vaccinium* produce edible berries, and these form an important crop in some parts of the world, particularly in the United States of America and to some extent in the Soviet Union. Several species and their hybrids are involved in the overall crop, among them *Vaccinium corymbosum* (highbush blueberry), *V. oxycoccus* (cranberry), *V. angustifolium* (lowbush blueberry) and *V. myrtillus* (bilberry).

Many species of Ericaceae are poisonous to stock and Man. Species of *Kalmia* are very poisonous to sheep, and are known as "lambkill" in various parts of the United States. The wood of some species is of local importance, particularly in the Himalayas.

J.C.

# EPACRIDACEAE

The Epacridaceae is a family of heath-like shrubs or small trees. Most members grow in rather open habitats and are distinctly light-demanding.

**Distribution.** The family is largely confined to Australia, but extends to New Zealand and has a few species in Indomalaysia and one in southern South America. It forms extensive heathlands similar to those dominated by *Erica* (heath) and *Calluna* (heather) in other parts of the world and, indeed, in Malaysia some species grow intermingled with *Erica*.

**Diagnostic features.** The leaves are usually alternate, simple, narrow, rigid, sessile and without stipules. The flowers are small, regular and usually bisexual; they have bracts and are borne in spikes, racemes or, rarely, panicles; occasionally they are solitary. There are four or five free, persistent sepals. The corolla is tubular, with four or five imbricate or valvate lobes. There are four or five stamens, borne on the corolla or rarely below the ovary, alternating with the corolla lobes, sometimes with glands or tufts of hair between them; the anthers have a single locule and dehisce longitudinally. The ovary is superior of two to five fused carpels, often subtended by a glandular disk, with one to 10 locules (usually five) and one to several ovules per locule, the placentation being

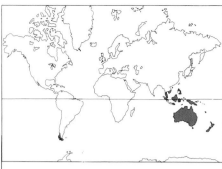

**Number of genera:** 30
**Number of species:** about 400
**Distribution:** mainly Australia with some species in Indomalaysia, New Zealand and S America.
**Economic uses:** some ornamental shrubs and occasional medicinal use.

axile or apical. The style is simple, the stigma capitate. The fruit is a loculicidal capsule with five valves or a drupe with one to several seeds; the seeds have a straight embryo and fleshy endosperm.

The species are mostly either insect-pollinated, for which the generally fragrant corolla with hairs inside seems to be an adaptation, or pollinated by birds, chiefly honeyeaters and parakeets.

**Classification.** The family is divided into three tribes:

STYPHELIEAE. Stamens borne at base of corolla; ovary with one ovule per locule; fruit indehiscent, usually fleshy. Eighteen genera, including *Styphelia* and *Trochocarpa*.

EPACRIDEAE. Stamens usually borne at base of corolla; ovary with several ovules per locule; fruit a loculicidal capsule. Ten genera, including *Richea, Dracophyllum* and *Epacris*.

PRIONOTEAE. Stamens borne below ovary, free; ovary with several ovules per locule; fruit a loculicidal capsule. Two monotypic genera (*Prionotes* in Tasmania, *Lebetanthus* in southern South America).

The family is most closely related to the Ericaceae, from which it differs in having palmate, open venation on the leaves and by the anthers which lack appendages and open by a slit rather than pores.

**Economic uses.** Several genera, notably *Dracophyllum, Epacris, Richea* and *Styphelia*, are grown as ornamental winter-flowering shrubs in the cool greenhouse. The

roots and leaves of *Styphelia malayana* are used locally for medicinal purposes while the inner bark of the stems is used for making canoes waterproof.                     D.M.M.

# EMPETRACEAE
*Crowberries*

The Empetraceae is a small family of evergreen shrubs of a heath-like habit, with small, close-set leaves, inconspicuous flowers without petals and fleshy or dry fruits.

**Distribution.** The family is interesting in that two of the genera show major geographical disjunctions. *Empetrum* (the crowberry) is an important member of heathlands in cool temperate regions of the Northern Hemisphere and southern South America, while the two species of *Corema* occur in eastern North America (*Corema conradii*) and southwest Europe (*C. alba*). The third genus, *Ceratiola*, has its only species in southeast North America.

**Diagnostic features.** The family consists of dwarf shrubs with small, linear, overlapping leaves, without stipules. The flowers are regular, usually unisexual with male and female on separate plants, rarely bisexual, one to three in number in the upper leaf axils

Epacridaceae. 1 *Richea sprengelioides* flowering shoot showing sheathing leaf bases and terminal inflorescences ($\times\frac{2}{3}$). 2 *R. gunnii* (a) cross section of ovary showing five locules ($\times12$); (b) dehisced fruit—a capsule ($\times6$). 3 *Epacris longiflora* flowering shoot ($\times\frac{2}{3}$). 4 *Styphelia laeta* (a) flowering shoot with non-sheathing leaves and inflorescences associated with abortive shoots ($\times\frac{2}{3}$); (b) half flower ($\times2\frac{2}{3}$). 5 *Dracophyllum rosmarinifolium* (a) flowering shoot ($\times1$); (b) half section of capsule ($\times4$). 6 *D. capitatum* half flower ($\times6$). 7 *Trochocarpa laurina* (a) flowering shoot ($\times\frac{2}{3}$); (b) fruit—an indehiscent drupe ($\times2\frac{2}{3}$); (c) cross section of fruit ($\times2$). 8 *Styphelia enervia* fruit ($\times4$).

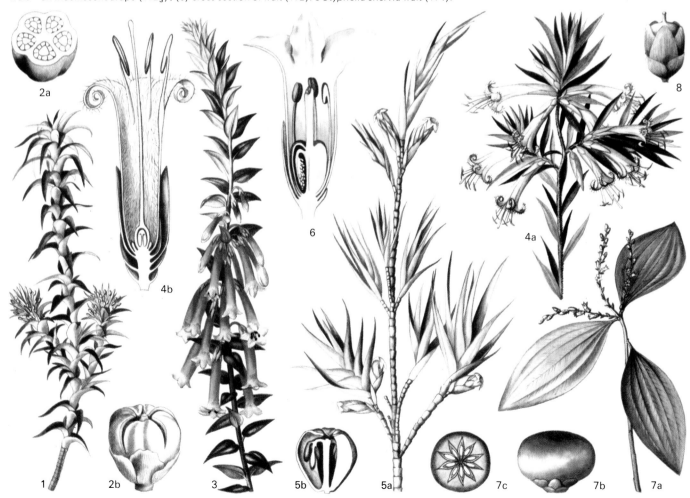

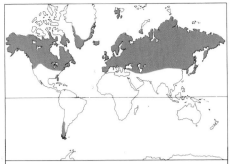

**Number of genera:** 3
**Number of species:** 4–6
**Distribution:** cool-temperate regions, SW Europe, E and SE N America.
**Economic uses:** limited uses as rock plants and for the edible berries.

or in terminal heads. There are four to six free perianth segments, usually in two similar whorls. The two to four stamens are free and episepalous; the anthers dehisce longitudinally; the disk is absent. The ovary is superior, has two to nine fused carpels and two to nine locules with one anatropous or campylotropous ovule per locule. There is a single short style, with two to nine fringed or lobed stigmatic branches. The fruit is a globose drupe, fleshy or dry with two or more pyrenes, each with one seed; the seeds have copious fleshy endosperm and the embryo is long and erect.

**Classification.** The three genera may be simply distinguished as follows: *Empetrum*, flowers solitary in leaf axils; three stamens. *Ceratiola*, two to three flowers in leaf axils; two stamens. *Corema*, flowers in terminal head; three or four stamens.

All genera are diploid except for *Empetrum*, which also contains tetraploids. The diploids characteristically have separate male and female plants; unisexual flowers and/or bisexual flowers on the same plant are a derived condition.

The Empetraceae is considered to be close to the Ericaceae, which it resembles both in certain well-marked embryological features and in the possession of several distinctive chemical compounds such as anthocyanidin, galactosides and some rare flavonols, in addition to general habit and morphology.

**Economic uses.** *Empetrum* and *Corema album* are cultivated for ornament in rock or heath gardens, and crowberry fruits (*Empetrum nigrum*) are used locally for jams and preserves.                D.M.M.

## PYROLACEAE
*Wintergreens*

The Pyrolaceae, commonly called the wintergreens, is a family of evergreen perennials with creeping rhizomes.

**Distribution.** The family is principally cool north temperate and Arctic, with some representatives as far south as Mexico and the West Indies.

**Diagnostic features.** The leaves are usually alternate, entire or dentate and without stipules. The flowers are terminal, in racemes, cymes or solitary. They are regular and bisexual, with five or four sepals and petals. The sepals are imbricate, the petals free or rarely shortly united. The stamens, eight to 10 in number, are in two whorls, the outer opposite the petals, the inner opposite the sepals. The anthers are free, opening by pores; the pollen is usually in tetrads. The stamens and petals are often situated at the edge of a nectariferous disk. The ovary is superior, of five or four fused carpels, and is incompletely five- or four-locular with thick fleshy axile placentas and numerous small anatropous ovules. The style is simple, and straight or declinate. The embryo is of a few cells without differentiation of the

Empetraceae. 1 *Ceratiola ericoides* (a) leafy shoot with flowers in leaf axils ($\times\frac{2}{3}$); (b) male flower with two anthers dehiscing lengthwise ($\times 8$); (c) ovary crowned by lobed stigma ($\times 12$); (d) cross section of fruit showing two seeds ($\times 6$). 2 *Corema conradii* (a) shoot with flowers in terminal heads ($\times\frac{2}{3}$); (b) head of flowers each with conspicuous stamens ($\times 4$). 3 *Empetrum rubrum* (a) shoot with solitary flowers in leaf axils ($\times 2$); (b) cross section of ovary with nine locules ($\times 10$); (c) male flower showing two whorls each of three perianth segments ($\times 8$); (d) gynoecium showing single, short style with six stigmatic branches ($\times 12$); (e) shoot bearing fruit ($\times\frac{2}{3}$); (f) fruit—a drupe ($\times 2$).

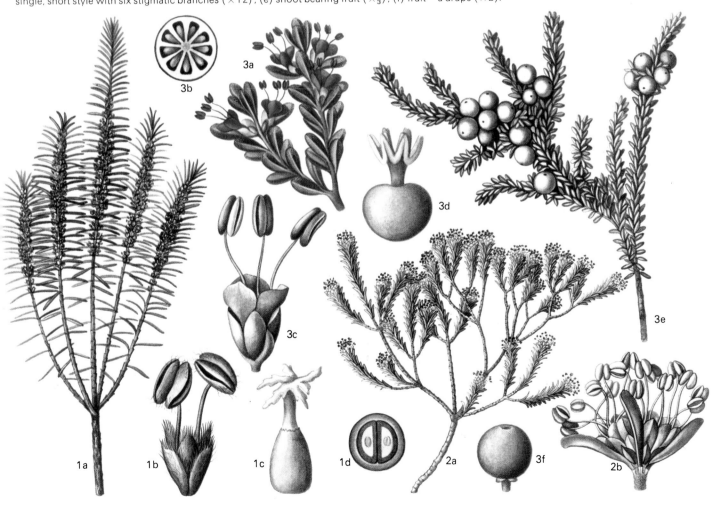

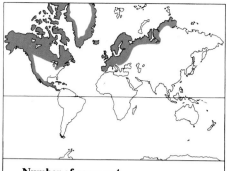

**Number of genera:** 4
**Number of species:** about 30
**Distribution:** temperate and Arctic.
**Economic uses:** medicinal.

cotyledons. The fruit is a loculicidal capsule with numerous very small wind-dispersed seeds, in a loose testa.

**Classification.** The family is represented by four genera, *Pyrola, Chimaphila, Moneses* and *Orthilia*. *Pyrola* contains 20 north temperate species. They are slender creeping plants with short, often distant aerial stems which are frequently reduced to a basal rosette of leaves. The inflorescence is a raceme of pink or white flowers. The characters of the style distinguish the commonest of the 20 species. *Pyrola minor* has a straight, included style, while that of *P. media* is long, straight and exserted, and that of *P. rotundifolia*, long, curved and exserted. The valves of the capsule of *Pyrola* are webbed at the edges. Nectar is secreted by the base of the petals.

The genus *Orthilia* is limited to one (possibly two) species found in circumpolar regions. It is similar to *Pyrola* but the greenish-white flowers of the racemose inflorescence are all arranged on one side. The disk consists of 10 small glands. The petioles are also shorter than those of *Pyrola*.

*Moneses* is represented by one boreal and Arctic species, and is distinguished from *Pyrola* and *Orthilia* by its opposite leaves, solitary flowers, the absence of webbing on the capsule and the absence of nectar. The disk is distinctly 10-lobed.

*Chimaphila*, represented by eight species of Eurasia and North and Central America, has strongly toothed, dark green leathery leaves, and pink flowers borne in an umbellate inflorescence.

The genus *Monotropa* is sometimes included in the Pyrolaceae, but in this work is considered to belong to the Ericaceae. The Pyrolaceae is related to the Ericaceae, but is distinguished from it by its herbaceous habit, incompletely septate ovary and undifferentiated embryo.

**Economic uses.** No economic uses for this family are recorded, except that the leaves of some species have been used to heal wounds.

S.A.H.

# DIAPENSIALES

## DIAPENSIACEAE
*Shortia and Galax*
The Diapensiaceae is a small family of herbs and dwarf shrubs.

**Distribution.** One of the species (*Diapensia lapponica*) is circumpolar, extending through North America, Greenland and northern Eurasia, as far south as South Korea. Three other genera, *Shortia, Pyxidanthera* and *Galax*, have species in North America, the latter two being endemic to the eastern United States. All other members of the family are east Asiatic, particularly in the Himalayan region, but extending east to Japan, where several species of *Shortia* and

Pyrolaceae. 1 *Pyrola rotundifolia* shoot and inflorescence ($\times\frac{2}{3}$). 2 *P. dentata* (a) flower ($\times 2$); (b) ovary ($\times 3$). 3 *Chimaphila umbellata* (a) shoot and inflorescence ($\times\frac{2}{3}$); (b) flower ($\times 1\frac{1}{3}$); (c) half flower ($\times 2$); (d) stamen side (left) and front (right) views ($\times 4$); (e) cross section of ovary ($\times 4$); (f) fruits ($\times\frac{2}{3}$); (g) dehisced fruit ($\times 3$). The following species are sometimes placed in the Pyrolaceae but are here included in the Ericaceae. 4 *Monotropa hypopithys* (a) habit ($\times\frac{2}{3}$); (b) flower ($\times 2$); (c) flower ($\times 2$); (d) gynoecium and stamens ($\times 3$); (e) stamen ($\times 12$); (f) vertical section of gynoecium ($\times 3$); (g) fruit ($\times 4$). 5 *Sarcodes sanguinea* (a) flower ($\times 1$); (b) half flower ($\times 2$); (c) gynoecium ($\times 2$); (d) stamen ($\times 3$).

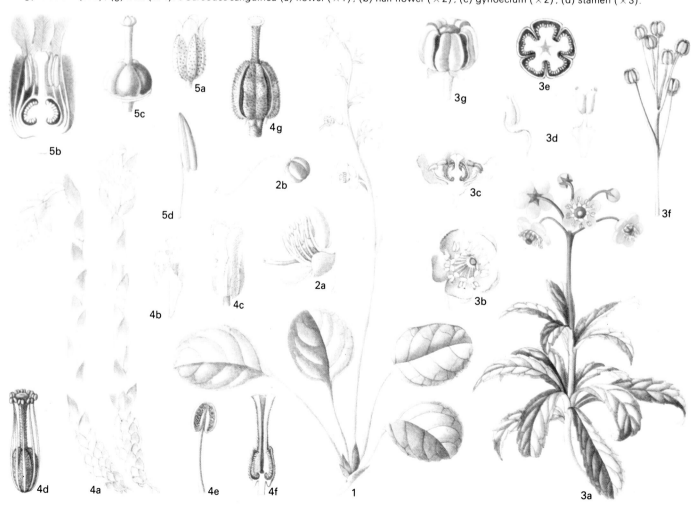

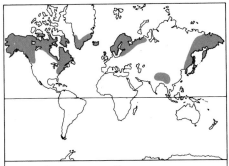

**Number of genera:** 7
**Number of species:** 20
**Distribution:** temperate and cold regions of the N Hemisphere.
**Economic uses:** cultivated ornamentals (*Shortia, Galax, Schizocodon*).

*Schizocodon* are found, and Taiwan, which has a number of endemic species of *Shortia*.

**Diagnostic features.** All members of the family are small shrubs or stemless herbs, usually with rosettes of simple leaves. The flowers are regular, bisexual, either solitary or in racemes, with five sepals, often partly fused, and five partly free petals, the latter forming a corolla tube. There are usually five stamens opposite the sepals and partly fused

to the corolla, and sometimes five staminodes opposite the petals. In *Diplarche* there are two whorls, each of five fertile stamens, the inner corresponding to the staminodes of other genera, these being joined to the base of the corolla tube. The ovary is superior, of three (sometimes five) fused carpels, usually with three locules (five in *Diplarche*), and with a simple style. There are few to numerous ovules on axile placentas. The fruit is a loculicidal capsule with many small seeds which have fleshy endosperm and a cylindrical embryo.

**Classification.** The family is usually considered to contain seven genera, which may be divided into three groups. One comprises *Diplarche*, a genus of two species from the eastern Himalayas, dwarf shrubs with branching racemes of pink flowers. *Galax aphylla*, the only species of its genus, forms a second group. It is a herbaceous perennial growing in woodlands, with creeping rhizomes bearing cordate leaves and producing racemose inflorescences of many small, white flowers on scapes. The other genera form a third group characterized by solitary flowers or few-flowered racemes. Two genera are cushion-like or creeping dwarf

shrubs with closely imbricate, linear leaves and solitary flowers: *Diapensia* (Arctic and montane, with several species in the Himalayas) and *Pyxidanthera* (two species on sandy sites in the eastern United States of America). The other three genera, *Shortia, Schizocodon* and *Berneuxia* are perennial herbs with stolons, mainly of montane woodlands, with well-developed lanceolate, ovate or cordate leaves and conspicuous white or pink flowers.

The whole family, *Diplarche* in particular, resembles some members of the Ericaceae, but in the latter family the stamens are always free and in a single whorl and the pollen is shed through pores, while the Diapensiaceae has anthers dehiscing by slits.

**Economic uses.** Several species are used horticulturally. Perhaps the most widely grown is *Galax aphylla*, which provides useful ground cover and decorative foliage. This and species of *Schizocodon* and *Shortia*, also attractive garden plants, are conspicuous in the autumn when their leaves turn bronze or crimson. *Diapensia* and *Pyxidanthera* species are sometimes grown in rock gardens.                T.T.E.

Diapensiaceae. 1 *Diapensia himalaica* (a) creeping shoot bearing small, simple overlapping leaves and solitary flowers ($\times\frac{2}{3}$); (b) fruiting shoot ($\times\frac{2}{3}$); (c) five-lobed perianth opened out to reveal five stamens fused to corolla ($\times 2$); (d) cross section of ovary ($\times 6$); (e) dehiscing fruit ($\times 3$); (f) detail of stamens ($\times 8$). 2 *Schizocodon soldanelloides* (a) habit ($\times\frac{2}{3}$); (b) part of corolla opened out to show fertile stamens inserted on corolla tube and linear staminodes at the base ($\times 2$); (c) fruit enclosed in persistent bracts and calyx ($\times 2$); (d) gynoecium ($\times 2$); (e) cross section of ovary ($\times 3$); (f) stamens, dorsal view (left) and ventral view (right) ($\times 6$). 3 *Galax aphylla* habit ($\times\frac{2}{3}$).

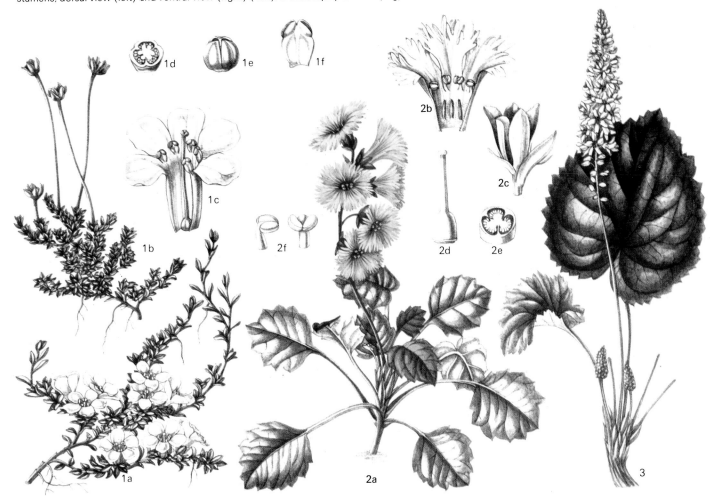

# EBENALES

## SAPOTACEAE
*Chicle, Gutta-percha and Sapodilla*

This large family of tropical trees yields timber and latex and edible fruits.

**Distribution.** The Sapotaceae occurs pantropically, mainly in lowland and lower montane rain forest.

**Diagnostic features.** White latex is present at least in the twigs, usually also in the bark, and is sometimes copiously produced. The leaves are simple, entire, spirally arranged, often crowded in false whorls and sometimes with stipules which soon drop. The flowers are borne in fascicles, often behind the leaves or on the trunk; they are bisexual, regular or irregular, scented, often white or cream and often nocturnal and bat-pollinated. The sepals are free, in two whorls of two, three or four, or one of five. The petals are usually equal in number to the sepals but usually in one whorl and are fused at the base. The stamens are epipetalous, either equal in number to and opposite the corolla lobes or more numerous, sometimes alternating with staminodes. The ovary is superior, of many fused carpels, with many locules each containing a single axile or basal ovule. The style is simple. The fruit is a berry, not articulated. The one or few seeds have an oily endosperm and bony testa and a large embryo.

**Classification.** The Sapotaceae is one of the

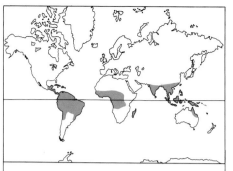

**Number of genera:** 35–75
**Number of species:** about 800
**Distribution:** pantropical in lowland and montane rain forests.
**Economic uses:** timber, gutta-percha, balata, chicle (chewing gum) and edible fruits (sapodilla plum, star apple, sapote).

families (another is the Lauraceae) within which generic limits are extremely difficult to perceive and opinions on subdivision vary considerably. From 35 to 75 ill-defined genera can be distinguished; there is no general agreement and new opinions continue to be published. *Sarcosperma*, sometimes considered a monogeneric family (Sarcospermataceae), represents an extreme of evolutionary trends. With this inclusion, the Sapotaceae stands as a close-knit family, the biggest member of the Ebenales.

**Economic uses.** Timber production is nowadays increasingly significant, for the Sapotaceae are an important component of many tropical rain forests (as in Malaya and Borneo), reaching 30m (100ft) tall and 2m (6.5ft) in girth. Some species have heavy timber, which is hard and naturally durable, but often siliceous; others have lighter timber, some without silica. Gutta-percha, obtained from the latex of *Palaquium* species (especially *Palaquium gutta* of Sumatra, Malaya, Java and Borneo) was once the premier product of the family. It is a polymer of isoprene, differing from rubber by having trans- instead of cis-isomerization, is almost non-elastic, a better insulator of heat and

Sapotaceae. 1 *Madhuca parkii* (a) tip of leafy flowering shoot with flowers in fascicles ($\times\frac{2}{3}$); (b) cross section of ovary showing eight locules each with an ovule on an axile placenta ($\times 3$). 2 *Mimusops zeyheri* var *laurifolia* (a) leafy shoot with axillary fascicles of flowers ($\times\frac{2}{3}$); (b) perianth opened out ($\times 3$); (c) petal with appendages ($\times 3$); (d) stamen ($\times 4$); (e) staminode ($\times 4$); (f) vertical section of ovary ($\times 3$). 3 *Sideroxylon costatum* (a) flowering shoot ($\times\frac{2}{3}$); (b) flower ($\times 3$); (c) corolla opened out ($\times 3$); (d) gynoecium ($\times 4$); (e) vertical section of ovary ($\times 4$). 4 *Achras sapota* (a) fruit—the sapodilla plum ($\times\frac{1}{3}$); (b) cross section of fruit ($\times\frac{1}{3}$).

electricity, becomes plastic on heating, and on cooling retains any shape given while hot. It was developed in the 19th century from a minor curiosity to a major industrial product, principally as an insulant for submarine telephone cables, but also used in golf balls and for temporary dental stoppings (only the last use still persists). The latex was mostly obtained by tapping trunks in a herringbone pattern. Early destructive tapping was followed by efforts for conservation and some plantations were established, as in Java and Singapore.

*Mimusops balata* (*Manilkara bidentata*) of northern South America also yields a latex, balata, formerly of considerable importance. Chicle, the elastic component of earlier chewing gum, is produced from the latex of *Achras zapota* (*Manilkara zapota*).

*Achras zapota* also yields the popular edible fruit chiku or sapodilla plum, and *Chrysophyllum cainito* the star apple; both are of American origin but now planted elsewhere in the humid tropics. Some of the fruits termed sapote are the product of *Calocarpum* species, notably *Calocarpum sapota*. The seeds of the north tropical African *Butyrospermum paradoxum*, the shea butter tree, yield an edible oil.

## EBENACEAE
*Persimmons and Ebonies*

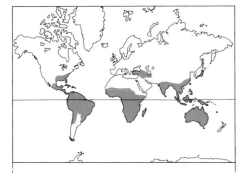

**Number of genera:** 2
**Number of species:** 400–500
**Distribution:** tropical, centered in Indomalesian rain forests; a few N temperate.
**Economic uses:** timber (ebony) and fruits (persimmon, date plum and Japanese date plum or kaki).

The Ebenaceae is a medium-sized family mainly of tropical trees, the source of ebony and persimmons.
**Distribution.** The greatest concentration is in the lowland rain forests of the Malay archipelago, with about 200 species, followed in abundance by tropical Africa, then Latin America. There are a few outlying species in the north temperate zone.
**Diagnostic features.** Members of the family are mostly small trees with a monopodial crown (ie with a single leading shoot) and flattened foliage sprays. Some are shrubs. The outer bark is usually black, gritty and

charcoal-like. There is no latex. The leaves are alternate (rarely opposite), simple, entire and without stipules.

The flowers are usually unisexual (with male and female on separate plants), rarely structurally bisexual. Inflorescences are short and determinate, in the leaf axils, sometimes reduced to a single flower, especially in the female. The flowers are jointed at the base and regular, with parts mostly in threes or fives (sometimes sixes or sevens). The sepals are fused, with lobes valvate or imbricate. The petals are fused into a tube with as many lobes as there are sepals, the lobes contorted, white, cream or suffused pink. The stamens are usually epipetalous, in two whorls, two to four times the number of the corolla lobes and fused in radial pairs. The female flowers usually bear staminodes. The ovary is superior (rarely inferior) and sessile, with as many locules as there are petals and sepals. Each locule has two pendent ovules attached at the apex, but is usually divided by a false septum with one ovule in each half, the halves being connate at the apex. The styles are fused at least at the base; there are as many stigmas as locules. The male flowers usually have a pistillode. The fruit is a berry, the pericarp pulpy to fibrous, with a stony inner part, only rarely dehiscent, seated on the persistent, often enlarged, calyx. The seeds are several, with endosperm, which is sometimes ruminate.
**Classification.** Nowadays nearly all species are put into the genus *Diospyros* (including *Lissocarpa* and *Maba*) with 400–500 species. *Diospyros* is pantropical with some outlying members. The only other genus, *Euclea* (14 species), is confined to eastern and southern Africa. *Diospyros* itself is closely knit and only a few small distinctive sections can be segregated from the main core.

Together with the Sapotaceae, the family comprises the bulk of the order Ebenales.
**Economic uses.** The family is best known for the black, hard heartwood, ebony of commerce, produced by most but not all species of *Diospyros*. *Diospyros reticulata* (Mauritius) and *D. ebenum* (Sri Lanka) are among the finest producers of ebony. The fruits of several species are eaten, throughout the range. A few have been brought into cultivation. Best known are the persimmons, a group of outlying warm north temperate species; *Diospyros kaki* of eastern Asia (kaki, Chinese or Japanese date plum, or persimmon), extensively cultivated in China and Japan, and known through to the Mediterranean; *D. lotus* of Eurasia (date plum) and *D. virginiana* of North America. Crushed seeds of certain Malesian species are used as fish poison. In all species the fruit is extremely astringent until it is very ripe.

## STYRACACEAE
*Silverbell and Snowbell Trees*
The Styracaceae is a family of shrubs and trees best known as the source of benzoin

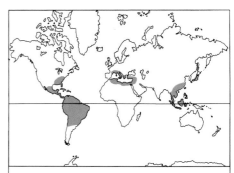

**Number of genera:** 12
**Number of species:** 180
**Distribution:** centered in E Asia, W Malesia, SE N, C and S America.
**Economic uses:** gum resins used medicinally and in perfumes, ornamentals (*Halesia* and *Styrax*).

(gum benjamin) and storax; it is also noted for several beautiful ornamentals.
**Distribution.** There are three centers of distribution: eastern Asia to western Malesia; southeastern North America to central and South America; and, with one species (*Styrax officinale*), in the Mediterranean.
**Diagnostic features.** The leaves are alternate, simple, without stipules and usually entire. The flowers are regular, usually bisexual and usually borne in racemes or panicles. The calyx is tubular with four or five persistent lobes, the corolla tubular at the base but often only very shortly so, with four to seven valvate lobes. The stamens are equal in number to and alternate with the corolla lobes, or double the number, usually adnate to the corolla tube or united as a tube. The ovary is superior or inferior, of three to five fused carpels with three to five locules each containing one to many anatropous ovules on axile placentas. The style is simple with a capitate or lobed stigma. The fruit is a drupe or capsule, with the calyx persistent. The one to few seeds with copious endosperm are eaten and dispersed by animals. The embryo is straight or slightly curved.
**Classification.** *Styrax* itself (130 species) is the most important genus. Most probably related to the Ebenaceae and Sapotaceae, the family is very heterogeneous in nature, unlike its relatives, and is perhaps not natural. *Afrostyrax*, a tree native to west tropical Africa, was formerly placed in this family but is now believed to be totally unconnected and is often placed together with *Hua* (another genus of tropical African trees) in the segregate family Huaceae. The affinities of the latter family are, however, entirely obscure.
**Economic uses.** The wood is mostly soft and of little use. Resin is the chief product. The tropical resins (chiefly from *Styrax benzoin*) are traded as benzoin (corrupted as gum benjamin), and used medicinally (in friar's balsam) and in incense. The resin is obtained by wounding the bark. Produced chiefly in Thailand, Sumatra and Bolivia, it consists

principally of two alcohols combined with cinnamic acid and free cinnamomic and benzoic acids. The trees are scattered and uncommon in lowland tropical rain forest, and production depends on continual and careful wounding. *Styrax officinale* (Mediterranean) yields the resin storax, used as an antiseptic, inhalant and expectorant. *Halesia* includes the silverbell or snowdrop trees, *Styrax* the snowbell trees; both groups are beautiful, distinctive ornamentals.

T.C.W.

# PRIMULALES

## PRIMULACEAE
*The Primrose Family*

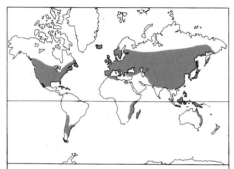

**Number of genera:** about 28
**Number of species:** nearly 1,000
**Distribution:** cosmopolitan, but chiefly in the N temperate zone with some alpine species.
**Economic uses:** many popular garden and house ornamentals (eg primroses, auriculas, cyclamens).

The Primulaceae is a family of perennial or annual herbs, including a number of popular garden ornamentals, such as primulas and cyclamens, and the familiar wild primrose, *Primula veris*.

**Distribution.** The family is cosmopolitan in distribution but with most members indigenous to the north temperate zone.

**Diagnostic features.** Most species perennate by means of sympodial rhizomes (as in *Primula*) or tubers (as in *Cyclamen*) and bear simple leaves without stipules (except in *Coris*). The leaves are opposite or alternate or in the form of a rosette arising from the stem base, and are usually entire (the submerged leaves of the aquatic genus *Hottonia* are pinnately dissected). Often the leaves and stems bear simple or compound glandular hairs.

The flowers are often borne on leafless scapes (as in *Primula veitchii, Soldanella alpina*), either solitary or in umbellate, racemose or paniculate inflorescences. The flowers have bracts and are usually regular (irregular in *Coris*), bisexual and often heterostylous. There are normally five sepals (occasionally six in *Lysimachia*) fused into a calyx tube which is persistent with four or five pointed segments. With the exception of

*Glaux*, which has no petals, the corolla consists of five (or occasionally four) petals usually fused into a tube but with the lobes reflexed in some genera (eg *Dodecatheon* and *Cyclamen*). There are five stamens fused to the corolla tube opposite the petals. In a few instances (including *Samolus* and *Soldanella*) there are also five staminodes alternating with the petals. This suggests that the antipetalous position of the stamens has resulted during the course of evolution from the loss of an ancestral outer whorl. The anthers are bilocular and dehisce longitudinally. The ovary is composed of five carpels fused to form a unilocular, superior or semi-inferior ovary with free central placentation and few to numerous ovules. There is a single style terminating in a head-like stigma. The fruit is a five-valved capsule or occasionally a pyxidium (as in *Anagallis*, where it opens by a cap-like cover). The fruit usually contains numerous small seeds with a small straight embryo surrounded by fleshy or hard endosperm.

**Classification.** The genera may be separated into a number of tribes on the basis of floral symmetry, position of the ovary and aestivation of the corolla:

PRIMULEAE. Ovary superior, corolla lobes imbricate in the bud, capsule with valvate dehiscence; includes *Primula* (primroses, about 500 species), *Androsace* (100 species), *Soldanella* (11 species), *Hottonia* (two species), *Dodecatheon* (about 50 species).

CYCLAMINEAE. Ovary superior, capsule with valvate dehiscence, flowers with reflexed petals, plants with tubers; only *Cyclamen* (15 species).

LYSIMACHIEAE. Ovary superior, capsule with valvate dehiscence or pyxidium, corolla lobes contorted in bud; includes *Lysimachia* (about 200 species), *Trientalis* (four species), *Glaux* (one species), *Anagallis* (28 species).

SAMOLEAE. Ovary semi-inferior; only *Samolus* (10–15 species, chiefly in the Southern Hemisphere).

CORIDEAE. Spiny calyx, flowers irregular; only *Coris* (two species, mainly Mediterranean). This is a curious genus of small thyme-like woody herbs, sometimes treated as a separate family, the Coridaceae.

The Primulaceae was once considered to be related to the Caryophyllaceae. The relationship was based on the nature of the gynoecium, and the vascularization and other anatomical features suggested the evolution of the primulaceous form from the caryophyllaceous type. However, the Primulaceae is more advanced in such features as the fusing of the corolla segments and the reduction of the androecium usually to five stamens. Here it is placed with the Myrsinaceae in a separate order, the Primulales, related to the Ebenales.

**Economic uses.** Although economically the Primulaceae is mainly of ornamental importance, it is worth noting that *Cyclamen purpurascens* (*C. europaeum*) (common cyc-

lamen) contains the poisonous glycoside cyclamin, while *Anagallis arvensis* was once an important medicinal plant and contains a poisonous glycoside similar to saponin. *Lysimachia vulgaris* yields a yellow dye and also has reputed uses as a febrifuge or fever-reducing agent. Flowers of *Primula veris* are used for home-made wine.

Many species of *Primula* are cultivated for their attractive flowers either as house pot plants (eg *Primula × kewensis* and *P. obconica*), in rock gardens (*P. auricula* and *P. allionii*) or in garden borders (*P. denticulata* and *P. bulleyana*).

A number of cyclamen species including *Cyclamen hederifolium* are grown in the open, but the progenitor of the popular winter flowering pot cyclamens is *C. persicum*, of which there are innumerable cultivars including 'Butterfly' with salmon-pink frilled petals; 'Cardinal' with scarlet petals, 'Silberstrahl' with red petals each with a narrow silvery-white margin and 'White Swan' with pure white petals.

Some species of *Anagallis* (eg *A. arvensis* and *A. linifolia*) have provided the source of garden varieties suitable for edging borders or for rock gardens.

*Lysimachia nummularia* is useful for ground cover and species of *Dodecatheon*, with their rose-purple flowers, are useful perennial border plants.

S.R.C.

## MYRSINACEAE

The Myrsinaceae is a medium-sized family of trees and shrubs, of little economic importance except for a few species grown as ornamentals.

**Distribution.** The family is mainly warm temperate, subtropical and tropical in distribution with representatives from New Zealand and South Africa in the south to Japan, Mexico and Florida in the north.

**Diagnostic features.** The leaves are alternate, simple, leathery, and without stipules and are usually dotted with glands or conspicuous resin ducts. The flowers are small, regular, bisexual or unisexual (then with the sexes on separate plants) and are normally borne in fascicles, either on scaly short shoots or on spurs in the leaf axils, but they may also be in terminal panicles, corymbs or cymes. There are four to six free or basally connate, small

Primulaceae. 1 *Dodecatheon meadia* habit showing basal rosette of leaves and flowers, with reflexed petals, borne on leafless stalks (×⅔). 2 *Primula veitchii* (a) habit (×⅔); (b) half flower showing epipetalous stamens and ovules on a free central placenta (×4); (c) cross section of unilocular ovary (×6). 3 *Samolus valerandi* half flower with staminodes and stamens (×8). 4 *Soldanella alpina* habit showing flowers with deeply divided petals (×⅔). 5 *Primula veris* dehisced fruit (a capsule) with part of persistent calyx removed (×3). 6 *Cyclamen hederifolium* (a) habit showing basal tuber (×⅔); (b) dehisced fruit (×4). 7 *Lysimachia punctata* leafy terminal inflorescence with yellow flowers (×⅔).

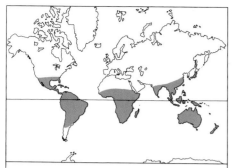

**Number of genera:** 32
**Number of species:** 1,000
**Distribution:** widely distributed from warm temperate to tropical regions.
**Economic uses:** several ornamentals and limited local uses as medicines.

sepals and the same number of petals which are connate, valvate or contorted into a four- to six-lobed corolla. The stamens are equal in number to, and usually opposite, the corolla lobes. The anthers have two locules which open by inwardly-facing longitudinal slits and are generally longer than the filaments which are often fused to the corolla. The ovary is either superior or semi-inferior, with one (sometimes four to six) locules and few to numerous ovules on axile or free central placentas. The fruit is a fleshy drupe. The seeds have a straight or slightly curved embryo and fleshy endosperm.

**Classification.** The Myrsinaceae is divided into two subfamilies: the MAESOIDEAE containing the genus *Maesa* characterized by its half-inferior ovary and many-seeded fruits; and the MYRSINOIDEAE with a superior ovary and a single-seeded fruit. The Myrsinoideae is further divided into two tribes: the MYRSINEAE with few ovules in one row (eg *Oncostemon*, *Embelia*, *Rapanea*) and ARDISIEAE with numerous ovules in many rows (eg *Aegiceras* and *Ardisia*).

The five genera *Theophrasta*, *Neomezia*, *Deherainia*, *Clavija* and *Jacquinia* with about 110 species of tropical American and West Indian trees and shrubs are sometimes placed in the Myrsinaceae, but are normally separated as the family Theophrastaceae. It differs from the Myrsinaceae in having anthers extrorse, five staminodes that are alternate with the corolla lobes and leaves that do not have large glands or resin ducts.

The Myrsinaceae is usually included in the order Primulales near to the Primulaceae but this classification is refuted by some authorities who consider the similarity between Myrsinaceae and the rest of the predominantly herbaceous Primulales to be superficial and place it in an order Myrsinales together with the Theophrastaceae, its most closely related family.

**Economic uses.** The Myrsinaceae is a family of little economic value. Species of the genera *Ardisia*, *Maesa*, *Myrsine* and *Suttonia* are sometimes grown as ornamentals. *Myrsine africana*, from Africa, China and India is cultivated in gardens of warmer areas for its attractive purple-blue fruits. Several species of *Maesa* from India, the Himalayas and Sikkim and about 16 species of *Ardisia*, especially *A. crispa*, with very persistent red fruits, are also used as garden or warm greenhouse subjects. The leaves of *Ardisia colorata* are used as an infusion in Malaya to treat stomach complaints and in the same area the fruits of *A. crispa* are eaten. In Java the sap of *A. fulginosa* is boiled with coconut oil and used to treat scurvy and in the Philippines the flowers and fruits of *A. squamulosa* are used to flavor fish dishes.

D.B.

Myrsinaceae. 1 *Ardisia humulis* (a) leafy shoot with flowers in axillary inflorescences ($\times\frac{2}{3}$); (b) half flower bud ($\times 4$); (c) corolla opened out ($\times 2$); (d) dehiscing stamen ($\times 3$); (e) fruit ($\times\frac{1}{3}$). 2 *Myrsine africana* (a) leafy shoot with fruits ($\times\frac{2}{3}$); (b) female flower ($\times 5$); (c) male flower ($\times 4$); (d) male flower opened out to show vestigial gynoecium ($\times 4$); (e) female flower opened out to show staminodes ($\times 6$). 3 *Embelia kraussii* (a) leafy shoot and axillary inflorescences ($\times\frac{2}{3}$); (b) flower ($\times 6$). 4 *Aegiceras corniculatum* (a) flower ($\times 1\frac{1}{3}$); (b) flower opened out ($\times 1\frac{2}{3}$); (c) dehisced anther showing transverse septa within the locules ($\times 6$); (d) leafy shoot and fruits ($\times\frac{2}{3}$). 5 *Maesa alnifolia* vertical section of ovary with calyx ($\times 6$).

# ROSIDAE

## ROSALES

### CUNONIACEAE

The Cunoniaceae is a family of trees and shrubs native to the Southern Hemisphere, which is important for its light timber.

**Distribution.** The main centers of distribution are Oceania and Australasia, but there are a few genera in South Africa and tropical America. The most important genus is *Weinmannia* which has 160 species distributed through Madagascar, Malaysia, the Pacific, New Zealand, Chile, Mexico and the West Indies. *Pancheria* has 25 species in New Caledonia. *Geissois* has 20 species in Australasia, New Caledonia and Fiji. The 20 species of *Spiraeanthemum* are native to New Guinea and Polynesia. *Cunonia* and *Lamanonia* are smaller genera, consisting of 15 and 10 species, respectively. The former has a discontinuous distribution in South Africa and New Caledonia, while *Lamanonia* is native to Brazil and Paraguay.

**Diagnostic features.** The leaves are leathery,

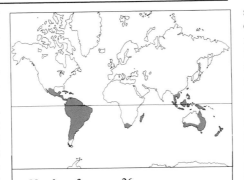

**Number of genera:** 26
**Number of species:** about 250
**Distribution:** S Hemisphere, chiefly Australasia and the Pacific.
**Economic uses:** timber.

often glandular, opposite or rarely in whorls, occasionally simple, but more often compound, trifoliolate or pinnate. Stipules are present and may be large and united in pairs.

The small flowers are regular, bisexual, but unisexual in a few species, with male and female on different plants. They are either solitary or borne in branched racemes or compact heads. The sepals are three to six in number and either free or fused together at the base. There may be four or five petals, free or united at the base. They are generally smaller than the sepals, and are absent in some species. Most species have numerous stamens, but some have four or five alternating with the petals and others have eight or ten. They are usually inserted by their free filaments on a ring-like, nectar-secreting disk, surrounding the ovary, which is superior, comprising two to five free or fused carpels. There are usually two (sometimes five) locules, each locule or free carpel containing numerous ovules set in two rows on recurved, axile or apical placentas. The styles are distinct, even in those species with fused carpels. The fruit is a capsule or nut and the seed has a small embryo surrounded by abundant endosperm.

**Classification.** This family has clear affinities with the Saxifragaceae, but differs in having a predominantly tree-like habit, opposite or whorled leaves and recurved placentas. A

Cunoniaceae. 1 *Pancheria elegans* (a) shoot with whorls of simple leaves and flowers in compact heads ($\times\frac{2}{3}$); (b) female flower showing three free sepals and petals and two free styles ($\times12$); (c) male flower with six stamens ($\times12$); (d) male flower opened out to show stamens with filaments of two lengths ($\times12$); (e) bilobed fruit ($\times12$). 2 *Cunonia capensis* shoot with pinnate leaf and flowers in a panicle ($\times\frac{2}{3}$). 3 *Weinmannia hildebrandtii* (a) shoot with trifoliolate leaves and flowers in panicles ($\times\frac{2}{3}$); (b) flower ($\times8$); (c) half flower ($\times8$). 4 *Geissois imthurnii* flower with four sepals, no petals and numerous stamens inserted on a nectar secreting disk ($\times2$). 5 *Davidsonia prunens*, fruit ($\times\frac{2}{3}$).

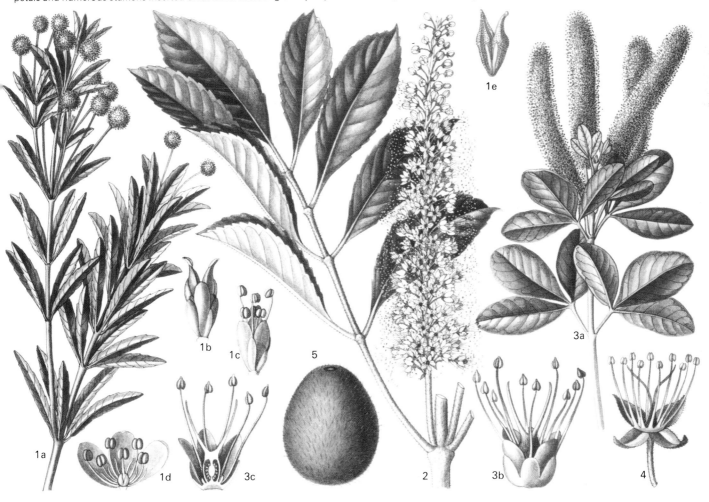

Pittosporaceae. 1 *Pittosporum crassifolium* (a) leafy shoot and inflorescences of male flowers (×⅔); (b) male flower (×1); (c) male flower with perianth removed showing large stamens and vestigial ovary (×1⅓); (d) gynoecium from female flower with vestigial stamens (×1⅓); (e) cross section of ovary (×1⅓); (f) dehiscing fruit—a capsule (×1⅓). 2 *Billiardiera mutabilis* (a) flowering and fruiting shoot (×⅔); (b) fruit—a berry (×1⅓). 3 *Sollya heterophylla* flowering shoot (×⅔). 4 *Marianthus ringens* (a) twining, leafy stem and inflorescence (×⅔); (b) flower (×1⅓); (c) androecium (×2); (d) stamen with flattened filament (×2⅔); (e) gynoecium (×2⅔).

further difference is that most species have flowers with numerous stamens.

**Economic uses.** The only member of the Cunoniaceae that is economically important is *Ceratopetalum apetalum*, a tall tree native to New South Wales. The timber (known as lightwood) is light brown to pinkish brown in color and is used in carpentry and cabinetmaking. It is also used extensively as flooring, paneling and skirting, and provides a plywood veneer as well as aircraft veneers.

S.R.C.

# PITTOSPORACEAE
*Parchment-bark*

The Pittosporaceae is a medium-sized family of evergreen shrubs and trees. *Pittosporum*, the type genus, contains a number of attractive ornamentals.

**Distribution.** The family is native mostly to the Old World tropics, eight of the nine genera being endemic to Australasia.

**Diagnostic features.** The plants are shrubs or small trees, sometimes climbers. The leaves are evergreen and leathery, typically entire, and without stipules. The flowers are bisexual, rarely tending towards unisexuality and polygamy (male, female and bisexual on the same plant), regular, but weakly irregular in one genus (*Cheiranthera*). There are five free sepals. There are five petals which are mostly united below and often clawed. The stamens are five, attached to the sepals. The ovary is superior, of two fused carpels (sometimes three to five) and one or many locules with placentas in two ranks, axillary or parietal. The style is simple, the fruit a loculicidal capsule or berry; the seeds are mostly numerous, sometimes winged, often (eg *Pittosporum*) smeared with a brownish resin-like mucilage (whence the name, the Greek word *pittos* meaning pitch); there is abundant endosperm. The bark is traversed by resin-containing canals.

**Classification.** Two tribes are recognized in the family, based on type of fruit:

PITTOSPOREAE. Fruit a capsule. *Pittosporum* (140–200 species, Canaries through West and East Africa and eastern Asia to Hawaii, Polynesia and, chiefly, Australasia); *Cheiranthera* (four species, Australia); *Hymenosporum* (one species, *H. flavum*, Australia and New Guinea); *Bursaria* (three species, Australia).

BILLARDIEREAE. Fruit a berry. *Sollya* (two species, Australia); *Citriobatus* (four species,

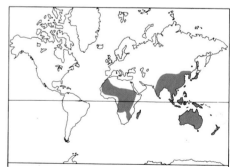

**Number of genera:** 9
**Number of species:** 200–240
**Distribution:** Australasia and Old World tropics.
**Economic uses:** ornamentals (*Pittosporum* and *Billardiera* species) and *Pittosporum* timber used locally.

Australia, one extending to Malaysia).

The genera may be divided by other characters:

Small trees, or shrubs without thorns; flowers not blue: seeds winged (*Hymenosporum*); seeds not winged (*Pittosporum*).

Shrubs with thorns: flowers in dense terminal panicles (*Bursaria*); flowers axillary (*Citriobatus*).

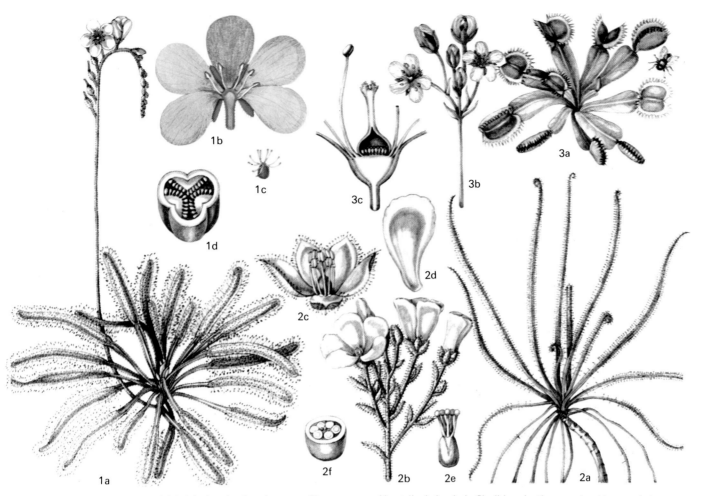

Droseraceae. 1 *Drosera capensis* (a) habit showing basal rosette of leaves covered in stalked glands ( ×⅔) ; (b) perianth opened out to reveal stamens ( ×2) ; (c) gynoecium ( ×2) ; (d) half section of ovary ( ×10). 2 *Drosophyllum lusitanicum* (a) habit ( ×⅔) ; (b) inflorescence ( ×⅔) ; (c) flower with petals removed ( ×2) ; (d) petal ( ×2) ; (e) gynoecium with free styles and rounded stigmas ( ×3) ; (f) half section of ovary ( ×4). 3 *Dionaea muscipula* (a) habit showing leaves modified to form a trap ( ×⅔) ; (b) inflorescence ( ×⅔) ; (c) vertical section of base of flower showing ovary with basal ovules ( ×4).

Subshrubs with climbing stems and slightly irregular flowers (*Cheiranthera*).

Subshrubs with climbing stems, flowers regular and often blue; anthers oblong, fruit a capsule (*Marianthus*), anthers oblong, fruit a berry (*Billardiera*); anthers linear and tapering, forming a closed cone around the style (*Sollya*); anthers linear, finally recurved from style (*Pronaya*).

The family is closely related to the Escallonioideae subfamily of the Saxifragaceae (sometimes regarded as a segregate family, the Escalloniaceae) but is distinguished by the presence of resin canals.

**Economic uses.** These are few and essentially confined to *Pittosporum*, whose wood is used locally and for inlay work. The chief use of *Pittosporum* (parchment-bark, Australian laurel) is ornamental. The following shrub and small tree species are grown for their flowers (purple, white or greenish yellow, sometimes very fragrant) and attractive foliage; they are hardy in warm temperate sheltered and "Riviera" sites: *Pittosporum crassifolium*, *P. ralphii*, *P. tenuifolium*, *P. eugenioides* (all from New Zealand), *P. tobira* (Japan and China) and *P. undulatum* (Victoria box, Australia). The Tasmanian

evergreen climber *Billardiera longiflora* (apple berry) is grown for its creamy white to purple flowers and blue edible berries.

F.B.H.

# DROSERACEAE
## *Sundews and Venus' Fly Trap*

The Droseraceae is a family of carnivorous annual or perennial herbs, sometimes with woody stems, comprising the genera *Drosera* (sundew), *Drosophyllum*, *Aldrovanda* and *Dionaea* (Venus' fly trap).

**Distribution.** *Drosera* is cosmopolitan, with concentrations in Australia and New Zealand. *Drosophyllum* is found in the Iberian peninsula, *Aldrovanda* throughout much of Eurasia southwards to Australia and tropical Africa, and *Dionaea* in southeastern United States of America. They commonly grow in bogs and other waterlogged soils, indeed *Aldrovanda* is aquatic, and the carnivorous habit may be an evolutionary response to their growth in habitats containing little or no available nitrogen.

**Diagnostic features.** The leaves are alternate, rarely in whorls, often in basal rosettes and may or may not have stipules. They are covered with sessile or stalked glands on the

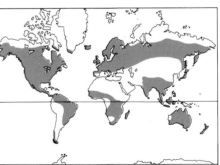

**Number of genera: 4**
**Number of species: about 83**
**Distribution: cosmopolitan, with centers in Australia and New Zealand.**
**Economic uses: ornamental curiosities.**

upper surface, and sometimes have median joints forming a trap. The flowers are regular, bisexual and are often in coil-like (circinnate) racemes, but sometimes solitary. There are five (sometimes four) sepals, more or less connate at the base, five (four) free petals and four up to 20 free stamens. The pollen is in tetrads. The ovary is superior, of two, three or five fused carpels with one locule, and three to numerous ovules on a basal placenta; the two to five styles are free or rarely connate. The fruit is a capsule which dehisces loculicidally into two to five valves; the seeds are usually numerous, containing endosperm and a small basal embryo.

*Drosophyllum* and *Drosera* have their leaves covered with long, red, gland-tipped hairs, motile in the latter genus, which entrap and digest insects by means of the proteolytic enzymes and ribonucleases which they secrete. This process is usually aided by bacterial activity, although this has not been detected in *Drosophyllum*, perhaps because of the antiseptic properties of the formic acid which is also secreted.

In *Dionaea* the leaves are divided longitudinally into two kidney-shaped halves which can come together like a trap. Each half-leaf has long marginal cilia (hair-like growths) and about three long sensitive slender hairs on the upper surface. When the sensitive hairs are touched, the halves of the leaf quickly swing on their common axis and the finger-like cilia intercross to form a barred cage. After subsequent movement the surfaces of the halves come into contact and any insect thus trapped is digested and absorbed by the secretory glands which cover the leaf surface. The trap opens and is active again in about 24 hours, but repeated stimulation greatly reduces the quickness of the motion.

In the aquatic *Aldrovanda* the leaves have a wedge-shaped basal part from which arise four to six long, narrow terminal segments and an orbicular lobe which is hinged along the midrib. This lobe can close rapidly (in about 1/50 second) to trap small aquatic animals, which are subsequently digested and absorbed by secretory glands on the leaf-surface.

**Classification.** The Droseraceae was formerly grouped with the other carnivorous families, Nepenthaceae and Sarraceniaceae, into a single order, Sarraceniales, but each family is now assigned to a different order on the basis of their individual affinities.

The carnivorous genera *Byblis* and *Roridula* bear considerable superficial resemblance to the Droseraceae and are sometimes placed in this family; other authorities recognize either two segregate families (Byblidaceae and Roridulaceae) or a single segregate family (Byblidaceae).

**Economic uses.** *Dionaea*, *Drosophyllum* and several species of *Drosera* are grown, usually in greenhouses, for ornament and interest.

D.M.M.

## BRUNELLIACEAE

The Brunelliaceae is a small family of tropical New World trees. There are about 45 species in *Brunellia*, the only genus in the family (not to be confused with *Brunella*, a synonym for the labiate generic name *Prunella*). *Brunellia* commemorates Gabriel

**Number of genera:** 1
**Number of species:** about 45
**Distribution:** New World tropics.
**Economic uses:** none.

Brunelli, onetime Professor of Botany and keeper of the Public Garden in Bologna, Italy. The family is restricted to the tropics of the New World from Mexico through the Caribbean into Peru. The leaves are simple opposite or in whorls, trifoliolate or compound and often densely hairy, with small caducous stipules. The flowers are regular, unisexual (sexes on separate plants) and borne in axillary or terminal panicles. They have no petals, but the calyx is divided into four to seven parts. There are eight to fourteen stamens in the male flowers. The female flowers have a superior ovary of two to five free carpels narrowing into elongate styles. These mature into dehiscent beaked fruits (follicles) which contain one or two seeds each. The seeds are black and shiny, containing flat cotyledons and a fleshy endosperm.

The family appears to be related to the Cunoniaceae and Simaroubaceae. The species have neither economic nor decorative value.

B.M.

## EUCRYPHIACEAE
*Eucryphias*

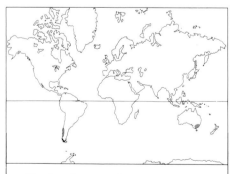

**Number of genera:** 1
**Number of species:** 5
**Distribution:** Chile, Tasmania, New South Wales.
**Economic uses:** hard timber and ornamental trees and shrubs.

The Eucryphiaceae is a family of evergreen (rarely half-evergreen or deciduous) trees or shrubs, consisting of a single Southern Hemisphere genus, *Eucryphia*.

**Distribution.** Of the five species in the genus, two are native to Chile, one to New South Wales in Australia, and two to Tasmania.

**Diagnostic features.** The leaves are opposite, simple or pinnately compound, sometimes trifoliolate, with odd terminal leaflets; the stipules are minute and connate. The flowers are bisexual and regular, large, white, axillary and usually borne singly. There are four imbricate sepals, cohering above, soon separating at the base and falling as one piece; the four (rarely five) petals are imbricate or convolute. The stamens are numerous and the filaments have basal tubular excrescences, inserted on an elongated axis below the (superior) ovary, which is formed of five to 12 fused carpels, with five to 12 (rarely 18) locules and as many styles. There are several pendulous ovules on axile placentas in each locule. The fruit is a woody or leathery capsule, dehiscing along the ventral sutures. The seeds are winged and the embryo embedded in endosperm.

**Classification.** In spite of its very disjunctive distribution, the family with its single genus is a very natural one. Phytochemical investigation of the flavanoids shows a close correlation with this distribution, the pattern in the three Australasian species being significantly very much simpler. Unfortunately, the flavanoid picture gives no direct information on the puzzling taxonomic position of the family, but indirectly supports a place in the order Rosales near the family Cunoniaceae.

**Economic uses.** In its native Chile, *Eucryphia cordifolia* makes a fine tree to some 24m (80ft), and the timber has been used for railway sleepers, telegraph poles, canoes, oars and cattle yokes. Indoor uses include furniture and flooring. The bark is a source of tannin. The timber of the Tasmanian species, *E. lucida*, is pinkish and used for general building as well as for cabinetmaking. *Eucryphia moorei* from New South Wales has similar applications.

F.B.H.

## BRUNIACEAE

The Bruniaceae is a small family of heath-like shrubby plants, some of which make attractive ornamentals.

**Distribution.** The family is almost confined to the Table Mountain sandstone areas of the Cape region of South Africa.

**Diagnostic features.** The slender twigs of the plants are densely covered in small, alternate leaves which are entire, needle-like, oblong and rigid, often with a black tip which is secretory on young leaves. There are no stipules. The more ornamental species have terminal spikes, panicles or heads of bisexual, regular flowers; others have small, solitary axillary ones. There are four or five sepals, free or sometimes joined forming a calyx tube which is more or less fused to the ovary. There are four or five petals, which may or may not be clawed, free or rarely

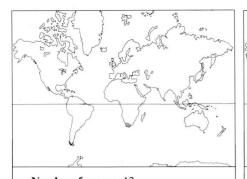

**Number of genera:** 12
**Number of species:** about 70
**Distribution:** Cape region of S Africa.
**Economic uses:** a few ornamentals.

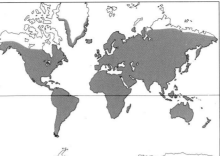

**Number of genera:** 122
**Number of species:** 3,370
**Distribution:** worldwide, centered in
N temperate regions.
**Economic uses:** major bush and tree
fruits of temperate regions (apples,
pears, cherries, plums etc) and many
valued ornamentals (roses, *Spiraea,
Filipendula, Sorbus, Cotoneaster* etc).

fused at the base. There are four or five
stamens alternating with the petals which
are generally free, or rarely partly fused to the
corolla; the anthers have two locules opening by long slits. The ovary is semi-inferior,
rarely inferior, of one to three fused carpels
and is crowned with two or three more or less
cohering styles. There are one to three
locules containing one to four pendulous
anatropous ovules. The fruits are either
capsules, splitting into two pieces containing
one to four seeds, or sometimes they are
indehiscent. The fruits are often decorated
with the remains of the sepals or petals. The
tiny oblong seeds contain copious fleshy
endosperm and a straight embryo.

**Classification.** The most important genera
are *Thamnea* (seven species), *Raspalia* (nine
species), *Staavia* (10 species) *Pseudobaeckea*
(10 species), *Berzelia* (seven species) and
*Brunia* (seven species). This is an isolated
family and is perhaps related to the Hamamelidaceae, but in this work is placed in the
Rosales.

**Economic uses.** Some species have been
grown for ornament both as cut flowers and
living plants and others are used for kind-
ling. B.M.

## ROSACEAE
*The Rose Family*

The Rosaceae is a large and important
family of woody and herbaceous plants. It is
valued both for its genera of bush and tree
fruits of temperate regions, including apples,
cherries, plums, peaches, raspberries and
strawberries, and for many popular horticultural ornamentals.

**Distribution.** The family is worldwide, but
with maximum development in the north
temperate area.

**Diagnostic features.** The Rosaceae includes
deciduous or evergreen trees, shrubs, shrublets or herbs (the last mostly perennial, but a
few annuals occur). There are few climbers
and no aquatics. The wood anatomy is
unspecialized and the vessels have simple
perforations. Branch thorns occur in *Crataegus, Prunus* and other genera, with surface
prickles in *Rosa* and *Rubus*. The leaves are
alternate (rarely opposite), simple or com-

pound, and typically bear a pair of stipules at
the base, although these may be difficult to
detect or even absent in a few genera
(*Exochorda, Spiraea* and allies). Glands are
commonly present, and are often paired at
the top of the petiole.

The flowers are characteristically insect-
pollinated and frequently large and showy: a
high percentage of all species are actual or
potential garden ornamentals. Usually the
flowers are regular and bisexual, showing
a series from hypogyny through perigyny to
epigyny, in which the carpels appear as if
swallowed up by the receptacle. *Rosa* is
unique in retaining free carpels although the
flower is epigynous. A common feature of
rosaceous flowers is the presence of an
epicalyx, that is, a second, smaller whorl of
five sepal-like organs below and alternating
with the regular calyx. These have been
interpreted as derived by the fusion of
stipules of adjacent sepals. The sepals and
petals are commonly five, although flower
doubling is common in cultivars developed
for ornament (*Kerria, Prunus, Rosa*, etc). It
arises by replacement of the numerous
stamens and sometimes also styles by petaloid organs. In extreme cases (some Japanese cherries and old-fashioned roses) the
center of the bloom may be green where the
carpels have been transformed into leaf-like
appendages. In such cases the flower is
usually completely sterile. The color range is
wide, but blue is almost completely absent.
The stamens are typically numerous and
whorled, not spirally arranged, and two,
three or more times the basic number of
petals. The anthers have two locules and
produce abundant pollen from longitudinal
splits. The carpels are normally numerous
and free, although varying degrees of fusion
occur, and in Prunoideae the number is
reduced to one. Each carpel contains usually
two anatropous ovules.

The fruits of Rosaceae are very diverse,
fleshy or dry, and provide important charac-

ters for dividing up the family. The fleshy-
fruited types, so important to European
horticulture, belong mainly to the Maloideae
(Pomoideae – the pome fruits), the Prunoideae (*Prunus* in the wide sense) and
Rosoideae (*Rubus* and *Fragaria*). The seeds
contain no endosperm, or only a trace.

Woody members of Rosaceae may pro-
pagate vegetatively by suckers, as in *Rubus*,
which also tip-roots in the brambles. Run-
ners (stolons) are a characteristic of some
herbaceous genera (*Fragaria*).

The flowers of the Rosaceae are mostly
among the simplest and least specialized for
pollination, relying on a large and wasteful
production of pollen which attracts a wide
range of insects, large and small. Some
genera, like *Rosa*, produce pollen only, but
most also secrete nectar from a disk surrounding the carpels. This disk may be freely
exposed (*Rubus*) or more or less screened by
the filaments (*Geum*). The latter flowers are
regarded as more highly evolved, eliminating
short-tongued visitors and attracting only
the longer-tongued flies and bees. Protandry
is the general rule, and self-compatibility is
exceptional. Several genera show a break-
down of the normal sexual cycle and may be
regarded as "evolutionary dead-ends." The
dog roses (*Rosa* section *Caninae*) are said to
be "subsexual" because half or more of the
chromosomes remain unpaired and are lost
prior to gamete formation. *Alchemilla,
Sorbus* and the brambles (*Rubus*) are more or
less completely apomictic.

A marked departure from insect pollination is found in the Sanguisorbeae, notably in the genera *Acaena* and *Poterium*.
These rely on wind-pollination. The flowers
are much reduced, in part unisexual, lacking
petals and nectar, and are massed together in
capitate or spicate heads.

**Classification.** The family can be divided into
groups, subfamilies and tribes as follows:

GROUP I: BASIC CHROMOSOME
NUMBER 7, 8 OR 9

SUBFAMILY SPIRAEOIDEAE

Two to five carpels (rarely one, or up to
twelve), whorled, not usually immersed in
the receptacle or elevated on a gynophore,
with two to many ovules; fruit usually
dehiscent.

SPIRAEEAE. Follicles with wingless seeds;
stipules almost or quite absent (*Aruncus,
Gillenia, Neillia, Physocarpus, Sibiraea,
Sorbaria, Spiraea, Stephanandra*).
EXOCHORDEAE. Capsules with winged seeds
(*Exochorda*).
HOLODISCEAE. Fruit indehiscent (*Holo-
discus*).

SUBFAMILY ROSOIDEAE

Carpels many (rarely few), on a usually
convex or conical gynophore, exposed or
enclosed in a hollow, persistent receptacle;
each carpel with one to two ovules; fruit
always indehiscent.

ULMARIEAE. Receptacle flat or weakly concave; filaments almost club-shaped, soon

1b

1a

2

3

4

5

6

7

8

9

10b

10a

10c

10d

15

11a

11b

16

14

13

12

Rosaceae. 1 *Rubus ulmifolius* (a) flowering shoot ($\times\frac{2}{3}$); (b) fleshy fruits ($\times\frac{2}{3}$). 2 *Rubus occidentalis* half flower showing hypogynous arrangement of parts ($\times 2$). 3 *Fragaria* sp vertical section of false fruit—a fleshy receptacle with the true fruits embedded in it ($\times 1$). 4 *Sanguisorba minor* leafy shoot and fruit ($\times\frac{2}{3}$). 5 *Agrimonia odorata* fruit comprising a receptacle covered with hooks enclosing the achenes (not visible) ($\times 4$). 6 *Rosa* sp, vertical section of hep showing urn-shaped receptacle enclosing the achenes ($\times 1\frac{1}{3}$). 7 *Potentilla agyrophylla* var *atrosanguinea* flowering shoot ($\times\frac{2}{3}$). 8 *Kerria japonica* flowering shoot ($\times\frac{2}{3}$). 9 *Rosa pendulina* flowering shoot clearly showing the stipules at the base of the leaf stalks ($\times\frac{2}{3}$). 10 *Chaenomeles speciosa* (a) flowering shoot ($\times\frac{2}{3}$); (b) vertical section of fruit comprising swollen receptacle and calyx enclosing the true fruits ($\times\frac{2}{3}$); (c) cross section of ovary ($\times 6$); (d) vertical section of flower showing epigynous arrangement of parts ($\times 2$). 11 *Cotoneaster salicifolius* (a) leafy shoot with fruit ($\times\frac{2}{3}$); (b) vertical section of fruit ($\times 2\frac{2}{3}$). 12 *Sorbus aria* flowering shoot ($\times\frac{2}{3}$). 13 *Prunus insititia* leafy shoot and fruits—drupes ($\times\frac{2}{3}$). 14 *Prunus* sp vertical section of fruit ($\times 1\frac{1}{3}$). 15 *Prunus avium* vertical section of flower with petals removed showing perigynous arrangement of parts ($\times 2\frac{2}{3}$). 16 *Spiraea cantoniensis* vertical section of flower ($\times 4$).

becoming deciduous (*Filipendula, Ulmaria*).

KERRIEAE. Receptacle flat or convex; carpels few, whorled; stamens many, narrowed above from a broad base (*Kerria, Rhodotypos*).

POTENTILLEAE. As Kerrieae, but carpels usually many, on a convex gynophore (*Dryas, Fragaria, Geum, Potentilla, Rubus*).

CERCOCARPEAE. Receptacle cylindrical, enclosing one carpel (*Cercocarpus*).

SANGUISORBEAE. Receptacle urn-shaped, usually hard, enclosing two or more achenes (*Acaena, Agrimonia, Alchemilla, Poterium, Sanguisorba*).

ROSEAE. Receptacle urn-shaped or cylindrical, soft in fruit, enclosing many free carpels (*Hulthemia, Rosa*).

### SUBFAMILY NEURADOIDEAE

Carpels five–10, united with one another and with the inner wall of the concave receptacle which is dry at maturity (*Neurada*).

### SUBFAMILY PRUNOIDEAE

Carpels one (rarely up to five), free, with terminal styles and pendulous ovules; fruit a drupe (*Prunus*, including *Amygdalus, Armeniaca, Cerasus, Laurocerasus, Padus, Persica*).

### GROUP II: BASIC CHROMOSOME NUMBER 17.

### SUBFAMILY MALOIDEAE

Carpels two to five, usually fused with the inner wall of the concave receptacle which together with the calyx enlarges to enclose the fruits as a pome (*Amelanchier, Aronia, Chaenomeles, Cotoneaster, Crataegus, Cydonia, Eriobotrya, Malus, Mespilus, Photinia, Pyracantha, Pyrus, Quillaja, Raphiolepis, Sorbus, Stranvaesia*).

The Rosaceae constitute the type family of the order Rosales, taken by some botanists in a narrow sense to include only two other minor families; by others in the broadest sense to include Leguminosae, Saxifragaceae and several other major families. This great discrepancy reflects perhaps no more than the artificiality of the order as a group in the taxonomic hierarchy.

Perhaps the closest ally of Rosaceae is the Saxifragaceae; it also has many common features with the Ranunculaceae, and the three large overlapping families can be separated only by taking a number of characters together – not by single absolute features. The Ranunculaceae differs in never having stipules and in having flowers almost entirely hypogynous, spirally arranged stamens, more carpels (usually) and endosperm in their seeds. The Saxifragaceae, on the other hand, is distinguished by being predominantly herbaceous, usually lacking stipules, and having a small number of whorled stamens and more or less united carpels. It, too, has endosperm in its seeds.

The fossil record shows the Rosaceae to be among the most ancient of dicotyledons, and its general structure and anthecology suggest that it is among the more primitive. Hutchinson regards it as an offshoot of the woody magnolian line of descent, and as being on a common line of evolution leading up to more specialized orders such as Leguminales (Fabales), Araliales (Umbellales) and the anemophilous tree orders Fagales and Juglandales. He points to the catkin-like, wind-pollinated inflorescences of *Poterium*, mentioned above, as giving a hint of how this last transformation might have occurred.

Taxonomic problems arise at all hierarchic levels in the Rosaceae: many of the controversies are of long standing. The tropical genera show far greater diversity in flower and fruit structure than the temperate, and include advanced features like irregular flowers, synandry, monoecy, apetaly and so on. Not surprisingly the subfamilies and most of the tribes have at one time or another been split off from the Rosaceae as separate families: Hutchinson lists the names of 26 segregate families. Of the subfamilies, the Spiraeoideae seems to be the least specialized in morphology, and has also contributed least to our gardens. By contrast, the well-defined subfamily Maloideae (Pomoideae) stands apart from all the rest not only by its unique fruit structure (pome) but by a basic chromosome number of 17, presumably derived from ancient amphidiploid hybrids between species with 8 and 9. Some botanists regard it as worthy of family status by itself: the Malaceae.

At the generic level controversy especially surrounds those of maximal economic interest, no doubt because their more intensive study has led to the discovery of more and finer taxonomic characters. Even the issue over whether or not the apple and pear should occupy the same genus or not is no nearer solution than it was two centuries ago! The subgenera within *Prunus* are raised to generic level by some botanists as *Amygdalus* (for the almonds), *Cerasus* (cherries), *Padus* (bird cherries) and *Laurocerasus* (Portugal and cherry laurels), etc.

As regards species, greatest problems concern those genera in which subsexual or asexual reproduction is the norm. These do not display the normal patterns of disjunct populations in the wild, and taxonomists are of two minds (or more!) on how to treat them. The single species *Rubus fruticosus* of Linnaeus has thus been split up into many hundreds of self-perpetuating microspecies by some.

**Economic uses.** Most of the important bush and tree fruits of temperate regions fall within the Rosaceae. Economically by far the most important is the apple (*Malus*), now grown in numerous hybrid cultivars of complex origin, with over 2,000 named varieties. Apples are grown mainly for dessert, but are also used for making cider. Annual world production is over 20 million tonnes. The next most important genus is *Prunus*, which produces almonds, apricots, cherries, damsons, nectarines, peaches and plums, all of which are grown extensively for consumption as fresh fruit and for canning and making into jams, conserves and liqueurs. Other major rosaceous fruits are blackberries, loganberries and raspberries (*Rubus*), loquats (*Eriobotrya*), medlars (*Mespilus*), pears (*Pyrus*), quinces (*Cydonia*) and strawberries (*Fragaria*).

Many *Prunus* species are also cultivated as ornamentals, notably the Japanese flowering cherries. However, it is the rose, the "queen of flowers," that overshadows all the other ornamentals, being probably the most popular and widely cultivated garden flower in the world, valued since ancient times for its beauty and fragrance. Modern roses are complex hybrids descended from about nine of the wild species. Rose-growing is now a large industry, with some 5,000 named cultivars estimated to be in cultivation.

Among other popular cultivated genera are herbaceous perennials such as *Alchemilla* (lady's mantle), *Geum* (avens), *Filipendula* (meadowsweet), and *Potentilla* (cinquefoil); and trees and shrubs such as *Amelanchier*, *Chaenomeles* (flowering quinces, including *C. lagenaria*, better known as the japonica), *Cotoneaster*, *Exochorda* (pearl bush), *Sorbus* (rowan, mountain ash), *Photinia* and *Pyracantha* (fire thorn).

Attar of roses is extracted from flowers of *Rosa damascena* and its production is a major industry in parts of western Asia and Bulgaria. The bark of *Moquila utilis*, the pottery tree of the Amazon, is used in making heat-resistant pots, and that of *Quillaja* species, the soap-bark tree, contains saponin, which is used as a substitute for soap in cleaning textiles; tannin is also extracted from the bark. G.D.R.

# CRASSULACEAE
*Stonecrops and Houseleeks*

The Crassulaceae is a family of succulent herbs and small shrubs.

**Distribution.** The Crassulaceae are distributed throughout the world, mainly in warm dry regions, but are centered in South Africa. Like the other two great families of succulents, Aizoaceae and Cactaceae, they are characteristic of hot, exposed, rocky habitats subjected to long periods of drought, but Crassulaceae have a wider range of adaptability: species of *Sempervivum* and some of *Sedum* are frost-hardy, and some species of *Crassula* and *Sedum* live in a plentiful supply of water, one *Crassula* being an aquatic.

**Diagnostic features.** The plants are perennial (rarely annual or biennial), soft-wooded and rarely larger than small shrublets. The leaves are always more or less fleshy and mostly entire, without stipules and commonly packed tightly in rosettes which may reach the extreme of surface reduction in a sphere (*Crassula columnaris*). A full range of xerophytic specializations is found, including surfaces covered in papillae, hairs, bristles or wax. Vegetative reproduction from offsets, adventitious buds and fallen leaves is common. The flowers are small but

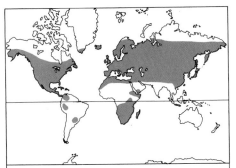

**Number of genera:** about 35
**Number of species:** about 1,500
**Distribution:** cosmopolitan, through the weedy species of *Crassula*, but centered in S Africa.
**Economic uses:** valued ornamentals for the rock garden and cool greenhouse, eg *Sedum* (stonecrops) and *Sempervivum* (houseleeks).

massed in showy corymbs or panicles. They are structurally simple, with equal numbers (normally five, but as few as three or as many as 30) of mostly perigynous free sepals, petals and carpels, and one (*Crassula*) or two whorls of stamens. The ovary is superior and the carpels (the same number as the petals) may be slightly joined at the base. The ovules

are numerous (rarely few) and inserted on the adaxial suture. The style is short or elongated. The fruit is a group of follicles, each with minute seeds, with little or no endosperm.

**Classification.** In general terms the family can be divided into two groups: those in which the stamens are as many as the petals and leaves are opposite, for example *Crassula* and *Rochea*; and those in which the stamens are twice as many as the petals and the leaves are alternate or opposite, for example *Kalanchoe* (flower parts in fours), *Umbilicus*, *Cotyledon* (corolla gamopetalous), *Sedum*, *Pachyphytum*, *Echeveria* (flower parts in fives) and *Monanthes*, *Sempervivum* and *Greenovia* (flower parts in sixes or more).

The Crassulaceae is closely related to the non-succulent Saxifragaceae.

**Economic uses.** The family is valued mainly as ornamentals: the hardy stonecrops (*Sedum* species) and houseleeks (*Sempervivum* species) for the rock garden or alpine house, the tender species for the succulent collection, *Echeveria* for summer bedding, and *Kalanchoe*, *Rochea* and a few others as house plants and florists' flowers. *Aeonium* is cultivated in warm regions.     G.D.R.

Crassulaceae. 1 *Echeveria nodulosa* (a) habit ($\times\frac{2}{3}$); (b) half flower showing fleshy petals and carpels joined at base ($\times 4\frac{1}{2}$). 2 *Pachyphytum longifolium* (a) shoot tip showing clusters of fleshy leaves and lateral inflorescence ($\times\frac{2}{3}$); (b) fruit ($\times 3$). 3 *Kalanchöe crenata* (a) habit showing opposite succulent leaves arranged on a fleshy stem ($\times\frac{2}{3}$); (b) tip of inflorescence bearing clusters of tubular flowers ($\times 1$).

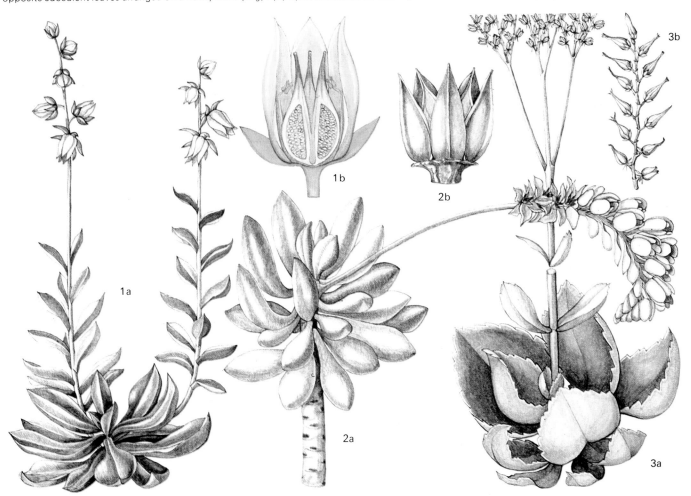

# CEPHALOTACEAE

*Flycatcher Plant*

The Cephalotaceae is a family represented by a single species of pitcher plants, *Cephalotus follicularis*, the Australian flycatcher plant.

**Distribution.** The species is native to Western Australia, where it grows in the drier parts of peaty swamps.

**Diagnostic features.** It is a perennial herb with short rhizomes. There are normal entire leaves and others consisting of a stalk which is broadened at the leaf base into a lid, and a blade which is modified into a pitcher decorated with fringed ribs up the sides. The mouth of the pitcher has a corrugated rim. Intermediate leaf forms also occur. The inflorescence, like the leaves, arises from soil level and is a leafless raceme of little cymes. The flowers are regular, bisexual and have a colorful calyx with six hooded lobes, but no petals. The 12 stamens are perigynous, six being longer than the others, and the filaments are free. The connective of the anther is swollen and glandular. The ovary is superior and consists of six free carpels in one whorl, each containing a single locule with one or rarely two basal, erect ovules,

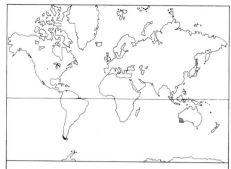

**Number of genera:** 1
**Number of species:** 1
**Distribution:** W Australia.
**Economic uses:** ornamental.

and crowned by a recurved style. The fruits are follicles, surrounded by the persistent calyx and contain a single seed with fleshy endosperm and a small, straight embryo.

The leaves start growth in July or August, are fully developed by October and are about 13cm (5in) long. The pitchers are mature by January and measure about 5cm (2in) long. It seems that a combination of partial digestion by pitcher-juices and bacterial decay provides nitrogenous materials from

trapped insects, but the plants can thrive without this supplement and are thus not obligate carnivores.

The inside of the pitcher consists of two recognizable zones. The upper is a smooth, glistening, opaque area with an overhang along its lower edge, the surface consisting of epidermis with downward-pointing hairs; this is contiguous with the lid. The lower smooth area is rich in secretory glands which have a digestive function, and has two lateral reddish-colored lumps of tissue endowed with large glands. The bottom lining of the pitcher is free of glands.

**Classification.** The curious evolutionary parallel between the pitcher-like leaves of *Nepenthes, Sarracenia* and *Cephalotus*, each of which occurs in a different geographical area, and family, has long been debated. Floral morphology has led botanists to think this family is related to the Saxifragaceae.

**Economic uses.** The plants are occasionally seen in glasshouse collections.          B.M.

# SAXIFRAGACEAE

*Currants, Hydrangeas and Saxifrages*

The Saxifragaceae is a large and widespread family consisting mainly of perennial herbs

Cephalotaceae. 1 *Cephalotus follicularis* (a) habit showing foliage leaves, pitchers and the leafless stalk to the inflorescence ($\times\frac{2}{3}$); (b) inflorescence—a raceme of small cymes ($\times\frac{2}{3}$); (c) pitcher and lid ($\times1\frac{1}{3}$); (d) half section of pitcher showing downcurved spikes at rim ($\times1\frac{1}{3}$); (e) section of flower showing hooded sepals, stamens of two lengths, a broad papillose disk and the gynoecium consisting of six free carpels ($\times8$); (f) long stamen with swollen glandular connective at apex ($\times24$); (g) short stamen ($\times24$); (h) vertical section of a carpel with single basal ovule ($\times24$); (i) flower from above ($\times4$); (j) flower ($\times6$); (k) fruit with part of wall removed to show single seed ($\times24$); (l) seed ($\times32$).

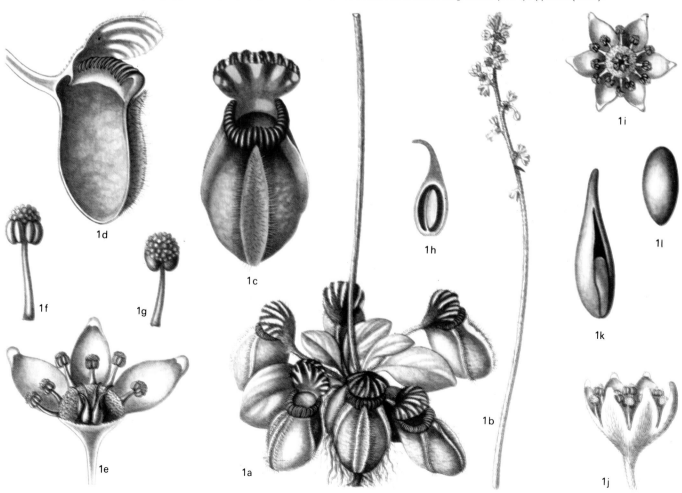

and shrubs, with a few annuals and a very few small trees. It contains the currants and gooseberries (both species of *Ribes*), as well as many popular garden flowers such as the hydrangeas and saxifrages. The family is here given a wide interpretation; some authors treat it much more narrowly and give family status to most of the subfamilies detailed below.

**Distribution.** On its wide interpretation the family is almost cosmopolitan. Its representation in the tropics, South Africa, Australia and New Zealand is, however, very scanty, and the great majority of the species are found in the north temperate zone, especially in eastern Asia and the Himalayas and in North America.

**Diagnostic features.** The uncertainty as to the limits of the family is due partly to its unspecialized nature and its "central" position, and consequently to the absence of any striking distinctive characters. The leaves are without stipules, and usually simple and alternate, but opposite or compound leaves are found in some genera. There are usually five valvate or imbricate sepals. The petals are always free, valvate or imbricate, and are usually four or five in number (occasionally

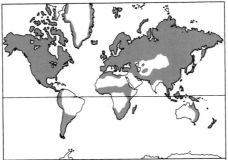

**Number of genera:** 80
**Number of species:** about 1,250
**Distribution:** almost cosmopolitan.
**Economic uses:** fruits (currants, gooseberries) and cultivated plants (eg saxifrages and hydrangeas).

absent); the stamens are typically twice as many as the petals, but may be more numerous. The ovary is superior or inferior with usually two or four carpels, united at least at the base, forming one to three locules, each with several rows of anatropous ovules on axile placentas. There are as many styles as carpels. The fruit is usually a small capsule, but in the Ribesoideae it is a berry with numerous seeds. The seeds contain

much endosperm and a small embryo.

**Classification.** The principal subfamilies into which the Saxifragaceae may be divided are as follows (some smaller ones are omitted):

ASTILBOIDEAE. Herbs with compound leaves; petals often reduced; carpels sometimes free. *Astilbe*; *Rodgersia*, with creeping rhizome and handsome, digitately compound leaves.

SAXIFRAGOIDEAE. Herbs; leaves usually simple; carpels usually two. *Saxifraga*; *Bergenia*, similar in flower, but with coarse, cabbage-like foliage; *Heuchera, Tellima, Tolmeia, Tiarella, Mitella*, all similar in habit to *Heuchera* but with less showy flowers, mostly woodland and damp meadow plants; *Chrysosplenium* (golden saxifrage), without petals.

FRANCOOIDEAE. Herbs; leaves simple but deeply lobed; four carpels. *Francoa*.

PARNASSIOIDEAE. Herbs; leaves entire; four carpels. *Parnassia* (grass of Parnassus), with curious staminodes, each bearing numerous shining yellow glands.

RIBESOIDEAE. Shrubs; ovary inferior; fruit a berry; leaves simple, alternate. *Ribes* (*Grossularia*).

HYDRANGEOIDEAE. Mostly shrubs; leaves simple, usually opposite; fruit a capsule.

Saxifragaceae. 1 *Parnassia palustris* (a) habit ($\times\frac{2}{3}$); (b) flower (viewed from above) showing fan-like staminodes alternating with the stamens ($\times 1\frac{1}{2}$); (c) fruit—a capsule with persistent staminodes at base ($\times 2$). 2 *Peltiphyllum peltatum* (a) leaf ($\times\frac{2}{3}$); (b) inflorescence ($\times\frac{2}{3}$); (c) flower with two petals removed to show stamens ($\times 2$); (d) fruits ($\times 2$). 3 *Bergenia crassifolia* (a) tip of shoot and inflorescence ($\times\frac{2}{3}$); (b) flower opened out showing central bicarpellate ovary ($\times 2$); (c) vertical section of ovary ($\times 2$). 4 *Itea virginica* (a) shoot and inflorescence ($\times\frac{2}{3}$); (b) fruit—a capsule ($\times 3$).

*Hydrangea, Philadelphus, Deutzia, Kirengeshoma.*

ESCALLONIOIDEAE. Shrubs; leaves simple, alternate; fruit a capsule. *Escallonia*; *Itea*, with evergreen, holly-like leaves and pendulous catkins of small, greenish flowers.

At various times some of these subfamilies and even groups within them have been separated from the Saxifragaceae and raised to the rank of family, eg Francoaceae, Parnassiaceae, Grossulariaceae, Hydrangaceae, Philadelphaceae, Escalloniaceae, Iteaceae.

The family is obviously closely related to the Rosaceae, and indeed *Astilbe* has been repeatedly confused with *Aruncus, Filipendula* and other members of the Rosaceae.

**Economic uses.** Gooseberries, black, red and white currants, all from the genus *Ribes*, are widely grown for their edible fruits. The most popular garden plants of the family are the hydrangeas and saxifrages. All the genera listed above are also widely cultivated garden ornamentals. D.A.W.

# CHRYSOBALANACEAE
## Coco Plum

The Chrysobalanaceae is a family of trees

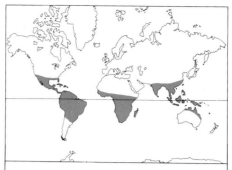

**Number of genera:** 17
**Number of species:** about 400
**Distribution:** tropical and subtropical lowlands.
**Economic uses:** fruits (coco plum), seed oil and local uses for timber.

and shrubs, some of which are locally important in the tropics as fruit trees.

**Distribution.** Most of the species are restricted to the lowlands of the tropics and subtropics. Large genera include *Parinari* (pantropical, 43 species), *Hirtella* (89 species in tropical America, one in Africa and Madagascar), *Couepia* (58 species, tropical and subtropical South America), *Licania*

(151 species in tropical America, one in Malaysia and New Caledonia), *Magnistipula* (10 species in tropical Africa) and *Maranthes* (10 species in Africa, one in Malaysia, one in Central America). There are some curious intercontinental distributions, for *Parinari excelsa* and *Chrysobalanus icaco* are said to occur in both tropical America and Africa, and the Asian *Maranthes corymbosa* is closely allied to the American species.

**Diagnostic features.** The leaves are simple, alternate and with stipules. The flowers are bisexual, less commonly unisexual, more or less irregular and markedly perigynous, with five sepals, up to five petals and two to 300 (*Couepia*) stamens. The anthers, borne on filiform filaments, dehisce introrsely. The ovary is superior and has three carpels, but only one develops, usually with two erect basal ovules. The style is simple, with a simple stigma. The fruit is a dry or fleshy drupe with a bony endocarp (stone). The seeds have no endosperm.

**Classification.** Although many of the genera seem superficially very similar, they are readily separable and characterized.

The family is closely allied to the Rosaceae and was formerly included in it, but differs,

Chrysobalanaceae. 1 *Moquilea canomensis* (a) shoot with axillary inflorescences ($\times\frac{2}{3}$); (b) flower with five free sepals and petals and numerous stamens ($\times 3$); (c) vertical section of ovary with two erect ovules ($\times 3$). 2 *Licania incana* half flower showing epipetalous stamens and ovary with single basal ovule ($\times 8$). 3 *Acioa pallescens* (a) flower with bundle of stamens and filamentous style ($\times 2$); (b) fruit ($\times\frac{2}{3}$). 4 *Hirtella zanzibarica* (a) flower opened out showing numerous epipetalous stamens with long filaments and globose anthers and lateral ovary ($\times 3$); (b) vertical section of ovary with single basal ovule and base of feathery style ($\times 6$); fruit (c) entire and (d) in vertical section ($\times 1$).

1b  1c  1a  2  4c  4d  4b  3b  3a  4a

among other things, in its erect ovules. The Australian shrubby genus *Stylobasium*, with two species, is sometimes made into a family of its own, Stylobasiaceae, perhaps allied to the Anacardiaceae.

**Economic uses.** The wood of *Licania terna-tensis* is hard and is used in construction under water and underground, as well as for charcoal. In the Solomon Islands, *Maranthes* and *Parinari* are two important genera of timber trees.

Several species are cultivated for their fruit. The most important is the coco plum, *Chrysobalanus icaco. Licania pyrifolia* (mere-cure) is grown in Venezuela. *Parinari excelsa* is the Guinea plum, which is rather dry and mealy, *P. macrophylla* is the ginger-bread plum, and *P. curatellifolia* is the straw-berry-flavored mobola plum, all eaten in Africa. The fruit of the last is used in beer-making and a red dye is extracted from the young leaves in West Africa. The fruits of the African *P. capensis* and *Magnistipula butayei* are edible and the latter tree is of importance in rain-making ceremonies. Oil may be extracted from the seeds of many species; *Licania rigida* (oiticica) is grown for this purpose in Brazil, the oil being used as a substitute for tung oil, while the oil of *L. arborea* is used in candle- and soap-making.

D.J.M.

# FABALES

## LEGUMINOSAE
*The Pea Family*

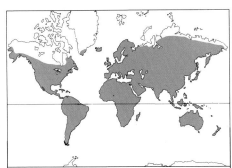

**Number of genera:** 700
**Number of species:** 17,000
**Distribution:** cosmopolitan.
**Economic uses:** important food crops (peas, beans, groundnut, soybean etc) forage crops (clover, lucerne) or ornamentals (broom, *Acacia*, sweet pea etc) and many other uses such as sources of dyes and timber.

The Leguminosae or Fabaceae is a very large family of herbs, shrubs and trees with a great variety of habit, including aquatics, xerophytes and climbers. Many species are of enormous importance to Man.

**Distribution.** The approximately 700 genera and 17,000 species have a cosmopolitan distribution in tropical, subtropical and temperate zones. Papilionoideae is the only

one of the three subgroups which is extensive in temperate regions.

**Diagnostic features.** The leaves are usually alternate, pinnately compound and with stipules. However, there are many exceptions: in gorse (*Ulex*) they are simple and small while in many species of *Acacia* the pinnate leaves are not developed in the young seedlings and the petioles become flattened into phyllodes. The stipules of some species of *Acacia* and *Robinia* develop into spines, while those of the garden pea (*Pisum*) are large and leaf-like. The number of leaflets on pinnate leaves varies greatly and can be of diagnostic value. For example, clover (*Trifolium*) and medick (*Medicago*) have leaves composed of only three leaflets while those of most species of vetch (*Vicia*) are composed of three to twelve pairs. Many species of Leguminosae have leaves which can alter their position at night; this usually involves the folding up of the leaflets. In some species of *Mimosa* (as in *Mimosa pudica*) the leaflets can be made to open merely by a shock stimulus, such as touch. The members of the family that can climb have twining stems (as in *Wisteria*) or tendrils or hooks which represent modified branches, leaves or leaflets.

A feature common to most Leguminosae is the presence of root nodules containing bacteria (*Rhizobium* species) which are capable of taking up atmospheric nitrogen and converting it into other nitrogenous compounds. Thus the leguminous plant can benefit from an augmented nitrogen supply and can grow well on relatively poor soils.

The inflorescence is usually an erect or pendulous raceme and sometimes, as in *Mimosa*, the flowers are arranged in tight clusters. The flowers are regular and unisexual or bisexual in Mimosoideae, and irregular and bisexual in Caesalpinioideae and Papilionoideae. There are five sepals which are more or less fused and and in some irregular flowers the calyx may be organized into two or four lobes. In Mimosoideae the five petals are small and equal, while in the Caesalpinioideae there is a range of irregularity in the corolla extending from the flowers of *Cassia* in which the five petals are nearly the same size, to those of the Judas tree (*Cercis siliquastrum*) which has flowers similar to those of the Papilionoideae. The five petals of the flowers of the Papilionoideae are organized into a butterfly shape with one upstanding dorsal petal (the standard), two lateral petals (the wings) and two lower ventral petals more or less fused along their contiguous margins to form a keel.

The stamens are usually numerous showing partial fusion of the filaments in Mimosoideae, while they are usually 10 or fewer in number, and free, in Caesalpinioideae. The 10 stamens in flowers of the Papilionoideae may be free as in *Sophora* but many have all the filaments fused together (monadelphous) as in *Ulex* or the filaments

of nine only may be fused with the dorsal stamen free (diadelphous) as in *Vicia*. The stamens are enclosed by the keel and the fused filaments form a tube around the ovary.

All the Leguminosae characteristically have a single carpel, the single ovary being superior, and surmounted by style and stigma. The ovules vary in number from two to many and are inserted in two alternating rows on a single placenta. Typically the fruit is a one-chambered pod (legume) which may be constricted between the seeds (lomentum). It is sometimes indehiscent as in the ground-nut (*Arachis*) or it may open explosively as in broom (*Cytisus*), gorse (*Ulex*) and lupin (*Lupinus*), as the two walls of the pod twist up violently to throw out the seeds. The legumes may be dry or fleshy, inflated or compressed, winged or not, greenish or brightly colored and range in size from a few millimetres to 30cm (1ft) or more. The seeds vary in number from one to many and often possess a tough coat. They contain a large embryo with little or no endosperm and in some genera they possess a colored appendage called a caruncle.

**Classification.** The chief characteristics of the three subfamilies (or families) are:

MIMOSOIDEAE (Mimosaceae). Mainly tropical and subtropical trees and shrubs (approximately 56 genera and 500–3,000 species). The leaves are often bipinnate and the flowers are regular with the petals valvate in bud, and with 10 or more stamens.

CAESALPINIOIDEAE (Caesalpiniaceae). Mainly tropical and subtropical trees and shrubs (approximately 180 genera and 2,500–3,000 species). The leaves are usually pinnate but sometimes bipinnate and the flowers are usually more or less irregular with the lateral petals (wings) covering the standard in the bud. There are 10 or fewer stamens, free or monadelphous.

PAPILIONOIDEAE (Papilionaceae). Temperate, tropical and sub-tropical in distribution, mostly herbs but some trees and shrubs amongst its 400–500 or so genera and upwards of 10,000 species. The leaves are usually pinnate but sometimes simple. The flowers are irregular with the lateral petals enclosed by the standard in the bud. There are 10 stamens, usually diadelphous but sometimes monadelphous or free.

The genera of Mimosoideae can be separated into five or six tribes on the basis primarily of the nature of the leaves and the number and degree of fusion of the stamens. Thus *Inga* (pinnate leaves), *Pithecellobium*, *Calliandra* (bipinnate leaves) and *Acacia* all have flowers with numerous stamens which are fused into a tube in the first three genera but free in the latter. Stamens are five to 10 in number in *Mimosa*, *Neptunia* and *Prosopsis*, the anthers being crowned with a gland in the latter two genera. An alternative classification divides the Mimosoideae into five groups based on their pollen types.

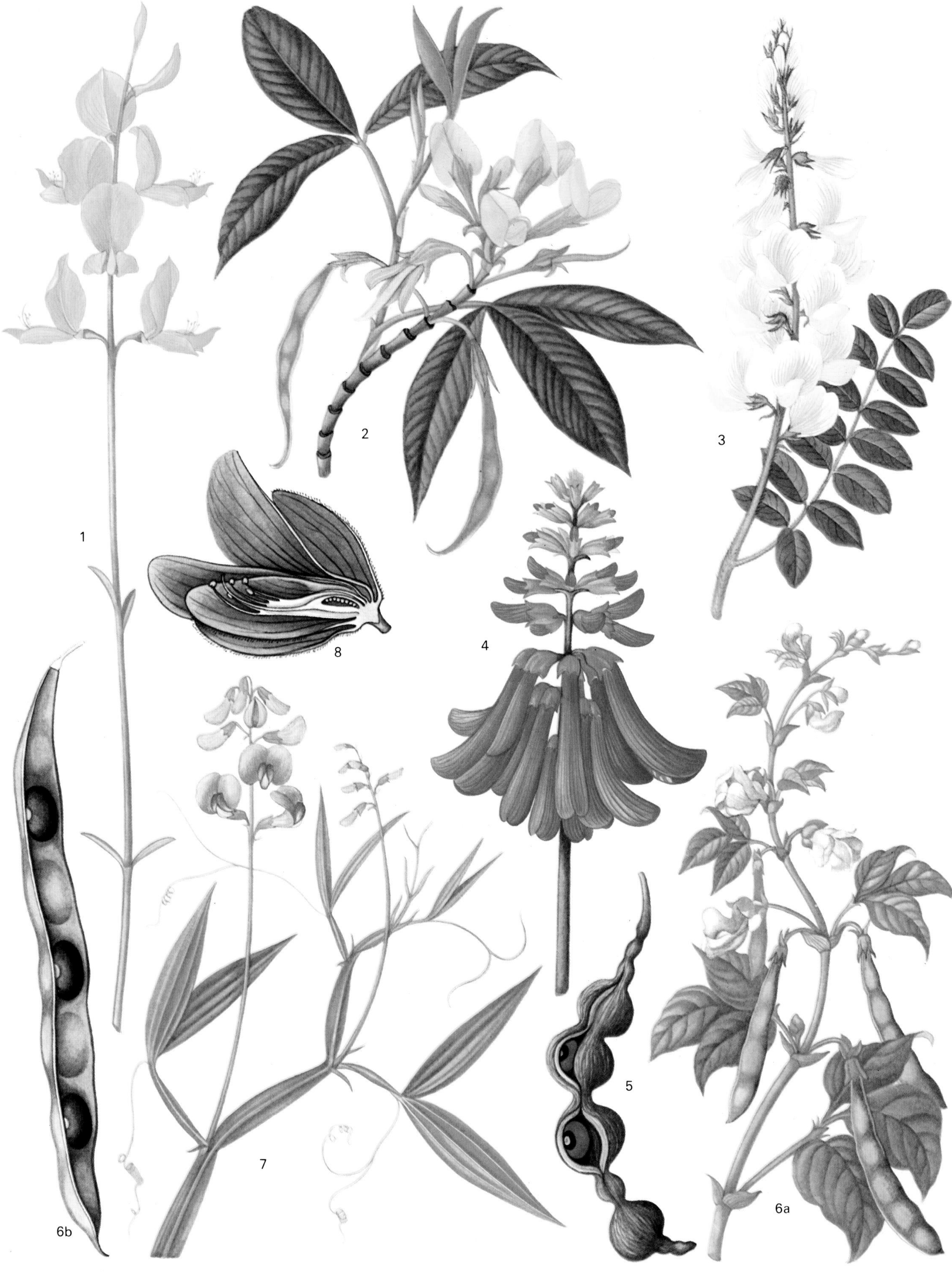
1
2
3
4
5
6a
6b
7
8

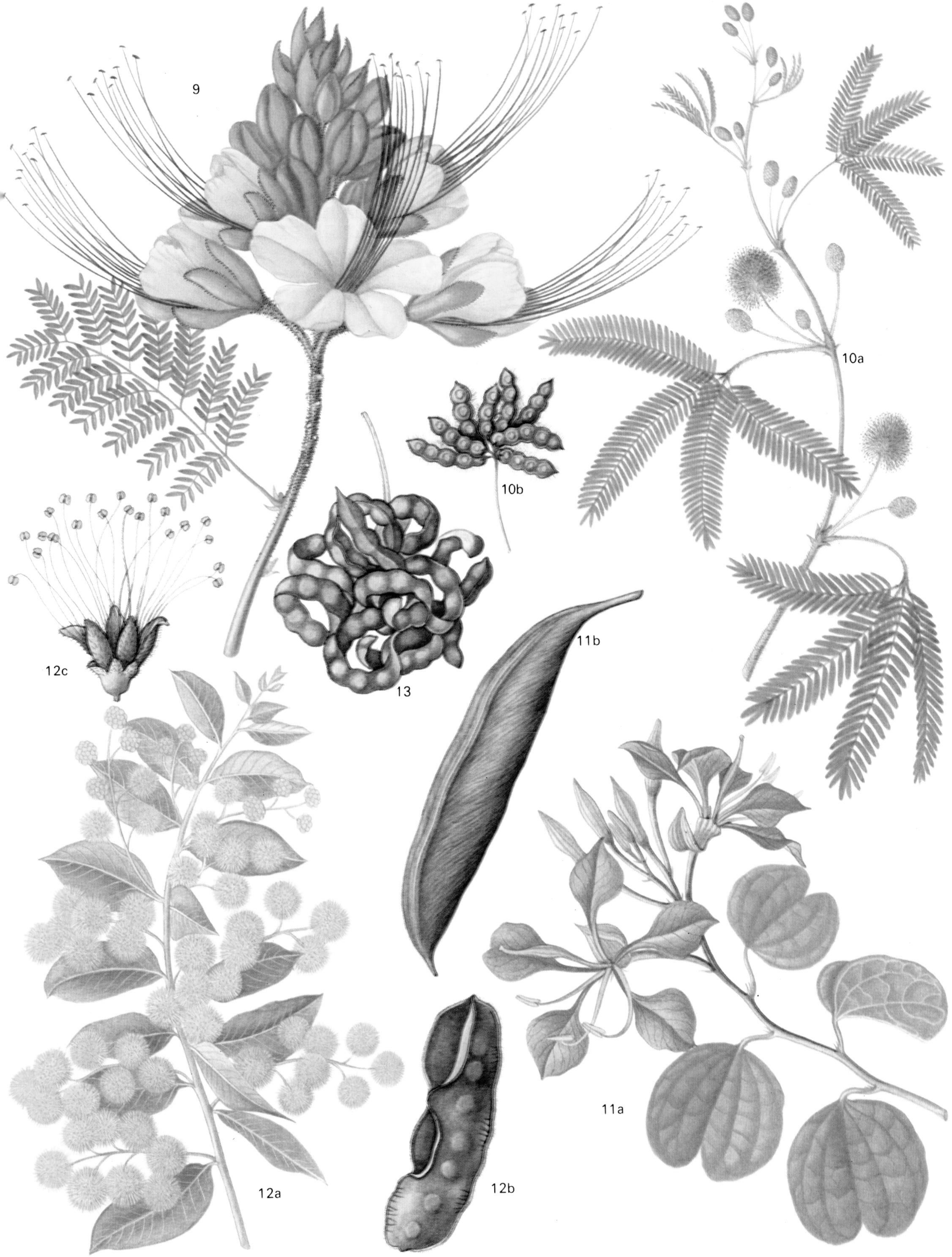

9
10a
10b
11b
12c
13
11a
12a
12b

Leguminosae. 1 *Spartium junceum* inflorescence – a raceme ( ×⅔). 2 *Piptanthus nepalensis* shooting bearing trifoliolate leaves with stipules, flowers and fruit ( ×⅔). 3 *Onobrychis radiata* inflorescence and pinnate leaf ( ×⅔). 4 *Erythrina humeana* inflorescence ( ×⅔). 5 *Erythrina abyssinica* dehiscing fruit (a pod or legume) and seeds ( ×⅔). 6 *Phaseolus vulgaris* (a) shoot bearing flowers and immature fruit ( ×⅔); (b) mature fruit with half of pod removed to show seeds ( ×⅔). 7 *Lathyrus sylvestris* shoot bearing leaves, tendrils and inflorescence ( ×⅔). 8 *Ulex europaeus*, half flower showing hairy sepals, upstanding standard petal, lateral wing petal and within it the keel petal which surrounds the stamens that have their filaments fused and ovary with numerous ovules ( ×2⅔). 9 *Caesalpinia gilliesii* shoot with bipinnate leaf and terminal inflorescence ( ×⅔). 10 *Mimosa pudica* (a) shoot with sensitive, bipinnate leaves with four secondary stalks and axillary tight clusters of flowers ( ×⅔); (b) clusters of mature fruit – compressed follicles ( ×⅔). 11 *Bauhinia galpinii* (a) shoot with simple, bilobed leaves and terminal inflorescence ( ×⅔); (b) mature fruit ( ×⅔). 12 *Acacia podalyriifolia* (a) shoot with simple leaves and globose inflorescences ( ×⅔); (b) mature fruit ( ×⅔); (c) flower which is regular and has numerous stamens. 13 *Dichrostachys cinerea* cluster of twisted follicles ( ×⅔).

The Caesalpinioideae can be divided into seven, eight or nine tribes or groups of genera on the basis of a number of characters, including the nature of the leaves, the irregularity of the flower, the degree of fusion of the sepals, and the mode of dehiscence of the anthers, but many of the groups are unsatisfactorily delimited or unnatural. Both *Poinciana* (calyx of five unequal lobes) and *Caesalpinia* (sepals free) have bipinnate leaves while *Cassia* (anthers dehiscing by pores) and *Cynometra* (anthers dehiscing by slits) have once-pinnate or simple leaves and free sepals. On the other hand both *Bauhinia* and *Cercis* possess a short-toothed calyx and simple entire or lobed leaves.

The position of the tribe Swartzieae is debatable. It is a small tropical African and South American group of 10 genera and is unusual in that the calyx is enclosed and entire in bud, becoming divided into lobes or slits as the flower opens. It is usually placed in the Caesalpinioideae as the last tribe, but other authorities place it in the Papilionoideae, or even recognize it as a separate subfamily.

The Papilionoideae is considered to consist of 10 or 11 tribes into which the genera can be grouped on features of habit, leaf form, and degree of fusion of the stamens. SOPHOREAE. Mainly trees or shrubs with leaves pinnate or simple; stamens free. PODALYRIEAE. Mostly shrubs (a few herbs) with simple or palmate leaves; stamens free. GENISTEAE. Mainly shrubs (a few herbs) with simple, palmate or pinnate leaves; stamens usually monadelphous; anthers some large and some small. TRIFOLIEAE. Mainly herbs (a few shrubs) with pinnate or palmate trifoliolate leaves; stamens usually diadelphous but occasionally monadelphous; anthers all alike. LOTEAE. Herbs or undershrubs with pinnate leaves or three or more entire leaflets; stamens monadelphous or diadelphous; anthers all alike. GALEGEAE. Herbs, shrubs (a few trees and climbing shrubs), with pinnate leaves, usually with five or more leaflets. HEDYSAREAE. Herbs, shrubs or woody climbers with usually diadelphous stamens and fruit a lomentum. VICIEAE. Herbs with pinnate leaves ending in a point or tendril; stamens usually diadelphous. PHASEOLEAE. Usually climbing herbs (a few shrubs and trees eg *Erythrina*) with pinnate (usually trifoliolate) leaves; stamens usually diadelphous. DALBERGIEAE. Trees or shrubs (a few climbers) with pinnate leaves of five to numerous pairs of leaflets; stamens monadelphous or diadelphous and fruit indehiscent.

There is little doubt of the close relationship to each other of the three subfamilies of the Leguminosae. The group as a whole has probably been derived from rosaceous ancestors, with perhaps the closest relationship with the Rosaceae (especially the segregate family Chrysobalanaceae) being shown by the Caesalpinioideae.

**Economic uses.** The family is of major economic importance. In the Mimosoideae, *Acacia* yields a number of valuable products. The Australian black wattle (*Acacia decurrens*) and golden wattle (*A. pycnantha*) are the source of wattle bark, which is used in tanning. A number of species, including the Australian blackwood (*A. melanoxylon*) and *A. visco*, are the source of useful timbers. Species including *A. stenocarpa* and *A. senegal* yield gum arabic while the pods and beans of the Mexican mesquite tree (*Prosopsis juliflora*) are ground up and used as an animal stock feed. A number of *Albizia* species are valuable timber trees.

Caesalpinioideae also contains a number of useful species, including *Cassia acutifolia* and *C. angustifolia* native to the Middle East, whose dried leaves are the source of the purgative senna. Several *Caesalpinia* species are sources of dyes and timber. The pods of the tamarind (*Tamarindus indica*) are used as a fresh fruit and for medicinal purposes in India, while a number of species, such as the flamboyant tree, *Delonix regia* (*Poinciana regia*) and species of *Caesalpinia* (eg *Caesalpinia pulcherrima*, pride of Barbados), are grown as ornamentals in the tropics and in greenhouses in temperate zones.

The Papilionoideae is especially important because the seeds and pods of many of the herbaceous species are sources of human and animal food. They are of particular value in the protein-deficient areas of the world because they are rich in protein as well as mineral content. Certain species such as clover (*Trifolium repens*) and lucerne (*Medicago sativa*) either can be used for feeding livestock or can be ploughed into the soil, functioning as excellent fertilizer and greatly increasing the nitrogen levels of the soil. Among the better-known species used as human food are the garden pea (*Pisum sativum*), chick pea (*Cicer arietinum*), French, haricot, snap, string, green or kidney bean (*Phaseolus vulgaris*), broad bean (*Vicia faba*), pigeon pea (*Cajanus cajan*), grass pea (*Lathyrus sativus*), lablab (*Dolichos lablab*), jack bean (*Canavalia ensiformis*), lima bean (*P. lunatus*), mung bean (*P. aureus*), scarlet runner (*P. coccineus*), lentil (*Lens culinaris*), soybean (*Glycine max*), and groundnut (*Arachis hypogea*). The cowpea (*Vigna sinensis*) clover and lucerne are widely used as forage plants.

Many genera contain species highly prized as ornamentals in both temperate and tropical countries. Some of the better-known in this category are lupin (*Lupinus*), broom (*Cytisus*), *Laburnum*, sweet pea ( *Lathyrus*), *Baptisia*, *Wisteria* and *Genista*.

The twigs, leaves and flowers of *Genista tinctoria* were the source of a yellow dye used for coloring fabrics. Species of *Indigofera* yield the dye indigo.       S.R.C.

# PODOSTEMALES

## PODOSTEMACEAE

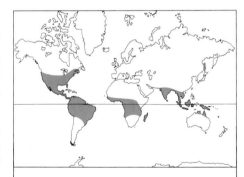

**Number of genera:** 45
**Number of species:** about 130
**Distribution:** mainly tropical, in streams and waterfalls.
**Economic uses:** none.

The Podostemaceae is a family of aquatic herbs resembling mosses.
**Distribution.** The plants are found widely in the tropics growing in rapidly flowing water, attached to rocks and stones.
**Diagnostic features.** These submerged freshwater herbs have a highly modified and varied vegetative structure, comprising a thallus and usually root-like organs (haptera) that anchor the plants. The flowers are regular, bisexual, either solitary or in a cymose inflorescence and when young are often enclosed in a spathe of enlarged, partially united bracts. The perianth comprises two to three sepals, united at the

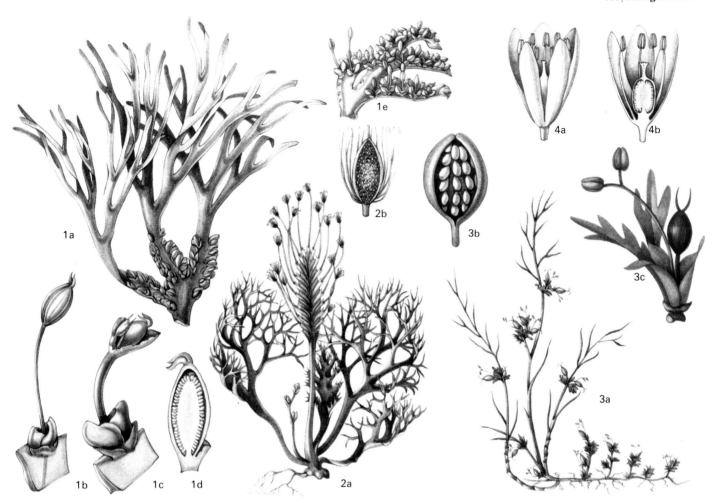

Podostemaceae. 1 *Dicraeia algiformis* (a) ribbon-like thallus (shoot) bearing flowers ($\times\frac{2}{3}$); (b) fruit—a capsule ($\times 3$); (c) expanded flower with basal involucre ($\times 3$); (d) vertical section of ovary showing numerous ovules on axile placentas ($\times 6$); (e) portion of thallus bearing flowers and fruit ($\times\frac{2}{3}$). 2 *Mourera weddelliana* (a) habit showing basal rhizoids and erect inflorescence ($\times\frac{2}{3}$); (b) vertical section of ovary surrounded by persistent stamens ($\times 4$). 3 *Podostemum ceratophyllum* (a) habit ($\times\frac{2}{3}$); (b) fruit with wall cut away ($\times 6$); (c) flower with two divided bracts, three green sepals, stalked ovary and two stamens with filaments fused at base ($\times 3$). 4 *Weddellina squamulosa* entire (a) and half (b) flower ($\times 9$).

base; petals are absent. There are one to four (sometimes numerous) hypogynous stamens; the filaments are either free or partially united and the anthers have two locules, dehiscing lengthwise. The ovary is superior, of two united carpels, forming two locules each with a large axile placenta bearing numerous ovules. The styles are free and slender. The fruit is a capsule containing numerous minute seeds without endosperm.

The seeds are dispersed during the dry season and do not germinate until they are submerged in the rainy season. A thallus grows from the seed and from it the haptera are produced. Flowers are not produced until the end of the rainy season and after fruiting the thallus usually dies.

**Classification.** The most common genera are *Apinagia* (50 species, South America), *Rhyncholacis* (25 species, South America), *Marathrum* (25 species, tropical America), *Podostemum* (17 species, pantropical), *Castelnavia* (nine species, Brazil), *Mourera* (six species, tropical America), *Dicraeia* (five species, Madagascar). *Weddellina* (one species tropical South America).

The peculiar structure of the family suggests no obvious close alliance with any other group, and the family is placed in its own order, Podostemales. It is generally felt that this is allied to Rosales and perhaps approaches Saxifragaceae or Crassulaceae (particularly the semiaquatic *Crassula* species) of that order. It may have affinities with the Hydrostachydaceae, another group of aquatics, which has sometimes been placed in the Podostemales but is nowadays referred to the Scrophulariales.

**Economic uses.** None are known.

D.J.M.

# HALORAGALES

## THELIGONACEAE

The Theligonaceae is a small family of annual and perennial herbs.

**Distribution.** The family occurs in the Mediterranean region, in China and Japan.

**Diagnostic features.** Members of the family have entire, ovate, fleshy leaves; the lower are opposite and the upper alternate, by the suppression of one leaf of each pair. There are peculiar united membranous stipules. Large club-shaped glands are present at the

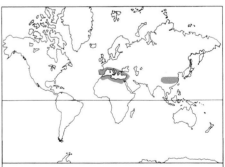

**Number of genera:** 1
**Number of species:** 2 or 3
**Distribution:** temperate N Hemisphere.
**Economic uses:** young shoots used as vegetable.

apex of the leaves. The flowers are unisexual (with both sexes on the same plant), in one- to three-flowered clusters. The male flowers have a valvate to globose perianth, splitting into two to five lobes when the flower opens. There are seven up to 12 (sometimes as few as two and as many as 30) stamens with filiform filaments and anthers that are erect in the bud, but pendulous later. The female flowers have a tubular, shortly toothed

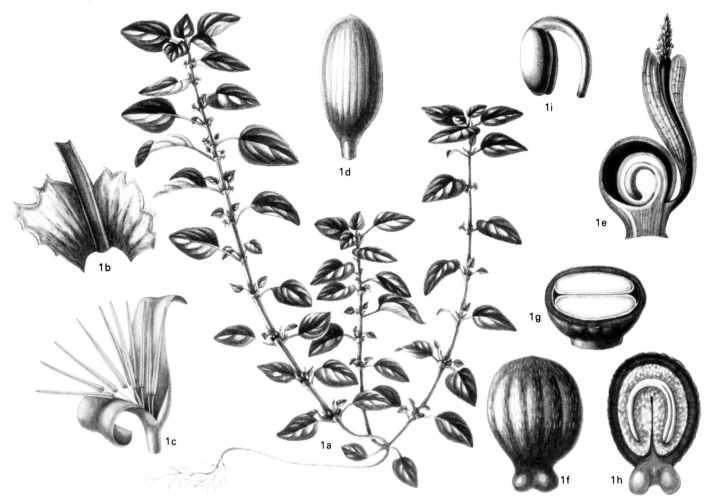

Theligonaceae. 1 *Theligonum cynocrambe* (a) habit showing leaves opposite at base of stem and alternate at the apex ( ×⅔) ; (b) united membranous stipules ( ×6) ; (c) male flower with bilobed perianth and numerous stamens with short, filiform filaments and long anthers ( ×10) ; (d) male flower bud ( ×10) ; (e) half female flower with lateral tubular perianth and style that arises from the base of the ovary ( ×10) ; (f) fruit—a nut-like drupe ( ×10) ; (g) cross section fruit ( ×10) ; (h) vertical section of fruit showing single seed with a curved embryo embedded in fleshy endosperm ( ×10) ; (i) embryo with two large, globose cotyledons ( ×12).

perianth and an ovary of a single carpel containing a single basal ovule. The style is simple and arises from the base of the ovary. The ovary enlarges irregularly on one side and the style becomes lateral at fruiting time. The fruit is a subglobose nut-like drupe containing one seed with fleshy endosperm.

The genus *Theligonum* exhibits myrmecochory (dispersal of seeds by the agency of ants). Ants feed on the oil body or elaiosome of various seeds and frequently carry the seed some distance from the parent plant. The oil body of the seeds is formed of a portion of the pericarp which remains attached to the base of the seed. The ants eat the oil body and then leave the seed undamaged.

**Classification.** The family is represented by one genus, *Theligonum*, with two or possibly three species. *Theligonum cynocrambe* grows in damp shady rock habitats in the Mediterranean region, as a glabrous somewhat succulent annual. *Theligonum japonicum* grows in the mountains and woods of western China and Japan, as a creeping, branched perennial herb with a line of short recurved hairs on the stems.

The relationship of the family Theligo-naceae (sometimes called the Cynocram-baceae) with other families is much disputed. Morphological observations have suggested a relationship with the Haloragaceae, Hippuridaceae or Portulacaceae, but there seems little anatomical evidence to support this. The Theligonaceae have also been regarded as part of the Urticaceae but this relationship is also doubtful.

**Economic uses.** Young shoots of *T. cynocrambe* are sometimes eaten as a vegetable.

S.A.H.

# HALORAGACEAE

## Gunneras, Water Milfoil, Haloragis

This family of aquatic or moist terrestrial herbs ranges from delicate aquatics, as in *Myriophyllum* (the water milfoil), to large robust species of *Gunnera* growing in forest margins, with leaves up to 6m (20ft) in circumference and stout inflorescences up to 1.5m (5ft) high.

**Distribution.** The family occurs in temperate and subtropical regions, chiefly in the Southern Hemisphere.

**Diagnostic features.** The aquatic representatives, like many of those of comparable habit in other families, show heterophylly,

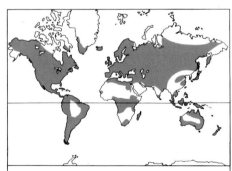

**Number of genera:** 7
**Number of species:** about 180
**Distribution:** temperate and subtropical, chiefly S Hemisphere.
**Economic uses:** garden ornamentals and ground cover (*Gunnera*) and limited use in tanning (*G. chilensis*).

with the submerged leaves pinnately divided into unbranched capillary segments and the aerial leaves normally simple, and toothed or entire.

The plants are annual or perennial herbs, aquatic or terrestrial, sometimes with a persistent woody base. The leaves are opposite to alternate or in whorls, entire to

Haloragaceae. 1 *Myriophyllum spicatum* shoot bearing submerged much divided leaves and aerial inflorescences ($\times\frac{2}{3}$). 2 *M. pedunculatum* (a) tip of inflorescence with separate male and female flowers ($\times 6$); (b) male flower ($\times 8$); (c) female flower with plumed stigmas ($\times 8$). 3 *Gunnera magellanica* (a) habit ($\times\frac{2}{3}$); (b) female inflorescence ($\times\frac{2}{3}$); (c) tip of male inflorescence each flower comprising only two stamens ($\times 4$); (d) female flower ($\times 4$); (e) fruit ($\times 4$). 4 *Haloragis cordigera* (a) flowering shoot ($\times\frac{2}{3}$); (b) flower ($\times 6$); (c) flower showing heart-shaped sepals, down-curved petals and large anthers ($\times 4$); (d) flower with petals and stamens removed to show styles and stigmas ($\times 8$). 5 *H. acanthocarpus* fruit ($\times 8$).

lobed, usually finely pinnate when submerged; stipules are present or absent. The flowers are bisexual or unisexual, regular, usually very small, solitary and axillary or in spikes or panicles. The calyx tube is adnate to the ovary, with two to four lobes, sometimes none. The petals are free, two, four or rarely three in number, or absent, concave, often hood-shaped. The stamens are eight or two to six in number, or absent, with filaments usually long and slender and basifixed anthers, dehiscing laterally. The ovary is inferior, of two to four fused carpels, with one, two or four locules and as many pendulous ovules as styles; there are one to four free styles with feathery or papillose stigmas. The fruit is usually very small, dry or succulent, indehiscent or separating into one-seeded nutlets; the seed has copious endosperm and an erect embryo.

**Classification.** The Haloragaceae is usually subdivided as follows (*Gunnera* is sometimes placed in a separate family, Gunneraceae):

SUBFAMILY HALORAGOIDEAE
Ovary two- to four-loculed; leaves without stipules; inflorescence small; fruit dry; petals present or absent.

HALORAGEAE. Fruit indehiscent. *Loudonia* (Australia), *Haloragis* (Australasia north to Himalayas), *Meziella* (Australia), *Laurembergia* (tropical Africa and Asia), *Proserpinaca* (North America).

MYRIOPHYLLEAE. Fruit separating into two or four nutlets. *Myriophyllum* (cosmopolitan in fresh waters).

SUBFAMILY GUNNEROIDEAE
Ovary unilocular; leaves stipulate; inflorescence large; fruit succulent; petals absent. *Gunnera.*

The family is most closely related to the Hippuridaceae (mare's tails), which has sometimes been included with it, and in floral features it resembles the Onagraceae, but the affinity is obscure.

**Economic uses.** Some species of *Gunnera* are used for ornament or ground cover in gardens. The stem of the Chilean *Gunnera chilensis* has been used on a small scale for tanning and dyeing. D.M.M.

# HIPPURIDACEAE
## Mare's Tail

The Hippuridaceae comprises the single species of aquatic herbs, *Hippuris vulgaris*, with a number of specialized races.

**Distribution.** *Hippuris* is found in wetlands

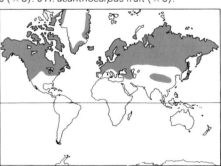

**Number of genera:** 1
**Number of species:** 1
**Distribution:** temperate and cold wetlands of the N Hemisphere.
**Economic uses:** young shoots eaten.

throughout the temperate and cold regions of the Northern Hemisphere.

**Diagnostic features.** *Hippuris* or mare's tail is a perennial herb that usually grows in shallow water. The stem is a perennial, creeping rhizome from which arise erect leafy shoots. The erect shoots bear flowers when they emerge above the surface of the water but after flowering they die down to the rhizome; the non-flowering submerged shoots usually remain green throughout the winter. The leaves are borne in whorls of four up to 12. The submerged leaves are linear, pale green and flaccid; the emergent leaves are shorter, obovate to linear, dark green and rigid. The flowers are inconspicuous and borne solitary in the axils of emergent leaves. Although the flowers are small and very reduced they are variable within a single plant and may be bisexual, unisexual or rudimentary and apparently sterile. The perianth is reduced to a small rim around the top of the ovary. There is one, relatively massive stamen: its size can probably be considered an adaptation to wind-pollination. The ovary is inferior formed of one carpel containing a single, pendulous ovule. The style is long and slender with stigmatic papillae throughout its length. The fruit is a small, ovoid, smooth one-seeded nutlet.

**Classification.** There are some ecologically specialized races of *Hippuris* found in the Arctic and the Baltic Sea, but as there is no clear-cut morphological differentiation it is usual to recognize only one species (*Hippuris vulgaris*). As with so many other reduced dicotyledonous aquatics, the relationships of the Hippuridaceae are disputed. It is generally considered to be related to the Haloragaceae, and is here placed in the Haloragales. Embryological and recent phytochemical investigations, however, suggest a relationship with the Tubiflorae (a large group including the Convolvulaceae, Labiatae, Scrophulariaceae, Solanaceae, Acanthaceae). However, no other Tubiflorae have a single, inferior carpel, and some recent workers have suggested that the family should be placed near the Cornaceae.

**Economic uses.** The submerged shoots of *Hippuris* remain green in winter and form an important winter food for many animals. The Eskimos also gather and eat the young shoots of *Hippuris*.                     C.D.C.

# MYRTALES

## SONNERATIACEAE

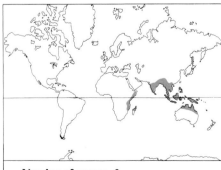

**Number of genera:** 2
**Number of species:** 8
**Distribution:** tropical.
**Economic uses:** wood used locally and fruit and leaves eaten locally.

This tropical family of trees and shrubs has two genera, *Sonneratia* and *Duabanga*.

**Distribution.** The family occurs from East Africa through Asia into Australasia and the western Pacific.

**Diagnostic features.** The leaves are opposite, simple, entire, and without stipules. The flowers are regular, bisexual or unisexual (sexes on the same plant), solitary or borne in threes and are conspicuous, having a bell-shaped, leathery calyx tube with four to eight lobes, four to eight free petals (sometimes absent), and many free stamens which are inflexed in bud and inserted on the calyx. The anthers are kidney-shaped and open lengthwise. The ovary is superior, of numerous fused carpels, sometimes partly fused to the base of the calyx tube; it contains four to many locules each with numerous ovules on axile placentas. There is a single long style with a head-like stigma. The fruits are capsules in *Duabanga*, berries in *Sonneratia*, and contain many seeds which lack endosperm and have short, leafy cotyledons.

**Classification.** The five *Sonneratia* species grow in tidal mudflats and possess erect conical breathing roots which project from the mud like those of certain species of mangrove. *Duabanga* contains three species of lowland and mountain forest trees growing to 30m (100ft). Their large flowers, which last only one night, are up to 7cm (3in) across, green outside, white inside, and in size resemble those of *Sonneratia*.

In many respects the flowers of both genera are morphologically close to those of *Lagerstroemia* in the family Lythraceae, to which the Sonneratiaceae is related, as it is to the Punicaceae (the pomegranates).

**Economic uses.** The wood of most species is used locally, and the fruit or leaves of some *Sonneratia* species may be eaten.       B.M.

## TRAPACEAE
*Water Chestnut*

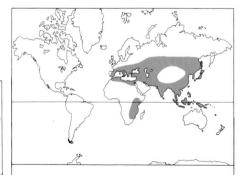

**Number of genera:** 1
**Number of species:** 1, 3 or about 30
**Distribution:** throughout Old World, naturalized in N America and Australia.
**Economic uses:** fruits (water chestnuts) a staple food in parts of Asia.

This family of floating aquatics has a single genus, *Trapa*, whose fruits, the water chestnut, are an important food in parts of Asia.

**Distribution.** Excepting the arctic regions, *Trapa* occurs almost throughout the Old World. It has become naturalized in North America and Australia.

**Diagnostic features.** *Trapa* is an aquatic annual or short-lived perennial. The stems are rooted in mud and float in water. Submerged leaves are opposite and linear, very short-lived and soon replaced by green, pinnately-branched roots, frequently mistaken for leaves. The floating leaves are alternate and form a rosette; the petiole often bears an ellipsoidal, spongy swelling which acts as a float; the leaf blades are rhombic with a toothed margin. The flowers are regular, bisexual, inconspicuous, solitary, axillary and short-stalked. The sepals are four, triangular, adnate to the ovary and develop into two, three or four hard horns or spines in fruit. The petals are four, white or lilac. The stamens are four. The ovary, of two fused carpels, is half-inferior, forming two locules, each with a single pendulous anatropous ovule on an axile placenta. The fruit is a large, woody, variously sculptured, spinose nut. The seeds have no endosperm and the cotyledons are very unequal in size and shape.

**Classification.** The sculpturing of the fruit is very variable and there is little agreement as to which variants should be recognized as species, the number of species being given variously as one, three or up to about 30. The Trapaceae is allied to the Onagraceae.

**Economic uses.** The fruits of *Trapa*, water chestnuts, contain much starch and fat and are a staple food in eastern Asia, Malaysia and India. Most tinned "water chestnut" is in fact *Cyperus esculentus*, *Trapa* grows very quickly and forms stable surface-floating mats which may hinder navigation in various parts of the world. The sturgeon breeding grounds in the Caspian Sea are threatened by *Trapa*.                     C.D.C.

## LYTHRACEAE
*Cupheas, Lythrums and Henna*

The Lythraceae is a small family of herbs, shrubs and trees including several ornamentals and species producing dyes, including henna (*Lawsonia inermis*).

**Distribution.** The family is distributed mainly in the tropics, but also in temperate regions. The temperate representatives are mostly annual and perennial herbs, often growing in damp habitats.

**Diagnostic features.** The leaves are opposite, whorled or spiral, simple and entire, with very small or no stipules. A constant anatomical feature is the presence of bicollateral vascular bundles (ie with phloem on both the inside and outside of the xylem). The flowers, borne in racemes, panicles or cymes, are usually regular and bisexual. They have four (or six) sepals and petals and twice as many stamens, although there is wide variation in number of parts. There is sometimes another whorl of appendages

Lythraceae. 1 *Lawsonia inermis* (a) leafy shoot with axillary and terminal inflorescences ($\times\frac{2}{3}$); (b) fruit ($\times 3$); (c) cross section of fruit ($\times 3$). 2 *Peplis portula* (a) habit showing adventitious roots ($\times\frac{2}{3}$); (b) vertical section of fruit ($\times 4$). 3 *Cuphea ignea* (a) leafy shoot with solitary axillary flower ($\times\frac{2}{3}$); (b) vertical section of flower ($\times 1\frac{1}{2}$). 4 *Lythrum salicaria* produces three types of flower (only one type on each individual) with the style and the two whorls of stamens at three levels in the floral tube (tristyly); seed-set is far higher when the stigma receives pollen from stamens of the same length as itself (shown as arrows) than when it is pollinated from longer or shorter stamens.

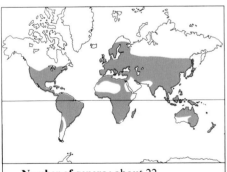

**Number of genera:** about 22
**Number of species:** about 450
**Distribution:** mainly tropical, sometimes temperate.
**Economic uses:** dyes, commercial timbers and ornamentals.

outside the sepals, forming an epicalyx. The petals, usually crumpled in bud, are free and borne on the rim of a deep calyx tube, alternating with the sepals. The stamens are inserted at different levels within the calyx tube, but rarely close to the petals. Heterostyly is quite a common family characteristic. The ovary is superior, sometimes stalked, and has two to six locules, each with two to many ovules on axile placentas; rarely there is one locule with parietal placentas. The style is simple and the stigma usually capitate. The fruit is a dry, dehiscent or indehiscent capsule. The seeds are numerous, with a straight embryo and no endosperm.

**Classification.** The main division of the Lythraceae is into two tribes, the Lythreae and Nesaeae, distinguished by a technical character: in the former the cross-walls (dissepiments) of the ovary are only partial so that the top of the fruit is unilocular, whereas in the latter the dissepiments are complete. The largest genera of the Lythreae are *Cuphea, Diplusodon, Rotala* and *Lythrum*, and of the Nesaeae, *Nesaea* and *Lagerstroemia*.

The Lythraceae, with their typically four-petaled flowers and internal phloem, belong to the Myrtales. The family has affinities with the Myrtaceae, Onagraceae, Punicaceae, Sonneratiaceae and Combretaceae, but it is a distinct group not linked with any other by intermediates. The families just mentioned differ from the Lythraceae in having inferior ovaries, fleshy fruits, and stamens which are generally inserted on the rim of the calyx tube with the petals.

**Economic uses.** The Lythraceae are known chiefly as the source of certain dyes. The most famous is henna, produced from *Lawsonia inermis*. The leaves of *Woodfordia fruticosa* yield a red color and the bark and wood of some *Lafoensia* species, including *Lafoensia pacari*, give a yellow dye. Several species have valuable timber, notably *Physocalymma scaberrima*, which has rose-pink wood, *Lafoensia speciosa*, and various members of *Lagerstroemia*. *Lagerstroemia indica* (crepe myrtle), *Lawsonia inermis* (mignonette tree) and *Woodfordia fruticosa* are grown as ornamental trees and shrubs in warm climates, and several *Cuphea* species are grown as pot-plants. *Lythrum salicaria* (purple loosestrife) and other *Lythrum* species are grown as perennials in temperate regions.                F.K.K.

## RHIZOPHORACEAE
*Mangroves*

The Rhizophoraceae is a tropical family of shrubs, climbers and trees, four genera of which are mangroves, half of the world's main mangrove genera.

**Distribution.** The family is found throughout, and is virtually confined to, the tropics. One of the genera, *Cassipourea* (not man-

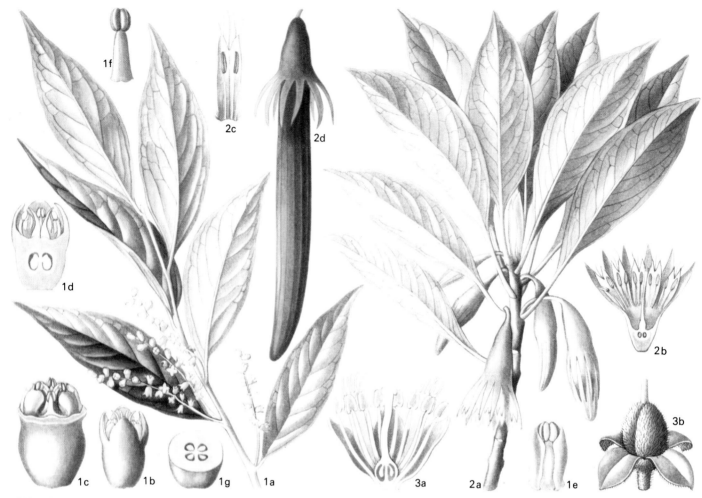

Rhizophoraceae. 1 *Anisophylla griffithii* (a) shoot with alternate leaves and axillary inflorescences ($\times \frac{2}{3}$); (b) flower with four sepals ($\times 4$); (c) flower with sepals removed ($\times 6$); (d) half flower showing inferior ovary with free styles ($\times 4$); (e) petal and stamen ($\times 14$); (f) stamen ($\times 14$); (g) cross section of ovary ($\times 5$). 2 *Bruguiera gymnorhiza* (a) shoot with alternate leaves and solitary axillary flowers ($\times \frac{2}{3}$); (b) half flower showing notched petals and inferior ovary ($\times \frac{2}{3}$); (c) petal and stamens ($\times 1\frac{1}{3}$); (d) fruit ($\times \frac{2}{5}$). 3 *Cassipourea rowensorensis* (a) half flower with superior ovary and single style ($\times 4$); (b) fruit ($\times 2\frac{2}{3}$).

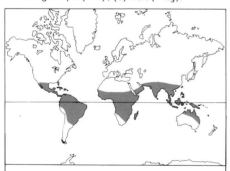

**Number of genera:** 16
**Number of species:** about 120
**Distribution:** tropical rain forests, and mangroves.
**Economic uses:** tannin (mangrove bark), timber for charcoal and fuel and local uses for food and medicine.

groves), contains well over half the species in the family, and is one of only four genera in the Americas.

**Diagnostic features.** Although so well known, the four genera of rhizophoraceous mangroves comprise only 16 species. The majority of species are shrubs or climbers (less often trees) of tropical rain forests. The leaves are simple and entire, usually opposite and with conspicuous, caducous stipules, rarely alternate and without stipules. The flowers are bisexual (rarely unisexual), regular, and hypogynous to epigynous, borne in cymes or racemes, rarely solitary, in the leaf axils. There are three up to 16 persistent, valvate sepals and the same number of petals, usually with a claw and jagged at the tip. There are eight to many stamens inserted on the edge of a disk, the anthers with four, rarely many, locules dehiscing by a valve. The female flowers bear staminodes sometimes adnate to the petals. The ovary is inferior or superior, of two to 12 fused carpels, and has two to 12 locules, each usually with two anatropous, pendulous ovules. There is usually a simple style. The fruit is a berry or drupe, or dry and indehiscent, rarely a dehiscent capsule or winged; the seeds, sometimes with an aril, and germinating on the plant in mangrove species, are with or without fleshy endosperm.

**Classification.** There are three tribes in the family, by far the best-known comprising four genera of mangroves: *Rhizophora*, which is pantropical; *Bruguiera* and *Ceriops*, found throughout tropical Asia and in Africa; and *Kandelia*, confined to Southeast Asia. These four genera, together with three genera of the Combretaceae and the genus *Avicennia*, are the most important mangroves throughout the world, and share many characteristics of ecology, growth-form and reproductive biology.

The Rhizophoraceae is quite closely related to the Combretaceae, and shares with it the woody habit and simple, usually entire leaves. It also has a similar range in ecology and associated growth-habit, but it has usually stipulate leaves and shows a greater range of basic floral construction.

**Economic uses.** Apart from many native uses in food and medicine, several species are valuable sources of timber. This is particularly true of the mangrove species, whose wood, hard and dense but not very durable, is used mainly for charcoal production and fuel. Mangrove bark is also widely used in the preparation of leather in the tanning industry.
C.A.S.

## PENAEACEAE

The Penaeaceae is a small family of heath-like shrubs which include some ornamentals.

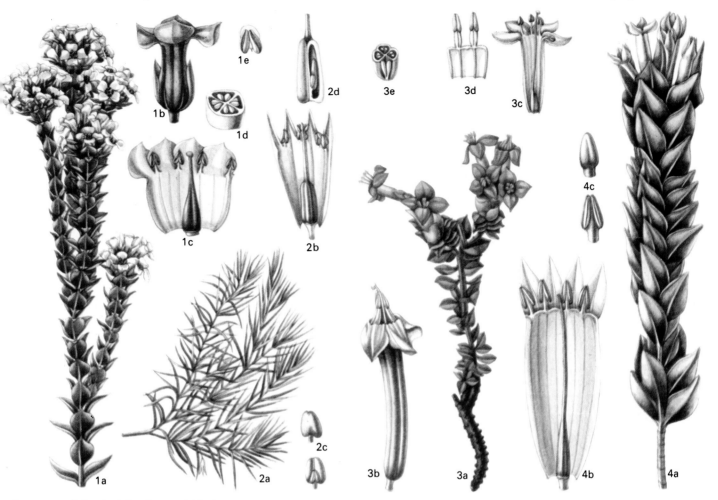

Penaeaceae. 1 *Brachysiphon fucatus* (a) leafy shoot and flowers ($\times\frac{2}{3}$); (b) flower ($\times 3$); (c) flower opened out showing stamens attached to the calyx ($\times 3$); (d) cross section of ovary ($\times 6$); (e) anthers ($\times 6$). 2 *Penaea ericifolia* (a) shoot ($\times\frac{2}{3}$); (b) flower opened out ($\times 4$); (c) stamens front (lower) and rear (upper) views ($\times 8$); (d) ovary with part of wall removed ($\times 6$). 3 *P. squamosa* (a) flowering shoot ($\times\frac{2}{3}$); (b) flower ($\times 2\frac{1}{2}$); (c) flower opened out ($\times 1\frac{1}{2}$); (d) stamens ($\times 3$); (e) cross section of ovary with part of vertical wall cut away to show basal ovules ($\times 4$). 4 *Glischrocolla formosa* (a) flowering shoot ($\times\frac{2}{3}$); (b) flower opened out ($\times 2$); (c) stamens back (upper) and front (lower) views ($\times 3$).

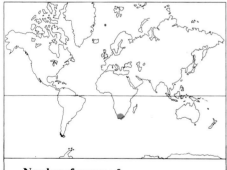

**Number of genera:** 5
**Number of species:** 27
**Distribution:** southern Africa.
**Economic uses:** some ornamentals.

**Distribution.** The family occurs only in southern Africa.

**Diagnostic features.** All the members of the family are shrubby and have a heath-like habit with small, opposite, sessile, entire leaves with or without minute stipules. The flowers are regular, bisexual and borne singly but often crowded in the upper leaf axils, and are subtended by two to four leafy and colored bracts. The tubular calyx has four lobes. There are no petals and the four

stamens are inserted on the throat of the calyx, alternating with the lobes. The anthers have only short filaments, two pollen sacs each, and open facing into the flower by long slits. The ovary is superior, with four locules, each opposite a calyx lobe, and has a terminal style with a four-lobed stigma. There are two to four basally inserted, or four axile, ovules in each locule. The fruit is capsular, four-valved on opening, with the calyx persistent. There are usually only one or two mature seeds in each locule of the fruit following abortion of the others, the seed having no endosperm and a thick embryo with small cotyledons. The seed coat is often smooth and glossy.

**Classification.** The largest genera are *Penaea* (12 species) and *Brachysiphon* (11 species), which together with *Saltera* (one species) are placed in the tribe PENAEEAE, characterized by basal placentation. In *Penaea* the style is four-angled, cross-shaped at the top with the four stigmatic surfaces in the angles. In *Brachysiphon* the style is cylindrical and the stigma capitate. *Saltera* has a calyx tube about three times as long as the free lobes, not twice as long as in *Penaea* and *Brachysiphon*. The genera with axile placentation,

the monotypic *Glischrocolla* and *Endonema* (two species), are placed in the tribe ENDONEMEAE. *Endonema* has a calyx tube about four times as long as the free lobes, and flowers borne laterally on branches, while *Glischrocolla* has a tube about three times as long as the free lobes, and flowers borne on branch tips.

The family is related to the Lythraceae by some botanists and to the Thymelaeaceae by others.

**Economic uses.** Certain species in the family are ornamental, eg *Brachysiphon fucatus* and *Endonema retzioides*. The gum (sarcocolla) yielded by *Penaea mucronata* and *Saltera sarcocolla* has been used locally in medicine. B.M.

## THYMELAEACEAE
### Daphne
The Thymelaeaceae is a medium-sized family, mainly of shrubs. *Daphne* includes attractive ornamentals.

**Distribution.** Although found in both temperate and tropical regions, the family is more diverse in the Southern Hemisphere than in the Northern Hemisphere, and is especially well represented in Africa. Many

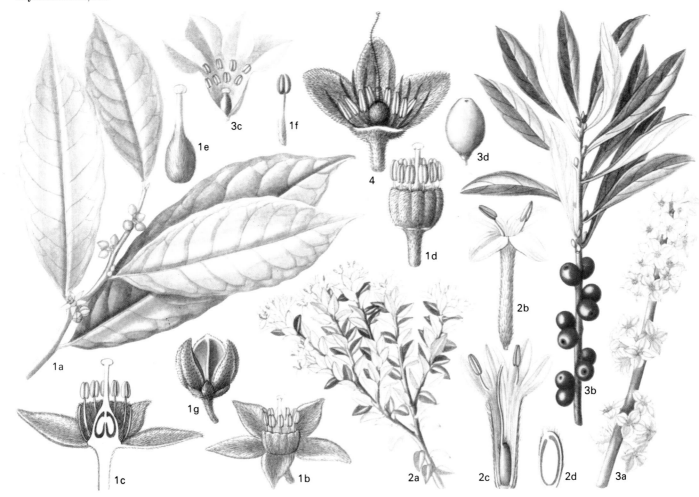

Thymelaeaceae. 1 *Octolepis flamignii* (a) leafy shoot with flowers and flower buds ($\times \frac{2}{3}$); (b) flower ($\times 3$); (c) half flower ($\times 4$); (d) hypanthial cup with stamens, and style and stigma ($\times 4$); (e) gynoecium ($\times 5$); (f) stamen ($\times 5$); (g) dehiscing fruit ($\times 1\frac{1}{3}$). 2 *Pimelea buxifolia* (a) leafy shoot with terminal inflorescences ($\times \frac{2}{3}$); (b) flower ($\times 3$); (c) flower opened out ($\times 3$); (d) vertical section of ovary ($\times 4$). 3 *Daphne mezereum* (a) flowering shoot ($\times \frac{2}{3}$); (b) leafy shoot with fruits ($\times \frac{2}{3}$); (c) flower opened ($\times 2$); (d) fruit ($\times 2$). 4 *Gonystylus augescens* flower with two sepals removed ($\times 4$).

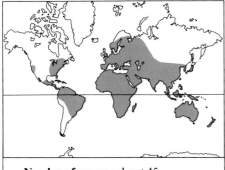

**Number of genera:** about 45
**Number of species:** about 500
**Distribution:** cosmopolitan, especially well represented in Africa.
**Economic uses:** some ornamentals and the bark is used locally to make paper.

genera are found in the Pacific Islands.

**Diagnostic features.** The leaves are alternate (occasionally opposite), entire and without stipules. The flowers are regular, usually bisexual, with parts normally in fours or fives and are grouped in racemes, capitula or fascicles. The flowers are basically cup-shaped, the hollowed-out receptacle forming a deep tube with the floral parts mostly arranged at the rim. The sepals are petaloid, appearing like a continuation of the tube, and the stamens, usually half to twice as many as the sepals, are inserted in the tube; the corolla is inconspicuous or absent. The ovary is superior, situated at the base of the receptacular cup, and comprises one or two (rarely three to eight) fused carpels, containing as many locules, each with one axial or parietal, pendulous ovule. The style is simple. The fruit is variable, an achene, berry, drupe or occasionally a capsule; the seed has little or no endosperm and the embryo is straight.

**Classification.** Although relatively homogeneous this family is usually divided into four subfamilies (which have even been given family status), drawing attention to a few genera which differ from the main bulk of members. Most genera belong to the subfamily THYMELAEOIDEAE, including the largest genus, *Gnidia*, with 100 species extending from Africa to Madagascar and through India to Ceylon. The other large genera of this group are *Pimelea* with 80 species in Australasia, *Wikstroemia* with 70 species in Australasia reaching South China, *Daphne* with 70 species in Australasia extending across Asia to Europe and North Africa, and

*Lasiosiphon* with 50 species covering about the same area as *Gnidia*. While Thymelaeoideae have a single pendulous ovule, the next largest subfamily, the AQUILARIOIDEAE, have two (rarely more). This group contains seven small genera from the Pacific area and Africa; *Octolepis*, from West Africa, has four to five locules in the ovary, the fruit being a capsule. The GONYSTYLOIDEAE, with three small genera from Southeast Asia and Borneo, includes *Gonystylus* which has numerous stamens. The fourth subfamily, GILGIODAPHNOIDEAE, contains one genus (*Gilgiodaphne*) from tropical West Africa, with four stamens and four staminodes united in a tube, the four fertile filaments arising as if axially from near the base, and the staminodes remaining united for most of their length.

The affinities of the Thymelaeaceae are not obvious. It has usually been placed, as here, in the Myrtales, although some authorities consider that it should be associated with the Flacourtiaceae.

**Economic uses.** Species of *Daphne* are cultivated as ornamental shrubs, usually fragrant-flowered. Some are evergreen. The bark of several genera, particularly *Wikstroemia*, yields fibers which are used locally

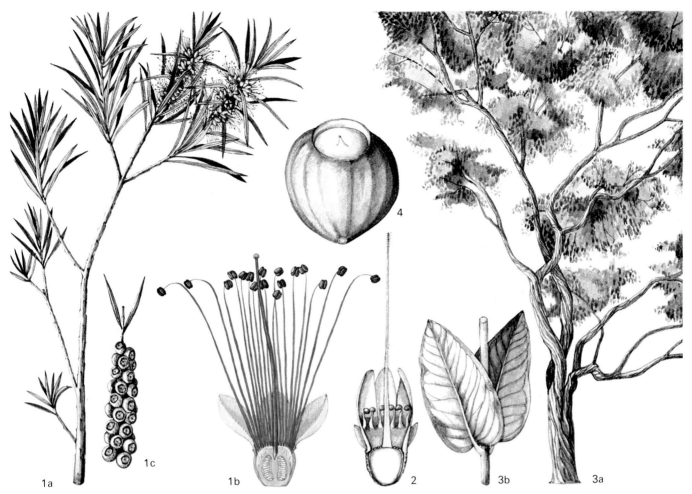

Myrtaceae. 1 *Callistemon subulatus* (a) leafy shoot and inflorescences ($\times\frac{2}{3}$); (b) half flower showing numerous stamens and inferior ovary crowned by a single style and containing ovules on parietal placentas ($\times 3$); (c) fruits ($\times\frac{2}{3}$). 2 *Darwinia citriodora* half flower with epipetalous stamens and a long style crowned by a hairy stigma ($\times 3$). 3 *Eucalyptus melanophloia* (a) habit; (b) adult leaves ($\times\frac{2}{3}$). 4 *Eugenia gustavioides* fruit—a berry with remains of the style ($\times 1$).

to manufacture paper. In some Mediterranean countries the bark of *Daphne* species is used to stupefy fish, and in the West Indies *Lagetto lintearia* yields the ornamental lace bark.                                                    I.B.K.R.

# MYRTACEAE

## Myrtles, Eucalyptus and Cloves

The Myrtaceae is a large family whose habit ranges from straggling and small shrubs to the lofty *Eucalyptus* trees which dominate the hardwood forests of Australia.

**Distribution.** The family is mostly tropical and subtropical, concentrated in America and eastern and southwestern Australia.

**Diagnostic features.** They are woody plants, mostly shrubs to large trees. The leaves are usually opposite (less often alternate), leathery, evergreen and typically entire, without stipules, and characteristically pellucid-dotted with subepidermal glands (also found on the young stem, floral organs and fruit) containing ethereal oils. The flowers are regular and bisexual, most frequently in cymose, less often in racemose, inflorescences, rarely solitary (*Myrtus communis*); they are generally epigynous, but varying degrees of perigyny are found. There are commonly four or five sepals, usually

free (sometimes more or less united, then sometimes forming a cap which drops off as the flower opens), sometimes much reduced to virtually absent. There are four or five petals which are typically free, small and round (in *Eucalyptus* more or less united, falling as one piece). The stamens are numerous (rarely few), sometimes in tufts opposite the petals (*Melaleuca*), typically free and with versatile anthers, the connective often gland-tipped. The ovary is commonly inferior with one to many (often two to five) locules, each with usually two to many ovules on axile (rarely parietal) placentas. The style is long and simple with a simple capitate stigma. The fruit is usually a fleshy berry (rarely a drupe) or dry (then a capsule or nut). There is little or no endosperm.

**Classification.** The family is divided into two subfamilies.

MYRTOIDEAE: Flowers epigynous, leaves always opposite; fruit fleshy, typically a berry, rarely a drupe.

LEPTOSPERMOIDEAE: Perigyny occurs but epigyny is the rule; leaves opposite or alternate; fruit dry, a capsule or nut-like.

The affinities of the Myrtaceae are with the Lythraceae and Melastomataceae.

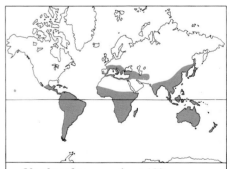

**Number of genera:** about 100
**Number of species:** about 3,000
**Distribution:** tropical and subtropical, chiefly America and Australia
**Economic uses:** timber, spices (cloves, allspice), essential oils, edible fruits and ornamentals

**Economic uses.** Economically the most important genus is *Eucalyptus*, chiefly for its timber. The family also yields some of the most valued spices – cloves (and clove oil) from *Syzygium aromaticum* (*Eugenia caryophyllata*); and allspice or pimento from *Pimenta dioica* (*P. officinalis*); *P. racemosa* provides oil of bay rum, and *Melaleuca leucadendron* cajeput oil. Eucalyptus oil, well known as a flavoring, expectorant and antiseptic, derives from several *Eucalyptus* species. Among the edible fruits are the tropical American and West Indian guava (*Psidium guajava*) and jaboticaba (*Myrciaria cauliflora*), and tropical American species of *Eugenia, Syzygium* and *Feijoa*. Ornamentals include the lillypilly (*Acmena smithii*), the bottlebrushes (*Callistemon* and *Melaleuca*), *Eucalyptus* (gumtrees, ironbarks, bloodwoods), the manuka or tea tree (*Leptospermum scoparium*), and the common myrtle (*Myrtus communis*, possibly the only European member of the family.    F.B.H.

## PUNICACEAE
*Pomegranate*

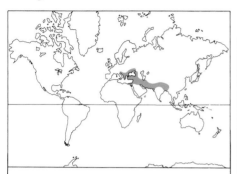

**Number of genera:** 1
**Number of species:** 2
**Distribution:** SE Europe to the Himalayas and Socotra.
**Economic uses:** fruit (pomegranate) eaten and used in grenadine drink, bark and fruit-skin used locally.

The Punicaceae contains a single genus with two species of shrubs and small trees.
**Distribution.** The family occurs from southeastern Europe to the Himalayas, and in Socotra.
**Diagnostic features.** The plants are deciduous, sometimes spiny, woody shrubs or small trees, with tapering cylindrical branches and simple opposite or subopposite entire leaves, without stipules. Four wings of epidermal and cortical parenchyma develop on young twigs but these are early deciduous. The flowers are solitary or in groups. They are bisexual, regular, with five to eight sepals, five to eight petals, numerous stamens and an inferior ovary of numerous fused carpels. Two whorls of carpels are laid down and then owing to peripheral growth are tilted from a vertical to horizontal position, so that the numerous locules are formed in two layers; numerous anatropous ovules are borne on parietal placentas in the upper locules and on axile placentas in the lower. This peculiar arrangement is also seen in the fruit, commonly called the pomegranate. The round fruit of up to 9cm (3.5in) diameter has a leathery brownish-red-yellow rind and characteristically a hard persistent calyx tube. The internal pulp contains numerous seeds, which lack endosperm.
**Classification.** The sole genus, *Punica*, is represented by two species, *Punica pro-*

*topunica*, native to Socotra, and *P. granatum*, which grows from the Balkans to the Himalayas and is also cultivated, mostly in southern Europe. The Punicaceae is taxonomically close to the families Lythraceae and Sonneratiaceae.
**Economic uses.** The chief economic use lies in production of pomegranates. The fruit ferments easily and is used in the production of the drink, grenadine. The fruit and bark have been used in Egyptian tanning processes and in medicinal preparations. Dwarf and double-flowered varieties of the red-flowered *P. granatum* are known in cultivation.    S.A.H.

## ONAGRACEAE
*Clarkias, Fuchsias and Evening Primroses*

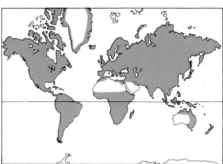

**Number of genera:** 18
**Number of species:** about 640
**Distribution:** cosmopolitan, centered in SW N America.
**Economic uses:** chiefly cultivated ornamentals (*Fuchsia, Oenothera*).

This widespread and botanically well-defined family of herbs and some shrubs contains a number of aquatics and popular ornamentals.
**Distribution.** Although virtually cosmopolitan, the family is most diverse in the western United States of America and Mexico, where all known genera occur. Most members occur in open habitats, either dry or moist, but *Ludwigia* is largely aquatic, while *Fuchsia* can inhabit wet woods.
**Diagnostic features.** The plants are herbs or rarely shrubs, often aquatic. The leaves are simple, opposite or alternate, usually without stipules. The flowers are solitary in axils or borne in racemes, and are bisexual or unisexual, then borne on separate plants (*Fuchsia*). Some are irregular, usually with a colored perigynous tube (hypanthium). The sepals are two, four or five in number, free and valvate. There are two, four or five free petals, rarely none. The stamens are usually eight in two whorls, but may be one, two, four, five or 10 in number. The anthers have two locules, dehiscing lengthwise. The ovary is inferior or rarely half-inferior, of four fused carpels, sometimes with one but usually with two, four or five locules, each containing one to many anatropous ovules on axile placentas. The single style is entire or has four lobes. The fruit is usually a

loculicidal capsule, sometimes a berry or indehiscent; seeds are numerous or solitary, without endosperm, with a plume of hairs at the chalaza in *Epilobium* and *Zauschneria*; the embryo is straight or almost so.

There has been a trend in flower structure from an original regular flower with parts in fours to more specialized, often irregular flowers with reduced numbers of petals. This can be associated with a change from cross-pollination to strict self-pollination, as in *Camissonia, Gaura* and *Ludwigia*. The earliest Onagraceae were presumably insect-pollinated, but the least specialized species of, for example, *Lopezia*, as well as many fuchsias, are bird-pollinated, insect-pollination being a derived condition. Many Onagraceae are pollinated by moths. Fly-pollination in *Lopezia coronata* results in a curious mechanism whereby the single stamen and mature stigmas are held under tension by the staminodes so as to snap apart when touched, flinging the pollen against the underside of the pollinating flies.
**Classification.** Developing from the intensive genetical work on the evening primroses (*Oenothera*) which started over 60 years ago, the Onagraceae has been the subject of considerable cytogenetical study. The most recent studies show it to comprise 18 genera, which are grouped into one large tribe, the ONAGREAE (*Calylophus, Camissonia, Clarkia* (including *Godetia*), *Gaura, Gayophytum, Gongylocarpus, Hauya, Heterogaura, Oenothera, Stenosiphon, Xylonagra*), and five smaller tribes: FUCHSIEAE (*Fuchsia*), LOPEZIEAE (*Lopezia*), CIRCAEEAE (*Circaea*), JUSSIAEEAE (*Ludwigia*) and EPILOBIEAE (*Boisduvalia, Epilobium, Zauschneria*).

The interrelationships are still a matter for debate, but Lopezieae and Jussiaeeae seem to have particularly close affinities, while the Onagreae stands apart as a distinctive group of genera, including all those in the family in which major chromosomal repatterning has been an important evolutionary mechanism.

Although the oldest fossils in the family belong to *Ludwigia*, it is *Fuchsia*, with fleshy fruits and unspecialized placentation, which appears to be the modern genus closest to the ancestors of all Onagraceae. The floral tube, once considered a derived condition, is present in the Lythraceae, Melastomataceae and Myrtaceae, which are presumed to have a common ancestry with the Onagraceae.
**Economic uses.** Many species are cultivated for ornament, as hardy or half-hardy annuals (eg *Clarkia, Oenothera* (evening primrose)) or as hardy or greenhouse shrubs (*Fuchsia*). Some *Ludwigia* species are grown as aquatics in greenhouses.    D.M.M.

## OLINIACEAE
The Oliniaceae is a family of only one genus (*Olinia*) with 10 species of trees and shrubs native to eastern and southern Africa.

The leaves are simple, without stipules and opposite on four-angled branches. The

Onagraceae. 1 *Fuchsia regia* var *alpestris* (a) flowering shoot ($\times\frac{2}{3}$); (b) half flower showing free sepals that are longer than the purple petals ($\times 1\frac{1}{2}$); (c) cross section of fruit ($\times 3$). 2 *Circaea cordata* (a) dehisced capsule ($\times 6$); (b) cross section of fruit ($\times 6$). 3 *Epilobium hirsutum* (a) flowering shoot ($\times\frac{2}{3}$); (b) fruit (a capsule) with part of wall cut away ($\times 2$); (c) ripe fruit dehiscing to release plumed seeds ($\times 2$). 4 *Lopezia coronata* flower with upper petals marked with blotches that resemble nectar, one erect fertile stamen and petaloid spoon-shaped staminodes—all adaptations to a specialized form of insect pollination ($\times 3$). 5 *Oenothera biennis* (a) flowering shoot ($\times\frac{2}{3}$); (b) partly opened flower with petals removed ($\times 2$).

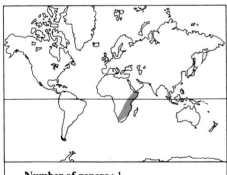

**Number of genera:** 1
**Number of species:** 10
**Distribution:** eastern and southern Africa.
**Economic uses:** none.

flowers are bisexual, regular and borne in cymes in the leaf axils or at the ends of stems. The four or five sepals are joined in a tube which is fused to the ovary and may have a limb of four or five teeth. The four or five petals are inserted on the margin of the calyx tube and alternate with as many incurved, colored, scales. The stamens are equal in number to the petals, and are also inserted on the margin of the calyx tube, immediately below the scales on very short filaments, expanding into a thickened connective. The ovary is inferior, of four or five fused carpels, with three to five locules, surmounted by a simple style ending in a club-shaped stigma. There are two or three pendulous ovules on axile placentas in each locule. The fruit is a false drupe, containing a single seed in each locule. The seed, without endosperm, contains a spiral or convoluted embryo.

Although relationships for this family have been suggested with the Cunoniaceae, the affinities are rather remote. It may be that the Thymelaeaceae are more closely related. There are no known economic uses for the family. S.R.C.

## MELASTOMATACEAE
### Dissotis and Medinilla

A relatively large family, the Melastomataceae is composed mainly of shrubs and small trees, but also of a few vines, herbs, marsh plants and, rarely, epiphytes. A number of species are cultivated for their showy flowers.

**Distribution.** The family is mainly tropical, rather uncommon in temperate zones. It is one of the largest families of South American plants and forms a particularly characteristic feature of the Brazilian flora.

**Diagnostic features.** The most useful diagnostic features are found in the leaf venation and stamen shape. The leaves are opposite and decussate (each pair at right angles to the next) but sometimes with one of each pair smaller than the other, or rarely alternate by abortion. Stipules are absent. The main veins (three to nine pairs) are usually palmate and parallel. The stem is often four-angled. The flowers are bisexual, regular and usually have four or five each of sepals and free petals. The stamens are usually twice as many as the petals. The stamen filaments are geniculate (elbow-shaped) with sterile appendages of various shapes: awl-shaped, spiny, club-like, curved or pronged. The anther lobes usually dehisce by a single, terminal pore. The ovary is superior or more commonly inferior by fusion with the receptacle. There are one to 14 carpels with four or five locules containing two to numerous ovules on axile, basal or parietal placentas. The style is simple. The fruit is a berry or loculicidal capsule. The seeds are small, numerous and lack endosperm.

**Classification.** The family is subdivided into

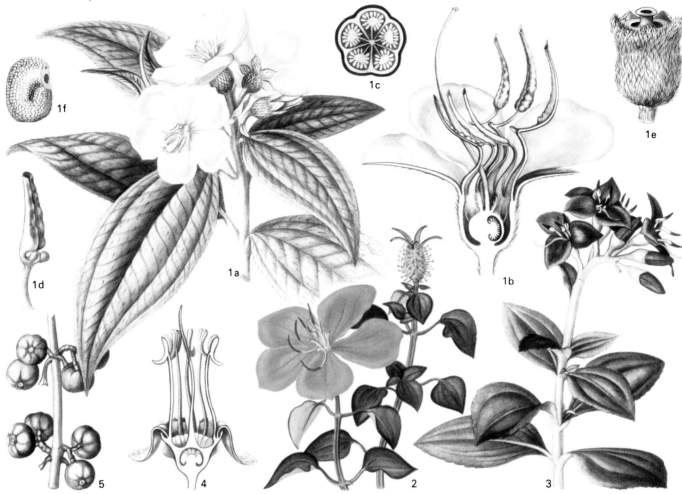

Melastomataceae. 1 *Melastoma malabathricum* (a) flowering shoot showing leaves with characteristic parallel veins ($\times\frac{2}{3}$); (b) half flower showing stamens with elbow-shaped filaments and anthers dehiscing by a single pore ($\times 1\frac{1}{3}$); (c) cross section of ovary ($\times 2\frac{2}{3}$); (d) stamen showing lobed appendages at base of the connective ($\times 3\frac{1}{3}$); (e) fruit—a capsule ($\times 2\frac{2}{3}$); (f) seed ($\times 8$). 2 *Medinilla speciosa* flowering and fruiting shoot showing four-angled stem ($\times\frac{2}{3}$). 3 *Sonerila grandiflora* leafy shoot and cymose inflorescence ($\times\frac{2}{3}$). 4 *Memecylon laurinum* half flower showing inferior ovary with sub-basal placentas ($\times 3\frac{1}{3}$). 5 *M. intermedium* fruits—berries ($\times\frac{2}{3}$).

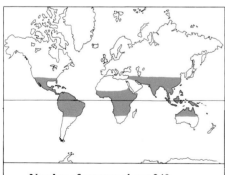

**Number of genera:** about 240
**Number of species:** about 3,000
**Distribution:** temperate but mainly tropical and subtropical, with center in S America.
**Economic uses:** local use of the hardwood and edible fruits of some species and some ornamentals.

three subfamilies and about 13 tribes. The largest subfamily, MELASTOMATOIDEAE, has an ovary of two to many carpels, either free from and superior to the receptacle, or more or less fused to the receptacle and inferior to the perianth parts; the numerous seeds have axile placentation; the fruit is a capsule or berry; principal genera *Tibouchina, Melastoma, Oxyspora, Sonerila* and *Miconia*. In the ASTRONIOIDEAE the ovary also comprises two to many carpels; the many seeds have subbasal or parietal placentation; the fruit is a capsule or berry; principal genera: *Astronia* and *Kibessia*. MEMECYLOIDEAE has ovaries of two to six carpels, sometimes reduced to one, with many seeds on subbasal or axile placentas; the fruit is a capsule or berry; principal genera: *Memecylon* and *Mouriri*. The Memecyloideae is considered a separate family, Memecylaceae, by some authorities.

The relationships of this family are indistinct and slender. One possibility is that some members of the Memecyloideae are related to some members of the Myrtaceae, since both have four to many carpels and, axile placentation of the two to many ovules in each locule. Another possible resemblance is with the Combretaceae, which like the Melastomataceae has twice as many stamens as petals and usually an inferior ovary.

**Economic uses.** None of the family is economically important, except in horticulture. The hardwood of some *Astronia* and *Memecylon* species is used locally in furniture and construction. The fruit and leaves of *Medinilla hasseltii* are eaten in Sumatra and the fruit of some species of *Melastoma* and *Mouriri* are eaten in tropical America. A few species yield dyes (yellow from *Memecylon edule*, Indian Ocean, Sumatra).

A number of shrub species are cultivated in gardens and greenhouses for their attractive flowers, which are often red, blue and purple, less commonly pink, white or yellow. Among them are *Dissotis grandiflora, Medinilla curtisii, M. magnifica, Tibouchina urvilleana (T. semidecandra), Melastoma malabathricum* and *Monochaetum alpestre*.

H.P.W.

# COMBRETACEAE

## Terminalia, Combretum and Quisqualis

The Combretaceae is a family of tropical trees, shrubs and lianas which includes a number of plants of economic and ornamental interest.

**Distribution.** The family is found throughout the tropics, scarcely ever extending beyond.

**Diagnostic features.** They are forest trees, 50m (165ft) or more high, to dwarf shrubs with subterranean rhizomes and short aerial

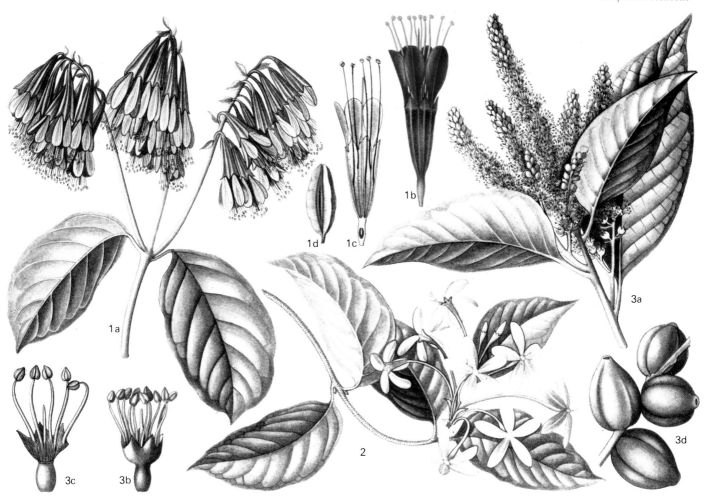

Combretaceae. 1 *Combretum grandiflorum* (a) leafy shoot with flowers in elongated heads ($\times\frac{2}{3}$); (b) flower with five-lobed, green calyx, five, red petals, ten stamens and a filamentous style ($\times 1$); (c) half flower showing inferior ovary with one locule and a pendulous ovule ($\times 1$); (d) winged fruit ($\times\frac{2}{3}$). 2 *Quisqualis indica* leafy shoot bearing flowers with a long perianth tube ($\times\frac{2}{3}$). 3 *Terminalia chebula* (a) leafy shoot and inflorescence ($\times\frac{2}{3}$); (b) flower with toothed calyx and numerous stamens ($\times 7$); (c) flower with half of calyx and stamens removed to show hairs on top of ovary ($\times 8$); (d) woody fruits ($\times\frac{2}{3}$).

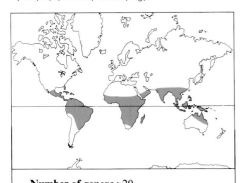

**Number of genera:** 20
**Number of species:** about 475
**Distribution:** mainly tropical, with a few species subtropical.
**Economic uses:** commercial wood, tropical and glasshouse ornamentals, tanning (fruits of *Terminalia* spp).

shoots which frequently become grazed or burnt off and may be more or less annual in appearance. In forests the large trees and lianas predominate, while in grassland the shrubby species are commoner. The leaves are entire, alternate or opposite and without stipules. The flowers are small, regular and bisexual (rarely unisexual), often clustered in globular or elongated heads, and nectar may be produced in abundance. There are usually five (sometimes four) valvate sepals, five (sometimes four) imbricate or valvate petals or the latter may be totally absent. The stamens are in one or two whorls of four or five (rarely they are numerous). The ovary is inferior, with one locule containing two to five pendulous ovules. The style is simple.

There is a tendency in many genera for the development of wind-dispersed winged fruits, dispersed aerially in the case of the lighter fruits but bowled along on the ground in the case of the heavier fruits, the rigid wings acting like spokes. In forests, however, where such methods are not practicable, most of the species have wingless fruits which are either fleshy and animal-dispersed or have spongy tissue and are water-dispersed. The mangrove genera have viviparous fruits which germinate on the parent plant. The seed has no endosperm, and the cotyledons are very variable in form.

**Classification.** Apart from one unusual African representative, the genera form three groups. One of these contains three genera, of which two are mangroves (*Lumnitzera* in East Africa, Asia and Australia; *Laguncularia* in West Africa and America).

The other two groups are centered on the large genera *Combretum* (200 species), mostly with petals, opposite leaves with glandular hairs and barely woody fruits, and *Terminalia* (150 species), mostly with no petals, spiral leaves without glandular hairs and strongly woody fruits. Both these genera are found virtually throughout the tropics (*Combretum* is absent from Australia), but the others related to them are much less widespread and most are restricted to one continent. *Conocarpus*, in the group related to *Terminalia*, is the third combretaceous mangrove genus.

The Combretaceae is probably most closely related to the Myrtaceae. All members of the family possess peculiar unicellular hairs which elsewhere in the angiosperms are known only in a few genera of the Myrtaceae.
**Economic uses.** Some of the trees of the genus *Terminalia* are important sources of timber for export; for example, idigbo is obtained from *Terminalia ivorensis* and afara from *T. superba*, both from West Africa. Many others are valued on a more local scale, eg assegai wood (*T. sericea*), and some of the larger trees are planted for shade.

Several of the climbers with attractive flowers are grown as ornamentals, either outdoors in the tropics or in glasshouses in temperate regions. The best-known are *Quisqualis indica* from Asia and various species of *Combretum*, notably *Combretum grandiflorum* from tropical West Africa.

Many of the Combretaceae are used locally as medicines and foods in a wide variety of situations. The Indian almond is the edible kernel of *Terminalia catappa*, from tropical Asia, which is now grown in many parts of tropical Africa and America as well. Myrobalans (species of *Terminalia*) yield fruit used in tanning.                        C.A.S.

# CORNALES

## NYSSACEAE
*Handkerchief and Tupelo Trees*

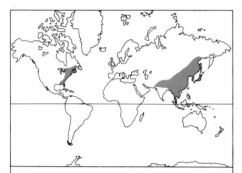

**Number of genera:** 3
**Number of species:** 8
**Distribution:** E N America, China and Tibet.
**Economic uses:** fruits, timber and valued ornamentals.

The Nyssaceae is a small family of trees and shrubs, including the ornamental handkerchief and tupelo trees.

**Distribution.** The family is represented by three genera, *Nyssa*, *Camptotheca* and *Davidia*. There are six species of *Nyssa*, four in eastern North America and two in east Asia. *Camptotheca* is represented by one species, *Camptotheca acuminata*, which grows only in China and Tibet. The single species of *Davidia* is native to China.

**Diagnostic features.** The leaves are alternate, entire or denticulate, and without stipules. The flowers are regular, unisexual and bisexual, the plants being polygamodioecious, with the males occurring in heads, racemes or umbels and the female and bisexual flowers occurring solitarily or in heads with two up to 12 flowers. The calyx is very small of five minute teeth or is totally absent. The petals usually number five, sometimes four or up to eight. There are usually ten stamens, sometimes eight to 16, each with a long narrow filament and small anthers. The ovary is inferior, consisting of one or two fused carpels, with a single locule (rarely more) containing a single apical,

pendulous ovule. The style may be simple or bifid, erect or spirally coiled. The fruit is drupe-like or samara-like. The single seed has thin endosperm and a fairly large embryo.

**Classification.** The three genera may be distinguished by a number of features. In *Davidia* the ovary has six to ten locules, the male flowers lack a perianth, the female and bisexual flowers have numerous perianth segments, the style is lobed and the fruit is a drupe. *Nyssa* and *Camptotheca* have a unilocular ovary, all flowers have sepals and petals, and the style is awl-shaped. In *Nyssa* the style is not divided and the fruit is a drupe, while in *Camptotheca* the style is forked and the fruit is a samara.

The family is related to the Cornaceae. *Davidia* is often considered to represent a separate family, the Davidiaceae.

**Economic uses.** Edible fruits and some timber are obtained from *Nyssa* (tupelo, black gum), but the greatest commercial value lies in the cultivation of the genus for its attractive autumn foliage. *Davidia involucrata* is the ornamental handkerchief tree. *Camptotheca acuminata* is also cultivated as an ornamental.                    S.A.H.

## GARRYACEAE

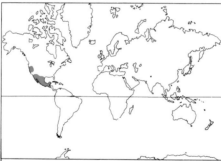

**Number of genera:** 1
**Number of species:** about 18
**Distribution:** SW N America, N Central America, W Indies.
**Economic uses:** winter-flowering ornamental shrubs.

The Garryaceae comprises a single genus (*Garrya*) of evergreen shrubs.

**Distribution.** The family is confined to western and southwestern North America, Mexico, Guatemala and the West Indies.

**Diagnostic features.** The leaves are opposite, oval to lanceolate, the margin being entire or slightly wavy. The flowers are unisexual, with sexes on separate plants in terminal or axillary catkin-like pendulous racemes. The male flowers have four introrse anthers with short filaments, surrounded by four valvate perianth segments. The female flowers are naked or with two to four small decussate bracts often united to form a cup and sometimes interpreted as sepals (or perianth segments). The ovary is inferior, of two to three united carpels, with one locule and two

pendulous ovules on parietal placentas. There are two slender styles. The fruit is a dryish, round, one- or two-seeded berry, with small seeds and abundant endosperm.

**Classification.** The catkin-like racemes and apetalous, unisexual flowers led 19th-century taxonomists to place the Garryaceae with other catkin-bearing plants, such as the Salicaceae and Myricaceae. There is now general consensus that the family is related to the Cornaceae, particularly through the genera *Griselinia* and *Aucuba*.

**Economic uses.** Several *Garrya* species are cultivated as winter-flowering ornamental shrubs.                                    F.B.H.

## ALANGIACEAE

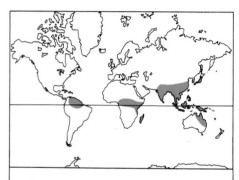

**Number of genera:** 2
**Number of species:** about 23
**Distribution:** Old World tropics and NW S America.
**Economic uses:** some ornamentals.

The Alangiaceae is a small, woody family of obscure affinities.

**Distribution.** Mainly tropical, the family comprises a single Old World genus, *Alangium*, with about 20 species from Africa to Japan and eastern Australia, and a single New World genus, *Matteniusa*, with three species in northern South America.

**Diagnostic features.** Both genera are trees or shrubs, sometimes spiny, with alternate, usually simple leaves without stipules. The flowers are bisexual and borne on bracteolate, jointed pedicels in axillary cymes. They are regular, with equal numbers (4–10) of sepals and petals, the sepals eventually bending backwards and downwards, often more or less twisted. The petals are sometimes united at the base, and are more or less hairy on the inner surface. The number of stamens varies from four to 40, and they are free or united with the petals, with often hairy filaments and usually basifixed, introrse anthers. A nectary disk is generally present at the base of the single style which surmounts the usually unilocular, inferior or superior ovary. The stigma is club-shaped, sometimes two- or three-lobed. One pendulous, anatropous ovule develops in the locule, and the drupaceous fruit is single-seeded with a hard endocarp, and usually crowned with the persistent calyx. The seed has a fleshy endosperm.

1e    1a    1f    1d    1c

Alangiaceae. 1 *Alangium salviifolium* (a) leafy shoot and axillary flowers ( ×⅔) ; (b) flower with short calyx tube, long, recurved petals, numerous stamens and single style with a lobed stigma ( ×2) ; (c) half section of gynoecium showing ovary with pendulous ovule ( ×6) ; (d) stamen with hairy filament and basifixed anther ( ×3) ; (e) fruiting shoot ( ×1) ; (f) cross section of fruit ( ×2).

**Classification.** The relationships of the family are obscure, although it is usually placed near the Cornaceae and related families.

**Economic uses.** Several species of *Alangium* are cultivated as half-hardy shrubs.

I.B.K.R.

## CORNACEAE
### Dogwoods

The Cornaceae is a small family of trees and shrubs, rarely herbs, known mainly for the dogwoods (various species of *Cornus*) and the several varieties of *Aucuba*, sometimes erroneously called laurels.

**Distribution.** The family is found growing mainly in north temperate regions, with a few species in the tropics and subtropics of Central and South America, Africa, Madagascar, Indomalaysia and New Zealand.

**Diagnostic features.** Members of the family are woody plants with opposite or occasionally alternate, simple leaves, sometimes evergreen (eg *Aucuba, Mastixia*). The inflorescences are usually corymbs or umbels, sometimes surrounded by large, showy bracts; flowers are small, bisexual or unisexual (the sexes are then on separate plants, eg *Aucuba, Griselinia*). The flowers are regular, with a four- or five-lobed calyx

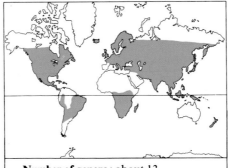

**Number of genera:** about 13
**Number of species:** over 100
**Distribution:** mainly N temperate; some species in tropics and subtropics.
**Economic uses:** popular ornamental shrubs (dogwoods), beverage and preserves from the fruit of *Cornus mas* and some useful timber.

tube (rarely absent). There are normally four or five free petals (rarely absent), and an equal number of stamens alternate with the petals. The short, bilocular anthers open lengthwise. The ovary is inferior, of two (rarely one, three or four) fused carpels, with one to four locules and axile placentation (parietal in *Aucuba*) with one anatropous, pendulous ovule in each locule. The style is simple and the stigma lobed. The fruit is a

drupe or berry with one to four locules and one or two stones.

**Classification.** *Cornus*, in the narrow sense (about four species), has a wide distribution in central and southern Europe as far as the Caucasus, eastern Asia and also California. *Cornus* in the wide sense includes species often separated into other genera such as *Afrocrania, Chamaepericlymenum, Benthamidia, Dendrobenthamia* and *Swida*. *Chamaepericlymenum*, the only herbaceous genus of this family, grows on mountains of North America, Europe, Asia and Japan. *Benthamidia* (three species) grows in east and west North America, Japan and the Himalayas. Thirty-six species of *Swida* grow in north temperate regions, three species in Mexico and one in the northern Andes. *Aucuba* is a genus of three species of hardy, evergreen, dioecious shrubs which is distributed from the Himalayas to Japan. *Toricellia* (three species), *Helwingia* (five species) and *Dendrobenthamia* (12 species) also range from the Himalayas to Japan. Two members of this family grow in tropical southern and East Africa, *Curtisia* and *Afrocrania*, a mountain tree. *Melanophylla* comprises eight species native to Madagascar. In the Indomalaysian region, there are some 25 species of *Mastixia*, which are medium to tall

Cornaceae. 1 *Curtisia faginea* (a) leafy shoot with terminal inflorescence ($\times\frac{2}{3}$); (b) flower ($\times 6$); (c) fruit ($\times 3$); (d) half-section of fruit ($\times 3$); (e) seed ($\times 3$). 2 *Chamaepericlymenum canadense* (a) leafy shoot with inflorescence of small flowers surrounded by large bracts ($\times\frac{2}{3}$); (b) fruits ($\times\frac{2}{3}$). 3 *Corokia buddleoides* (a) flowering shoot ($\times\frac{2}{3}$); (b) flower-bud ($\times 3$); (c) perianth opened out ($\times 3$); (d) vertical section of gynoecium ($\times 3$); (e) fruit, entire (left) and in vertical section (right) ($\times 3$); (f) fruits ($\times\frac{2}{3}$). 4 *Aucuba japonica* (a) leafy shoot and female flowers ($\times\frac{2}{3}$); (b) male flower ($\times 2$); (c) female flower ($\times 2$); (d) fruits ($\times\frac{2}{3}$); (e) cross section of fruit ($\times\frac{2}{3}$).

evergreen trees. *Griselinia* contains six species of trees or shrubs, sometimes epiphytic, which have a most disjunct distribution, with four species in Brazil and two in New Zealand.

The family is divided into two subfamilies and several tribes. The subfamilies are: CURTISIOIDEAE. Ovules with ventral raphe; flowers bisexual; fruit a drupe with uni- or quadrilocular endocarp. *Curtisia, Mastixia.* CORNOIDEAE. Ovules with dorsal raphe; flowers bisexual or unisexual; ovary of one to four parts, always unilocular. *Toricellia, Helwingia, Cornus, Aucuba, Griselinia, Melanophylla.*

Some taxonomists put *Curtisia, Mastixia, Toricellia, Helwingia, Aucuba, Griselinia* and *Melanophylla* each into separate families.

The Cornaceae is related to the Alangiaceae and Nyssaceae, but these families have stamens more numerous than the petals (except some Alangiaceae), and leaves strictly alternate. The secretory ducts and extrorse micropyle of *Mastixia* are thought to link the Cornaceae with the Araliaceae.

**Economic uses.** *Cornus, Aucuba* and *Griselinia* species are widely grown as ornamental shrubs. Fruits of *Cornus mas* can be used for preserves and, in France, the alcoholic beverage vin de cornouille. The wood of several species of *Cornus* and of *Curtisia* (assegai wood) is used for furniture, agricultural implements and bobbins and shuttles for weaving. H.P.W.

# PROTEALES

## ELAEAGNACEAE
*Oleaster and Sea Buckthorn*

The Elaeagnaceae is a small family of much-branched shrubs, covered with silvery or golden scales. It contains several ornamentals, such as oleaster (*Elaeagnus angustifolia*) and the sea buckthorn (*Hippophaë*).

**Distribution.** The family is mainly distributed in North America, Europe and southern Asia and Australia, frequently in coastal regions or steppes.

**Diagnostic features.** A considerable number of species (eg those of *Hippophaë*) are thorny. The stems and leaves are covered with silvery, brown, or golden hairs which are either peltate or scaly. The leaves are alternate, opposite, or in whorls, and are leathery in texture, simple, entire and without stipules.

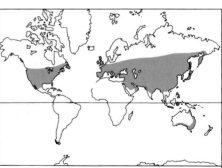

**Number of genera:** 3
**Number of species:** about 50
**Distribution:** N America, Europe, S Asia, Australia, mainly coasts and steppes.
**Economic uses:** ornamentals and limited uses of fruits and wood.

The flowers are regular and either solitary or borne in clusters or racemes. They are bisexual or unisexual, the male and female usually being borne on different plants (dioecious) in *Shepherdia* and *Hippophaë*. There are no petals, the perianth comprising a single whorl of two to eight fused sepals. In the male flower the receptacle is often flat,

Elaeagnaceae. 1 *Shepherdia argentea* (a) flowering branch showing thorns and male flowers ( ×⅔); (b) leafy shoot with female flowers ( ×⅔); (c) female flower ( ×6); (d) male flower ( ×6); (e) fruit—a drupe-like structure cut away here to reveal the single seed ( ×3). 2 *Hippophaë rhamnoides* (a) thorny, leafy, shoot bearing fruits ( ×⅔); (b) female flower with bilobed calyx-tube ( ×3); (c) gynoecium ( ×3); (d) fruit ( ×2); (e) fruit with part of fleshy calyx cut away ( ×2); (f) male flower ( ×3); (g) male flower opened out ( ×2). 3 *Elaeagnus multiflora* (a) shoot bearing fleshy fruits ( ×⅔); (b) shoot bearing bisexual flowers ( ×⅔); (c) flower opened out with vertical section of unilocular ovary ( ×2).

while in the bisexual and female flowers it is tubular. There are four to eight stamens, with free filaments and bilocular anthers. The ovary is superior, with one carpel containing a single, erect, anatropous ovule. The style is long and bears a simple stigma. The fruit is an achene or drupe-like structure enclosed by the thickened lower part of the persistent calyx. It contains a single seed with little or no endosperm and a straight embryo with thick, fleshy cotyledons.

**Classification.** The three genera, *Elaeagnus* (45 species), *Hippophaë* (two or three species) and *Shepherdia* (three species) may be distinguished by the following characters:

*Shepherdia* (North America): leaves opposite, plants dioecious, with four sepals and eight stamens in the male flower.

*Hippophaë* (Europe and Asia): leaves alternate, plants usually dioecious, female flower with elongated receptacle and short two-lobed calyx tube, male flower with two large sepals and four stamens.

*Elaeagnus* (Europe, Asia, North America, Australia): leaves alternate, flowers bisexual or unisexual, male and female on the same plant; four-lobed calyx tube is elongated beyond the ovary; four stamens present.

This family is considered to be related to the Thymelaeaceae or possibly to the Rhamnaceae but is distinguished by the golden or silvery hairy indumentum, by the nature of the fruits and by the presence of a basally inserted ovule in the ovary. It is here allied with the Proteaceae, sharing with it such features as perigynous flowers, reduced or absent petals and a single carpel.

**Economic uses.** A number of species are grown as ornamental shrubs, notably *Elaeagnus angustifolia* (oleaster), *E. pungens*, *E. umbellata* and *E. macrophylla*, which are mainly grown as deciduous or evergreen shrubs for their attractive foliage, and *Hippophaë rhamnoides* (sea buckthorn) whose female plants produce bright, orange berries in autumn and winter.

The fruits of a number of species are edible, for example those of *Shepherdia argentea* (silver buffalo berry). Its fruits are used as jelly and are also eaten dried with sugar in various parts of the United States of America and Canada. The berries of *S. canadensis* (russet buffalo berry) when dried or smoked are used as food by Eskimos. The berries of *H. rhamnoides* are made into a sauce in France, and into jelly elsewhere. The

wood of this species is fine-grained and is used for turnery. The fruits of the Japanese shrub *Elaeagnus multiflora* (cherry elaeagnus) are used as preserves and are used in an alcoholic beverage.               S.R.C.

## PROTEACEAE
### Proteas, Banksias and Grevilleas
The Proteaceae is one of the most prominent families of the Southern Hemisphere. It provides numerous examples of links between the floras of South America, South Africa and Australasia. One such example is the genus *Gevuina*, of whose three species one is native to Chile and the other two to Queensland and New Guinea. Many of the species can live in near-drought conditions but a few ancestral species are rain forest dwellers.

**Distribution.** The family is found in southern Africa, Asia, Australasia and Central and South America, especially in areas with long dry seasons.

**Diagnostic features.** Almost all of the species are trees or shrubs with alternate, entire or divided leaves which are without stipules, leathery and often hairy to some extent. The flowers are borne in sometimes showy

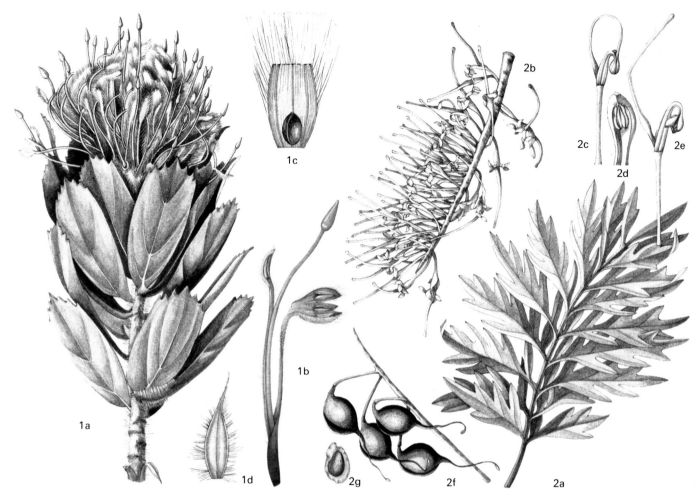

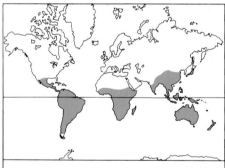

Proteaceae. 1 *Leucospermum conocarpodendron* (a) leafy shoot with terminal cone-like inflorescence, each flower with a conspicuous arrow-shaped stigma ($\times \frac{2}{3}$); (b) flower with stamens fused to perianth segments and long style with arrow-shaped stigma ($\times 1\frac{1}{2}$); (c) fruit opened out to show seed ($\times 10$); (d) hairy bract ($\times 4$). 2 *Grevillea robusta* (a) deeply-divided leaf ($\times \frac{2}{3}$); (b) inflorescence each flower with a projecting stigma ($\times \frac{2}{3}$); (c) young flower with stigma retained in bud ($\times 2$); (d) petal with anthers directly attached ($\times 5\frac{1}{2}$); (e) mature flower with extended style and stigma ($\times 2$); (f) fruits ($\times \frac{2}{3}$); (g) winged seed ($\times \frac{2}{3}$).

**Number of genera:** 62
**Number of species:** over 1,000
**Distribution:** Asia, southern Africa, Australasia, and C and S America, particularly dry regions.
**Economic uses:** ornamentals, edible seeds, honey and timber.

racemes, spikes or heads with a ring of bracts, in *Banksia* with up to 1,000 flowers. The flowers are normally bisexual but sometimes unisexual, with male and female on separate plants. They are irregular and have four perianth lobes (which some botanists regard as sepals) with reflexed tips. Two to four scales (which some regard as petals) alternate with the perianth lobes. There are four stamens inserted on the perianth lobes with often only the anther free and conspicuous, the fused filament not at all evident. The ovary may be stalked and is superior, with one carpel, and one to many ovules in the single locule. The style is long and terminal, often bent inwards and sometimes fleshy or wiry. The fruit is a follicle, drupe or nut; the seeds are often winged and have no endosperm.

The flowers tend to be protandrous, the male parts becoming functional before the female, with pollen exposed to the air on non-receptive stylar surfaces for transfer to pollinators such as birds or insects, a feature not found in groups from wetter climates.

**Classification.** There are two subfamilies: GREVILLEOIDEAE and PROTEOIDEAE. The former, generally considered the more primitive, has flowers in pairs and the latter has single flowers. The inflorescences of *Protea* and *Leucadendron* resemble genera of the Compositae and pine families respectively. The Proteaceae is generally regarded as taxonomically isolated and of uncertain affinities. It is probably most closely related to the Elaeagnaceae, sharing such common features as strongly perigynous flowers, with the petals reduced or lost, and the presence of only one carpel.

**Economic uses.** Primarily ornamental, this family has many genera (eg *Banksia*, *Embothrium*, *Grevillea* and *Telopea*) which have been successfully cultivated in tropical, subtropical and temperate climates. Examples are *Grevillea robusta* (silk oak or golden pine), *Protea cynaroides* (giant protea), *P. neriifolia*, the rare *P. stokoei*, *P. grandiceps*, *P. barbigera*, *Embothrium coccineum* (the Chilean fire bush) and *Hakea laurina* (pincushion flower). *Gevuina avellana* and the macadamia nut (*Macadamia integrifolia*) both yield edible seeds, the latter being farmed in Australia and Hawaii. Species of *Grevillea* (among others) provide timber.

B.M.

# SANTALALES

## SANTALACEAE
### Sandalwood

The Santalaceae is a family of tropical and temperate herbs, shrubs and trees. Most, if not all, are semiparasites; they are able to manufacture their own complex food

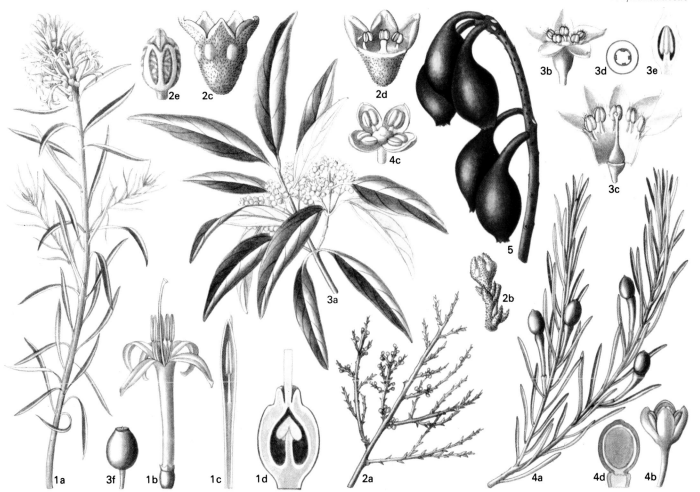

Santalaceae. 1 *Quinchamalium majus* (a) leafy shoot and terminal inflorescence ($\times\frac{2}{3}$); (b) flower ($\times 3$); (c) stamen attached to perianth-segment ($\times 4$); (d) vertical section of ovary ($\times 12$). 2 *Thesium lacinulatum* (a) flowering shoot ($\times\frac{2}{3}$); (b) shoot tip showing scale-like leaves ($\times 3$); (c) flower ($\times 6$); (d) flower with two perianth segments removed ($\times 6$); (e) fruit ($\times 6$). 3 *Santalum yasi* (a) flowering shoot ($\times\frac{2}{3}$); (b) flower ($\times 6$); (c) flower opened out ($\times 8$); (d) cross section of ovary ($\times 18$); (e) vertical section of ovary ($\times 18$); (f) fruit ($\times\frac{2}{3}$). 4 *Anthobolus foveolatus* (a) leafy shoot and fruits ($\times\frac{2}{3}$); (b) partly open flower ($\times 4$); (c) flower ($\times 4$); (d) vertical section of fruit ($\times 1\frac{1}{3}$). 5 *Scleropyrum wallichianum* fruits ($\times\frac{2}{3}$).

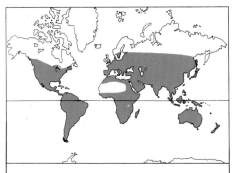

**Number of genera:** about 35
**Number of species:** about 400
**Distribution:** widespread in the tropics and temperate regions.
**Economic uses:** timber (sandalwood), oil and edible fruits.

substances by photosynthesis but require the presence of a host plant from which they absorb water and minerals through haustorial connections. They are mostly root-parasites but a few are epiphytic branch-parasites.

**Distribution.** The family is widespread throughout the tropics and in temperate regions, being concentrated in relatively dry areas.

**Diagnostic features.** The leaves are usually simple, entire, without stipules and spirally arranged, although a few genera, eg *Santalum*, have opposite leaves. However, members of several genera, including *Anthobolus*, *Exocarpos* and *Phacellaria*, have scale-like leaves and the plants then superficially resemble a *Cytisus* or *Cupressus*. *Exocarpos phyllanthoides* and other species have flattened branches (cladodes) which imitate true leaves. The flowers, borne in spikes, racemes or heads, are generally small and inconspicuous. The perianth comprises a single united whorl of three to six segments (tepals) which may be greenish or colored. The stamens are adnate to the perianth, one opposite each lobe. The ovary is inferior or semi-inferior, with a single locule containing one to five naked ovules (ie lacking an integument) suspended from a central placental column. The style is more or less simple. Only one ovule develops, and the fruit is a nut or drupe containing a single seed which has no testa and copious endosperm.

**Classification.** The Santalaceae has been divided into three tribes:

SANTALEAE, Ovary inferior; receptacle shallowly saucer- or cup-shaped, lined with a nectar-secreting disk. Twenty-seven genera including *Acanthosyris*, *Okoubaka*, *Osyris*, *Phacellaria* and *Santalum*.

THESIEAE. Ovary inferior; receptacle deeply cup-shaped, without a disk. Five genera including *Arjona*, *Quinchamalium*, *Thesium*.

ANTHOBOLEAE. Ovary superior to inferior. Ovules not fully differentiated from the placenta. Pedicel becoming swollen and fleshy as the fruit develops. Three genera including *Anthobolus* and *Exocarpos*.

The Santalaceae is related to the Loranthaceae which differs in the presence of a calyx (though very reduced) and in having viscid fruits adapted to dispersal by birds.

**Economic uses.** The best-known and most useful member of the Santalaceae is *Santalum album*, the sandalwood tree. The sapwood of this medium-sized tree yields a fragrant white timber suitable for fine carvings and carpentry, while sandal oil, used in eastern countries for anointing the body and in the manufacture of soap and perfumes, is distilled from the yellow heartwood and roots. Sandalwood is also used as a form of incense in Hindu, Buddhist and

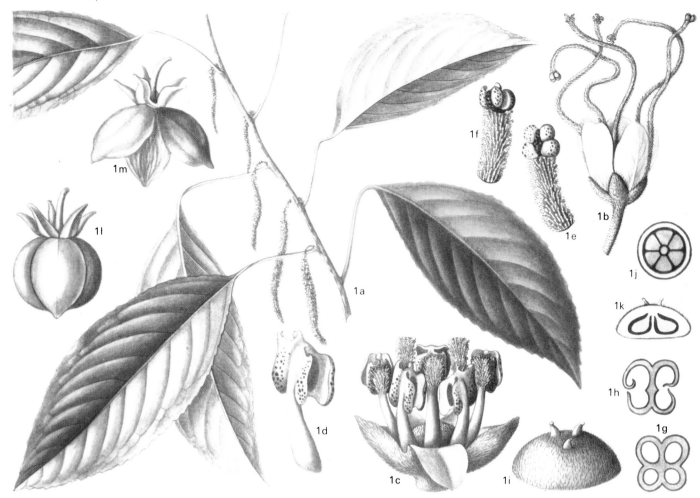

Medusandraceae. 1 *Medusandra richardsiana* (a) leafy shoot bearing axillary, pendulous racemes of flowers ( ×⅔) ; (b) flower showing the five long, hairy staminodes ( ×6) ; (c) flower with petals removed to show the five short, fertile stamens and bases of the staminodes ( ×18) ; (d) stamen with dehisced anthers ( ×26) ; (e) tip of staminode ( ×26) ; (f) tip of staminode with vestigial anthers split open ( ×26) ; (g) cross section of anther ( ×26) ; (h) cross section of dehisced anther ( ×26) ; (i) gynoecium showing three short styles ( ×26) ; (j) cross section of ovary ( ×10) ; (k) vertical section of ovary showing pendulous ovules ( ×14) ; (l) fruit—a capsule ( ×1) ; (m) dehisced fruit showing the three valves ( ×1).

Muslim ceremonies, and used as joss sticks.

*Exocarpos cupressiformis*, known as the Australian cherry, is one of the few species with edible fruits. The fruit of *Acanthosyris falcata* is also edible. F.K.K.

# MEDUSANDRACEAE

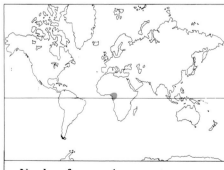

**Number of genera:** 1
**Number of species:** 1
**Distribution:** Cameroon, possibly Nigeria.
**Economic uses:** none.

The Medusandraceae is a curious family of trees from equatorial West Africa. It is generally accepted that the family is repre-sented by only one genus, *Medusandra*, containing a single species (*M. richardsiana*).

**Distribution.** *Medusandra richardsiana* grows in the rain forest of Cameroon and possibly Nigeria.

**Diagnostic features.** *M. richardsiana* is a smooth-barked understory tree of the rain forest, attaining a height of about 18m (60ft). The wood is pinkish and close-grained, with a white sapwood. The leaves are alternate, elliptic-ovate, leathery, slightly toothed 10cm–30cm (4in–12in) long, with eight lateral nerves on each side of the midrib. The young leaves are pinkish-green. The long petioles have a swelling at either end. The flowers are small, bisexual and are borne in dense pendulous racemes 3cm–15cm (1in–6in) long, which are solitary or paired in leaf axils. There are five small sepals and petals. Five short, fertile stamens with large four-locular anthers are borne opposite the petals, while five densely hairy staminodes are borne opposite the sepals; they are long and white and make the otherwise insignificant flowers conspicuous. The ovary is superior, with three fused carpels forming one locule, with six pendulous ovules attached to the roof. There are three short styles. The fruit is a three-valved, pale yellow or green capsule, with a silky-haired interior, subtended by five persistent sepals and containing a single seed. The fruits become brown and brittle on ripening, and are readily eaten by parrots and baboons, although few appear to be produced. The seeds have copious endo-sperm and a small straight embryo.

**Classification.** Two morphological features seem to distinguish the Medusandraceae from other families, the staminode and ovary structure. The staminodes appear short in the bud, but are in fact folded, and unfold to become much longer than the petals. The ovary is peculiar in having a central column. *Medusandra richardsiana* is not known to have any very close relatives. The leaves have curious bent hairs on their lower surfaces and a very complex vascular structure in the petiole, and also secretory cells. The Dipterocarpaceae exhibits similar secretory cells and petiolate vascular struc-tures, and the Lacistemataceae possesses similar hairs, but neither family is considered closely related to the Medusandraceae. Affinities with the Olacaceae, Icacinaceae and Euphorbiaceae have been suggested.

**Economic uses.** None are known. S.A.H.

Olacaceae. 1 *Heisteria parvifolia* (a) leafy shoot with small flowers ($\times\frac{2}{3}$); (b) half flower ($\times7$); (c) cross section of ovary ($\times7$); (d) fruit ($\times\frac{2}{3}$). 2 *Ximenia caffra* (a) fruits ($\times\frac{2}{3}$); (b) vertical section of fruit ($\times1$). 3 *Olax obtusifolia* (a) flowering shoot ($\times\frac{2}{3}$); (b) flower with calyx represented by a small rim below the recurved petals ($\times3$); (c) part of flower showing gynoecium and stamens ($\times3$). 4 *Schoepfia vacciniflora* (a) flowering shoot ($\times\frac{2}{3}$); (b) flower ($\times2$); (c) flower opened out showing disk around the ovary with below it the reduced calyx ($\times2$); (d) vertical section of ovary which is partly sunk in the disk ($\times4$); (e) cross section of ovary ($\times4$); (f) fruits ($\times\frac{2}{3}$).

# OLACACEAE
*American Hog Plum and African Walnut*

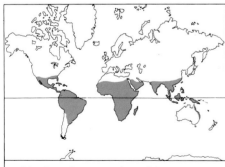

**Number of genera:** about 25
**Number of species:** about 250
**Distribution:** mainly tropical Africa, Asia and America.
**Economic uses:** timber, fruits eaten locally and a few ornamentals.

The Olacaceae is a family of shrubs, trees, climbers and lianas, most of which are native to the Old World tropics. Some have useful timber or edible fruits.

**Distribution.** The family is pantropical, but principally African and Asian, with a secondary concentration in America and some species in Australia and the Pacific Islands.

**Diagnostic features.** The leaves are alternate and entire, without stipules, and have a characteristically rough and parchment-like texture. The flowers are regular, green or white, each with a much-reduced calyx of four to six shallow lobes or teeth. The petals are valvate, equal in number to the calyx segments. The stamens are up to twice as many as the petals and opposite to them, with anthers opening often by pores. Excluding *Strombosia*, the ovary is never truly inferior, but usually partially sunk in a disk with one to three locules, each containing one ovule. There is a single style and a two- to five-lobed stigma. The fruit is a single-seeded nut or drupe. In *Harmandia* the drupe is on a conspicuous thick fleshy disk formed from the flattened calyx. The seed has a small, straight embryo and copious endosperm.

**Classification.** With their small, reduced flowers the Olacaceae are included in the order Santalales, many of whose members are adapted to a semiparasitic way of life. There are also affinities between the Olacaceae and members of the order Celastrales. Closely related and sometimes placed within the Olacaceae are *Opilia* and seven other genera which are more often separated into the family Opiliaceae.

**Economic uses.** Several genera are well known in the parts of the world where they grow and are worthy of note. *Olax* is an Old World genus represented by several climbers. Some species have leaves and fruits smelling of garlic, eg *Olax subscorpioidea, O. gambecola.* Seeds of the latter are used as condiments in parts of West Africa. In India *O. nana* is well known as one of the first species to emerge after forest fires, the shoots growing directly from buried roots. *Scorodocarpus* (literally "garlic-fruit") is another genus which smells strongly of garlic. Despite the odor, the timber is utilized in heavy construction work because of its great strength.

*Ximenia americana* is known as tallow wood or the hog plum. The wood is hard, yellow-pink in color and is used as a substitute for sandalwood in South America. The fruits of *Ximenia* are extremely bitter owing to the presence of prussic acid in the flesh.

*Heisteria* is predominantly American. Its timber is hard and is used in construction work. *Harmandia coccinea* is occasionally grown as an ornamental hothouse plant. *Coula edulis*, the African walnut, is so named because of its walnut-like nuts. They are eaten fresh, boiled or roasted. The timber also is useful; the wood is very strong and is used in house-building.                    S.W.J.

## LORANTHACEAE
### Mistletoes

The Loranthaceae is a family of parasites with green leaves, most of which are anchored to a host plant by means of suckers usually regarded as modified adventitious roots. Many species do not require a particular host species, often even tolerating plants from different families. However, some are host-specific. A few root into the earth, and the Western Australian *Nuytsia* is a small tree.

**Distribution.** The family is widely distributed in wooded areas of the tropics and extends into temperate regions; some of the groups tend to have a restricted distribution, often being either Old or New World.

**Diagnostic features.** The stem of these usually shrubby parasites is sympodial, often dichasial. There is often a large outgrowth

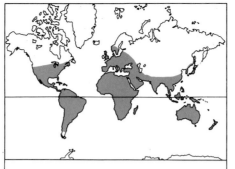

**Number of genera:** at least 35
**Number of species:** about 1,300
**Distribution:** mainly tropical, extending into temperate areas.
**Economic uses:** none, except for the decorative and symbolic value attached to mistletoe.

where the parasite's root enters the host tissue, and the root often branches considerably within the host. The leaves are usually evergreen, leathery, opposite, and without stipules. The flowers are regular, in cymes which are often reduced to three (rarely two) flowers. More rarely the inflorescence is a spike, in which case flowers

will then also occur on the internodes. The cup-shaped receptacle bears perianth segments at its rim; the segments may be green or petaloid. The flowers can be bisexual or unisexual, with sexes on the same or separate plants. The stamens are equal in number to the perianth segments, and are fused to them; the pollen is often arranged in an unusual manner, in numerous cells within the anthers. The ovary has a single locule and is sunk into the receptacle. The ovules are usually numerous and more or less immersed in the placenta. The style is simple or absent. The fruit is usually a berry or drupe, and there is a characteristic layer of sticky viscin surrounding the seeds, which adheres to the beak of birds feeding on the fruits.

**Classification.** The family is readily divisible into two main subfamilies, Loranthoideae and Viscoideae – sometimes regarded as two separate families, Loranthaceae and Viscaceae. The LORANTHOIDEAE, with the exception of the aberrant *Nuytsia*, have a characteristic rim or fringe (calyculus) below the perianth on the receptacle. The pollen is usually three-lobed, and the fruit has its viscous layer outside the vascular strands leading to the perianth segments. *Nuytsia* is

Loranthaceae. 1 *Nuytsia floribunda* (a) shoot with flowers in threes ( ×⅔) ; (b) flower ( ×4) ; (c) dry three-winged fruit ( ×1). 2 *Dendrophthora cupressoides* (a) flowering branch ( ×⅔) ; (b) male inflorescence ( ×4) ; (c) male flower from above ( ×8) ; (d) part of female inflorescence ( ×4) ; (e) female flower with sessile stigma ( ×8). 3 *Viscum album* (mistletoe) (a) shoot with berries ( ×⅔) ; (b) leaf base ( ×2) ; (c) male flowers ( ×6) ; (d) stamen adnate to perianth segment ( ×8) ; (e) female flowers ( ×6) ; (f) ovary ( ×10) ; (g) fleshy fruit ( ×2). 4 *Loranthus kimmenzae* (a) terminal inflorescence ( ×⅔) ; (b) flower with rim (calyculus) at base ( ×2) ; (c) half of flower base with detail of calyculus ( ×5) ; (d) epipetalous stamen ( ×4).

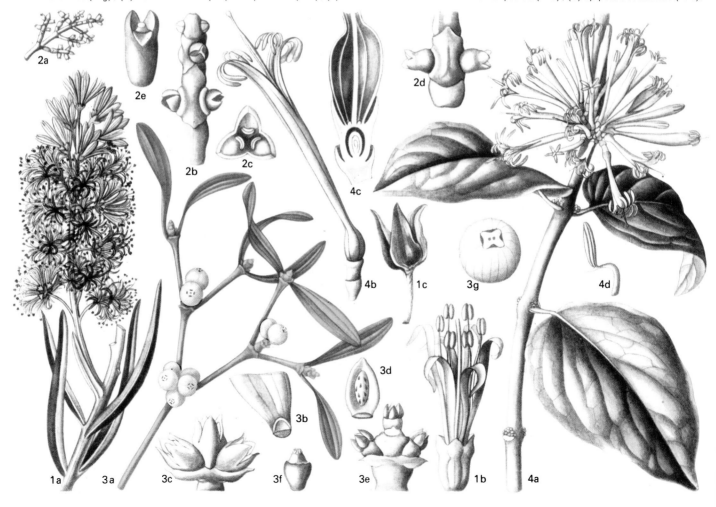

placed in a tribe on its own, the NUYTSIEAE; it has a dry, three-winged fruit. The other species belong to the tribe LORANTHEAE, and number about 850 in some 24 genera, the largest being *Loranthus* with 600 species. In the VISCOIDEAE, there is no calyculus, the pollen is spherical and the viscous layer is between the vascular strands leading to the perianth and the inner strands leading to the ovary. The ovules are reduced to sporogenous cells embedded in the tissue of the placenta. About 450 species are divided among some 11 genera, and four tribes are recognized (EREMOLEPIDEAE, PHORADENDREAE, ARCEUTHOBIEAE and VISCEAE).

The affinities of the family are rather obscure but there is a strong superficial resemblance in foliage and inflorescence to the Myrtaceae; probably the Santalaceae and Misodendraceae are the nearest-related families.

**Economic uses.** Of little economic value, the family provides the popular Christmas mistletoes (*Viscum album* in Europe, *Phoradendron flavescens* in North America). There is increasing evidence that some species can considerably threaten the establishment of indigenous hardwood trees in parts of the tropics, notably in eastern India; native conifers of North America, and *Citrus* plantations of Central America are now also recognized to be significantly affected.

I.B.K.R.

## MISODENDRACEAE

The Misodendraceae is a family of semiparasitic shrubs comprising a single genus, *Misodendrum*, of mistletoe-like plants up to about 60cm (2ft) in diameter. All grow from the trunks and branches of the southern beeches (*Nothofagus*).

**Distribution.** The family is confined to the *Nothofagus* forests of the Andes south of about 38°S and southern Tierra del Fuego.

**Diagnostic features.** The primary root is transformed into a haustorium. The stems are woody, apparently dichotomously or trichotomously branched, and often articulated at the nodes, with numerous whitish lenticels. The leaves are alternate, deciduous, green or small, brown and scale-like. Stipules are absent. The flowers are unisexual with the sexes on separate plants. The parts of the flowers are arranged in threes (rarely twos) and the flowers themselves are minute, sessile or with short pedicels, and without bracts or bracteoles. They are borne in spikelets or clusters which are sometimes in pairs or solitary, and the whole inflorescence is racemose. The male flowers have two or three stamens and the anthers have one locule; the ovary is totally absent. The female flowers have three staminodes partly included in grooves at the corners of the ovary, which later extend into very long barbed bristles. The ovary is superior, formed of three fused carpels, with three locules at first, later only one. The placenta is

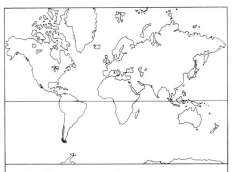

**Number of genera:** 1
**Number of species:** about 11
**Distribution:** southern S America.
**Economic uses:** none.

free and central, and three naked ovules hang from the apex of the placenta. The styles are very short (sometimes almost absent), with three stigmas, each bearing short hairs. The fruit is an achene with three barbed awns. There is a single albuminous seed with a short, erect embryo, bearing a minute sticky disk at the radical end.

**Classification.** The genus is normally divided into two subgenera: *Misodendrum*, which has leafy inflorescences and flowers with three stamens, and *Gymnophyton*, which has bracteate inflorescences and flowers with two stamens. The family is related to the Loranthaceae and Santalaceae.

**Economic uses.** The plants do not have any economic importance.

D.M.M.

## CYNOMORIACEAE

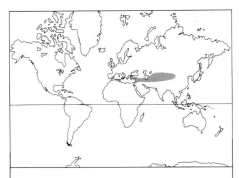

**Number of genera:** 1
**Number of species:** 1 or 2
**Distribution:** temperate Eurasia and the Mediterranean region.
**Economic uses:** occasional use as a condiment.

The Cynomoriaceae is a small family of obligate parasites of temperate Eurasia and the Mediterranean region.

**Distribution.** The family comprises the single genus *Cynomorium*, which has one or two species found very locally in dry coastal habitats from the Canary Islands and the Mediterranean basin to the steppes of Central Asia and Mongolia.

**Diagnostic features.** The whole plant is reddish-brown to purplish-black, the bulk

underground and comprising a thick rhizome bearing many haustoria and thick, simple, flowering stems, with many scales and club-shaped inflorescences of numerous epignynous flowers. The plants are polygamous (male, female and bisexual flowers on the same plant). The males have one to five or rarely up to eight linear petals and one stamen with a four-locular anther, while the females have one to five petals and one inferior carpel with a single ovule, which has a thick integument. The style is terminal. The fruit is nut-like, with a single seed which has abundant endosperm and a very small embryo.

*Cynomorium coccineum* of the Mediterranean is known to parasitize many saltmarsh plants, eg *Obione* and *Salsola* (Chenopodiaceae), *Inula* (Compositae), *Tamarix* (Tamaricaceae), *Melilotus* (Leguminosae) and *Limonium* (Plumbaginaceae). It is so unlike flowering plants in its appearance that in medieval times it was known as *Fungus melitensis*, or Maltese mushroom.

**Classification.** The family is often included in the Balanophoraceae, *Cynomorium* being linked to that family by the African *Mystropetalon*, which is the only balanophoraceous genus with an inferior ovary. Nevertheless, *Cynomorium* differs from the rest of that family in having a sculptured pollen wall (exine) and a well-developed ovule integument.

**Economic uses.** The roots of *Cynomorium coccineum* are used as a condiment by some African peoples, but only locally.

D.J.M.

## BALANOPHORACEAE

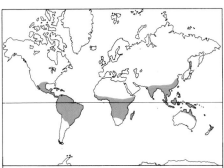

**Number of genera:** 18
**Number of species:** 120
**Distribution:** pantropical, particularly moist upland forests.
**Economic uses:** waxes burnt for lighting in Java.

The Balanophoraceae is a family of obligate parasites notable for their reduced structure, curious development and obscure affinities.

**Distribution.** The Balanophoraceae (excluding Cynomoriaceae) is a pantropical family, usually found in moist upland forest. The hosts, of which only roots are attacked, are varied and no specificity has yet been demonstrated. The hosts are normally trees.

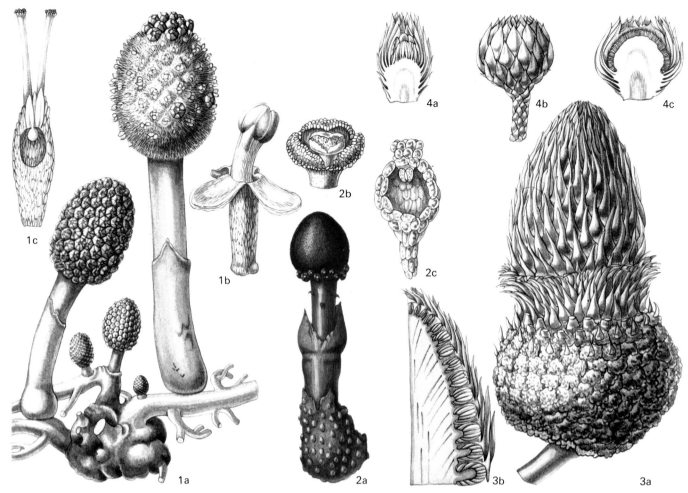

Balanophoraceae. 1 *Helosis mexicana* (a) habit showing irregular, underground tuber, spreading rhizomes and aerial, club-shaped inflorescences ( ×2) ; (b) male flower with stamens fused into a tube ( ×1,000) ; (c) female flower with part of wall removed to show pendulous ovule ( ×1,000). 2 *Balanophora involucrata* (a) tuber and inflorescence ( ×⅔) ; (b) male flower ( ×1,000) ; (c) vertical section of ovary ( ×1,000). 3 *Lophophytum weddellii* (a) habit ( ×⅔) ; (b) vertical section of male inflorescence ( ×⅔). 4 *Thonningia sanguinea* (a) male inflorescence ( ×⅔) ; (b) female inflorescence ( ×⅔) ; (c) vertical section of female inflorescence ( ×⅔).

**Diagnostic features.** The overground parts are usually fleshy, club-shaped inflorescences or "capitula" which have a very fungus-like appearance and are pale yellow to brown, pink or purplish, bearing many flowers which are amongst the smallest known. The underground parts, to which the host is attached, appear tuberous, and may reach the size of a baby's head; only in *Lophophytum* do they bear scale leaves.

The tuber may be entirely composed of parasite tissue as in *Dactylanthus* (New Zealand), *Helosis, Lophophytum* and *Scybalium* (all American). In others, the tuber is a "corpus intermedium," ie part parasite, part host, as noted by the Dutch botanist, Karl Blume, as early as 1827, when it was generally thought that these plants were fungi. The tuber is a chimerical system unknown elsewhere in higher plants.

The inflorescences develop inside the tuber, rupturing the tuber tissue which remains as a "volva" at the base; in *Chlamydophytum*, they mature completely before rupturing the tuber. The inflorescence may have spirally arranged branches as in *Sarcophyta*, or the capitula may be un-branched and flattened or club-shaped. The

flowers are unisexual with the sexes either on separate plants or the same plant; in the latter case there may be separate male and female inflorescences or mixed ones with the males towards the base. In *Helosis* the inflorescence is covered with startlingly geometrical hexagonal scales, each of which is surrounded by two concentric rings of female flowers, while the males occupy corners under the scales. In the male flowers the perianth is either lacking or may be valvate and three- to eight-lobed. The stamens are one or two in male flowers without petals and equal in number to the segments in those with a perianth. The anthers have two, four or many locules. In *Hachettea*, the filament is contracted and the single stamen has one locule. In *Helosis* and *Scybalium* the lower parts of the stamens are united into a tube with discrete anthers, and in other genera there is merely a tube tipped with pollen sacs. The pollen is much reduced as in many parasitic plants, and the outer wall (exine) is not sculptured. The female flowers lack a perianth and are so reduced that ovules, placentas and carpels are not easily recognizable. The ovary is normally super-ior, but in *Mystropetalon* it is inferior. There

are one to three locules, each with a single, usually pendulous ovule. The ovules are represented by embryo sacs without integu-ments. There are one or two styles with a terminal stigma; occasionally the stigma is sessile. The fruits are nut- or drupe-like, containing seeds with abundant endosperm and a small embryo.

Small flies are thought to visit those with a sickly sweet odor, though *Juelia*, which is largely subterranean, may be apomictic. The "pedicels" of some female flowers become elaiosomes, attractive to ants, which are known to disperse the seeds of *Mystro-petalon*.

**Classification.** At different times the Bal-anophoraceae has been split into six discrete families, now treated as tribes. Five of these have a storage substance resembling starch, while species of the sixth (Balanophoroideae) accumulate a waxy substance called bal-anophorin. The pollen of *Mystropetalon* is unique in flowering plants in being tri-angular, square or pentagonal when viewed end on, but almost always square when viewed from the side. This is so character-istic as to suggest that *Mystropetalon* may be only distantly related to the rest of the genera.

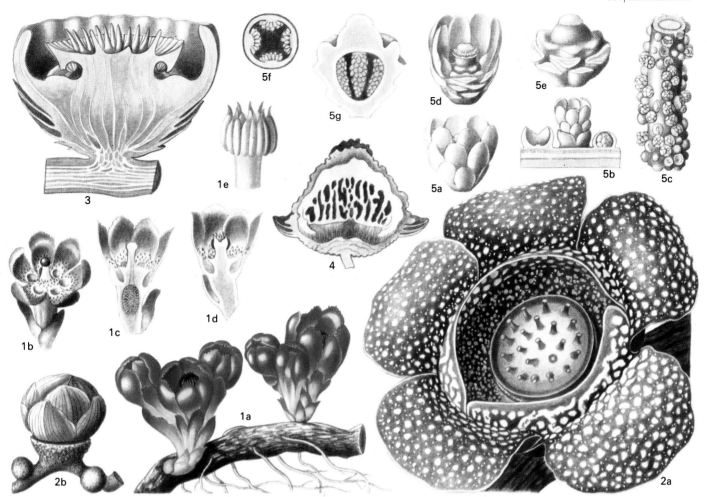

Rafflesiaceae. 1 *Cytinus sanguineus* (a) inflorescences on surface of host stem ($\times\frac{2}{3}$); (b) female flower ($\times\frac{2}{3}$); (c) half female flower ($\times\frac{2}{3}$); (d) half male flower ($\times\frac{2}{3}$); (e) staminal column ($\times 2$). 2 *Rafflesia manillana* (a) male flower ($\times\frac{1}{3}$); (b) flower buds ($\times\frac{1}{4}$). 3 *R. patma* half male flower bud showing "mycelia" ramifying through the host tissue ($\times\frac{1}{4}$). 4 *R. rochussenii* vertical section of fruit ($\times\frac{1}{3}$). 5 *Apodanthus welwitschii* (a) male flower ($\times 4$); (b) section of host branch showing flowers and flower buds ($\times 2$); (c) host branch bearing flowers ($\times\frac{2}{3}$); (d) male flower with part of calyx removed ($\times 4$); (e) female flower with perianth removed ($\times 3$); (f) cross section of ovary ($\times 3$); (g) vertical section of ovary ($\times 4\frac{1}{2}$).

The Balanophoraceae is closely related to the Cynomoriaceae, which is often placed in it. As these plants are clearly reduced and simplified in structure, it is difficult to decide whether they represent a "natural" assemblage or are end products of convergent evolution; specialists tend to agree on the former idea. They are probably unrelated to other parasitic groups though they are often said to be related to Hydnoraceae and Rafflesiaceae. They also have affinities with other families of the Santalales, such as Santalaceae, Olacaceae and Opiliaceae. In the opinion of recent workers, the true relationship, albeit a distant one, may be with *Gunnera* (Haloragaceae).

**Economic uses.** These plants are sometimes considered to have aphrodisiac properties. The only place where real use is made of them is Java, where waxes are extracted and burnt for lighting. D.J.M.

# RAFFLESIALES

## RAFFLESIACEAE
The Rafflesiaceae are total parasites, invading the stems or roots of other flowering

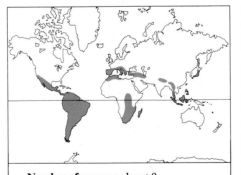

**Number of genera:** about 9
**Number of species:** about 500
**Distribution:** tropics and subtropics.
**Economic uses:** none.

plants. The flowers of *Rafflesia* are the largest known, some being up to 1m (3ft) across.

**Distribution.** The family is found mainly in the tropics and subtropics.

**Diagnostic features.** Only the reproductive parts of the Rafflesiaceae are recognizably angiospermous. Apart from a few scaly bracts below each flower, the vegetative tissues are represented by a kind of "mycelium" which ramifies through the host

cambium. The flower buds begin their development inside the host and then push through to the surface. In some species the "mycelium" penetrates into the growing points of the host's aerial or subterranean parts, and the parasite may then develop synchronously with the host. Thus in the Iranian *Pilostyles haussknechtii*, which infects *Astragalus* stems, the parasite's flowers are borne very regular in pairs at the base of leaves (belonging to its host) which were produced at the beginning of that growing season. Again, a Japanese species of *Mitrastemon*, which parasitizes the roots of *Quercus*, produces its flowers in annual "fairy rings" corresponding to a zone a few centimeters within the perimeter of the host's root system. In cases such as these the parasite eventually becomes fragmented into many parts occupying the terminal twigs or roots.

The flowers have no petals but a fleshy petaloid calyx with four to six lobes. They are usually unisexual (then borne on the same or separate plants) and rarely bisexual. In the center of the flower is a stout column with a grooved margin; in male flowers the numerous stamens occur on the column, under the margin, while in females the corresponding area is stigmatic. The ovary is inferior (rarely superior) of four to numerous fused carpels and either comprises one locule with four to 14 parietal placentas bearing very many ovules, or (in *Rafflesia*) consists of a honeycomb of irregular chambers lined with ovules. The fruit is fleshy, and contains numerous tiny hard seeds.

**Classification.** The Rafflesiaceae has been divided into four tribes:

MITRASTEMONEAE. Flowers bisexual and solitary, ovary superior. *Mitrastemon*, Southeast Asia, Central America.

APODANTHEAE. Flowers unisexual and solitary, small; stamens in two to four rings on the central column; ovary inferior with four placentas or one continuous placenta. *Apodanthes, Pilostyles, Berlinianche*, tropical America, tropical Africa and Iran.

RAFFLESIEAE. Flowers unisexual and solitary, large; stamens in one ring; ovary inferior with many irregular chambers. *Rafflesia, Sapria, Rhizanthes*, Southeast Asia from India to Indonesia.

CYTINEAE. Flowers unisexual and in racemes; stamens in one ring; ovary inferior with 8–14 placentas. *Cytinus, Bdallophyton*, Mediterranean region, South Africa, Madagascar, Mexico.

The closest relatives of the Rafflesiaceae are undoubtedly the Hydnoraceae, which differ in having root-like structures, no bracts below the flowers, bisexual flowers and stamens not borne on a column. Some botanists prefer to isolate *Mitrastemon* as a separate family placed between the Rafflesiaceae and Hydnoraceae. The affinities of this group are very uncertain. Most authorities link the Rafflesiaceae with the Aristolochiaceae because they have a similar perianth, but the evidence for this association is not conclusive.

**Economic uses.** None are known.        F.K.K.

## HYDNORACEAE

Members of the Hydnoraceae are parasites which feed from the roots of the host plant. They are found in Madagascar and tropical Africa (the 12 species of *Hydnora*) and in South America (the six species of *Prosopanche*).

The plant is leafless and rootless; the large, solitary, bisexual flowers (the only aerial parts) are borne on the thick, creeping underground "stem." There are no petals, but three to five sepals (or perianth segments) are fused into a tube at the base, and open as valves above. Many anthers are united in a ring-like or undulating mass on the calyx tube with many parallel longitudinal or transverse locules, there being no stamen filaments. The ovary is inferior (usually well below soil level) and consists of three to five fused carpels with a single locule containing very many reduced ovules on placentas which are leaf-like or parietal or pendulous from the apex of the ovary. The

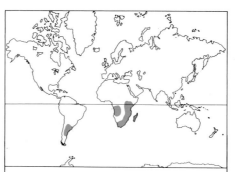

**Number of genera:** 2
**Number of species:** 18
**Distribution:** Madagascar, tropical Africa, S America.
**Economic uses:** none.

stigma is button-like and crowns the ovary, which matures into a large, thick-walled fleshy berry containing many minute seeds with copious endosperm. The plants are thought to attract pollinating beetles by generating a smell of carrion.

The Hydnoraceae is most closely related to its fellow parasite family, the Rafflesiaceae, from which it differs chiefly in having bisexual flowers. No economic uses are known.        B.M.

# CELASTRALES

## GEISSOLOMATACEAE

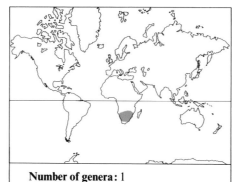

**Number of genera:** 1
**Number of species:** 1
**Distribution:** S Africa.
**Economic uses:** none.

The Geissolomataceae is a monotypic family with a single species (*Geissoloma marginatum*), native to South Africa.

*Geissoloma* is a small xerophytic shrub with entire sessile leaves, without stipules, arranged in opposite pairs in four rows down the stem. The flowers are bisexual, regular and arranged in a much reduced axillary racemose type of inflorescence. The flower is subtended by six persistent bracts. There are four petaloid sepals which are fused together at the base to form a short tubular structure. There are no petals and the eight free stamens with slender filaments are attached to the base of the calyx. The stamens are in two whorls of four, lying opposite and alternate with the sepals. The ovary is

superior, formed of four fused carpels, and has four locules, with two pendulous ovules inserted at the apex of each of the locules. The ovary is surmounted by four tapering styles which are separate at the base but coherent at their apexes. The fruit is a four-locular capsule with normally only a single seed in each locule. The seed contains a straight embryo and a little endosperm.

The relationships of Geissolomataceae are disputed. It has been related to the Thymelaeaceae and the Penaeaceae (order Myrtales), sharing with them such features as woody habit, apetalous flowers, a petaloid calyx, rather leafy bracts and a straight embryo in the seed. But it is barely perigynous and lacks internal phloem. Its pollen resembles that of the Celastraceae. No economic uses are known.        S.R.C.

## CELASTRACEAE
*Spindle Tree*

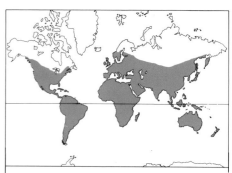

**Number of genera:** 55
**Number of species:** 850
**Distribution:** mainly tropical and subtropical regions.
**Economic uses:** source of Arabian tea (khat leaves), cultivated as ornamentals, timber used for carving and yields a seed oil and yellow dye.

The Celastraceae is a family of trees and shrubs, many of which are climbing (eg *Salacia*) or twining (eg *Hippocratea*) in habit.
**Distribution.** The family is widespread, but concentrated in sub- and tropical regions.
**Diagnostic features.** The leaves may be opposite or alternate even in a single genus as in *Cassine*. They are simple, often leathery, with or without small stipules. The flowers are small, greenish, regular, bisexual or unisexual and arranged usually in cymose inflorescences. The sepals and petals are inserted on or below the margin of a distinctive glandular fleshy disk. There are three to five sepals, free or united at the base, and three to five free petals (rarely none). The stamens alternate with and are equal in number to the petals and are inserted on the disk. The anthers have two locules and dehisce longitudinally (transverse dehiscence in *Hippocratea*). The ovary is superior and consists of two to five fused carpels with two to five locules, each usually with two (sometimes one, rarely many) erect ovules on axile placentas. There is a single, very short

Celastraceae. 1 *Euonymus myrianthus* (a) leafy shoot bearing fruits—loculicidal capsules (×⅔); (b) flower with four distinct petals inserted on a fleshy disk (×2); (c) half flower showing stamens inserted on the disk and gynoecium with ovules on axile placentas (×4); (d) stamen (×12). 2 *E. vagans* cross section of ovary (×14). 3 *Hippocratea welwitschii* (a) shoot with dehiscing fruit (×⅔); (b) seed with aril (×⅔). 4 *Celastrus articulatus* (a) leafy shoot and cymose inflorescences (×⅔); (b) flower (×2); (c) two stamens and gynoecium (×2); (d) fruit—a capsule (×2). 5 *Elaeodendron aethiopicum* (a) flower (×6); (b) half flower (×8).

style terminated by a capitate or two- to five-lobed stigma. The fruit is a loculicidal or indehiscent capsule, a samara, berry or drupe. The capsule of some species of *Euonymus* may possess spiny outgrowths. The seeds possess a large, straight embryo surrounded by fleshy endosperm and are often covered by a brightly colored aril which aids in dispersal by birds.

**Classification.** The chief genera are *Maytenus* (225 tropical species), *Salacia* (200 tropical species), *Euonymus* (176 species mostly from the Himalayas, China and Japan), *Hippocratea* (120 species, tropical South America, Mexico and southern United States of America), *Cassine* (40 species, South Africa, Madagascar, tropical Asia and the Pacific), *Celastrus* (30 subtropical and tropical species), *Elaeodendron* (16 subtropical and tropical species), *Pachystima* (five species, North America) and *Gyminda* (three species, Central America, Mexico and Florida).

The Celastraceae is probably related to the Aquifoliaceae but is distinguished from it by the presence of the glandular disk surrounding the ovary and by the brightly colored aril. The absence of aril and endosperm and

the unusual form of anther dehiscence in *Hippocratea* have been considered by some taxonomists to warrant separate family status (Hippocrateaceae).

**Economic uses.** The small khat tree, *Catha edulis*, is cultivated in the Middle East and Ethiopia for its leaves, used to make an infusion of tea (Arabian tea) or in the making of a honey wine (in Ethiopia). The seeds of *Kokoona zeylanica* (kokoon tree) are used as a source of oil in Ceylon. A number of species of *Euonymus* yield useful products, including the spindle tree (*Euonymus europaeus*) whose fine-grained wood is used for turnery and figure carving and as a source of charcoal. The seeds of this species yield an oil used in soap manufacture and a yellow dye for coloring butter. The heavy, durable, close-grained wood of the Japanese shrub, *E. hians*, is also used for turnery and for making printing blocks. Some species, such as *E. japonicus* and *E. sieboldiana* yield a rubber-like latex from the stems and roots. Extracts from *E. purpureus* and *E. americanus* are used in native medicines in North America. Species of other genera which yield medicinal extracts include *Elaeodendron glaucum*, *Maytenus boania*, *M. ilicifolia* and *M.*

*senegalensis*, and *Hippocratea acapulcensis*.

Species of *Celastrus*, *Euonymus*, *Elaeodendron*, *Pachystima* and *Maytenus* are cultivated as ornamentals. *Celastrus orbiculatus* and *C. scandens* are attractive, vigorous climbing shrubs. Although the flowers of *Euonymus* are inconspicuous, a number of deciduous species including *E. alatus*, *E. europaeus*, *E. latifolius* and *E. yedoensis* are grown for their attractive autumn tints and distinctive bright-colored fruits. The evergreen *E. japonicus* has an upright, bushy habit which makes it suitable for hedges.

S.R.C.

## STACKHOUSIACEAE

The Stackhousiaceae is a small family of three genera of more or less xerophytic herbs with a branched rhizome system.

**Distribution.** The family occurs chiefly in Australasia, but also in Malesia.

**Diagnostic features.** The leaves are alternate, simple, with stipules, and are often leathery or succulent. The flowers are regular, bisexual and borne in racemes or spaced clusters, and consist of a five-lobed calyx tube on which the five petals and five alternating stamens (often three long and two

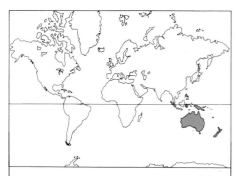

**Number of genera:** 3
**Number of species:** 27
**Distribution:** chiefly Australasia.
**Economic uses:** none.

short) are inserted. The petals may be linear or spoon-shaped, and free or partly fused together into a tube in the middle portion but not at the base. The ovary is superior, comprising two to five locules with one erect ovule at the base of each locule. The styles are the same number as locules and are partly fused. The fruit is a schizocarp which splits into two to five segments, each indehiscent and with one seed. The seeds have fleshy endosperm and a large, straight embryo.

**Classification.** *Macgregoria* differs from *Stackhousia* and *Tripterococcus* in having free petals, not partly fused, and stamens of equal length and is sometimes placed in a separate subfamily the MACGREGORIOIDEAE, the latter two being placed in the subfamily STACKHOUSIOIDEAE.

The only species of *Macgregoria*, *M. racemigera* (arid Australia), is a slender, glabrous annual with star-like, white, terminal racemes of flowers. *Stackhousia* comprises some 25 mostly Australian species. *Stackhousia intermedia* (Malesia) is a lank herb to 50cm (20in) in open grassy situations; *S. minima* (New Zealand) is a slender herb to 5cm (2in) tall with solitary flowers; *S. monogyna* (eastern Australia, including South Australia) has white to cream, densely flowered racemes. *Tripterococcus* contains a single species from southwestern Australia. The family is not clearly allied to any other but has been related to the Euphorbiaceae, Celastraceae and the order Sapindales.

**Economic uses.** None are known.          B.M.

## SALVADORACEAE
The Salvadoraceae is a small family of trees and shrubs.

**Distribution.** The family is native to arid, often saline, areas in Africa, Madagascar, Arabia, India and Asia.

**Diagnostic features.** The leaves are opposite and simple, with minute stipules. Some species of *Azima* have axillary spines. The flowers are borne in dense axillary clusters or in branched racemes. They are regular, bisexual or unisexual (the male and female then being borne on separate plants). The calyx comprises two to four fused sepals. The

petals are free or partially fused, four or five, with teeth or glands on the inner side. There are four or five stamens which alternate with the petals. The filaments may be fused into a tube and are often inserted at the base of the petals. The anthers have two locules and open lengthwise. A disk in the form of separate glands may be present between the filaments. The ovary is superior and consists of one or two carpels with one or two erect ovules in each locule. The style is short and bears a forked stigma. The fruit is a berry or drupe and contains a single seed without endosperm and an embryo with thick cotyledons.

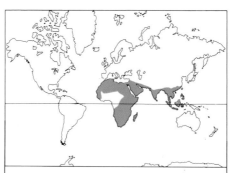

**Number of genera:** 3
**Number of species:** 11
**Distribution:** Africa and Madagascar to Arabia and Asia.
**Economic uses:** shoots and leaves used as salad, local medicinal uses, fat for candles, Kegr salt and perfume oil.

**Classification.** The three genera differ in habit and floral structure: *Azima* (four species), thorny shrubs whose flowers possess free petals and stamens and a bilocular ovary, *Dobera* (two species) and *Salvadora* (five species), spineless trees and shrubs. *Dobera* has free petals and the stamen filaments united at the base. The petals of *Salvadora* are united at the base and the stamens are epipetalous.

The relationships of this family are doubtful; it is here placed in the Celastrales.

**Economic uses.** The shoots and leaves of *Salvadora persica* (the toothbrush tree or salt bush) are used in salads or as food for camels. Kegr salt is obtained from the plant ash. The flowers of *Dobera roxburghii* provide an essential oil used by Sudanese women as perfume.          S.R.C.

## CORYNOCARPACEAE
The Corynocarpaceae is a small family of Pacific trees and shrubs.

**Distribution.** The family is native to Polynesia, New Zealand and Australia and some other areas of the southwestern Pacific.

**Diagnostic features.** The leaves are alternate, simple, entire and without stipules. The flowers are bisexual, regular and borne on indeterminate branching racemes. The sepals are five in number, and are distinctly imbricate. The five petals are fused to the base of the sepals. The filaments of the five

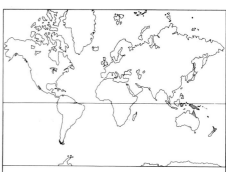

**Number of genera:** 1
**Number of species:** 5
**Distribution:** Australasia and Polynesia.
**Economic uses:** seeds eaten by Maoris and canoes made from trunks.

stamens are in turn attached to the base of the petals. The stamens are inserted opposite the petals (antipetalous) and not alternate with them as in most other families. In addition, there are five alternating, clawed, petaloid staminodes, opposite to each of which is a large disk-shaped gland. The ovary is superior, of one or two fused carpels containing one or two locules, surmounted by one or two free styles. When there are two locules, only one of them is fertile; each locule contains a single pendulous ovule. The fruit is a drupe, and the seed it contains is without endosperm, while the embryo has a tiny radicle.

**Classification.** There is only one genus, *Corynocarpus*, with five species. The affinities of the family are obscure, and it has been placed in a number of orders, including the Ranunculales and the Sapindales, but it is here referred to the Celastrales, having some features in common with the Celastraceae.

**Economic uses.** The fruits of *Corynocarpus laevigatus* are used as food by the Maoris and the seeds form a staple food. Some natives of the southwestern Pacific use the trunks of the trees for making canoes.          S.R.C.

## ICACINACEAE
This family comprises trees, shrubs and lianas, almost all of which inhabit tropical rain forests.

**Distribution.** The family occurs mainly in Malaysia and tropical regions of India, Africa and Central America, decreasing rapidly toward the subtropics. Among the few species of tall trees are *Apodytes* (Malaysia, northeastern Australia and southern Africa), found in primary rain forest on steep slopes, ravine and stream edges, and the Malaysian *Stemonurus* found in peatlands (occasionally coastal mangroves), swamp forest, in lowlands or even dry hilly land. Among the tallest of southern African Icacinaceae are *Poraqueiba* and *Dendrobangia*. Smaller trees and shrubs such as *Gonocaryum* (Taiwan, Southeast Asia and Indomalaysia) and *Gomphandra* (Southeast

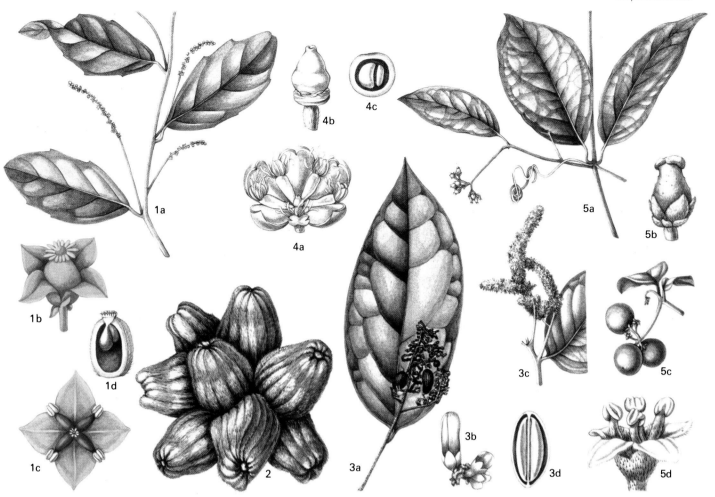

Icacinaceae. 1 *Pyrenacantha volubilis* (a) leafy shoot with axillary inflorescences ($\times\frac{2}{3}$); (b) female flower ($\times 6$); (c) male flower ($\times 6$); (d) vertical section of ovary ($\times 6$). 2 *Phytocrene bracteata* fruits ($\times\frac{2}{3}$). 3 *Citronella suaveolens* (a) leaf and fruits ($\times\frac{2}{3}$); (b) flower bud ($\times 3$); (c) inflorescence ($\times\frac{1}{3}$); (d) cross section of fruit ($\times 5$). 4 *Stemonurus vitensis* (a) flower ($\times 6$); (b) ovary ($\times 16$); (c) cross section of ovary ($\times 16$). 5 *Iodes usambarensis* (a) leafy shoot with tendril and inflorescence ($\times\frac{2}{3}$); (b) female flower ($\times 6$); (c) fruits ($\times 1$); (d) male flower ($\times 6$).

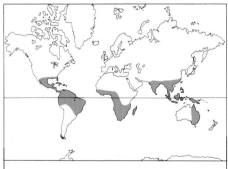

**Number of genera:** about 60
**Number of species:** about 400
**Distribution:** chiefly tropical
rain forest.
**Economic uses:** timber, starches, oils
and tea substitute.

Asia to Solomon Islands) grow in the forest understory in low montane zones. Some *Citronella* species (South America) have also been found on dry, barren soil as well as dense, moist forests. Most of the rain forest lianas are dioecious, such as *Miquelia* (Indochina and Indonesia) and *Phytocrene* (mainly Malaysia but also Asia),
**Diagnostic features.** The leaves are usually alternate, entire, and leathery, always with-

out stipules. The inflorescence is usually a cyme or thyrse. The flowers are small, regular, not fragrant, bisexual or unisexual, the sexes then borne on separate plants. There are four or five each of sepals, petals and stamens (no sepals in *Pyrenacantha*). The stamens alternate with the petals and the anthers have two locules, rarely four lobes. The ovary is superior, initially of three (sometimes two or five) carpels; there is usually one locule, rarely two or more. There are two anatropous ovules, pendulous from the top of each locule. The style is simple and short, with three (rarely two or five) stigmas. The fruit is usually a one-seeded drupe, sometimes a samara; the endocarp is often laterally flattened and ornamented with depressions or pits and the seeds usually have endosperm and a small, straight embryo.

**Classification.** The Icacinaceae is subdivided into four tribes:

ICACINEAE. Trees or shrubs, rarely climbers; flowers bisexual; vessels in wood with ladder-like perforations; chief genera *Citronella*, *Gonocaryum*, *Gomphandra*, *Poraqueiba* and *Icacina*.

IODEAE. Climbers or twining shrubs; plants dioecious; vessels in wood with simple

perforation plates; chief genus *Iodes*.

SARCOSTIGMATEAE. Climbing shrubs, inner surface of endocarp weakly wrinkled; chief genus *Sarcostigma*.

PHYTOCRENEAE. Climbing or twining shrubs; endocarp inner wall always warty; chief genera *Pyrenacantha* and *Phytocrene*.

The Icacinaceae shows connections with the Celastraceae, Aquifoliaceae and Olacaceae. The differences are in such characters as presence or absence of stipules, petals and floral disk, aestivation of perianth parts and number of locules in the fruit.

**Economic uses.** The hard, heavy, fragrant timber of *Cantleya corniculata* is exported from Sarawak and Brunei and is used for house- and shipbuilding, and as a sandalwood substitute. The wood of *Apodytes* is used in Indochina for cabinet work. The tubers and seeds of *Humirianthera* have abundant starch. *Poraqueiba* fruits and seeds yield starch and oil. Leaves of *Citronella* species are used as a substitute for yerba maté tea. The seeds of *Sarcostigma kleinii* yield a useful oil in India used to treat rheumatism. The leaves and bark of *Cassinopsis madagascariensis* yield an antidysenteric. Cut stems of *Miquelia* and *Phytocrene* yield drinkable water.   H.P.W.

Aquifoliaceae. 1 *Ilex aquifolium* (a) leafy shoot and fruits ($\times\frac{2}{3}$); (b) female flower with four staminodes ($\times 4$); (c) male flower with four stamens alternating with the petals ($\times 4$); (d) corolla of male flower opened out showing stamens attached at the base ($\times 4$); (e) gynoecium with sessile stigma ($\times 4$). 2 *I. anomala* (a) flowering shoot ($\times\frac{2}{3}$); (b) bisexual flower ($\times 3$); (c) perianth opened out to show stamens fused to the base of the perianth tube and alternating with the lobes ($\times 3$); (d) fruit—a berry ($\times 4$). 3 *I. paraguaensis* (a) leafy shoot with fruits ($\times\frac{2}{3}$); (b) fruit with wall cut away to show four hard pyrenes each containing a single seed ($\times 4$).

## AQUIFOLIACEAE
*Hollies and Yerba Maté*

This family of trees and shrubs comprises the large genus *Ilex* (holly) with about 400 species, *Nemopanthus* with two species and *Phelline* with ten.

**Distribution.** The family is widely distributed in both temperate and tropical regions, though relatively poorly represented in Africa and Australia. *Nemopanthus* is restricted to northeastern North America and *Phelline* to New Caledonia.

**Diagnostic features.** The leaves are leathery, sometimes evergreen and usually alternate; stipules are present (except in *Phelline*) but may fall soon after enlargement of the leaf. The inconspicuous, greenish-white flowers are bisexual or unisexual (in which case male and female are normally borne on different plants). They are regular and usually in bundles or cymes (spikes or panicles in *Phelline*), rarely solitary, with minute bracts. The sepals and petals, usually four of each (four to six in *Phelline*), are imbricate (petals valvate in *Phelline*); the petals are united at the base (free in *Nemopanthus* and *Phelline*), and the stamens, usually equal in number to and alternating with the petals, are freq-

**Number of genera:** 3
**Number of species:** about 400
**Distribution:** widespread in tropical and temperate regions.
**Economic uses:** valued ornamental trees and shrubs, useful hardwood and tea (yerba maté).

uently fused to the petals; otherwise they are free, and are often represented as staminodes in female flowers. The anthers have two locules and open inward. The ovary is superior, of three or more united carpels, comprising three or more locules with a single terminal, sometimes minute, style; placentation is axile, with one or rarely two pendulous, usually anatropous, ovules in

each locule. The fruit is a berry with usually four hard pyrenes each with one seed with copious endosperm.

**Classification.** The genus *Phelline* is sometimes separated as the family Phellinaceae. Closely allied to the Celastraceae, the family is distinguished mainly by the absence of a distinct, annular nectary disk at the base of the stamens, and by the usually solitary ovule in each locule.

**Economic uses.** The family is important for the hard, white wood of *Ilex* species, used in carving, inlay and many other ways. Holly has long been used as decoration, for its attractive foliage and berries. There is a Christmas trade in holly in several countries and many species are grown as ornamentals. The leaves of *Ilex paraguensis* are used as a tea (yerba maté). I.B.K.R.

## DICHAPETALACEAE

This is a family of tropical shrubs and a few climbers and small trees, some of which are very poisonous. It is sometimes incorrectly known as the Chailletiaceae.

**Distribution.** There are about 200 species in four genera: *Dichapetalum* (150 or more tropical species, notably in Africa); *Stepha-*

Dichapetalaceae. 1 *Stephanopodium peruvianum* (a) leafy shoot and axillary inflorescences united to the leaf stalk (×⅔); (b) half flower with sessile epipetalous anthers (×8); (c) vertical section of ovary showing pendulous ovules (×14); (d) fruit—a hairy drupe (×⅔); (e) flower (×8). 2 *Dichapetalum mombongense* (a) flowering and fruiting shoot (×⅔); (b) half flower (×8); (c) cross section of ovary (×21); (d) cross section of fruit (×1½). 3 *Dichapetalum toxicarum* (a) flower with gynoecium removed (×14); (b) gynoecium (×14). 4 *Tapura ciliata* (a) leafy shoot with small stipules and axillary inflorescences (×⅔); (b) flower (×4); (c) corolla opened out (×6); (d) hypogynous gland (×12); (e) ovary (×12).

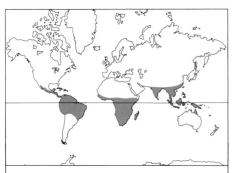

**Number of genera:** 4
**Number of species:** about 200
**Distribution:** throughout the tropics, with one S African species.
**Economic uses:** a few ornamentals and leaves and seeds used as a poison.

*nopodium* (seven species, tropical South America); *Tapura* (20 species, tropical South America and the West Indies, and four species in tropical Africa); and *Gonypetalum* (five species in tropical South America). The only extra-tropical species is a South African *Dichapetalum*.

**Diagnostic features.** The leaves are simple, alternate, with stipules, and often covered

with fine gray hairs. The inflorescence is an axillary cyme or fascicle, its stalk often united to the leaf petiole below as in some species of *Dichapetalum* and *Tapura*. The flowers are regular or irregular, bisexual or unisexual (then with both types on the same plant). They have five imbricate sepals and four or five petals; the petals are forked or bilobed and often dry black. The five stamens are borne on the petals or are free. The ovary is superior with two or three united carpels, each with two pendulous ovules; it is surrounded by a cup-shaped or lobed disk and topped with two separate or united styles. The fruit is a lobed drupe, which is usually pubescent, containing one to three locules each with a single seed. The seeds often bear a caruncle, have no endosperm but a large, straight embryo.

**Classification.** Both *Dichapetalum* and *Stephanopodium* have five fertile stamens in each bisexual flower, those of the latter being sessile. *Tapura* and *Gonypetalum*, united by some authors, have only three fertile stamens. The Madagascan *Falya* is now placed in the same genus as the West African *Carpolobium* (Polygalaceae), but was at one time placed in the family Dichapetalaceae.

The affinities of the Dichapetalaceae are a matter for debate, and it has been referred to the Rosales, Celastrales or Euphorbiales.

**Economic uses.** A few species are grown as ornamentals in the tropics, and the fruits of some East African species are said to be edible. The leaves and seeds of all species of *Dichapetalum* tested are poisonous. The leaves of *Dichapetalum stuhlmannii* are used to poison wild pigs, monkeys and rats in East Africa, where an extract is said to have been used in arrow poisons. The seeds of *D. toxicarium* are similarly used in West Africa, particularly as an effective rat poison. *D. cymosum* of the high veldt of southern Africa begins growth before the veldt grasses, and is therefore eaten by cattle, giving rise to "gifblaar" poisoning with disastrous effects. The toxic principle is fluoracetic acid, which disrupts the tricarboxylic acid cycle of respiration. D.J.M.

# EUPHORBIALES

## BUXACEAE
*Box*

The Buxaceae is a small family of six genera of evergreen shrubs (rarely trees or

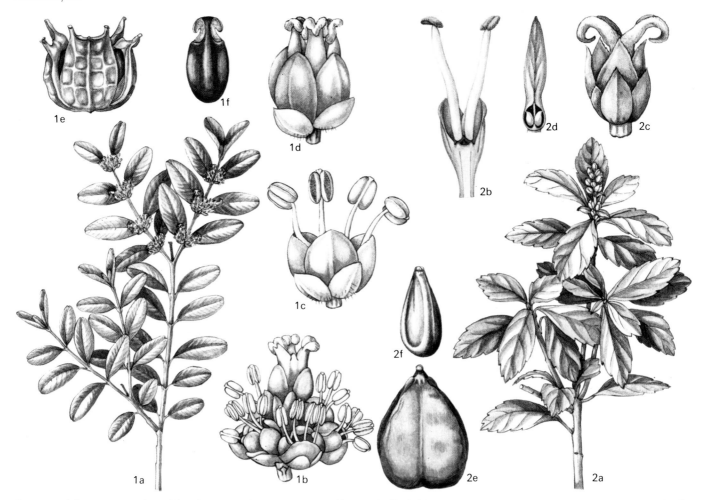

Buxaceae. 1 *Buxus sempervirens* (a) leafy shoot with axillary clusters of flowers ($\times\frac{2}{3}$); (b) female flower surrounded by cluster of male flowers ($\times 6$); (c) male flower with four stamens having introrsely dehiscing anthers ($\times 8$); (d) female flower with three styles crowned by convolute stigmas ($\times 8$); (e) fruit—a capsule dehiscing by three valves ($\times 2$); (f) seed ($\times 4$). 2 *Pachysandra terminalis* (a) leafy shoot and inflorescence ($\times\frac{2}{3}$); (b) half male flower ($\times 3$); (c) female flower ($\times 6$); (d) vertical section of ovary ($\times 6$); (e) fruit ($\times 4$); (f) seed ($\times 4$).

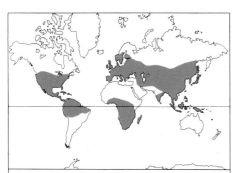

**Numbers of genera:** 4 or 6
**Number of species:** about 100
**Distribution:** temperate, subtropical, tropical.
**Economic uses:** or namentals (box, *Sarcococca*) and high quality timber.

herbs) whose best-known member is the box tree.

**Distribution.** The family is scattered throughout temperate, subtropical and tropical regions.

**Diagnostic features.** The leaves are alternate or opposite, simple and without stipules, and have a rather leathery texture. The flowers are regular, unisexual, and borne on the same plant or separate plants in spikes, racemes or fascicles. They are bracteate and have no petals. There are usually four imbricate sepals fused together at the base, sometimes six or more. The male flowers usually possess four or six stamens (rarely seven to ten) but in two genera (*Simmondsia* and *Styloceras*) they are numerous. When there are four stamens they are opposite the sepals, and when there are six, two of the pairs are opposite the inner two sepals. The anthers are often large and are either sessile or on long filaments. They have two locules, dehiscing by valves or longitudinal slits. Sometimes a rudimentary ovary is present. The female flowers are fewer (sometimes solitary) and possess a superior ovary of three fused carpels with three locules (four to six in *Styloceras*). The placentation is axile with normally one or two ovules in each locule. The ovary is surmounted by three styles which may be separate or fused at the base. The fruit is a loculicidal capsule or a drupe, containing black, shiny seeds, sometimes with a caruncle. They possess fleshy endosperm surrounding a straight embryo with flat cotyledons.

**Classification.** Two of the genera, *Simmondsia* (native to California) and *Stylo-ceras* (native to South America) are sufficiently distinct to be sometimes considered as separate families (Simmondsiaceae and Stylocerataceae). *Styloceras* (three species, South America) has alternate leaves, six to 30 stamens and male flowers lacking a perianth. *Simmondsia* (one species, California) has opposite leaves and numerous stamens. In both genera the rudimentary ovary is absent from the male flowers. All other genera have flowers with either four stamens, ie *Sarcococca* (16–20 species, China and Indomalaysia) which has alternate leaves and *Buxus* (70 species, temperate Northern Hemisphere, southern and tropical Africa, West Indies) which has opposite leaves; or six stamens, ie *Notobuxus* (seven species, southeast Africa) and *Pachysandra*. The four species of *Pachysandra* (eastern Asia and the United States of America) are distinguished by their herbaceous habit and their alternate, coarsely toothed leaves.

This family is thought to be related to the Euphorbiaceae and the Celastraceae but shows a number of clearly distinguishing features, eg absence of milky sap, apetalous flowers, the black, shiny, carunculate seeds and leaves always without stipules.

**Economic uses.** Economically the Buxaceae is important for its ornamentals. The best-known is *Buxus sempervirens* (common box), a slow-growing species with ovate, glossy, dark-green leaves, which is often used as hedges, border edgings etc. *Pachysandra procumbens* (eastern North America) and *P. terminalis* (Japan) are spreading subshrubs, grown as ground-cover plants. Species of *Sarcococca* have small but strongly fragrant flowers in winter. The wood of common box and of Cape box (*B. macowani*) is particularly valuable for carving, inlaying furniture and for making rulers and instruments and has been prized by wood engravers.

S.R.C.

# PANDACEAE

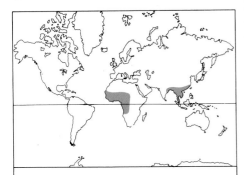

**Number of genera:** 3
**Number of species:** about 28
**Distribution:** W Africa, Asia, Indomalaysia.
**Economic uses:** seeds eaten locally.

The Pandaceae is a small family of tropical trees, comprising about 28 species in three genera.

**Distribution.** The genus *Panda* is restricted to tropical West Africa, *Microdesmis* to tropical West Africa and Asia and *Galearia* to Indomalaysia.

**Diagnostic features.** All members of the family have alternate, simple, often serrate leaves arranged in two rows on shoots which have buds in their axils. The leaves have no axillary buds, the terminal bud of the leafy shoot is reduced and non-functional, and the stipules which would normally subtend the shoot are reduced or absent. Thus the whole leafy shoot resembles a pinnate leaf. A similar arrangement is found in some members of the Euphorbiaceae.

The flowers are regular, unisexual with male and female on different trees in closely crowded inflorescences in the axils of shoots (*Microdesmis*) or directly on the stems of older wood (*Galearia* and *Panda*). They have five fused sepals and five petals. There are one, two or three series of five stamens and a rudimentary ovary in the male flowers, and a superior ovary of two to five locules each with one apical, pendulous ovule in the female flowers. The fruit is a drupe, the innermost layer of the ovary wall being stony. The seeds are flattened, and contain copious oily endosperm. The embryo has cordate cotyledons.

**Classification.** Although the Pandaceae are close relatives of the Euphorbiaceae the similarities are not sufficient to merit their inclusion in that already large family. The stony fruit of the Pandaceae is a character not found anywhere in the Euphorbiaceae.

**Economic uses.** *Panda oleosa* yields edible, oily seeds which are used locally for cooking, but is of no economic importance.

S.W.J.

# EUPHORBIACEAE
*The Spurge Family*

The Euphorbiaceae is a large family of flowering plants, including some 300 genera and over 5,000 species of dicotyledonous herbs, shrubs and trees. Some of the genera are very large, such as *Phyllanthus*, *Euphorbia*, *Croton* and *Acalypha*, but a great number are monotypic. Important products of the family include cassava, rubber (from *Hevea brasiliensis*) and tung oil.

**Distribution.** The family is predominantly tropical in its distribution, although there are strong local concentrations, particularly of the genus *Euphorbia*, in such extra-tropical regions as the southern United States of America, the Mediterranean basin, the Middle East and South Africa; the greatest number of the genera, however, are entirely tropical. In the tropics, the richest concentration of the family is perhaps in the Indomalaysian region, with the New World tropics coming a close second; the genus *Croton*, for example, is remarkably well developed in South America, with about 300 species in Brazil alone. Although it is a large family in Africa, it is not quite as rich and varied there as in the other two tropical realms.

**Diagnostic features.** The leaves are alternate or rarely opposite, and have stipules. They are usually simple, and when they are compound they are always palmate and never pinnate.

The flowers are regular, unisexual, and may occur either on the same plant (monoecious), as in *Euphorbia*, or on different plants (dioecious), as in *Mercurialis*. The flowers usually have five perianth segments, but in some genera (eg *Jatropha*, *Aleurites* and *Caperonia*) petals are also present and in others the perianth is lacking altogether. There are one to very numerous stamens, and anthers with two (sometimes three or four) locules usually opening lengthwise, rarely by pores. A pistillode (non-functional ovary) is often present in male flowers. The ovary is superior and usually consists of three fused carpels having three locules with one or two ovules on axile placentas in each locule; the styles are free or variously united.

The fruit is usually a schizocarp, sometimes a drupe; the type of schizocarp commonly found in the family is known as a regma, where the mericarps themselves

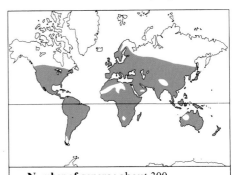

**Number of genera:** about 300
**Number of species:** over 5,000
**Distribution:** mostly tropical, but with some temperate species.
**Economic uses:** rubber, castor oil, cassava, tung oil, vegetable tallow, timber, purgatives, dyes, many ornamental species, including the poinsettia.

dehisce after having separated from each other. In a number of genera, the seeds are carunculate, eg *Euphorbia*, *Jatropha* and *Ricinus*; they usually have copious endosperm.

Latex is present in the tribe Euphorbieae, but is generally absent from the other tribes; it is usually poisonous, especially in *Hippomane mancinella*, the mancineel tree of the West Indies, and can cause temporary blindness.

Glands are a noteworthy feature of many Euphorbiaceae; they are generally associated with the flowers, where they either encircle or are situated among the stamens or encircle the base of the ovary, or else they are associated with the inflorescence, as in the tribe Euphorbieae, or with the vegetative organs, as in many other tribes; the leaf glands of *Macaranga aethiadenia* from Borneo are very large and shaped like volcanic cones.

Pseudanthial states are characteristic of the tribe Euphorbieae and the genus *Dalechampia*, where the close aggregation of the tiny simple flowers in the inflorescence, in conjunction with bracts, glands and appendages which may be petaloid in appearance, often produces an effect akin to that produced by a true flower. Thus *Dalechampia roezliana* has large bright pink bracts, *Euphorbia fulgens* has bright scarlet glands and *E. corollata* has white gland appendages, and all of these serve to heighten the likeness of the inflorescence to a flower, as do the florets in the capitulum of the Compositae.

The highly specialized inflorescence of *Euphorbia* is called a cyathium. This is usually a small cup-shaped structure consisting of a turbinate involucre which bears a number of glands (usually four or five) of varying shape around its rim. Within the involucre are numerous male flowers, each of very simple structure, surrounding a single central female flower. The inflorescences are commonly protogynous, and cross-pollination is effected chiefly by flies of the

1

2a

2b

3a

3b

4a

4b

4c

5a

5b

order Diptera, which are attracted to the cyathia by the flat gland-platforms which produce copious nectar. After fertilization, the female pedicel elongates, pushing the developing fruit up and out into the open above the gland-platform. It then commonly bends over at a point where there is a gap between the glands, the fruit then occupying a position beside or below the involucre. The fruit is almost invariably an explosively dehiscent three-lobed regma.

**Classification.** The family can be divided into the following major tribes:

PHYLLANTHEAE, eg *Phyllanthus, Breynia.*

BRIDELIEAE, eg *Bridelia.*

CROTONEAE, eg *Croton, Chrozophora, Caperonia, Mallotus, Macaranga, Mercurialis, Dalechampia.*

ACALYPHEAE, *Acalypha.*

RICINEAE, *Ricinus.*

JATROPHEAE, eg *Jatropha, Aleurites, Manihot, Codiaeum, Ricinodendron, Hevea.*

SUREGADEAE, *Suregada.*

EUPHORBIEAE, eg *Hippomane, Hura, Sapium, Sebastiania, Euphorbia.*

Some genera traditionally referred to the Euphorbiaceae, although rather anomalous with respect to the "main mass" of the family, have in recent years been segregated therefrom, usually into small families of their own specially erected to receive them. Notable in this respect are *Buxus* and its allies into the Buxaceae and *Panda, Microdesmis* and *Galearia* into the Pandaceae. Other genera often similarly treated include *Aextoxicon, Androstachys, Antidesma, Bischofia, Centroplacus, Daphniphyllum, Hymenocardia, Pera* and *Uapaca.*

The family has links with several other families, such as the Flacourtiaceae, Malvaceae and Urticaceae, but these are regarded by other authorities as being of secondary importance; here the family is placed in a separate order close to the Celastrales.

Although the family takes its name from the genus *Euphorbia,* the largest of the family, *Euphorbia* is not a typical member on account of the extreme simplicity of the flowers. Apart from some other members of the tribe Euphorbieae, the flowers in the family, although simple, are not as reduced as in the type-genus.

**Economic uses.** A number of genera include species of considerable economic importance. *Hevea,* for example, has *Hevea brasiliensis,* the para rubber tree which is the source of most of the world's natural rubber. *Manihot* also includes rubber-producing species such as the ceara rubber, *Manihot glaziovii,* but this genus is best known on account of the manioc, cassava or tapioca plant, *M. esculenta,* source of a staple foodstuff of poorer people in many tropical countries. Castor oil comes from *Ricinus communis. Aleurites moluccana* is the source of candlenut oil, used in the manufacture of soap, paints and varnishes; candles shaped from the paste of the kernels were formerly used for illumination, hence the common name. *Vernicia* is a genus related to *Aleurites*; it comprises three species, from each of which is obtained an oil of commercial value known as tung oil, used mainly in varnishes and paints. The large genus *Sapium* includes the Chinese tallow tree, *Sapium sebiferum,* which has a greasy tallow surrounding the seeds, used in making soap and candles; the leaves yield a black dye.

The genus *Jatropha* includes the physic nut, or purging nut, *Jatropha curcas,* from the seeds of which a powerful purgative is obtained. The most drastic of all purgatives, however, comes from the seeds of *Croton tiglium*; it is now generally considered unsafe to use, and has been dropped from the British and certain other pharmacopoeias. A red dye is obtained from the regmata of *Mallotus philippinensis,* and *Chrozophora tinctoria* yields purple and blue dyes. The timber of *Ricinodendron* species is known in the trade as African oak. The unripe fruits of the sandbox tree, *Hura crepitans,* were formerly used as containers of sand for blotting ink, or, filled with lead, as paperweights. The seeds of the jumping bean plant, *Sebastiania pringlei,* contain larvae of the moth *Carpocapsa saltitans,* and they show characteristic jumping movements when warmed up. Some species of the genus *Colliguaja* are also jumping bean plants.

A number of genera include species of horticultural merit, eg *Euphorbia* (including the poinsettia, *Euphorbia pulcherrima*), *Breynia, Jatropha, Codiaeum, Acalypha, Ricinus* and *Dalechampia.* A.R.-S.

# RHAMNALES

## RHAMNACEAE
*Buckthorn and Jujube*

The Rhamnaceae is a large family of temperate and tropical trees and shrubs with some climbers. The American genus *Ceanothus* provides the finest of the family's cultivated ornamental species, while its

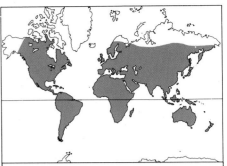

**Number of genera:** 58
**Number of species:** about 900
**Distribution:** cosmopolitan
**Economic uses:** medicinal (chiefly purgative), some fine ornamentals, dyes, charcoal and a soap-substitute.

better-known products include cascara and the jujube.

**Distribution.** The family is widely distributed throughout the world.

**Diagnostic features.** Most species are trees or shrubs, but several are adapted to climbing by the use of twining stems (*Berchemia*), tendrils (*Gouania*) or hooks (*Ventilago*). Some are armed with spines, eg *Rhamnus* and *Paliurus* (Christ's crown of thorns was reputedly made from *Paliurus spina-christi*) *Colletia* is odd, having not one but two buds in each leaf axil, the upper one developing into a thorn, the lower into a shoot. The leaves are alternate or opposite, simple and with stipules. Throughout the family the flowers are similar. They are small, inconspicuous, sometimes without petals, bisexual (rare unisexual), regular and usually borne in cymes. There are four or five valvate sepals and four or five small, incurved petals that are often closed over the four or five stamens. The ovary is superior and free or sometimes embedded in a prominent disk. There are three or two locules (or one by abortion), each with a single (rarely two) basal, anatropous ovule. The style is simple or divided.

The fruits vary in form according to their means of dispersal. Some are dry, dehiscent and wind-dispersed (eg *Paliurus*), but most are fleshy drupes or nuts, dispersed by the mammals and birds which eat them.

**Classification.** The chief genera include *Rhamnus, Hovenia, Zizyphus, Ceanothus, Ventilago, Phylica* and *Frangula.*

Various classification schemes have associated the Rhamnaceae with the families Vitaceae and Celastraceae. The Vitaceae are probably the closest relatives, differing only in their larger petals, in small features of the receptacle and fruit, and in their leaves which are never simple but lobed or compound. The Rhamnaceae shows some relationships with the Celastraceae, but differs in having stamens opposite the petals not opposite the sepals.

**Economic uses.** Some *Rhamnus* species are used in the manufacture of dyes. Sap-green is derived from the berries of *Rhamnus cathar-*

Euphorbiaceae. 1 *Euphorbia stapfii* a cactus-like species ($\times\frac{2}{3}$). 2 *Phyllanthus* sp (a) shoot with flat green phylloclades (modified stems) that bear flowers on their margins ($\times\frac{2}{3}$); (b) female flower with single perianth whorl and three-lobed stigma ($\times 12$). 3 *Acalypha* sp (a) leafy shoot and lateral inflorescence ($\times\frac{2}{3}$); (b) female flower with large, branched styles ($\times 6$). 4 *Euphorbia amygdaloides* (a) flowering shoot showing inflorescences (cyathia) condensed to resemble a single flower ($\times\frac{2}{3}$); (b) the cyathium consisting of an outer cup-shaped structure bearing horseshoe-shaped glands on the rim, within which is a ring of male flowers each consisting of a single stamen and in the center the female flower which consists of a stalked ovary and branched stigmas ($\times 6$); (c) three-lobed fruit ($\times 4$). 5 *Croton fothergillifolius* (a) flowering shoot ($\times\frac{2}{3}$); (b) fruit ($\times 4$).

Rhamnaceae. 1 *Zizyphus jujuba* (a) shoot bearing inconspicuous flowers and leaves with thorny stipules ( ×⅔) ; (b) fruits ( ×⅔) ; (c) cross section of fruit ( ×⅔). 2 *Gouania longipetala* (a) inflorescence, leaf and coiled tendril-like stipule ( ×⅔) ; (b) half flower ( ×6) ; (c) section of ovary ( ×12) ; (d) winged fruit ( ×4). 3 *Phylica nitida* (a) flowering shoot ( ×⅔) ; (b) flower ( ×6) ; (c) vertical section of gynoecium and receptacle ( ×12) ; (d) cross section of fruit ( ×3). 4 *Paliurus virgatus* winged fruits ( ×⅔). 5 *Ceanothus veitchianus* (a) flowering shoot ( ×⅔) ; (b) flower ( ×8). 6 *Colletia cruciata* (a) flowering shoot ( ×⅔) ; (b) flower opened out ( ×4) ; (c) vertical section of flower base ( ×4) ; (d) cross section of flower base ( ×6).

*tica*, yellow from the berries of *R. infectoria* and Chinese green indigo from the bark of *R. chlorophora*. Other species are used in medicine, notably for their purgative properties. The best-known of these is *R. purshiana*, a North American species whose bark yields cascara sagrada (sacred bark), a well-known purgative. In the West Indies the bark of *Gouania domingensis* is chewed as a stimulant. *Hovenia dulcis* (Japan, China) has pink fleshy flower stalks which are dried and used locally in medicine. Leaf and bark extracts of African *Gouania* species are often applied to sores and wounds. *Ventilago oblongifolia* is similarly used in Malaya as a poultice to cure cholera. Chemical analysis has shown that many members of the Rhamnaceae contain substances related to quinine. This may account for their wide use in medicine.

In the Philippines the root extract of *Gouania tiliifolia* is used as a substitute for soap. The roots contain saponin, which froths readily when mixed with water. *Zizyphus jujuba* is the jujube or Chinese date. *Zizyphus lotus* is believed to be the lotus fruit of antiquity. The timber of members of this family is not particularly strong or useful, and few species are of any importance.

Several genera of the Rhamnaceae are well-known ornamentals. The best-known is *Ceanothus* which contains many beautiful flowering shrubs. Other genera which are occasionally cultivated include *Pomaderris*, *Phylica*, *Noltea*, *Rhamnus* and *Colletia*.

S.W.J.

# VITACEAE
*Grapevine and Virginia Creeper*

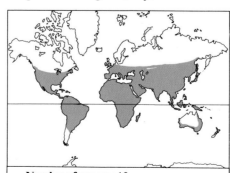

**Number of genera:** 12
**Number of species:** about 700
**Distribution:** tropics and subtropics.
**Economic uses:** wine, grapes, dried grapes (raisins, sultanas, currants), ornamentals.

The Vitaceae is a family of mainly climbers and some shrubs, celebrated on account of one species, *Vitis vinifera*, the grapevine, age-old provider of wine and fruit. Several genera are grown as ornamental creepers, eg *Cissus*, *Parthenocissus* and *Vitis* itself.
**Distribution.** The Vitaceae is mainly found in the tropics and subtropics. *Vitis vinifera* is widely cultivated in temperate climates and as far north as the lower Rhineland.
**Diagnostic features.** Most members of this family are climbing shrubs with tendrils, but there are also some small erect shrubs and trees, the nodes of which are often jointed or swollen. The tendrils are either modified shoots or inflorescences and may end in disk-like suckers. In the case of the grapevine (*Vitis*) the tendril is negatively phototropic and thus forces its way into crevices in the supporting structure. Here the tendrils expand into large balls of tissue, which become sticky and mucilaginous and help the tendril to adhere firmly to its support.

The leaves are alternate and simple or pinnately or palmately compound, often possessing stipules, and with pellucid dots on the blade. The flowers are very small, regular and bisexual or unisexual (usually

Vitaceae. 1 *Vitis thunbergii* (a) inflorescence ($\times\frac{2}{3}$); (b) flower bud ($\times 3$); (c) flower with petals removed showing cup-like calyx and five stamens ($\times 3$); (d) gynoecium ($\times 3$); (e) part of shoot with leaf and immature and mature fruits—berries ($\times\frac{2}{3}$). 2 *Tetrastigma obtectum* (a) leafy shoot with axillary inflorescences ($\times\frac{2}{3}$); (b) flower bud ($\times 4$); (c) flower viewed from above showing four petals and four stamens ($\times 3$); (d) stamens ($\times 6$). 3 *Cissus velutinus* (a) leafy shoot with axillary inflorescence and unbranched, coiled tendrils ($\times\frac{2}{3}$); (b) flower ($\times 4$); (c) vertical section of gynoecium showing erect ovules and short style with a discoid stigma ($\times 4$).

with male and female on the same plant). They are usually arranged in cymose or racemose inflorescences, arising opposite a leaf. There are four or five sepals, fused into a cup-like structure which is toothed or lobed, and four or five petals which may be free but are often united at the tips. In the latter condition they often fall off as a hood on the opening of the flower bud. The stamens are equal in number to the petals and opposite to them. They are inserted on a ring-like or lobed disk, the anthers being free or joined. The ovary is superior and consists usually of two fused carpels, with two to six locules each usually containing two erect ovules. The style is short, ending in a discoid or rarely four-lobed stigma (as in *Tetragyna*). The fruit is a berry; the seeds have a straight embryo surrounded by copious endosperm.

**Classification.** The largest genus is *Cissus* whose 350 species are almost entirely tropical in distribution. The flowers of nearly all the members of the genus are bisexual with four free petals and four stamens. Some of the species of *Cissus* are erect and others tendril-climbing as in *Vitis* (50 species in the subtropics and warm temperate regions), but the latter usually has flower-parts in fives and

united petals. Other important genera include *Ampelopsis* (about 25 species in Asia and North America) which also normally has a climbing habit but has flower-parts in fives and free petals, and *Parthenocissus* (about 15 species of tendril-bearing climbers in Asia and North America) whose flowers lack the ring-like or lobed disk found in most genera. The family also includes *Pterisanthes* (about 20 species in Burma and West Malaysia) and *Leea* (about 70 species throughout the tropics). The latter genus is sometimes separated as the family Leeaceae.

The Vitaceae is most closely related to the Rhamnaceae with which it shares the predominantly woody climbing habit, bisexual or unisexual flowers, a single whorl of stamens opposite the petals, and only one or two ovules per locule in the syncarpous ovary. Both families also contain many species whose flowers possess a ring-like or lobed disk.

**Economic uses.** Economically the Vitaceae is important because of the grapevine (*Vitis vinifera*) which originates from the Orient and northwest India. More than 25 million tonnes of wine are made annually from the fruit of this species, and viticulture is now a

scientific study. When dried, the fruits are termed raisins, or sultanas if the grape is of the seedless variety. Currants are the dried fruits of the Corinthian variety. Grapes of the Muscatel variety are used to make the wine of that name, as well as raisins. The fruits of some other species of *Vitis*, eg *V. aestivalis* and *V. labrusca*, are also used for wine-making. These are North American species which are resistant to the insect *Phylloxera*. On account of the devastation caused by this pest, most European vines are now grafted on to American root stocks. The stems of some species, eg *V. papillosa* (Java) and *V. sicyoides* (Mexico) are used locally as cordage.

Some other members of the family are prized as ornamentals, notably the 10 species of *Parthenocissus* (Virginia creeper). All are climbers, *Parthenocissus quinquefolia* or "true" Virginia creeper, possessing leaves with three or five coarsely serrated leaflets which turn crimson in autumn. This and other species of *Parthenocissus*, eg *P. inserta* and *P. himalayana*, are suitable for covering walls, fences and pergolas, as are some species of *Vitis* such as *V. amurensis* and *V. davidii*, and of *Cissus*.                    S.R.C.

Staphyleaceae. 1 *Tapiscia sinensis* (a) leaflet ($\times\frac{2}{3}$); (b) inflorescence ($\times\frac{2}{3}$); (c) flower showing fused sepals ($\times 4$); (d) vertical section of flower ($\times 6$); (e) indehiscent fruits ($\times\frac{2}{3}$). 2 *Staphylea holocarpa* (a) shoot showing axillary inflorescence and trifoliolate leaves with paired stipules ($\times\frac{2}{3}$); (b) flower with free sepals ($\times 2$); (c) cross section of ovary ($\times 7$); (d) dehiscing fruit ($\times\frac{2}{3}$). 3 *Turpinia insignis* (a) inflorescence and unifoliolate leaf with stalk having a pair of stipules at base and a pair of stipels part of the way up ($\times\frac{2}{3}$); (b) flower ($\times 2$); (c) flower with petals and sepals removed to show stamens with flattened filaments ($\times 3$); (d) gynoecium ($\times 3$); (e) cross section of ovary ($\times 3$).

# SAPINDALES

## STAPHYLEACEAE
*Bladder Nuts*

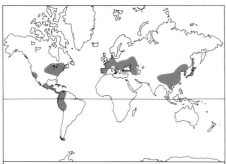

**Number of genera:** 5
**Number of species:** about 60
**Distribution:** N temperate, Cuba, Hispaniola, S America and Asia.
**Economic uses:** ornamental shrubs, local uses of timber and of fruits in medicine.

The Staphyleaceae is a family of five genera of temperate and tropical trees and shrubs. Some *Staphylea* and *Euscaphis* species are cultivated as ornamental garden plants.

**Distribution.** *Staphylea* is found throughout the north temperate region. *Tapiscia* is native to China, *Huertea* to Cuba, Hispaniola, Colombia and Peru, *Euscaphis* to east Asia and *Turpinia* to tropical and temperate Asia and America.

**Diagnostic features.** The leaves are opposite or alternate, trifoliolate or pinnate, with paired stipules. The flowers are regular, bisexual or sometimes unisexual with male and female on the same plant, rarely on separate plants, and are borne in paniculate clusters. There are five imbricate sepals, five imbricate petals and five stamens alternating with the petals, the filaments sometimes being flattened. The ovary is superior and consists of two to four fused carpels (or three to four free in *Euscaphis*), and each locule contains one or a few ovules on axile placentas. The two to four styles are variously free to completely fused together, and the fruits are either berry-like or inflated capsules with an open top. There are few seeds and these have straight embryos, flat cotyledons and a fleshy or horny endosperm. The seeds of *Euscaphis* are arillate and the fruits follicular.

The name *Staphylea* derives from the Greek word *staphyle*, a cluster, relating to the arrangement of the flowers, which may be pink or white in these shrubs.

**Classification.** The family has no clear and obvious affinities with other members of the Sapindales. Some authorities suggest a relationship with the Cunoniaceae or Celastraceae.

**Economic uses.** *Staphylea pinnata* is native to Europe and is known as the bladder nut on account of the inflated fruits. It has long been used as a garden ornamental. The Caucasian *S. colchica* has better flowers from a gardener's viewpoint, but poorer fruits than *S. pinnata*, and attempts have been made to combine the better attributes of each plant in hybrids, *S. × coulombieri* and *S. × elegans* being reputed to be such hybrids. One of the best bladder nuts is the Chinese *S. holocarpa* var *rosea*, a pink-flowered variety of the species introduced from central China.

*Euscaphis* comprises four species in Japan, China and Vietnam. Known as hung-liang, *Euscaphis japonica* is a common tree or shrub in Japan and central China, where its fruits are used as a drug: in Japan it is called gonzui zoku.

Melianthaceae. 1 *Melianthus pectinatus* (a) shoot with pinnate leaves, small stipules, and inflorescence with flowers and immature fruits ($\times\frac{2}{3}$); (b) half flower with irregular sepals and petals and swollen nectar-secreting disk ($\times 1$); (c) capsule ($\times\frac{2}{3}$). 2 *Bersama tysoniana* (a) leafy shoot with stipules in the axils ($\times\frac{2}{3}$); (b) inflorescence ($\times\frac{2}{3}$): (c) mature flowers with long stamens ($\times 3$); (d) young androecium with four short stamens fused at the base and ovary crowned by simple style and lobed stigma ($\times 4\frac{1}{2}$); (e) cross section of ovary with four locules and ovules on axile placentas ($\times 3$); (f) fruits ($\times\frac{2}{3}$); (g) seed with aril ($\times 1$).

*Turpinia* species number between 30 and 40, of which *Turpinia nepalensis* is a common tree in western China with a useful tough wood. *T. occidentalis* is the cassada wood of the West Indies. B.M.

# MELIANTHACEAE

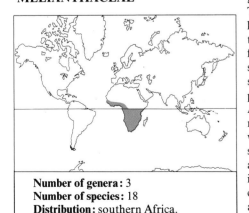

**Number of genera:** 3
**Number of species:** 18
**Distribution:** southern Africa.
**Economic uses:** limited use of timber and as ornamentals and in medicine.

The Melianthaceae is a small family consisting of three genera and about 18 species of shrubs and small trees of little economic use.

**Distribution.** The family is native to southern Africa.

**Diagnostic features.** The leaves are alternate, usually pinnately compound but occasionally simple. The stipules are within the petiole (intrapetiolar). The flowers are irregular, borne in terminal or axillary racemes. They are usually bisexual but occasionally have no stamens or no carpels. The stalks of the flowers twist through 180° at the time of flowering. There are four or five unequal sepals, fused at their bases and frequently swollen at one side. There are four or five petals, unequal in size and distinctly clawed. An annular or crescent-shaped disk, bearing nectar glands, lines the inside of the calyx, within which are inserted four, five or 10 stamens. They alternate with the petals and are often bent forward. The ovary is superior, of four or five fused carpels, and comprises either four or five locules with axile placentation, or one locule with five parietal placentas. There are one or several ovules on each placenta, either erect or pendulous. The ovary is surmounted by a single style which is divided at the tip into four or five stigmatic lobes (sometimes flattened or toothed). The fruit is a papery or woody capsule which dehisces at the apex. The seed contains a straight embryo surrounded by endosperm and in some species it has an external outgrowth (aril).

**Classification.** Two of the genera (*Melianthus* and *Bersama*) display axile placentation in the ovary while the third, *Greyia*, has parietal placentation. This, and a number of other floral and vegetative features, are considered by some botanists sufficient to place *Greyia* in a separate family (Greyiaceae).

*Melianthus* (six species) is a genus with pinnate leaves, four stamens and two to four ovules in each locule of the ovary, while *Greyia* (three species) has simple leaves and 10 stamens in the flower. *Bersama* (two polymorphic species) has pinnate leaves, four or five stamens and only one ovule per locule.

The family is closely related to the Sapindaceae, sharing with it such features as leaf form, insertion of stamens within a disk and a superior ovary. However, the Melianthaceae differs in the twisting of the flowers on their stalks before maturation, and in having seeds with copious endosperm.

**Economic uses.** This family is not economi-

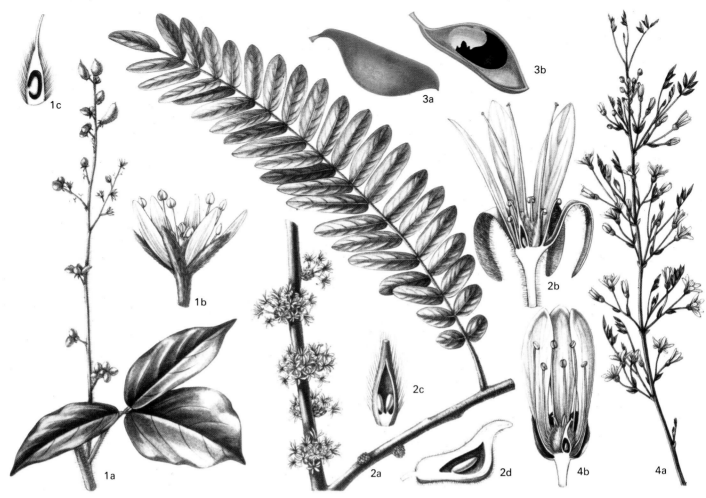

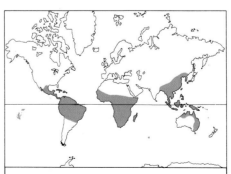

Connaraceae. 1 *Agelaea hirsuta* (a) flowering shoot showing trifoliolate leaf and flowers in terminal panicles ($\times\frac{2}{3}$); (b) flower showing five hairy sepals fused only at the base, five free petals, eight stamens and free styles and stigmas ($\times 6$); (c) vertical section of a single carpel ($\times 12$). 2 *Cnestis laurentii* (a) flowering shoot ($\times\frac{2}{3}$); (b) half flower ($\times 6$); (c) vertical section of single carpel showing two erect ovules ($\times 12$); (d) vertical section of fruit—a follicle ($\times\frac{2}{3}$). 3 *Connarus monocarpus* fruit (a) entire ($\times\frac{2}{3}$) and (b) in vertical section showing large seed with an aril ($\times\frac{2}{3}$). 4 *Rourea foenum-graecum* (a) flowering shoot ($\times\frac{2}{3}$); (b) half flower ($\times 6$).

cally important, but species of all three genera are grown as ornamentals in warm regions. Species of *Melianthus* emit a strong scent, for example *Melianthus major*, a shrub with large, reddish-brown flowers, and the smaller *M. comosus*, with long racemes of orange, red-spotted flowers. The cultivated species of *Greyia* is *Greyia sutherlandii*, a small tree with light-colored bark and conspicuous scarlet flowers in large racemes. The root, bark and leaves of *M. comosus* are used in South Africa for treating snake bites, while a decoction of the leaves of *M. major* is used for healing wounds. *Bersama abyssinica*, a medium-sized tree, produces a hard, heavy wood used for house construction in West Africa.　　　S.R.C.

# CONNARACEAE
## Zebra Wood

The Connaraceae is a dicotyledonous family of tropical trees or twining shrubs.

**Distribution.** The family is pantropical. The most important genera are *Byrsocarpus* (about 20 species in Africa and Madagascar), *Connarus* (about 100 species distributed in Africa, Asia, the Pacific, Australasia, and tropical America), *Rourea* (about 90 species

widely distributed in Australia and the Pacific), *Cnestis* (about 40 species in Africa, Madagascar and Malaysia), and *Agelaea* (50 species in tropical Africa, Madagascar, Southeast Asia and Malaysia).

**Diagnostic features.** The leaves are alternate and without stipules, and are pinnate or trifoliolate, a few species being unifoliolate.

**Number of genera:** 16
**Number of species:** about 350
**Distribution:** pantropical.
**Economic uses:** timber (zebra wood), some species yield medicines, tannins and fibers are used locally.

The flowers, which are produced in panicles, are generally bisexual and either regular or slightly irregular. The calyx has five sepals, which are either free or fused. The five petals are free or slightly fused. There are generally five or 10 stamens, rarely four or eight, often bent downwards with their filaments joined below. Sometimes a thin, nectar-secreting disk is present. The ovary is superior, with one to five free carpels containing two erect ovules in each locule. The fruit is generally a follicle containing a single seed which may or may not contain endosperm. The seeds often have an outer appendage (aril).

**Classification.** This family is sometimes regarded as being related to the Dilleniaceae, Crossosomataceae, and Brunelliaceae. It shares with these families its tree or shrublike habit, pinnate leaves, bisexual flowers, superior, free carpels and arillate seeds. In evolutionary terms it is considered to be rather more advanced than the Leguminosae, possibly on a line leading to the Oxalidaceae. In this work its traditional place among the Sapindales is retained.

**Economic uses.** The family is economically important for zebra wood, which is obtained

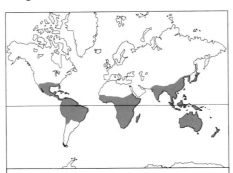

Sapindaceae. 1 *Dodonaea bursarifolia* (a) flowering shoot ( ×⅔) ; (b) male flower ( ×4) ; (c) female flower ( ×4) ; (d) fruit ( ×1½) ; (e) cross section of fruit ( ×1½) ; (f) vertical section of part of fruit ( ×4). 2 *Litchi chinensis* fruits entire and in section ( ×⅔). 3 *Serjania exarata* shoot tip with coiled tendrils, winged fruits and inflorescences ( ×⅔). 4 *Paullinia thalictrifolia* (a) flowering shoot ( ×⅔) ; (b) flower ( ×4) ; (c) flower with two petals removed ( ×4) ; (d) gynoecium ( ×6). 5 *Cupaniopsis anacardioides* (a) pinnate leaf ( ×⅔) ; (b) inflorescence ( ×⅔) ; (c) flower viewed from above ( ×2) ; (d) gynoecium surrounded by disk and calyx ( ×2) ; (e) cross section of ovary ( ×4).

from *Connarus guianensis*, a native of Guyana. The seeds of the African *C. africanus* are made into a flour which is effective as an anthelmintic. The leaves of the West African tree *Agelaea villosa* are used to treat dysentery, while those of *A. emetica* (native to Madagascar) yield an essential oil which promotes vomiting. The West African species, *Cnestis corniculata* and *C. ferruginea*, have leaves which are the source of an astringent and a laxative respectively. The bark of *Rourea glabra* (native to Central America) is used for tanning animal skins a dark blue or purple color. The roots of this plant yield a strong fiber used for rope-making. Its seeds and the fruits of the Pacific *R. volubilis* are used for poisoning dogs.                                    S.R.C.

## SAPINDACEAE

*Akee, Litchi and Rambutan*

The Sapindaceae contains about 150 tropical and subtropical genera and 2,000 species. About 300 species are lianas, and the rest are trees and shrubs including the economically important food plants in such genera as *Litchi* and *Blighia* and the ornamentals in *Koelreuteria*, *Xanthoceras* and *Dodonaea*.

**Distribution.** The family is found throughout the tropics and subtropics.

**Diagnostic features.** The leaves are normally alternate, simple or compound, and without stipules. The flowers may be regular or irregular and are often unisexual, or func-

**Number of genera:** about 150
**Number of species:** about 2,000
**Distribution:** tropical and subtropical.
**Economic uses:** edible fruits (akee, litchi and rambutan), stimulating beverages from *Paullinia* species (guarana) and some cultivated ornamentals.

tionally so, borne in cymose inflorescences. There are usually five free or fused sepals, and five free petals, which may be absent, and a well-marked disk between the petals and stamens. The stamens are in two whorls of five, often with two stamens missing and so appearing as eight. The filaments are free and often hairy. The ovary is superior, of three fused carpels, may be lobed, divided or simple and has one to four locules, each with one or two (rarely many) ovules on a central axis or rarely on parietal placentas. The style is terminal, simple or divided; rarely there are two to four styles. The fruits are various: capsules, nuts, berries, drupes, samaras or schizocarps, often red, containing seeds which are often arillate. They lack endosperm, and the embryos are folded or curved.

The woody lianas in genera such as *Serjania* or *Paullinia* make up an important ecological component of tropical woodland, climbing by tendrils which are modified inflorescence axes, and having anomalous stem anatomy in comparison with other woody plants, brought about by the climbing habit and the consequent internal stem stresses experienced. The tissues of many sapindaceous plants contain resinous or laticiferous compounds secreted by specialized cells.

**Classification.** The family is related to the Aceraceae, Hippocastanaceae and Melianthaceae. The monotypic Australian genus *Akania* is sometimes placed in the Sapindaceae, although other authorities place it in its own family, the Akaniaceae.

**Economic uses.** *Blighia sapida*, akee in the West Indies, or akye in Africa, is native to the forests of West Africa, where the aril is eaten, tasting rather like scrambled egg when cooked; it is poisonous if not eaten at the correct stage of ripeness. The plant was introduced to the West Indies and in particular has become naturalized in Jamaica where it is the national fruit. Litchi or lychee, *Litchi chinensis* (*Nephelium litchi*), is a native of southern China, but is widely grown in all tropical regions for its sweet acid aril. Its relative the mamoncillo, *Melicocca bijuga*, is grown in America. *Nephelium lappaceum* is the rambutan, a much-prized fruit eaten in the Old World tropics.

*Paullinia cupana*, one of 180 American species of liana, is the source of guarana, much drunk in Brazil. *Paullinia* yields yoco, a drink similarly rich in caffeine. *Sapindus saponaria*, from Florida, the West Indies and South America, has berries rich in saponins which form a lather with water and are used as a soap substitute. *Schleichera trijuga* is the source of macassar oil, used in ointments and for illumination.

Of sapindaceous ornamentals perhaps *Koelreuteria* is the most important, making good street trees. Also with bladder fruits is *Cardiospermum halicacabum*, the balloon vine of the tropics and subtropics. The tree *Xanthoceras sorbifolia* is cultivated for its attractive flowers.                    B.M.

## SABIACEAE

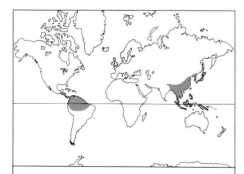

**Number of genera:** 4
**Number of species:** about 80
**Distribution:** tropical and subtropical Asia and America.
**Economic uses:** a few species are cultivated as ornamentals.

The Sabiaceae is a small tropical and subtropical family of trees or shrubs and a few climbers of limited use as ornamentals.
**Distribution.** The four genera of the family are *Sabia* (20–30 species from Southeast Asia and Malaysia), *Meliosma* (50–60 species from tropical and subtropical Asia and America), *Phoxanthus* (one species from the Amazon basin) and *Ophiocaryon* (one species from Guyana).

**Diagnostic features.** The leaves are alternate and are odd-pinnate or simple, without stipules. The flowers are regular, bisexual or unisexual with both sexes on the same plant, and are borne in terminal or axillary cymes or panicles. They have five (rarely three or four) imbricate, unequal sepals which are opposite five (rarely four) imbricate petals, the inner two of which are often smaller. There are five stamens (rarely four or six) which are opposite the petals, but the outer three are often reduced to staminodes. The ovary is superior, of two fused carpels, and usually has a disk at the base. It has two locules (rarely three) and each locule has one ovule (rarely two). The fruit is indehiscent and drupaceous, and the seeds have little or no endosperm.

**Classification.** *Sabia* has five fertile stamens. The other three genera, which have only two fertile stamens, are sometimes regarded as constituting a separate family, the Meliosmaceae. The position of the sepals, petals and stamens opposite to each other is most unusual and some authorities regard the family as closely related to the Menispermaceae (Ranunculales), although others place it in the Sapindales.

**Economic uses.** *Sabia latifolia* and *S. schumanniana* are cultivated for their attractive blue fruits. Several species of *Meliosma* are cultivated for ornament, eg *Meliosma beaniana*, western China, which is very free-flowering with drooping panicles of flowers 20cm (8in) long.                    M.C.D.

## JULIANIACEAE

The Julianiaceae is a small family comprising two genera and five species of resinous trees and shrubs not unlike the genus *Rhus* in habit.

**Distribution.** The family is found in Central America, from Mexico to Peru.

**Diagnostic features.** The leaves are alternate, pinnate (rarely simple), covered in fine hairs, and are without stipules. The leaflets have serrated edges. The flowers are green, small and unisexual, the separate sexes occurring on different plants; male flowers are numerous in pendulous or erect panicles and consist of a three- to nine-lobed calyx, no petals, and as many stamens as sepals in alternate arrangement; female flowers are in stalked clusters of three or four subtended by a collar of bracts, and lack sepals and petals, having only a superior ovary, with one locule, and a single three-lobed style. A single ovule is inserted at the base of the locule. The fruits are dry and club-shaped, do not open on the tree and are enclosed by the enlarged collar of bracts. The fruit stalk may become broad and flattened in *Amphi-*

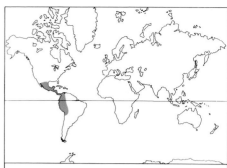

**Number of genera:** 2
**Number of species:** 5
**Distribution:** Central America.
**Economic uses:** some medicinal uses and produce tannins and a red dye.

*pterygium* (*Juliania*), or not so in the monotypic *Orthopterygium*.

**Classification.** On the basis of anatomy, pollen and habit this family is closely related to the Anacardiaceae.

**Economic uses.** The Mexican *Amphipterygium adstringens*, known variously as "quetchalalatl" or "cuachalala," has a bark used medicinally as an astringent, for malaria and hardening the gums. The bark also contains tannins and a red dye.

B.M.

## HIPPOCASTANACEAE
*Horse Chestnuts and Buckeyes*

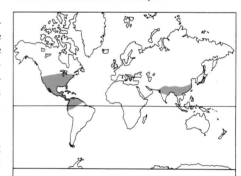

**Number of genera:** 2
**Number of species:** 15
**Distribution:** north temperate (*Aesculus*) and S Mexico and tropical S America (*Billia*).
**Economic uses:** ornamentals, timber and local medicinal uses.

The Hippocastanaceae is a small family of trees comprising two genera, *Aesculus* (horse chestnuts and buckeyes) with 13 deciduous species and *Billia* with two evergreen species.

**Distribution.** *Aesculus* is widespread in north temperate regions, while *Billia* is restricted to southern Mexico and tropical South America.

**Diagnostic features.** Characteristic of the family are the large winter buds covered with resinous scale leaves. The leaves are opposite, palmate and without stipules. The inflorescence is usually a raceme with lateral cymes, but in *Billia* it is paniculate. The

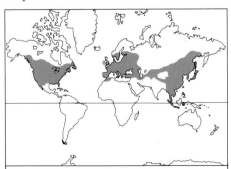

Hippocastanaceae. 1 *Billia hippocastanum* (a) leaf (×⅔); (b) sepal (×2); (c) flower with sepals removed to show five slightly unequal petals and six stamens (×1½); (d) gynoecium with curved style and simple stigma (×4½); (e) cross section of trilocular ovary with ovules on axile placentas (×10). 2 *Aesculus hippocastanum* (a) leafless mature tree; (b) digitate leaf and inflorescence (×⅔); (c) half flower with fused sepals and unequal petals (×1); (d) dehiscing fruit (a capsule) exposing seeds (×½); (e) seed with large hilum (×½); (f) cross section of ovary (×4); (g) vertical section of ovary (×4).

upper flowers, which are functionally male, open first, to be followed by the protogynous bisexual flowers below. The flowers are irregular, comprising five sepals, united at the base (free in *Billia*), and a large whitish, yellowish or red corolla of four or five free petals. There is an irregular disk between the petals and the five to eight stamens. The ovary is superior, consisting of three fused carpels and three locules (rarely two or one by abortion), each locule containing two ovules on axile placentas. The style is elongated, with a simple stigma. The fruit is a leathery capsule, usually three-valved and single-seeded. The seed is large and without endosperm.

**Classification.** The family is probably related to the Sapindaceae, but is readily distinguished by the palmate leaves, leathery capsule and the relatively large, usually solitary seed.

**Economic uses.** Species of *Aesculus* have various medicinal uses, and extracts from some have been used by North American Indians to stupefy fish. The wood is light but not very durable, and is used for making boxes and charcoal. The genus is best known for its ornamental trees, notably the horse

chestnut (*Aesculus hippocastanum*), which is grown for its winter buds, large leaves and striking inflorescences; its shiny brown seeds are the "conkers" much prized by children.

I.B.K.R.

# ACERACEAE
## Maples

**Number of genera:** 2
**Number of species:** 102–152
**Distribution:** cosmopolitan, centered in China.
**Economic uses:** ornamentals (maples), timber and maple sugar.

The Aceraceae is a family of predominantly medium to small, deciduous trees of north-

ern temperate regions. It includes some large trees and evergreens, and a few subtropical species. There are only two genera, *Acer*, the maples, and *Dipteronia*.

**Distribution.** The two *Dipteronia* species are found in central and southern China. Maples have an enormous geographical range centered in China, where 100–150 species are found. A few species occur through southern China to Malaysia and Java, and these are evergreen, as are some of the many ranging to and along the Himalayas. There are 19 maples in Japan and two on Formosa. Eight species are native to Asia Minor and the Caucasus Mountains, and there are 10 in southern Europe, one of which, *Acer campestre*, is the only species to occur in Africa and also the only native species in the British Isles. In North America there are four maples native to the western states only, eight native to the eastern and southern states while one, *A. negundo*, is found from Montreal to southern California.

**Diagnostic features.** The branches grow in opposite pairs. The leaves are opposite and highly variable in shape: simple or compound, entire or deeply toothed, unlobed or deeply lobed. Stipules are absent. The

Aceraceae. 1 *Acer platanoides* (a) shoot with opposite, palmately-lobed leaves and fruits comprising pairs of winged samaras ($\times\frac{2}{3}$); (b) shoot with terminal inflorescence, young leaves and bud scales ($\times\frac{2}{3}$); (c) male flower with four sepals, four petals, eight stamens and central vestige of the ovary ($\times 3$); (d) half bisexual flower showing winged ovary with forked style and short stamens on a lobed disk ($\times 3$); (e) silhouette of a leafless tree showing the much branched habit.

flowers are regular, with five free sepals and five free petals, the latter often absent. Species are andromonoecious (male and bisexual flowers on same plant), androdioecious (male and bisexual flowers on separate plants) or dioecious (male and female flowers on separate plants). The male and bisexual flowers have 4–10 stamens, normally eight, and in male flowers a vestigial ovary is often present. The ovary is superior, of two fused carpels and two locules each containing two ovules on axile placentas. The paired fruits are samaras, each with a membranous wing. The seeds are solitary and without endosperm.

**Classification.** *Acer* is distinguished from *Dipteronia* by having obovate wings attached to the seeds; *Dipteronia* seeds are surrounded by a circular membrane. The family is related to the Sapindaceae and Hippocastanaceae.

**Economic uses.** Many maples are grown for ornamental purposes, often in towns, and are prized for their beautiful foliage and spectacular autumn colors. Many species produce good timber, particularly the sycamore (*Acer pseudoplatanus*), while the sugar maple (*Acer saccharum*) and some other species yield maple sugar.     A.F.M.

# BURSERACEAE
*Frankincense and Myrrh*

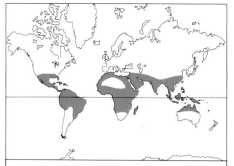

**Number of genera:** about 17
**Number of species:** about 500
**Distribution:** tropical, especially Malaysia, Africa and America.
**Economic uses:** aromatic resins (including frankincense and myrrh), used for making incense, perfumes, soap and paints.

The Burseraceae is a family of tropical trees and shrubs; important products include frankincense and myrrh.

**Distribution.** The family is found throughout the tropics, but especially in Malaysia, South

America and Africa. Trees of the Burseraceae are a common constituent of the main story of lowland dipterocarp forests of central and southern Malaya. *Canarium*, *Dacryodes* and *Santiria* are the three main genera, growing particularly in the lowlands. One of the principal members of this family in Africa and Madagascar is *Commiphora*. Various different species abound in hot, dry bushveld, desert, thorn scrub and throughout vast areas of savannah. Other species grow on alluvial slopes and in coastal belts along tidal mangrove swamps and in tropical rain forests. Large trees such as *Dacryodes* and *Canarium* and smaller trees such as *Santiria* are quite common in Africa, but are different from the species found in Malaya. *Boswellia serrata* is a component of deciduous forests and often gregarious on dry hillsides in India. Trees and shrubs of the genera *Bursera* and *Protium* are particularly well represented in South America. *Tetragastris* and *Dacryodes* also grow in various parts of South America and the West Indies.

**Diagnostic features.** All parts of the plants, especially the bark, contain resin. The leaves are spirally arranged, usually crowded at twig-tips, imparipinnate (pinnate with terminal leaflets) and may or may not have

Burseraceae. 1 *Boswellia popoviana* (a) twig with leaves crowded at tip ( ×⅔) ; (b) flower ( ×4) ; (c) cross section of ovary with five locules ( ×12). 2 *Commiphora marlothii* (a) terminal cluster of fruits ( ×⅔) ; (b) fruit ( ×1). 3 *Canarium hirtellum* (a) inflorescence ( ×⅔) ; (b) flower ( ×2) ; (c) flower with perianth removed to show stamens and globose stigma ( ×3) ; (d) section of trilocular ovary ( ×3). 4 *Boswellia papyrifera* habit. 5 *Protium guianense* (a) inflorescence ( ×⅔) ; (b) section of ovary with five locules ( ×6) ; (c) shoot with imparipinnate leaf and fruits ( ×⅔).

stipules. The flowers are grouped in panicles, also usually crowded at twig-ends. They are small, often unisexual (with each sex on separate plants), and usually greenish or cream, with parts in threes to fives. The sepals are fused, and either imbricate or valvate; the petals are usually free, also either imbricate or valvate. The stamens are equal to or double the number of petals. The ovary is superior and has three to five carpels and two to five locules, with two ovules (rarely one) on axile placentas in each locule. The style is simple. The fruit is usually a drupe, sometimes a capsule. The seeds lack endosperm.

**Classification.** The family can be divided into three tribes:

PROTIEAE. Drupe with two to five, free or adhering parts but not fused in the stony endocarp; exocarp occasionally splitting by valves. Six genera, including *Protium* and *Tetragastris*.

BURSEREAE. Drupe with an endocarp of fused parts; exocarp always dehiscing by valves. Five genera, including *Boswellia*, *Bursera* and *Commiphora*.

CANARIEAE. Drupe with an endocarp of fused parts. Six genera, including *Canarium* and *Santiria*.

This family is distinguished from the Rutaceae and the Simaroubaceae by the presence of resin ducts in the bark, by the distinct stamens and short single style, and by the usual absence of transparent glandular dots on the leaves.

**Economic uses.** The wood of *Canarium littorale*, *Dacryodes costata*, *Santiria laevigata* and *S. tomentosa* in Malaya and *Aucoumea* and *Canarium schweinfurthii* in Africa is used for general building construction and carpentry. Frankincense comes from *Boswellia carteri* (Somaliland) and some other species. Myrrh, used in incense and perfumes, is obtained from *Commiphora abyssinica*, *C. molmol* and some other species which are cultivated in Arabia and Ethiopia. Varnish is obtained from several species of *Bursera* in Mexico and probably elsewhere.                    H.P.W.

# ANACARDIACEAE
*Cashew, Mango, Sumacs and Poison Ivy*

The Anacardiaceae contains about 600 species of trees, shrubs and lianas including some popular ornamental trees as well as species producing commercially valuable fruit and nuts, such as the cashew, pistachio, Dhobis, hog plum, Jamaican plum, mango.

**Distribution.** The family is mainly tropical and subtropical, and is equally well represented in South America, Africa and Malaysia. A few genera are native to temperate North America and Eurasia.

**Diagnostic features.** Most members of the family have resinous tissues, although the leaves are not gland-dotted. Sometimes the resinous exudate is poisonous, causing severe irritation of the skin, as in poison ivy (*Rhus radicans*). The irritant substances may be distributed throughout the plant body or concentrated in particular organs, eg in the fruit wall of the cashew, *Anacardium occidentale*. The leaves are alternate (rarely opposite) and usually pinnately compound, although simple leaves occur, for example in *Cotinus*, *Anacardium* and *Mangifera*. Stipules are absent. The flowers are regular and bisexual (or sometimes unisexual), typically with five fused sepals, five free petals and five to 10 or more stamens. Between the stamens and ovary is a fleshy disk of tissue, the torus. The ovary is usually superior and comprises one to five carpels, usually united, very rarely free, each containing a single pendulous ovule. There are one to three styles, often widely separated. The fruit is usually a drupe; the solitary seed may have very thin

Anacardiaceae. 1 *Pistacia lentiscus* (a) shoot with imparipinnate leaves and male inflorescences ($\times\frac{2}{3}$); (b) male flower with short, lobed calyx and stamens with very short filaments ($\times 10$); (c) female flower with three spreading stigmas ($\times 14$); (d) vertical section of fruit ($\times 4$). 2 *Anacardium occidentale* (a) shoot with inflorescence and fruits, the latter with swollen, a pear-shaped stalk and receptacle with the kidney-shaped fruit below ($\times\frac{2}{3}$); (b) simple leaves ($\times\frac{2}{3}$); (c) male flower with a single stamen protruding ($\times 3$); (d) bisexual flower with petals removed showing all stamens except one to have short filaments ($\times 4$); (e) vertical section of fruit ($\times 1$). 3 *Rhus trichocarpa* habit.

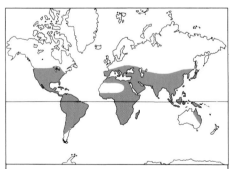

**Number of genera:** about 77
**Number of species:** about 600
**Distribution:** mainly tropical and subtropical with a few temperate representatives.
**Economic uses:** sources of tannins (*Cotinus, Pistacia, Schinopsis* and *Rhus*), fruits (eg cashew, pistachio, mango), and grown as ornamentals (eg sumac, smoke tree).

endosperm, or none, and fleshy cotyledons.
**Classification.** The family, which comprises about 77 genera, has been classified into four tribes largely on the basis of differences in the number and degree of fusion of the carpels.

ANACARDIEAE. Flowers with five free carpels, and compound leaves; or flowers with a solitary carpel and simple leaves. Old World tropics (eight genera, including *Mangifera*) and South America (*Anacardium*).

SPONDIEAE. Flowers with four or five united carpels each containing one ovule which develops into a seed. Pantropical (21 genera, including *Spondias*).

SEMECARPEAE. Flowers with an inferior ovary of three united carpels. Only one ovule develops. Old World tropics (six genera, including *Semecarpus*).

RHOIDEAE. Flowers with a superior ovary of three united carpels. Only one ovule develops. Pantropical, temperate Eurasia, South Africa and North America (42 genera, including *Cotinus, Pistacia, Rhus, Schinopsis* and *Schinus*).

The Anacardiaceae is closely related to several other families in the order Sapindales, notably the Sapindaceae, Aceraceae, Hippocastanaceae, Burseraceae and Juliianaceae. The latter two share with the Anacardiaceae the possession of specialized resin ducts.

**Economic uses.** Several of the more important uses of the Anacardiaceae are related to their resinous properties. Species of *Cotinus, Pistacia, Schinopsis* and *Rhus* are major sources of tannins for the leather industry. The resin of *Rhus verniciflua*, native to China, is the basis of lacquer. Mastic and pistachio turpentine are produced from species of *Pistacia*.

The family yields some important fruits and nuts, eg cashew nuts and cashew apples (*Anacardium occidentale*), pistachio nuts (*Pistacia vera*), Dhobi's nut (*Semecarpus anacardium*), the mango (*Mangifera indica*) and the Otaheite apple, hog plum and Jamaica plum (fruits of *Spondias* species). Some species produce useful timber: examples are *Schinopsis quebracho-colorado* (quebracho) and the cashew tree. The Anacardiaceae also includes some commonly grown ornamental trees, such as the sumacs (*Rhus* species), the smoke or wig tree (*Cotinus coggygria*) and the pepper tree or American mastic tree (*Schinus molle*). A mildly alcoholic drink is made from the latter species in its native Peru and the ground seeds may be used as a condiment and/or an adulterant for pepper, and the tree also produces a gum resin of the mastic type. The smoke tree yields a yellow dye.          F.K.K.

Simaroubaceae. 1 *Quassia amara* (a) shoot with trifoliolate leaves and inflorescence ($\times \frac{2}{3}$); (b) flower with petals removed to show numerous stamens ($\times 1\frac{1}{2}$); (c) calyx and gynoecium with disk at base ($\times 1\frac{1}{2}$); (d) fruits ($\times \frac{2}{3}$); (e) cross section of fruit ($\times \frac{2}{3}$). 2 *Harrisonia abyssinica* (a) flowering shoot with pinnate leaf ($\times \frac{2}{3}$); (b) flower bud ($\times 3$); (c) flower ($\times 3$); (d) calyx and gynoecium crowned by five styles ($\times 3\frac{1}{3}$); (e) stamen with hairy scale at the base ($\times 6$); (f) fruiting shoot ($\times \frac{2}{3}$); (g) fruit ($\times 2$). 3 *Ailanthus excelsa* (a) part of pinnate leaf ($\times \frac{2}{3}$); (b) fruits—twisted samaras ($\times \frac{2}{3}$); (c) half section of fruit showing single seed ($\times \frac{2}{3}$).

# SIMAROUBACEAE

*Quassia and Tree of Heaven*

The Simaroubaceae is a medium-sized family of trees and shrubs which include the medicinal genus *Quassia* and the ornamental genus *Ailanthus*.

**Distribution.** The family is found throughout the tropics and subtropics.

**Diagnostic features.** The leaves are alternate, pinnate, rarely simple, and usually without stipules. The often numerous small flowers are regular, bisexual or unisexual, and are borne in cymose spikes or dense panicles. There are three to seven free or united sepals and petals, the petals rarely being absent altogether. A ring or cup-like disk occurs between the petals and the stamens which are free and as many as or double the number of petals. The ovary is superior, of two to five carpels which are fused or free below and united above by the style or stigma. There are two to five styles, and each of the one to five locules contains a single ovule (rarely two) inserted on an axile placenta. The fruit is a samara, schizocarp, or capsule, the seeds with or without endosperm, having thick cotyledons and a straight or curved embryo.

**Classification.** The family is closely related to

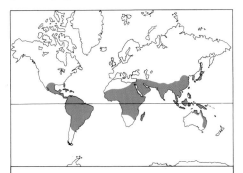

**Number of genera:** about 20
**Number of species:** about 120
**Distribution:** tropical and subtropical.
**Economic uses:** quassia wood and other medicinal species, ornamental trees, and timber.

the Rutaceae, but differs conspicuously in lacking pellucid glands on the leaves.

**Economic uses.** *Ailanthus* (10 species) is the most important decorative genus in the family, with *Ailanthus altissima*, the tree of heaven, being widely grown. The male trees smell offensively in flower, so that females are best grown in urban environments.

*Quassia amara*, one of 40 species from the tropics, has been widely cultivated both as an attractive shrub or small tree with red flowers, and for its bitter wood which has medicinal value in the form of the glucoside, quassiin. The properties of quassia wood are like those of the genus *Picrasma*, the Old World genus of six species, and as a consequence *Picrasma* has been used as a quassia substitute.

The genus *Picramnia*, comprising about 50 species in tropical America, and often encountered in montane woodland, also has medicinal value. *Picramnia antidesma*, from the West Indies and Central America, has bitter leaves and bark with a flavor some compare to liquorice. It is still employed in rural medicine, and was once exported, in bark form, to Europe as a treatment for erysipelas and venereal disease.

*Kirkia* comprises eight southern African species of tree, *Kirkia acuminata*, the white syringa, being the most attractive and commonest. It is a graceful, 18m (60ft) tall tree with corky bark, pinnate leaves to 45cm (18in) long, wood which is well colored and figured when worked into furniture, flooring or ornaments, and swollen roots which store

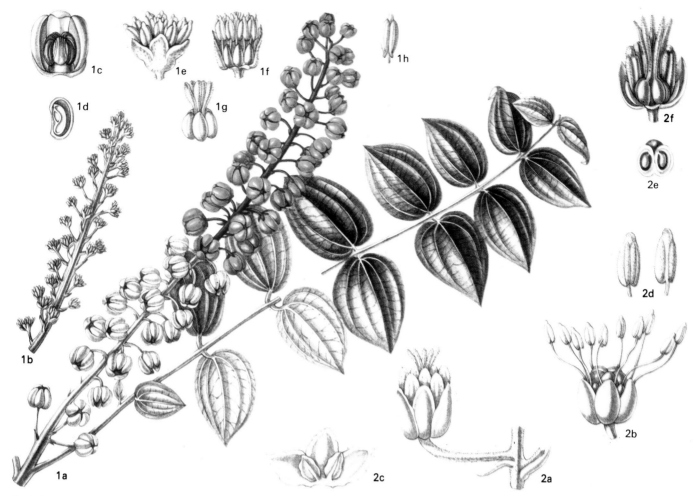

Coriariaceae. 1 *Coriaria terminalis* (a) leafy shoot with fruits (pseudodrupes) surrounded by fleshy petals ($\times\frac{2}{3}$); (b) inflorescence ($\times\frac{2}{3}$); (c) fruit with two petals removed ($\times 2$); (d) vertical section of achene ($\times 3$); (e) flower showing large anthers ($\times 2$); (f) flower with two sepals removed to show small petals ($\times 2$); (g) gynoecium with ovary surrounded by five petals ($\times 3$); (h) stamen ($\times 3$). 2 *C. ruscifolia* (a) young protogynous flower with stigmas fully emerged ($\times 6$); (b) fertilized flower with mature stamens ($\times 6$); (c) part of perianth showing three sepals and two small petals ($\times 6$); (d) anthers ($\times 12$); (e) vertical section of ovary ($\times 8$); (f) young flower with some of petals and sepals removed to show free carpels ($\times 6$).

liquid which is readily tapped in times of drought.

The "berries" of certain species of the genus *Brucea* are used in the treatment of dysentery.      B.M.

## CORIARIACEAE
### Coriarias

The Coriariaceae is a small family of warm temperate shrubs comprising a single genus, *Coriaria*.

**Distribution.** The family is found in warm temperate areas in central and western South America, the Mediterranean region, the Himalayas, and into eastern Asia, New Guinea and New Zealand. It is notably absent from Africa and Australia. *Coriaria* is a good example of a genus with a discontinuous geographical distribution pattern. The Northern Hemisphere coriarias belong to one section of the genus, the Southern to another.

**Diagnostic features.** The family are shrubs, sometimes spreading, with opposite or whorled angular branches which are often frond-like in appearance. The leaves are opposite, ovate to lanceolate, entire, with three or more veins arising from the base,

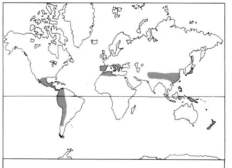

**Number of genera:** 1
**Number of species:** 8
**Distribution:** warm temperate zones.
**Economic uses:** *Coriaria myrtifolia* yields fly poison, tannin and ink, and some species are cultivated as ornamentals.

and without stipules. The flowers are regular small, green, bisexual or unisexual (with both sexes on the same plant), often borne on the previous year's wood, axillary (or terminal when on current growth) and either solitary or racemose. There are five persistent sepals, five shorter, keeled, fleshy petals, 10 stamens with large anthers, either

all free or five fused to the petal keels. The ovary is superior, comprising five to 10 free carpels and five to 10 locules, each with a single pendulous, anatropous ovule. The styles are elongated and conspicuous. The often purplish petals become succulent after fertilization, and being intruded between the carpels enclose them, forming a pseudodrupe before the compressed seeds are released. The seed has a thin endosperm when mature. The infructescences are often attractive, although several species are poisonous and all should be regarded as such in the context of their value as garden ornamentals.

**Classification.** The family is difficult to place satisfactorily in any taxonomic group. In the system adopted in this work it is placed in the Sapindales.

**Economic uses.** Redoul, *Coriaria myrtifolia*, is native to the Mediterranean region, and the fruits when crushed in water make a good fly poison. The leaves are rich in tannins and used for making ink and curing leather. It is a deciduous shrub up to 2m (6ft) high, with small greenish flowers in 3cm (1in) long racemes from the joints of the previous year's wood.

A popular ornamental species is the

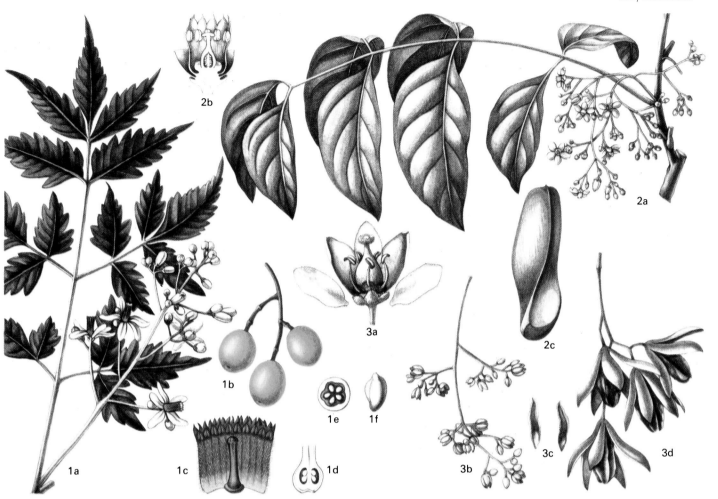

Meliaceae. 1 *Melia azedarach* (a) bipinnate leaf and axillary inflorescence ($\times\frac{2}{3}$); (b) fruits ($\times\frac{2}{3}$); (c) staminal tube opened out to show anthers attached and gynoecium with basal disk ($\times 2$); (d) vertical section of ovary with ovules on axile placentas ($\times 6$); (e) cross section of fruit ($\times\frac{2}{3}$); (f) seed ($\times 4$). 2 *Swietenia mahagoni* (a) shoot with pinnate leaf and axillary inflorescence ($\times\frac{2}{3}$); (b) half flower showing staminal tube ($\times 8$); (c) winged seed ($\times\frac{2}{3}$). 3 *Cedrela australis* (a) flower opened out showing free stamens arising from the disk and superior ovary crowned by discoid stigma ($\times 3\frac{1}{2}$); (b) part of inflorescence ($\times\frac{2}{3}$); (c) winged seeds ($\times 1$); (d) fruits surrounded by persistent sepals ($\times\frac{2}{3}$).

rhizomatous and spreading *C. terminalis* from the Sino–Himalayas, which has good autumnal leaf-tints, and black or yellow fruiting clusters about 22cm (9in) long, each fruit being about 1cm (0.4in) in diameter. It grows to about 1.3m (4ft). Another is *C. japonica* from Japan, this being only about 60cm (2ft) high, also with good autumnal leaf-color and coral-red to black fruits.

Several authorities state that the succulent fruits of several species other than *C. myrtifolia* are edible and taste rather like bilberries, but as the seeds and other plant parts are reputed to be poisonous, caution is recommended. Convulsions like those produced by strychnine characterize *Coriaria* poisoning.        B.M.

## MELIACEAE
*Mahoganies*

The Meliaceae is a family of mostly trees and shrubs, economically important chiefly because of its high-quality timbers, including the true mahoganies. It also includes a number of fruit trees.

**Distribution.** The family is restricted to the tropics and subtropics and comprises about 50 genera and some 550 species, which are

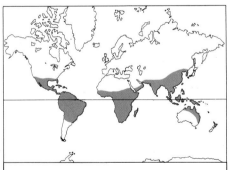

**Number of genera:** about 50
**Number of species:** about 550
**Distribution:** tropical and subtropical.
**Economic uses:** important timber, including mahogany, some edible fruits, and some ornamentals.

particularly common as understory trees in rain forests.

**Diagnostic features.** The family consists of trees or shrubs, sometimes unbranched "tuft trees," rarely herbaceous plants with woody stocks. The leaves are alternate, usually pinnate, sometimes simple and rarely bipinnate, and are without stipules. The flowers are often in cymose panicles and may be borne on the trunk or branches or in the axils of undeveloped leaves, or may be terminal or, rarely, borne on the leaves (*Chisocheton*). They are regular, and are usually bisexual, but often unisexual (the sexes borne on separate plants), although they may appear bisexual. There are three to five united or free sepals, three to five (rarely up to 14) usually free petals and five (rarely three) to ten (rarely up to 23) stamens, which may be free but are usually united into a staminal tube. The style may be absent and the stigma is often disciform or capitate. The ovary is superior with two to six (rarely one or up to 20) locules with one, two or more pendulous ovules in each locule. The fruit is a capsule, berry or drupe, or, rarely, a nut. The seeds are often winged while others have a fleshy aril or fleshy testa (sarcotesta); they are with or without endosperm and have a straight or curved embryo.

**Classification.** The family is divided into five subfamilies, of which three are small and restricted to Madagascar, while the other two, Melioideae and Swietenioideae, are pantropical.

The MELIOIDEAE have unwinged seeds

and usually have a fleshy sarcotesta or aril. They include *Turraea* (about 65 species, Old World), *Melia* (five species, Old World), *Azadirachta* (two species, Indomalaysia and the Pacific), *Guarea* (35 species, Africa and America), *Chisocheton* (52 species, Indomalaysia), *Dysoxylum* (60 species, Indomalaysia and Pacific) and *Xylocarpus* (two or three species in the mangrove and coastal forests of the Old World).

The SWIETENIOIDEAE have winged seeds and include the commercial mahoganies, *Swietenia* (seven or eight species, America), *Entandrophragma* (11 species, Africa), *Khaya* (eight species, Africa), *Cedrela* (nine species, America), *Toona* (15 species, Asia and Australasia) and *Lovoa* (two species, Africa).

The Meliaceae is undoubtedly allied to the pinnate-leaved tree families in the Sapindales such as the Anacardiaceae, Burseraceae, Sapindaceae, Simaroubaceae and Rutaceae, from all of which Meliaceae may be distinguished by the androecium.

**Economic uses.** The true mahoganies are Swietenioideae, particularly *Swietenia* in the Americas and *Khaya* in Africa. All are highly prized for their excellent color, working properties and finish. Important meliaceous timbers besides these include sapele *Entandrophragma cylindricum*, utile (*E. utile*), omu (*E. candollei*) and species of *Melia, Carapa, Azadirachta, Guarea, Cedrela, Toona, Soymida, Chukrasia, Dysoxylum, Lovoa, Aglaia* and *Owenia*.

Oils for soap-making have been extracted from the seeds of *Trichilia emetica* in Uganda, and oil from the Malayan *Chisocheton macrophyllus* has been used as an illuminant. Insecticides have been derived from *Melia* and *Azadirachta* species. The flowers of *Aglaia odorata* are used in the East in flavoring tea. The fruit of some *Aglaia* and *Lansium* species is important locally, and the most important commercial one, langsat from *Lansium domesticum*, is popular in Southeast Asia, as is the santol, *Sandoricum koetjape*. Many genera are very ornamental and are increasingly seen in cultivation, particularly *Aglaia, Chisocheton, Dysoxylum, Melia* and *Turraea*.          D.J.M.

# CNEORACEAE
*Spurge Olive*

The Cneoraceae is a family containing a single genus, *Cneorum*, which has two species, *Cneorum tricoccum* (the spurge olive) and *C. pulverulentum*. The latter has been placed in a separate genus *Neochamaelea* by some botanists.

**Distribution.** *Cneorum tricoccum* is native to the western Mediterranean region, *C. pulverulentum* to the Canary Islands.

**Diagnostic features.** *Cneorum* are evergreen shrubs with alternate gray-green, narrow, leathery leaves without stipules. The yellow flowers are solitary (*C. pulverulentum*) or in small corymbs (*C. tricoccum*) in the axils of

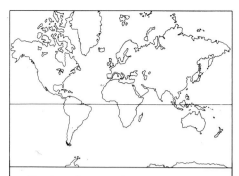

**Number of genera:** 1
**Number of species:** 2
**Distribution:** western Mediterranean and Canary Islands.
**Economic uses:** local medicinal use as a purgative and rubefacient.

terminal leaves. They are regular, bisexual and usually have three (sometimes four) free sepals, petals and stamens, a single style and an elongate receptacle or disk. The ovary is inferior, of three or four fused carpels, and lobed, with two pendulous ovules in each locule. The hard, red-brown fruit is composed of three (sometimes four) globose segments each with two seeds which contain endosperm.

**Classification.** The Cneoraceae is considered to be closely allied to the Zygophyllaceae but differs in having only one whorl of stamens, no ligules and no stipules, and in having oil glands in the leaves. The Cneoraceae shares this latter feature with other families in the Sapindales such as the Rutaceae.

**Economic uses.** The leaves and fruits of *C. tricoccum* are used locally as a purgative and a skin rubefacient.          M.C.D.

# RUTACEAE
*The Citrus Fruit Family*

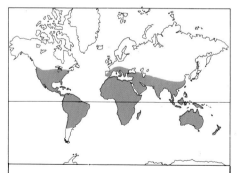

**Number of genera:** 150
**Number of species:** about 900
**Distribution:** tropical and warm temperate regions particularly S Africa and Australia.
**Economic uses:** citrus fruits (lemon, orange, grapefruit etc) and oils in perfumery and medicine (*Ruta, Galipea, Toddalia*).

The Rutaceae is a large family of shrubs and trees, and occasionally herbs, which are of great economic importance yielding the

citrus fruits of commerce (lemons, oranges, mandarins, tangerines, limes, grapefruit) as well as some attractive ornamental species such as *Choisya* and *Skimmia*.

**Distribution.** The Rutaceae has a more or less cosmopolitan distribution, but are largely centered in the tropics and temperate regions of the Southern Hemisphere, especially Australia and South Africa. Commercially, citrus fruits are grown in the tropical and warm temperate "citrus belt" which spans the whole globe, notably in the Mediterranean region, southern United States of America, Mexico, South Africa and Australia.

**Diagnostic features.** The family owes its name to the rue (*Ruta graveolens*), a small, hardy, evergreen, aromatic shrub which for centuries has been grown in herb gardens as a medicinal plant. Like most other members of the family, the crushed leaves of rue produce a strong foetid aroma from oil glands which can be seen as small, translucent black dots on the leaves. This is the most distinctive feature of the Rutaceae.

*Ruta* is somewhat typical for the family and can be characterized as a genus of 60 hardy shrubs, subshrubs and perennials, with strong-smelling, alternate, trifoliolate or compound, pinnate leaves without stipules and terminal inflorescences (corymbs or panicles) that are subtended by leafy bracts. The flowers are insect-pollinated, greenish-yellow in color and composed of a four- or five-lobed persistent calyx, four or five overlapping, toothed or ciliate (having fine hairs or projections) petals, a thick basal disk of eight to ten glands or pits from which nectar is secreted and eight to ten stamens. The ovary is superior, deeply lobed, with five or four fused carpels forming four or five locules each with numerous ovules. The fruit is a berry and the seeds have endosperm. Of course, there are many deviations from this pattern and some of the more obvious ones are: the tree growth habit in various prominent genera such as *Citrus, Poncirus,* and *Phellodendron*; simple leaves in *Diosma, Boronia* and *Skimmia*; leaves reduced to thorns in various species of the orange subfamily (Aurantioideae); leaves not dotted with glands in *Leptothyrsa* and *Phellodendron*; stamens more numerous than 10 in *Peltostigma, Citrus* and *Asterolasia*; and the

Rutaceae. 1 *Ruta graveolens* (a) shoot with bipinnate leaves and cymose inflorescences ($\times\frac{2}{3}$); (b) flower with four sepals, four petals, eight stamens and a superior, lobed ovary with a basal disk and crowned by a single style ($\times 2\frac{2}{3}$); (c) vertical section of ovary ($\times 6$); (d) cross section of ovary showing four locules and ovules on axile placentas ($\times 4$). 2 *Citrus aurantium* (sweet orange) (a) half flower with numerous stamens and prominent disk at the base of the ovary ($\times 2$); (b) fruit—a pulpy berry ($\times\frac{2}{3}$). 3 *Ptelea trifoliata* winged fruit—an unusual feature for the family ($\times 1\frac{1}{2}$). 4 *Citrus limon* (lemon) flowering shoot ($\times\frac{2}{3}$). 5 *Crowea saligna* flowering shoot ($\times\frac{2}{3}$).

flowers epiphyllous, ie emerging from the leaves, in some species of *Erythrochiton*. The flowers are irregular in *Cusparia* and unisexual in *Toddalia*. The ovary is semi-inferior in *Platyspermatica* and many species have two ovules in each locule. The fruits are very variable between different subfamilies and tribes, being schizocarps, drupes or berries. The seeds may lack endosperm.

**Classification.** The Rutaceae can be divided into four subfamilies as follows:

RUTOIDEAE. Ovaries with two to five deep lobes, the carpels quite separate and connected only by the styles and stigmas. The fruit is a berry. Tribe RUTEAE: herbs and shrubs of the Northern Hemisphere (*Ruta*, *Dictamnus* and *Thamnosma*). Tribe ZANTHOXYLEAE: South American and Australian trees and shrubs (*Melicope*, *Pelea*, *Choisya*, *Euodia*, *Fagara* and *Zanthoxylum*). Tribe BORONIEAE: perennial herbs and shrubs from Australia (*Eriostemon*, *Phebalium*, *Asterolasia*, *Correa*, *Boronia* and *Dipholaena*). Tribe DIOSMEAE: mostly perennial herbs or shrubs and rarely trees of South Africa (*Diosma*, *Calodendrum*, *Barosma*, *Agathosma* and *Macrostylis*). Tribe CUSPARIEAE: shrubs and trees of South America (*Flindersia*, *Esenbeckia*, *Galipea*, *Cusparia* and *Ravenia*).

TODDALOIDEAE. Ovaries entire (not lobed) or slightly two- to five-lobed with two to five incompletely or completely united carpels. The fruit consists of two to four drupelets or is a thick-skinned drupe. They are predominantly Old World tropical and temperate trees and shrubs (*Phellodendron Ptelea*, *Amyris*, *Vepris*, *Toddalia*, *Skimmia*).

RHABDODENDROIDEAE. Ovaries two- to five-lobed and the carpels united but distinguished by the possession of unique bowl-shaped receptacles. Comprising a single tree genus, *Rhabdodendron*.

AURANTIOIDEAE. Ovaries entire, and the fruit a large, pulpy berry (*Aegle*, *Citrus*, *Atalantia*, *Glycosmis*, *Murraya*, *Clausena* and *Micromelum*).

The Rutaceae belongs to an order of plants known as the Sapindales, a group of 16 families characterized, with few exceptions, as woody plants with mostly compound leaves. The principal character separating the Rutaceae from the other families is the presence of oil glands in the leaves.

**Economic uses.** The Rutaceae is important for a large number of crop and garden plants. It includes the citrus fruits of the subfamily Aurantioideae. The commonly cultivated fruits belong to three genera, *Citrus*, *Fortunella* and *Poncirus*. *Citrus* is undoubtedly the most important genus in the whole family and of the 60 known species most are cultivated. Botanically, the citrus fruit is a berry with a tough leathery skin containing aromatic oil glands and a flesh composed of enlarged cells filled with juice. The most widely cultivated species include the lemon (*Citrus limon*), the citron (*C.*

*medica*), the sour or Seville orange (*C. aurantium*), the edible or sweet orange (*C. sinensis*), the mandarins, satsumas and tangerines (*C. reticulata*), the limes (*C. aurantifolia*) and the grapefruit (*C. paradisi*). Lesser-known fruits include the kumquats belonging to the genus *Fortunella* and the inedible trifoliolate orange, *Poncirus trifoliata*.

Many species are cultivated for their essential oils (fragrant aromatic compounds). The bergamot, a dwarf variety of the Seville orange, produces the valuable "bergamot oil" (for use in perfumes) from its fruit and "oil of neroli" (for use in Eau de Cologne) from its blossom. The Mexican *Choisya ternata* is a very beautiful shrub frequently grown in hothouses and sheltered gardens. Popular park plants include: shrub species of *Skimmia*, notably *Skimmia japonica* and *S. reevesiana* and their hybrid, *S. × foremannii*; the highly scented hop tree of North America (*Ptelea trifoliata*); and the "cork" trees *Phellodendron japonicum*, *P. amurense* and *P. sachalinense*. Desirable house plants include the heavily scented shrubs of the genus *Diosma* (which means divine scent), various species of *Agathosma* and *Barosma*, and the scented indoor heather-like shrubs *Coleonema album* and *C. pulchrum*.

The burning bush (*Dictamnus albus*) is covered with oil glands that exude a strong, spice-scented secretion. Its oil production is so high that on really hot days the plants can easily be ignited without damaging the bush. The extremely poisonous rue plant (*R. graveolens*) was an ancient herbal remedy for faintness, cramp, hysteria and diseases of the womb, as well as its bizarre use for treating croup in poultry.

Other valuable medicinal oils have been obtained from the bark glands of species of the genera *Galipea* and *Toddalia*. Only one species of the Rutaceae has any value as timber, yielding valuable hardwoods, namely the West Indian silkwood (*Zanthoxylum flavum*). Jaborandi (the source of the alkaloid pilocarpine) is obtained from the dried leaves of the South American shrubs, *Pilocarpus jaborandi* and *P. microphyllus*.

C.J.H.

# ZYGOPHYLLACEAE
## Lignum Vitae

The Zygophyllaceae is a largely tropical and subtropical family of shrubs, some herbs and a few trees, many of which are adapted to dry or salty habitats. It includes some valuable timber trees such as lignum vitae (*Guaiacum* species).

**Distribution.** The family is found widely in the tropics and subtropics, often in drier areas, forming a conspicuous element of scrub vegetation, with some temperate representatives.

**Diagnostic features.** The branches are sometimes jointed at the nodes, The leaves are fleshy or leathery, usually opposite, rarely

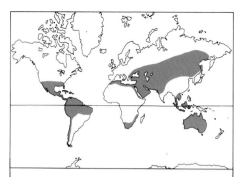

**Number of genera:** about 25
**Number of species:** about 240
**Distribution:** mainly tropical and subtropical, mostly in dry areas.
**Economic uses:** timber (lignum vitae), medicinal resins, some edible fruits and dyes.

alternate, with stipules which are sometimes spiny or become spiny as they develop.

Apart from one genus (*Neoluederitzia*), the flowers are bisexual, usually regular and borne solitary, paired or in cymes. There are four or five free, imbricate sepals, and four or five free petals also generally imbricate, rarely absent. The stamens are in one, two or three whorls of five, the outer series lying opposite the petals. The filaments are separate and often have a scale at the base. A disk is usually present. The ovary is superior, usually of five fused carpels (rarely two to 12) often with wings. There are usually five locules with one to numerous pendulous ovules on axile placentas. The lobed or flattened stigmas are sessile or borne on a short style. The fruit is generally a capsule, sometimes splitting into five portions, rarely berry-like or drupe-like. The seeds have a straight or slightly curved embryo usually surrounded by endosperm.

**Classification.** The family is divided into five or six subfamilies largely on the basis of fruit structure. The most important genera are *Guaiacum*, *Bulnesia*, *Nitraria*, *Peganum*, *Balanites*, *Neoschroetera*, *Zygophyllum* and *Tribulus*. The family clearly belongs in the complex of orders Sapindales-Geraniales-Polygalales, but any closer relationships are difficult to determine.

**Economic uses.** Several species of *Guaiacum*, especially *Guaiacum officinale* and *G. sanctum* (from tropical America and the West Indies) are the source of lignum vitae, a heavy, durable wood which resists splitting. The fruits of some *Balanites* and *Nitraria* species are edible. The tropical American trees, *Bulnesia arborea* (Maracaibo lignum vitae) and *B. sarmienti* (Paraguay lignum vitae), provide useful timber as well as an essential oil used in perfumes. *Neoschroetera tridentata*, the creosote plant from Mexico and adjacent areas, is used medicinally, and its flower buds are used as a caper substitute, as are the buds of *Zygophyllum fabago*. The seeds of the Mediterranean *Peganum harmala* produce the dye, turkey red. S.R.C.

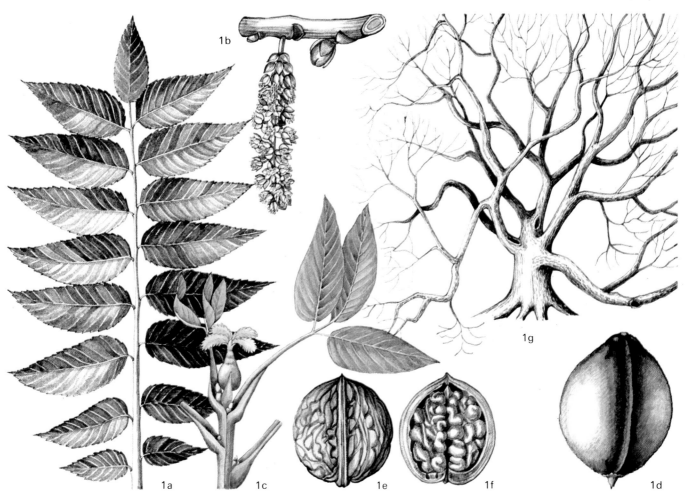

Juglandaceae. 1 *Juglans regia* (a) imparipinnate leaf ($\times\frac{2}{3}$); (b) male, catkin-like inflorescence borne on old wood ($\times\frac{2}{3}$); (c) tip of shoot with female flower with plumose stigmas ($\times\frac{2}{3}$); (d) fruit ($\times\frac{2}{3}$); (e) fruit with fleshy husk removed to show hard, sculptured endocarp (inner fruit wall) ($\times\frac{2}{3}$); (f) fruit with endocarp removed to show seed with contorted cotyledons ($\times\frac{2}{3}$); (g) habit of old tree.

# JUGLANDALES

## JUGLANDACEAE
*Walnuts, Hickories and Pecan Nuts*

The Juglandaceae is a small family of deciduous trees, whose best-known members are the walnuts (*Juglans*) and hickories (*Carya*).

**Distribution.** The family is basically north temperate and subtropical, with extensions southwards to India, Indochina and down the Andes in South America.

**Diagnostic features.** In winter, the shoots bear brown, hairy buds. The leaves are alternate (rarely opposite), pinnate and without stipules. The flowers are bracteate and unisexual, male and female being borne on the same plant, the males usually in catkin-like, pendulous inflorescences forming on the previous year's twigs, the females in smaller erect spikes forming on the new twigs. The perianth is typically four-lobed, but is often reduced or absent by abortion. The male flowers have three to 40 free stamens in two or more series, with short filaments and bilocular anthers opening lengthwise. Pollination is by wind. The ovary of the female flowers is inferior and consists of two fused carpels forming a single locule containing one erect, orthotropous ovule.

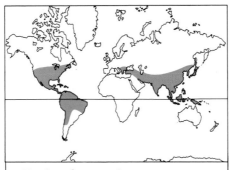

**Number of genera:** 7
**Number of species:** about 50
**Distribution:** mainly N temperate and subtropical.
**Economic uses:** nuts and oil (walnut, hickory and pecan), timber and ornamentals.

The style is short, with two stigmas. The fruit is a drupe or nut; the characteristic "boats" from a walnut shell do not correspond to the two carpels, but the suture is along their midribs. The seeds have no endosperm.

**Classification.** The family has been divided into two subfamilies, the JUGLANDOIDEAE with two genera, *Juglans* and *Carya* and the OREOMUNNEOIDEAE with six genera, *Pterocarya*, *Engelhardtia*, *Oreomunnea*, *Platycarya* and *Alfaroa*. The subfamilies are character-

ized by their fruits, being drupes and winged nuts, respectively. The two best-known genera, *Juglans* and *Carya* are distinguished by the former having an indehiscent, sculptured "nut" and the latter a dehiscent, smooth "nut" inside the fleshy part of the drupe.

The relationships of the family are obscure, although there may be a connection with the Anacardiaceae. The family Rhoipteleaceae represented by a single Chinese tree species *Rhoiptelea chiliantha*, is often regarded a primitive type of the order Juglandales leading to the more advanced Juglandaceae. It differs from the Juglandaceae in having bisexual and female flowers, a superior ovary, stipulate leaves and fruit a samara.

**Economic uses.** The family is best known for its edible nuts, the walnuts (*Juglans regia* and other species), pecan nuts (*Carya pecan*, *C. illinoensis*) and hickory nuts (*C. ovata*). The oil from the nuts is also used in foods and in the manufacture of cosmetics and soap and as a drying agent in paints.

Both *Juglans* and *Carya* produce valuable timber and are much prized for their fine grain and their toughness.

Species of walnut, hickory and *Pterocarya* (wing nut) are grown for their ornamental value, particularly in autumn.     I.B.K.R.

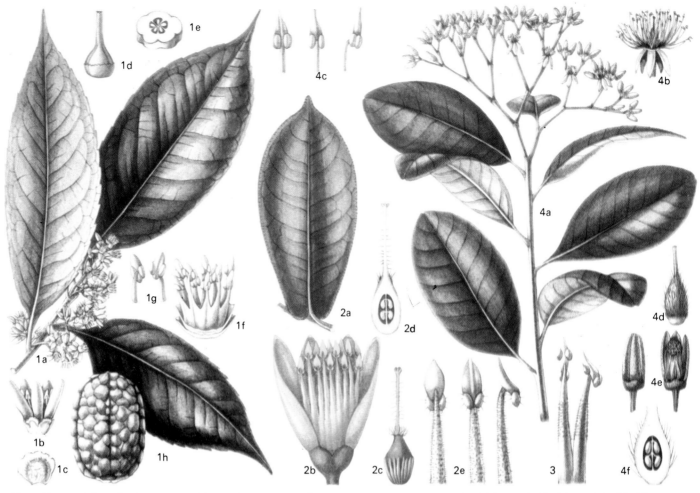

Houmiriaceae. 1 *Sacoglottis amazonica* (a) part of shoot with leaves and axillary inflorescences ($\times\frac{2}{3}$); (b) flower ($\times 2$); (c) sepal ($\times 4\frac{1}{2}$); (d) gynoecium ($\times 4$); (e) cross section of ovary ($\times 8$); (f) androecium ($\times 4$); (g) views of anthers ($\times 7$); (h) fruit—a drupe ($\times\frac{2}{3}$). 2 *Houmiria balsamifera* (a) leaf ($\times\frac{2}{3}$); (b) flower with one petal removed ($\times 6$); (c) gynoecium with toothed, ring-like disk ($\times 6$); (d) vertical section of gynoecium ($\times 6$); (e) stamens ($\times 14$). 3 *Duckesia verrucosa* stamens each with four anther lobes ($\times 14$). 4 *Vantanea parviflora* (a) leafy shoot ($\times\frac{2}{3}$); (b) flower ($\times 2$); (c) stamens ($\times 14$); (d) gynoecium ($\times 7$); (e) flower bud and bud with one petal removed ($\times 1\frac{1}{2}$); (f) vertical section of gynoecium ($\times 10$).

# GERANIALES

## HOUMIRIACEAE
### Bastard Bullet Tree

The Houmiriaceae is a family of mainly tropical American trees and shrubs.

**Distribution.** The family is native to Central and tropical South America, with only two species (of *Sacoglottis*) also occurring in West Africa.

**Diagnostic features.** The leaves are alternate, simple, entire or crenate, without stipules. In some members of the family, such as *Sacoglottis*, the petiole is swollen at the base where it joins the stem. The flowers are regular and bisexual, borne in axillary or terminal cymes. There are five sepals, free or slightly connate in the lower part, sometimes covered with fine hairs. There are five free petals, persistent or caducous shortly after opening. There are 10, 20, 30 or more stamens in one, two or several whorls, with filaments more or less connate in the lower part. The anthers are versatile (attached near the middle and moving freely), with two or four locules opening by lengthwise slits. The flower possesses a ring-like disk, often toothed or

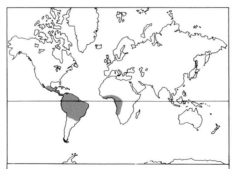

**Number of genera:** 8
**Number of species:** about 50
**Distribution:** Central and tropical S America with two species also in W Africa.
**Economic uses:** wood and fruit used locally.

made up of separate glands between the stamens and surrounding the base of the ovary. The ovary is superior, consisting of five fused carpels and bearing a simple undivided style with a capitate stigma; it has five locules, each with one or two anatropous ovules, pendulous from the apex. The fruit is drupaceous, with a rather thin fleshy pericarp and a hard woody endocarp often with

resin-filled cavities. It has five locules but usually only one or two seeds. The seeds contain fleshy, oily, copious endosperm surrounding a straight embryo with short cotyledons.

**Classification.** The family includes *Houmiria* (*Humiria*) (three or four species), *Endopleura*, *Duckesia* and *Hylocarpa* (each with one species), *Vantanea* (14 species), *Schistostemon* (seven species) and *Humiriastrum* (12 species). Distinguishing characters include the number of stamens and the nature of the anthers and ovary. *Vantanea* flowers have numerous stamens with four locules in the anthers, while *Houmiria* and *Sacoglottis* flowers have 10 or 20 stamens with two locules in the anthers. The ovary of flowers of *Houmiria* contains two ovules per locules, that of flowers of *Sacoglottis* only one ovule per locule.

The family is related to the Linaceae.

**Economic uses.** The only member of this family of economic importance is *Houmiria floribunda* (bastard bullet tree); the durable red-brown hardwood is used locally for construction purposes. The fruit of *Sacoglottis gabonensis* is used locally to make a fermented beverage. S.R.C.

Linaceae. 1 *Linum grandiflorum* (a) leafy shoot with cymose inflorescence ($\times\frac{2}{3}$); (b) petal ($\times 2$); (c) flower with petals removed to show five blue stamens and five pink staminodes ($\times 3$); (d) vertical section of ovary showing axile, pendulous ovules ($\times 4$). 2 *Hugonia castaneifolia* (a) leafy shoot showing hook-like modifications at the base of the inflorescence and flower buds ($\times 1$); (b) flower ($\times 3$); (c) stamens in two whorls of five surrounding the gynoecium which has three styles ($\times 4$); (d) vertical section of fruit ($\times 1\frac{1}{2}$). 3 *Reinwardtia sinensis* (a) leafy shoot and inflorescence ($\times\frac{2}{3}$); (b) fruit—a capsule ($\times\frac{2}{3}$); (c) cross section of ovary ($\times 3$); (d) whorl of stamens and small staminodes ($\times 3$); (e) calyx and four styles ($\times 3$).

# LINACEAE

*Flax and Linseed*

The Linaceae is a small but widespread family of herbs and some shrubs, of which flax (*Linum usitatissimum*) is the most economically important member.

**Distribution.** The family is chiefly distributed in temperate zones, with some tropical representatives.

**Diagnostic features.** The leaves are usually alternate, small, entire and with or without stipules. The inflorescence is cymose, bearing regular bisexual flowers. In the tropical genus *Hugonia* the lower parts of the inflorescence are modified as hooks for climbing. The calyx is composed of five (occasionally four) sepals which are either free or united at their base. It is usually persistent through to the stage of seed liberation. The petals are equal in number to the sepals, usually free but sometimes joined at the base, and wither away and fall early. The stamens usually equal the number of petals and are alternate with them. Their filaments are usually short and fused at their bases to form a glandular ring. There may also be five staminodes alternating with the stamens. The ovary is superior and consists

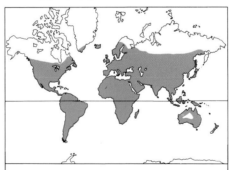

**Number of genera:** 13
**Number of species:** about 300
**Distribution:** chiefly temperate, with some tropical representatives.
**Economic uses:** fibers, linseed oil and cattle feed from flax, some fruits and locally used timber, and garden and greenhouse ornamentals.

of two to five fused carpels. It may, however, have four to ten locules owing to the development of additional (false) septa. The placentation is axile with one or two pendulous ovules in each locule. The two to five styles are separate. The fruit is usually a capsule but in some species drupaceous. The

seed contains a straight embryo and endosperm ranges from none to copious. The seed of flax (*Linum*) has a mucilaginous coat which swells upon wetting.

**Classification.** The genera can be distinguished on the basis of habit and floral characteristics such as presence or absence of staminodes, number of locules in the ovary, number of styles and type of seed. The most important genera are *Linum* (230 species, mainly Mediterranean), *Hugonia* (40 species, Africa and the Far East), *Reinwardtia* (two species of shrubs, northern India and China), *Anisadenia* (two species, China), *Roucheria* (eight species, tropical South America) and the monotypic genus *Radiola* (Europe and North Africa). The genus *Ctenolophon* is considered by some authorities to constitute a separate family (Ctenolophonaceae).

This family is characterized by the usually early-falling petals and the fusion of the short stamen filaments. It is placed in the same order as the Geraniaceae with which it shares such features as regular flowers, two whorls of stamens or staminodes, and a gynoecium of two to five fused carpels with axile placentation. The genus *Anisadenia* is considered to show features which link the

Geraniaceae. 1 *Geranium malviflorum* (a) shoot with compound leaves and inflorescences ($\times\frac{2}{3}$) : (b) vertical section of flower showing bilobed petals ($\times 1\frac{1}{3}$). 2 *G. sanguineum* fruit with persistent calyx and with one awn separating from the central axis to disperse a seed ($\times 1\frac{1}{3}$). 3 *Erodium romanum* (a) tip of leafy shoot with inflorescence and fruits ($\times\frac{2}{3}$) ; (b) fruit before dehiscence ($\times 1\frac{1}{3}$). 4 *Sarcocaulon patersonii* (a) fleshy stem with thorns (remains of leaf stalks) and solitary flowers ($\times 1$) ; (b) flower with petals removed showing five pointed sepals and fifteen stamens of two lengths and with fused filament bases ($\times 3$).

Linaceae with the more advanced Plumbaginaceae.

**Economic uses.** *Linum* is the most important genus as it includes the flax and linseed plant, *Linum usitatissimum*, an annual herb cultivated for both its stem fibers and its seeds. The fibers are durable and have great tensile strength. They are used in the manufacture of linens, fine writing paper and cigarette paper. Linseed oil is prepared by extraction from the pressed seed and is chiefly used in the manufacture of paints, varnishes and printing ink; the residual oilcake makes valuable cattle food. Europe is the chief center of fiber production, Argentina the world's largest producer of linseed oil.

The African species of *Hugonia* (*Hugonia obtusifolia* and *H. platysepala*) provide edible fruits while the Malaysian tree *Ctenolophon parvifolium* provides a hard durable timber used in house construction. A number of other *Linum* species are grown as ornamentals, either as rockery plants (eg *L. arboreum*) or as border plants (*L. flavum*), while *Reinwardtia trigyna*, with bright yellow flowers, is an attractive winter-flowering shrub grown in greenhouses in temperate zones.                                    S.R.C.

# GERANIACEAE
*Geraniums and Pelargoniums*

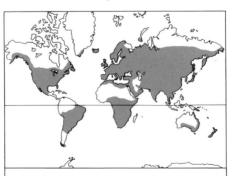

**Number of genera:** 11
**Number of species:** about 750
**Distribution:** mainly temperate and subtropical.
**Economic uses:** geranium oil from *Pelargonium* species, and garden and greenhouse ornamental, including *Pelargonium* (geranium), *Geranium* (cranesbill) and *Erodium* (storksbill).

The Geraniaceae is a family of mostly temperate and subtropical annual or perennial herbs and a few small shrubs. It includes the garden geraniums (*Pelargonium* species)

which should not be confused with the genus *Geranium* (cranesbill) in the same family.

**Distribution.** The family is widely distributed in temperate and subtropical regions of both Northern and Southern Hemispheres. Some species of *Geranium* are found within the Arctic while others occur in Antarctica.

**Diagnostic features.** The stems often have jointed nodes and these, and the leaves, are frequently covered with glandular hairs. The leaves are opposite or alternate, simple or compound, often with stipules. The flowers are regular (rarely slightly irregular), bisexual and borne either solitary or in cymose inflorescences. There are usually five sepals and five, often large and brightly colored petals surrounding one, two or three whorls each of five stamens (and/or staminodes), which usually have the filaments fused at the base. If there are two whorls, the outermost lies opposite the petals. Nectaries are often present at the base of the stamens. The ovary is superior and comprises usually five carpels fused around a central axis (carpophore); there are one or two ovules on axile placentas in each of the five locules. Each of the five styles terminates in a separate stigma. The fruit is a schizocarp (rarely a capsule as in

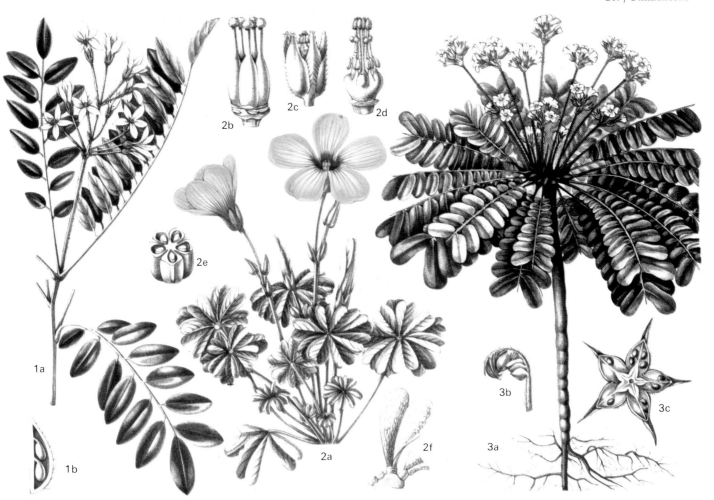

Oxalidaceae. 1 *Eichleria blanchetiana* (a) flowering shoot showing imparipinnate leaves and flowers in a cymose inflorescence ($\times\frac{2}{3}$); (b) vertical section of part of ovary ($\times$14). 2 *Oxalis adenophylla* (a) habit showing palmate leaves ($\times\frac{2}{3}$); (b) gynoecium comprising five united carpels each with a capitate stigma ($\times4\frac{1}{2}$); (c) flower with petals removed ($\times$2); (d) androecium and gynoecium showing trimorphic heterostyly (ie stamens in two rows each at different levels to the stigmas) ($\times$3); (e) section of base of fruit ($\times$2); (f) leaflet and short petiolule ($\times$2). 3 *Biophytum sensitivum* (a) habit showing pinnate leaves with the terminal leaflet reduced to a bristle ($\times\frac{2}{3}$); (b) leaf unfolding ($\times\frac{2}{3}$); (c) dehiscing fruit ($\times$3).

*Viviania*), and the carpels with their long persistent styles (awns) split off elastically at the base of the central axis. The seeds contain a curved embryo and little or no endosperm.

**Classification.** The 11 genera can be disposed into five subfamilies:

GERANIOIDEAE, characterized by the presence of a beaked ovary, contains nearly half of the genera: *Geranium*, *Erodium*, *Pelargonium*, *Monsonia* and *Sarcocaulon*. Three other subfamilies (BIEBERSTEINIOIDEAE, VIVIANIOIDEAE and DIRACHMOIDEAE) lack a beaked ovary and each contain a single genus: *Biebersteinia* (ovary contains only one seed), *Viviania* (fruit a capsule) and *Dirachma* (eight carpels present), respectively. *Viviania* (30 South American species) is sometimes separated off as a separate family, the Vivianiaceae. The other subfamily, WENDTIOIDEAE (sometimes with a beaked ovary), contains the other three genera, *Balbisia*, *Wendtia* and *Rhynchotheca* and is separated by some authorities as the family Ledocarpaceae.

The family is closely related to the Tropaeolaceae, Oxalidaceae, Linaceae and possibly the Balsaminaceae.

**Economic uses.** The so-called geraniums grown so extensively in gardens and greenhouses in fact belong to the South African genus *Pelargonium*. Most of the horticultural pelargoniums are hybrids, as in the case of zonal pelargoniums, which are the result of crosses between *Pelargonium zonale* and *P. inquinans*, or the large-flowered regal pelargoniums, which are the result of crosses between *P. cucullatum*, *P. fulgidum* and *P. grandiflorum*. Other species of *Pelargonium*, notably *P. graveolens*, *P. odoratissimum*, *P. capitatum* and *P. radula*, are cultivated for geranium oil, which can be distilled from the leaves and shoots, and finds wide uses in the perfume and oil industries. A number of species of *Geranium* (cranesbill) and *Erodium* (storksbill) are cultivated as border and as rockery plants.          S.R.C.

## OXALIDACEAE

### Wood Sorrel and Bermuda Buttercup

The Oxalidaceae is a family of mainly tropical and subtropical annual and perennial herbs including a number of ornamentals.

**Distribution.** Most of the family is native to tropical and subtropical Asia and Africa, and tropical America. Some are temperate.

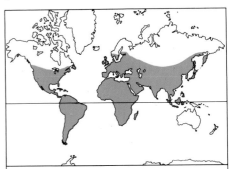

**Number of genera:** 3
**Number of species:** about 900
**Distribution:** centered in the tropics and subtropics, but widespread in temperate regions.
**Economic uses:** tubers and leaves of some *Oxalis* species eaten and some rock garden ornamentals.

**Diagnostic features.** The leaves are alternate and without stipules, sometimes simple but often pinnately or palmately compound. Many species of *Oxalis* possess leaflets which fold downward at night and in cold weather; in some species of *Biophytum* the leaflets bend when touched. In a few species of

*Oxalis*, such as *Oxalis bupleurifolia*, the ordinary leaves are replaced by phyllodes (leaf-like petioles).

The flowers are regular, bisexual, and borne either solitary or in cymose inflorescences. They have five, free, persistent sepals and five petals which may be free or fused just at the base. There are 10 stamens arranged in two whorls and connate at the base, the outer whorl of five lying opposite the petals. Sometimes the outer stamens are sterile. The anthers are two-celled and dehisce lengthwise. The ovary is superior, consisting of five free (as in *Biophytum*) or fused (as in *Oxalis*) carpels with five free styles and capitate stigmas. There are five locules, each with one or two rows of ovules on axile placentas. Many species of *Oxalis* display trimorphic heterostyly, ie flowers with long styles and medium and short stamens, medium styles and long and short stamens, and short styles and long and medium stamens. Fertile crosses are only possible between different flower-types. The Eurasian species *Oxalis acetosella* (wood sorrel) has flowers which in cold conditions exhibit cleistogamy (pollination and fertilization before flower-opening).

The fruit is a capsule. The seeds have a straight embryo surrounded by fleshy endosperm. The seeds of some species of *Biophytum* and *Oxalis* may have a fleshy aril at the base. The turgid inner cell layers of the aril turn inside out rapidly, separate from the testa, and the seed is explosively flung out.

**Classification.** The three genera are *Oxalis* (about 800 species), with leaves with one to 20 or more leaflets; *Biophytum* (70 species), whose leaves possess a bristle representing the end leaflet; and *Eichleria* (two species), whose leaves possess a terminal leaflet. Classification within *Oxalis* is based on features such as the number, shape and size of the leaflets and the form of inflorescence and color of flowers.

The family is related to the Geraniaceae and Linaceae but is distinguished for example from the Geraniaceae by the five distinct styles and by the possession of arillate seeds. The tree genera *Averrhoa* and *Connaropsis* were at one time included in the Oxalidaceae but are now sometimes separated as the family Averrhoaceae.

**Economic uses.** The Oxalidaceae is of minor economic use. *Oxalis crenata*, a small Peruvian perennial herb, has tubers which are boiled and eaten as a vegetable and the leaves used in salads. The leaves of *O. acetosella* are sometimes used in salads instead of sorrel and the bulbous stems of *O. pes-caprae* (*O. cernua*, Bermuda buttercup) are sometimes used as a vegetable in southern France and North Africa. The tubers of the Mexican species *O. deppei* are also used as food and are cultivated in France and Belgium.

A number of *Oxalis* species are cultivated as rock garden plants and some are troublesome weeds. S.R.C.

# ERYTHROXYLACEAE
*Coca/Cocaine*

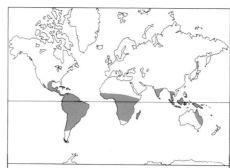

**Number of genera:** 4
**Number of species:** about 260
**Distribution:** tropical and subtropical centered in S America.
**Economic uses:** cocaine, widely used in medicine, and local uses of wood, bark dye and essential oils.

This tropical and subtropical family of trees and shrubs includes the important cocaine-producing coca plant.

**Distribution.** The family is centered in the Andes and Amazonian basin of South America and also occurs in Africa, Madagascar, Southeast Asia and tropical Australia.

**Diagnostic features.** The plants have simple, alternate (rarely opposite), ovoid leaves and stipules within the petioles (intrapetiolar). The terminal or axillary flowers are very small, occurring in fascicles. They are usually regular and bisexual, rarely unisexual (and then the sexes are on separate plants). The calyx is bell-shaped with five imbricate lobes and the corolla consists of five free, imbricate, deciduous petals. The ten stamens are in two series joined at their bases. The ovary is superior and consists of three or four fused carpels containing three or four locules, only one or two of which are fertile, each containing one or two pendulous, anatropous ovules. There are three fused or free styles. The fruit is an ovoid drupe borne beyond the persistent calyx. The seeds have endosperm and a straight embryo.

**Classification.** The family comprises four genera, readily distinguished by leaf and floral characteristics. *Aneulophus* (two species, parts of tropical Africa) has opposite leaves. *Erythroxylum* (about 250 species, tropical South America, Africa and Madagascar) has alternate leaves and filaments fused in a tube. *Nectaropetalum* (six species, tropical and southern Africa) has alternate leaves, filaments free and sessile flowers in fascicles in the leaf axils. *Pinacopodium* (two species, parts of tropical Africa) has alternate leaves, filaments free and stalked flowers in axillary or terminal fascicles.

The Erythroxylaceae is a well-defined group closely allied to and often included in the family Linaceae.

**Economic uses.** The leaves of *Erythroxylum*

*coca* and *E. novagranatense* (coca) yield the important alkaloid cocaine, a narcotic widely used in modern medicine. They are chewed as a stimulant by South American Indians and are cultivated in South America, Sri Lanka and Java for the cocaine, used as a local anaesthetic and in some medicines. Other species are of local importance for their wood, bark dye, wood tar, essential oil or medicinal uses.

C.J.H.

# LIMNANTHACEAE
*Poached Egg Flower*

**Number of genera:** 2
**Number of species:** 11
**Distribution:** N America, chiefly California.
**Economic uses:** limited as cultivated ornamentals.

The Limnanthaceae is a small North American family of two genera of delicate, sometimes attractive, annual herbs.

**Distribution.** Most of the 11 species are Californian, and all are North American, growing in moist habitats.

**Diagnostic features.** These fragile, sappy, annual herbs have alternate, pinnatifid leaves without stipules. The solitary flowers are regular, usually white, bisexual and borne in leaf axils. There are three or five valvate sepals and an equal number of contorted petals. There are six or ten stamens. The ovary is superior, of three or five carpels, which are almost free but have a common gynobasic style. Each carpel locule contains one ascending ovule, and at maturity the fruit separates into three or five tuberculate nutlets; the seeds lack endosperm and have a straight embryo.

**Classification.** In *Limnanthes* the flower parts are in fives. In the monotypic *Floerkea* they are in threes and the petals are shorter than the sepals. Although generally similar to the Polemoniaceae and Hydrophyllaceae, the family is more closely related to the Geraniaceae.

**Economic uses.** Of the 10 species of *Limnanthes* (Greek: marsh flower) the best-known is *Limnanthes douglasii* (poached egg flower). Its striking flowers have yellow petals with white tips. Borne in profusion, they are very attractive to bees. It is grown in gardens and in glasshouses. B.M.

Balsaminaceae. 1 *Hydrocera triflora* (a) irregular flower (×1); (b) section of fruit—a berry (×1). 2 *Impatiens walleriana* leafy shoot and flowers (×⅔). 3 *I. balsamina* (a) fruit (×1⅓); (b) leafy shoot and axillary flowers (×⅔). 4 *I. glandulifera* (a) irregular flower with posterior spurred sepal and small anterior sepals (×1); (b) stamens fused around the ovary (×3); (c) anterior sepal (×3); (d) lateral petal (×1); (e) anterior petal (×1); (f) fruit—a capsule (×1½).

# BALSAMINACEAE
## Balsams

The Balsaminaceae is a family of annual and perennial herbs with watery, translucent stems. The major genus is *Impatiens*, several species of which are cultivated.

**Distribution.** The family is represented throughout temperate and tropical Eurasia, Africa, Madagascar, and Central and North America.

**Diagnostic features.** The leaves are alternate or opposite, toothed, and usually without stipules. The flowers are bisexual and very irregular. There are usually three free sepals, the posterior one petaloid and usually spurred, and in addition sometimes two small or aborted anterior sepals. There are five unequal petals, the four lower ones connate in lateral pairs. There are five stamens with short, flat filaments and introrse anthers which are more or less united to form a cap over the ovary. The ovary is superior, consisting of five fused carpels, with five locules containing numerous ovules on axile placentas. There are one to five, more or less sessile stigmas. The fruit is a capsule or rarely a berry, and the seed has no endosperm and a straight embryo.

**Classification.** The genus *Impatiens* has

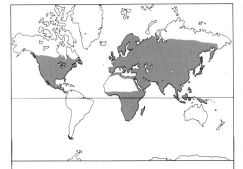

**Number of genera:** 4
**Number of species:** 500–600
**Distribution:** temperate and tropical regions.
**Economic uses:** valued greenhouse and pot plants of the genus *Impatiens*.

leaves that are oval to lanceolate, toothed and petiolate, and alternate or opposite. The flowers appear to have only four sepals but the anterior one, which is notched at the apex, probably originated as two. The posterior one, which may appear anteriorly as the flower hangs on the pedicel, is the largest and forms a spurred sac. The brightly colored corolla of pink, red, purple, white or yellow has five petals, the laterals united in

pairs. The fruit is an explosive capsule, with a fleshy pericarp, the outer cells of which are highly turgid and create an intense pressure on the whole fruit. The explosive method of dehiscence is termed septifragal dehiscence and is initiated by touch, when the fruit is ripe, and valves roll inwards from the base with a violent explosion which is clearly audible to the human ear, and the black seeds are scattered in all directions. *Impatiens* exhibits dimorphism, bearing small cleistogamic flowers which are fertilized in bud and never open, and large showy ones which rarely produce seeds that ripen.

*Impatiens* (with 500–600 species) is distinguished from the genus *Hydrocera* by two characters: the lateral petals are connate in pairs and the fruit is a capsule in *Impatiens*, while in *Hydrocera* the lateral petals are free and the fruit is a berry. *Hydrocera* is represented by one species from Indomalaysia, *Hydrocera trifolia* (*H. angustifolia*), an erect marsh plant, about 1m (3ft) tall with an angled stem and alternate, narrow leaves. The flowers are borne one to three together. The outer petals are large and concave, and the sepals are colored. The fruit is a red berry. Two further genera of the Balsaminaceae, of somewhat doubtful tax-

Tropaeolaceae. 1 *Tropaeolum majus* (a) prostrate stem bearing peltate leaves, solitary, spurred flowers and fruit—a schizocarp comprising three mericarps (×1); (b) vertical section of mericarp containing a single seed (×2). 2 *Magallana porifolia* (a) stem with deeply palmate leaves, flowers and fruits (×⅔); (b) irregular flower with two of the petals differing from other three and eight stamens (×2); (c) base of flower opened out to show free stamens (×3); (d) gynoecium (×10); (e) winged fruits (×10).

onomic status, are *Semeiocardium*, represented by one species in Indomalaysia, and *Impatientella* with one species in Madagascar.

The presence of a spur in the flowers of the Balsaminaceae has led people to associate the family with the Geraniaceae and Tropaeolaceae, but the spur is strictly an outgrowth of the calyx, whereas in the other families there is evidence that receptacle tissue is involved in spur formation. Therefore the spurs are of different origins and it is doubtful whether the supposed relationship is as close as often stated.

**Economic uses.** The only economic value of the Balsaminaceae lies in the cultivation of *Impatiens* species as greenhouse or pot plants or garden ornamentals. The plants known commercially as "busy lizzies" are hybrids between *Impatiens holstii* and *I. sultanii*, and have white, pink, red or orange flowers and green, red or bronze foliage.                S.A.H.

# TROPAEOLACEAE
### Nasturtiums and Canary Creeper
This small family of climbing succulent herbs includes the cultivated *Tropaeolum majus* (garden nasturtium – not to be confused with the genus *Nasturtium*, family Cruciferae).

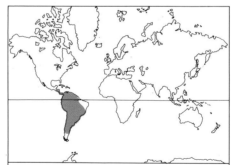

**Number of genera:** 2
**Number of species:** about 90
**Distribution:** mountainous C and S America.
**Economic uses:** ornamentals (garden nasturtium), and pickled seeds of *Tropaeolum majus* are a caper substitute.

**Distribution.** The family is native mainly to the mountains from Mexico to central Chile and Argentina.

**Diagnostic features.** The plants are usually succulent herbs with an acrid mustard oil present in the sap, as in the family Cruciferae. Sometimes root tubers are produced. The stems are prostrate, though frequently climbing by means of sensitive petioles,

which twine around any support in a similar manner to those of *Clematis*. The leaves are alternate, peltate, sometimes deeply lobed, and without stipules. The showy flowers are bisexual, irregular and spurred, and usually borne solitary in the axils of leaves. The perianth consists of a calyx with five distinct sepals, one modified to form a long nectar-spur, and a corolla with five distinct, usually clawed petals, the upper two smaller than the lower three. There are eight stamens. The ovary is superior, formed of three fused carpels, with three locules each containing one axile, pendulous ovule; the single apical style has three stigmas. The fruit is a three-seeded schizocarp, each mericarp separating to become an indehiscent "seed," lacking endosperm. The embryo is straight and has thick fleshy cotyledons.

**Classification.** There are only two genera; *Tropaeolum*, with about 90 species, and *Magallana*, with only one. *Magallana* is native to Patagonia and is named after the Portuguese navigator Fernando Magellan (1480–1521). It differs from *Tropaeolum* in having winged fruits.

The family was at one time placed in the Geraniaceae but is now kept separate, as it differs in having distinct stamens and no

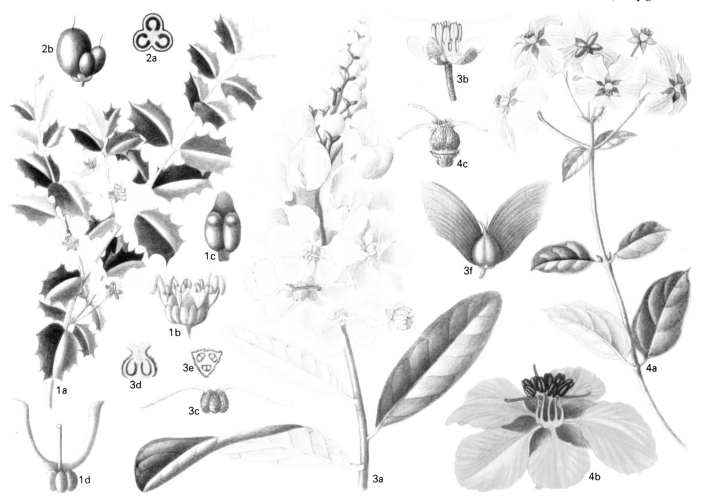

Malpighiaceae. 1 *Malpighia coccigera* (a) flowering shoot ($\times\frac{2}{3}$); (b) flower with petals removed ($\times2\frac{2}{3}$); (c) sepal dorsal view ($\times5\frac{1}{3}$); (d) gynoecium ($\times5$). 2 *M. heterantha* (a) cross section of ovary ($\times6$); (b) fruit ($\times2$). 3 *Acridocarpus natalitius* (a) inflorescence ($\times\frac{2}{3}$); (b) flower with petals removed ($\times2$); (c) gynoecium ($\times2\frac{2}{3}$); (d) vertical section of ovary ($\times3$); (e) cross section of ovary ($\times2\frac{2}{3}$); (f) winged fruit ($\times\frac{2}{3}$). 4 *Sphedamnocarpus pruriens* (a) leafy shoot and terminal inflorescence ($\times\frac{2}{3}$); (b) flower with filament bases united in a ring ($\times2$); (c) gynoecium ($\times3$).

beak on the fruit. Other relationships have been suggested with the Limnanthaceae and Sapindaceae.

**Economic uses.** About eight species are cultivated for ornament; most commonly met with are *Tropaeolum majus*, the garden nasturtium, and *T. peregrinum* (*T. canariense*), canary creeper or canary-bird flower. The unripe seeds of *T. majus* are occasionally pickled and used like capers. The leaves or tubers of some species (eg *T. tuberosum*) are eaten locally.                    S.L.J.

# POLYGALALES

## MALPIGHIACEAE

The Malpighiaceae is a family containing numerous tropical climbers as well as shrubs and trees. The fruits as well as the flowers of many species are highly attractive.

**Distribution.** Members of this family are found in the tropics, especially in South America.

**Diagnostic features.** The leaves are simple, usually opposite, glandular below or on the stalk, and with or without stipules. The flowers are borne in racemes, and are regular or irregular, bisexual (rarely male, female

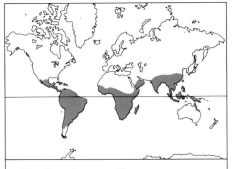

**Number of genera:** 60
**Number of species:** 800
**Distribution:** tropics, especially S America.
**Economic uses:** some ornamentals, a few edible fruits, cordage and occasional local uses.

and bisexual all on the same plant). There are five imbricate sepals each of which often has paired glands at the base, five imbricate, usually clawed petals, and 10 (rarely fewer) stamens with filaments often fused at the base. The ovary is superior, formed of three fused carpels, with only one pendulous ovule on an axile placenta in each of the three locules, and is set obliquely to the central axis

of the flower. The styles are usually distinct. The mature schizocarpic fruit-parts are often winged when they fall apart, or the fruit may be a fleshy or woody drupe. The seeds have no endosperm. The family is often recognizable by a special type of single-celled branched hair.

**Classification.** There are two groups within the family: PYRAMIDOTORAE (torus pyramidal and fruits winged, chief genera *Tetrapteris*, *Banisteriopsis*, *Heteropterys*, *Acridocarpus*, *Stigmaphyllon*); PLANITORAE (torus flat or concave and fruits not winged, chief genera *Malpighia*, *Bunchosia*, *Byrsonima*).

The genus *Stigmaphyllon* comprises some 65 tropical American species. They are woody vines, the leaves having two glands at the top of the leaf stalk. The showy yellow flowers are borne in umbel-like corymbs. There are six anther-bearing stamens and four infertile ones situated opposite each of the four sepals, each of which has two glands. The five hairless petals are unequal in size. The stigmas are leaf-like. The fruits are samaras. *Tristellateia* comprises about 20 Old World species ranging from Africa through Madagascar, southeast Asia and into Australasia. *Banisteria* has been split up by some botanists into the genera *Banisteri-*

*opsis* and *Heteropterys*, both with about 100 species.

The African genus *Acridocarpus* comprises about 50 species, and extends into Madagascar, Arabia and New Caledonia. In the shrubby genus *Camarea*, some eight species from Brazil tending to grow in dry savanna regions, a tough bulbous rootstock has evolved which affords the plant some protection from periodic drought or fire hazards.

The Malpighiaceae has been associated with the Trigoniaceae and Tremandraceae within the order Polygalales and with the Houmiriaceae (Geraniales) and Zygophyllaceae (Sapindales) outside this order. It is usually considered a primitive family within the Polygalales.

**Economic uses.** A number of species make fine ornamentals, particularly in warm temperate and tropical gardens, for example *Stigmaphyllon ciliatum*, *Tristellateia australasiae*, *Hiptage benghalensis*, *Acridocarpus alternifolius*, *A. smeathmannii* and *A. natalitius*, *Heteropterys beecheyana*, *Banisteria laevifolia*, and *Malpighia glabra*. Cordage is made from some species of *Banisteria*. The leaves and shoots of *Banisteria caapi* yield a hallucinatory drug. The leaves of *Hiptage benghalensis* are used in India to treat skin diseases. The fruits of some species of *Malpighia* are edible, being eaten raw or in jellies.                                          B.M.

# TRIGONIACEAE

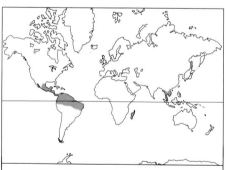

**Number of genera:** 4
**Number of species:** about 35
**Distribution:** tropical S America, Malaysia and Madagascar.
**Economic uses:** wood used locally for furniture.

The Trigoniaceae is a small family of shrubs and trees of no real economic importance.
**Distribution.** The family is native to tropical South America, Malaysia and Madagascar.
**Diagnostic features.** The leaves are opposite or alternate, simple, and generally with small stipules. In some species the bases of the stipules on opposite leaves are fused.

The flowers are irregular, bisexual and borne in racemose inflorescences. There are five overlapping sepals which are either free or united at the base, and five or three unequal petals, the posterior (dorsal) one being the largest. There is a variable number of stamens and staminodes (3–12) with the filaments fused at the base. There is often a disk gland opposite the posterior petal. The ovary is superior and comprises three fused carpels, with three locules (rarely one), each with two to numerous ovules on axile placentas. It is surmounted by a single style. The fruit is a capsule, sometimes winged, rarely a samara, containing seeds which are usually hairy and have a straight embryo and no endosperm.

**Classification.** Criteria used to separate the four genera include the leaf arrangement, the presence or absence of wings on the fruit, the number of ovules per fruit and the number of stamens per flower. The genera are *Euphronia* (three species, tropical America), *Trigonia* (30 species, tropical America), *Trigoniastrum* (one species, Malaysia) and *Humbertiodendron* (one species, Madagascar). Examples of the generic differences can be seen in *Euphronia* which has three to five stamens and two ovules per locule, and *Trigonia* which has up to 12 stamens and an ovary with numerous ovules. *Euphronia* has opposite leaves, while *Trigonia* has alternate leaves.

The family is related to the Polygalaceae and the Vochysiaceae and possibly to the Sapindaceae.
**Economic uses.** *Trigoniastrum* provides wood used locally for furniture.          S.R.C.

# TREMANDRACEAE

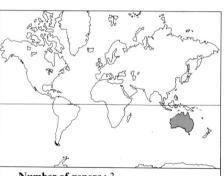

**Number of genera:** 3
**Number of species:** 43
**Distribution:** Australia.
**Economic uses:** ornamentals.

The Tremandraceae is a family of small shrubs, bearing flowers which sometimes combine moderate size with extravagant color.
**Distribution.** The family is confined to Australia.
**Diagnostic features.** With occasionally winged stems and reduced foliage, these shrublets often have glandular or rarely star-shaped hairs. The leaves are alternate, sometimes opposite or whorled, simple, entire or toothed, without stipules and often narrow. The flowers are regular, bisexual, frequently red or purple, solitary and axillary, with three to five free sepals and petals. There are twice as many stamens as petals, the anthers having two or four locules and opening by a single apical pore. The receptacle is sometimes swollen and glandular, lobed between the stamens and petals. The ovary is superior and comprises two fused carpels forming two locules, each with one to three apical, pendulous ovules. The style is slender with a small stigma. The fruit is a compressed capsule opening by slits which follow the locule partitions or are at right angles to them. The seeds have an outgrowth at one end, a straight embryo and copious endosperm, and are sometimes hairy.

**Classification.** The three genera are *Tremandra* (two species), *Tetratheca* (39 species) and *Platytheca* (two species). The family is related to the Pittosporaceae by some botanists, and to the Polygalaceae by others. The habit of growth of many species resembles that of certain species in the Rutaceae or Ericaceae.
**Economic uses.** One or two species of *Tetratheca* are occasionally cultivated in Australian gardens, but the family has no other economic use.                                          B.M.

# VOCHYSIACEAE

The Vochysiaceae is a small family of trees, shrubs, and climbers.
**Distribution.** This family is native to tropical Central and South America and West Africa.
**Diagnostic features.** The leaves are simple and either opposite, in whorls or alternate; stipules are either small or lacking. The flowers are bisexual, slightly irregular, and usually arranged in compound racemes. There are five imbricate sepals, connate at the base, the outer one often the largest and swollen or spurred at the base. The petals vary in number from one to five and are unequal in size. There is only one fertile stamen and two to four staminodes. The anthers have two locules and open longitudinally. The ovary is superior or sometimes attached to the calyx and inferior. It consists of either three fused carpels and three locules or one carpel and one locule; one to numerous axile ovules occur in each locule. There is a single style terminating in a simple stigma. The fruit is a capsule or samara-like, containing one or more seeds which are often winged and sometimes covered with soft hairs. The embryo is straight and the seed is usually without endosperm.

**Classification.** One of the six genera (*Erismadelphus* with three species) is native to tropical West Africa, while the others (*Erisma* with 20 species, *Callisthene* with 10 species, *Qualea* with 60 species, *Salvertia* with one species, and *Vochysia* with 105 species) are all native to tropical Central and South America. Two genera (*Erisma* and *Erismadelphus*) have an inferior ovary and a unilocular fruit, with the sepals adherent to it and persisting as wings. *Erisma* has an ovary containing two ovules, surmounted by a slender thread-like style, while the corolla is reduced to a single petal. *Erismadelphus* has

Vochysiaceae. 1 *Vochysia divergens* (a) leafy shoot and inflorescence ($\times\frac{2}{3}$); (b) winged fruit ($\times 1\frac{1}{3}$). 2 *V. guatemalensis* (a) vertical section of flower ($\times 1\frac{1}{3}$); (b) stamen ($\times 2$); (c) staminode ($\times 4$); (d) cross section of ovary ($\times 2$); (e) vertical section of ovary ($\times 2$). 3 *V. obscura* winged seed ($\times 1$). 4 *Salvertia convallariodora* part of inflorescence ($\times\frac{2}{3}$). 5 *S. convallariodora* flower showing single fertile stamen ($\times\frac{2}{3}$). 6 *Erismadelphus exsul* var *platiphyllus* (a) flower ($\times 4$); (b) vertical section of flower base ($\times 6$); (c) winged fruit ($\times\frac{2}{3}$).

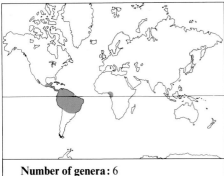

**Number of genera:** 6
**Number of species:** about 200
**Distribution:** tropical C and S America, and W Africa.
**Economic uses:** timber and a fat from the seeds used for making soap and candles.

a single ovule in the ovary, which is surmounted by a very short, stubby style, while the corolla consists of five subequal petals. All other genera have flowers with a superior three-locular ovary which develops into a three-chambered capsule with winged seeds. *Salvertia* has flowers with five subequal petals, and a club-shaped style with a lateral stigma, while *Vochysia* has flowers with one to three or more petals and a style with a terminal stigma. Both *Qualea* and *Callisthene* have flowers with only one petal and an ovary containing numerous ovules. These two genera can be distinguished on the basis of a number of characters including the nature of the axis of the capsule.

The family is related to the Trigoniaceae and Polygalaceae, sharing with them the similarity of the tree, shrub or climbing habit, leaf arrangement, floral structure, one- to three-locular ovary and hairy, non-endospermic seeds.

**Economic uses.** Some members of this family are economically valuable, eg *Vochysia tetraphylla* whose timber is used in furniture-making. The timber of *Vochysia hondurensis* is used in Brazil for boat construction and for fence posts. The seeds of *Erisma calcaratum* are the source of jaboty butter which is used for making candles and soap.

S.R.C.

## POLYGALACEAE

The Polygalaceae is a family of herbs, shrubs, small trees, climbers and even saprophytes. The family is remarkable for the superficial resemblance of the flowers to the well-known papilionaceous flower of the Leguminosae. Over 500 out of about 1,000 species belong to the genus *Polygala*.

**Distribution.** The family is almost cosmopolitan, being absent only from New Zealand and many of the southern Pacific Islands, and the extreme northern parts of the Northern Hemisphere.

**Diagnostic features.** The leaves are usually alternate, always simple, and usually without stipules. The flowers are bisexual, irregular, each subtended by a bract and two bracteoles, and are arranged in spikes or racemes or are solitary. The calyx of five (rarely four to seven) sepals is variously modified, most commonly either with the two lowermost united or with the two inner (lateral) enlarged and often petaloid. The corolla is usually reduced to three petals, with the lowest (median) petal often saucer-shaped and sometimes with a fringed crest. The usually eight stamens are generally joined to the very base of the corolla, with their united filaments forming a split sheath; the anthers are basifixed, usually dehiscing by an apical pore; the pollen grains have a distinctive pattern on their outer wall. A ring-shaped disk is sometimes present inside

Polygalaceae. 1 *Xanthophyllum scortechinii* (a) leafy shoot and irregular flowers (×⅔); (b) flower with petals removed showing free stamens (×1½); (c) petal (×1½); (d) gynoecium (×2); (e) cross section of ovary (×2); (f) globose fruit (×⅔). 2. *Polygala apopetala* (a) inflorescence (×⅔); (b) flower with lateral sepals removed (×2); (c) androecium with filaments united in a split sheath (×3); (d) stamens (×4); (e) gynoecium (×3); (f) vertical section of ovary (×8). 3 *Carpolobia lutea* (a) leaves and fruit—a drupe (×⅔); (b) fruit entire and in cross section (×⅔). 4 *Bredemeyera colletioides* flowering shoot (×⅔). 5 *Securidaca longipedunculata* winged fruits—samaras (×⅔).

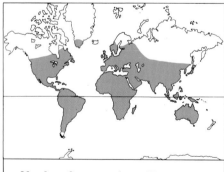

**Number of genera:** about 17
**Number of species:** about 1,000
**Distribution:** almost cosmopolitan.
**Economic uses:** limited local uses.

the base of the staminal whorl. The ovary is superior, usually of two united carpels with a single pendulous ovule on an axile placenta in each of the two locules, although there are various exceptions to this general structure. The style is simple. The fruit is usually a loculicidal capsule, but several smaller genera have exceptions to this; the seeds, sometimes hairy, generally have an aril, and contain a straight embryo and fleshy endosperm (sometimes absent).

**Classification.** The family has been divided into three tribes: the POLYGALEAE (about 12 genera), the MOUTABEAE (four genera) and the XANTHOPHYLLEAE (one genus). *Xanthophyllum*, with about 40 species from the Indomalaysian region, has been treated by some authorities as a distinct family, the Xanthophyllaceae, on account of the almost free stamens and close superficial resemblance to the Caesalpinioideae of the Leguminosae. The other tribes and genera are based primarily on fruit differences: while *Polygala* and the tufted-seeded *Bredemeyera* have capsular fruit, *Securidaca* has a samara, *Monnina* and *Carpolobia* have a drupe and *Atroximia* a nut, one of the carpels being rudimentary in the latter cases.

The nearest family to the Polygalaceae is the Krameriaceae, with compound leaves and five petals. The Leguminosae are not related; similarities are due to parallel evolution; their ovary with its single carpel and the stipulate leaves readily distinguish them.

**Economic uses.** The family is of little importance. Local medicines are extracted from several species; the best-known is snake-root in eastern North America from *Polygala*

*senega*, the constituent glucoside seregin being used by the Indians to cure snake bites. A few species of *Polygala* produce dyes; *P. butyracea* from tropical Africa yields a fiber.

I.B.K.R.

# KRAMERIACEAE

The Krameriaceae is a monogeneric family of shrubs and perennial herbs.

**Distribution.** The family is native to the United States of America, the West Indies and Central and South America.

**Diagnostic features.** The stems are often covered with short, soft or silky hairs, as are the leaves, which are alternate, entire, simple or trifoliolate and without stipules.

The flowers are bisexual and irregular, borne either in the axils of leaves or in terminal racemes. The pedicels bear two opposite leaf-like bracteoles. There are four or five free, imbricate, unequal sepals and five petals, of which the three upper are large with long claws, while the two lower are much smaller and sometimes broad and thick. The three larger petals may be partially fused below. There are three or four stamens, inserted either on the receptacle or on the claws of the three upper petals. The

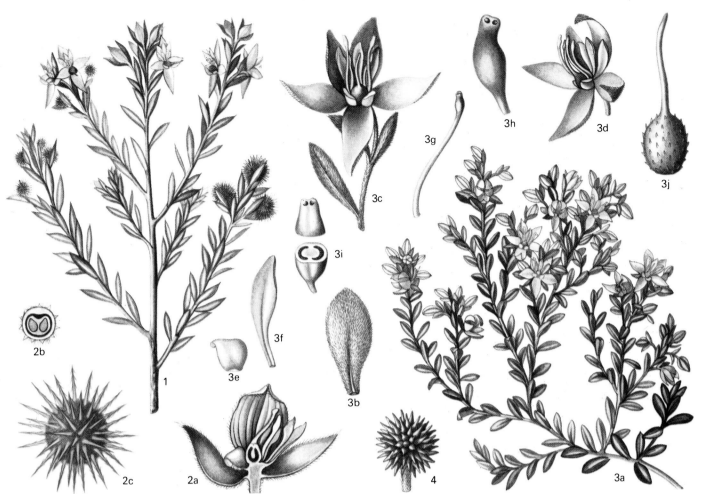

Krameriaceae. 1 *Krameria triandra* shoot with alternate leaves, flowers and fruits (×⅔). 2 *K. tomentosa* (a) half flower (×2); (b) cross section of ovary (×4); (c) bristly, barbed indehiscent fruit (×2⅔). 3 *K. cistoidea* (a) flowering shoot (×⅔); (b) leaf (×3); (c) flower with two opposite subtending bracts, five large sepals, five unequal petals, four stamens and a simple style (×2); (d) flower side view showing petals of two sizes (×2); (e) small anterior petal (×4); (f) large posterior petal (×4); (g) stamen (×4); (h) anther showing porose dehiscence (×12); (i) cross section of anther (×12); (j) gynoecium (×6). 4 *K. argentea* fruit (×1⅓).

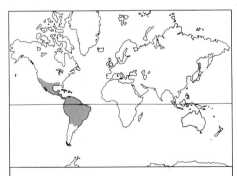

**Number of genera:** 1
**Number of species:** 25
**Distribution:** USA, W Indies, C and S America.
**Economic uses:** roots yield medicinal extracts, tannins and dyes.

anthers have two locules, opening by terminal pores. The ovary is superior, comprising one free carpel with one locule which contains two pendulous, anatropous ovules. The cylindrical style bears a disk-shaped stigma. The fruit is globose and indehiscent, bearing numerous often barbed bristles, and containing a single non-endospermic seed which has a straight embryo.

**Classification.** There is only one genus in this family: *Krameria*. In earlier taxonomic systems it was placed in either the Polygalaceae or in the Leguminosae (subfamily Caesalpinioideae). These relationships were argued partly on the basis of anatomical features and partly on vegetative and floral characters. For example, *Krameria* shares with members of the Polygalaceae the herb and shrub habit, bisexual, irregular flowers and seed with a straight embryo. On the other hand, apart from the absence of stipules, *Krameria* shows many similarities to the Caesalpinioideae, such as irregular flowers with five petals, porose dehiscence of the anthers, and a unilocular superior ovary. However, the distinctive features of *Krameria* such as the form of the petals and the number of stamens probably justify its classification as a separate family.

**Economic uses.** Economically, *Krameria triandra*, a low-growing shrub native to Bolivia and Peru, is important as the source of a medicinal extract made from its dried roots. This extract has astringent properties and was used at one time as a tooth preservative. The root contains a tannin which is used locally for tanning. The root of the tropical American *K. tomentosa* is also used for tanning, while a dye obtained from the roots of *K. parvifolia* (western United States of America and Mexico) is used to color fabrics.

S.R.C.

# UMBELLALES

## ARALIACEAE
*Ivies and Ginseng*

The Araliaceae is a medium-sized family of tropical and temperate herbs, shrubs and trees. Its best-known members are ivy and ginseng.

**Distribution.** Species are distributed throughout the world in both temperate and tropical regions, although the family is mainly tropical, the chief centers being Indomalaysia and tropical America.

**Diagnostic features.** The leaves are usually alternate and often large and compound; they have small stipules and are frequently covered with stellate hairs. In those species with a climbing habit, aerial roots are modified for clinging to the supporting structures. The flowers are regular, small,

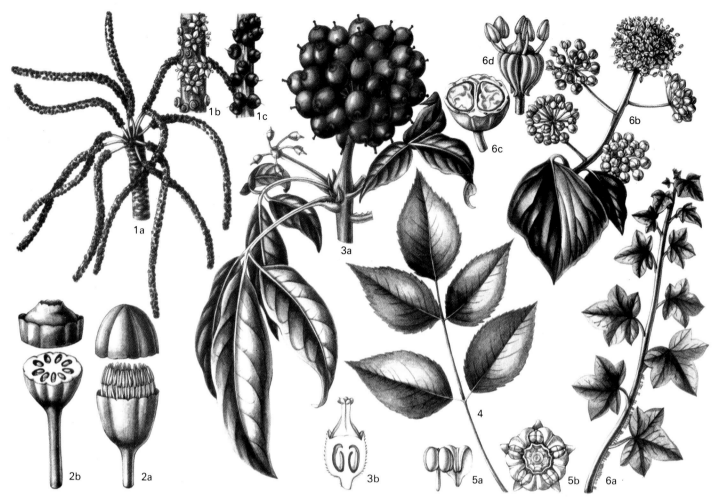

Araliaceae. 1 *Cussonia kirkii* (a) portion of stem crowned by fruiting head ($\times\frac{1}{8}$); (b) part of inflorescence ($\times\frac{2}{3}$); (c) fruit ($\times\frac{2}{3}$). 2 *Tetraplasandra hawaiensis* (a) flower with cap-like corolla which falls off ($\times 2$); (b) cross section of fruit ($\times 2$). 3 *Acanthopanax henryi* (a) shoot with stipulate trifoliolate leaves, and young and mature fruits ($\times\frac{2}{3}$); (b) vertical section of ovary ($\times 3$). 4 *Aralia scopulorum* pinnate leaf ($\times\frac{2}{3}$). 5 *Mackinlaya macrosciadea* (a) petal and stamen ($\times 12$); (b) flower from above ($\times 12$). 6 *Hedera helix* (a) climbing shoot with juvenile leaves and adventitious roots ($\times\frac{2}{3}$); (b) shoot with adult leaves and flowers in umbels ($\times\frac{2}{3}$); (c) cross section of fruit ($\times 2$); (d) flower ($\times 3$).

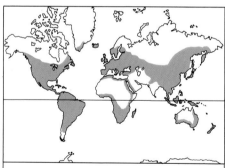

**Number of genera:** about 55
**Number of species:** about 700
**Distribution:** cosmopolitan.
**Economic uses:** ginseng and some medicinal products, rice paper, ivies (*Hedera*) and other species grown as ornamentals.

often greenish or whitish, and aggregated into clusters (compound umbels). They are bisexual or unisexual (then having sexes on separate plants but occasionally male, female and bisexual on the same plant). They have a very small calyx of four or five teeth, fused to the ovary. There are generally five, but occasionally three petals which may be free or partially fused. The stamens are free, equal in number to the petals, and alternate with them. They are attached to a disk which lies on top of the ovary. The latter is inferior and surmounted by five (occasionally fewer or 10) free or fused styles, the bases of which are confluent with the disk. The ovary, of five fused carpels, has five locules, each with a single pendulous ovule. The fruit is a drupe containing five seeds, each with copious endosperm and a small embryo.

**Classification.** Various vegetative and floral characters can be used to separate the most important genera. For example *Kalopanax*, *Fatsia*, *Hedera* and *Tetrapanax* all have simple (but often deeply lobed) leaves, while *Aralia*, *Polyscias*, *Dizygotheca*, *Acanthopanax*, and *Panax* all have compound leaves. In the latter group *Aralia* and *Polyscias* have pinnate leaves, while in the other three genera the leaves are all palmate. *Dizygotheca* has flowers with 10 styles, while *Panax* and *Acanthopanax* have flowers with five styles or fewer.

The family shows relationships with the Umbelliferae and the Cornaceae. Features such as the umbellate inflorescence, small simple flowers with one ovule in each locule in the inferior ovary are common to most members of these three families, although the Cornaceae is now regarded as belonging to a separate order.

**Economic uses.** Economically the family is important for the ginseng (*Panax quinquefolia*) from the roots of which can be obtained an extract with stimulant and supposed aphrodisiac properties. The Chinese also obtain a tonic medicinal product from the roots of *P. repens*. The thin "rice paper" is obtained from the pith of *Tetrapanax papyrifera*. Medicinal extracts have also been obtained from a number of species of *Aralia*, eg *A. cordata* and *A. racemosa*.

There is a number of attractive ornamentals. Many ivies are grown as house plants, in particular the ornamental cultivars of *Hedera helix*. The Canary Island ivy *H. canariensis*, grows to a height of 5m (15ft) in sunny situations. *Fatsia japonica* is also an attractive house and outdoor plant with its glossy green leaves. The bigeneric hybrid of *Fatsia* and *Hedera* (*Fatshedera*) is often grown as a ground-cover plant. Shrubby *Polyscias* and *Acanthopanax* species are used as garden ornamentals. Several ivies are grown outdoors as ground cover. **S.R.C.**

# UMBELLIFERAE

## The Carrot Family

The Umbelliferae or Apiaceae is one of the best-known families of flowering plants, because of its characteristic inflorescences and fruits and the distinctive chemistry reflected in the odor, flavor and even toxicity of many of its members. Several umbellifers were known to the ancient Chinese and Mexican Indian civilizations, as well as to the Mycenaeans Greeks and Romans of the Mediterranean basin. The family was recognized under the name of *Narthekodes* by Theophrastus and the Greek word *Narthex* was replaced by *Ferula* in Latin, the name applied to the dried stalks of umbellifers such as fennel (*Foeniculum*) or *Ferula*. In Greek art Dionysus is often shown bearing a *Ferula* or ferule in his hand. Herbs or condiments such as anise, cumin, coriander, dill and fennel were known to Theophrastus and characterized by their naked seeds and herbaceous stems. The Umbelliferae seems to be the first flowering plant family to be recognized as such by botanists about the end of the 16th century, although only the temperate Old World species were then known. It was also the first group of plants to be the subject of a systematic study published by Robert Morison in 1672.

**Distribution.** The Umbelliferae contains about 300 genera and 2,500 to 3,000 species. The family is found in most parts of the world, although commonest in temperate upland areas and relatively rare in tropical latitudes. The three subfamilies into which it is divided (see below) have characteristic distributions: the largest, the Apioideae, is bipolar but mainly developed in the Northern Hemisphere in the Old World; the Saniculoideae is also bipolar but better represented in the Southern Hemisphere than the Apioideae; the third subfamily, the Hydrocotyloideae, is predominantly a Southern Hemisphere group. About two-thirds of the species of Umbelliferae are native to the Old World but the distribution of the subfamilies in the Old and New Worlds is different, 80% of the Apioideae being found in the Old World, and 60% of the Hydrocotyloideae in the New World, almost 90% of these occurring in South America, where they form a significant component of the flora of temperate southern zones. The subfamily Saniculoideae is almost evenly split between the Old and New worlds. This pattern reflects the long history of evolution and differentiation of this almost cosmopolitan family. Many curious distributions are found in the Umbelliferae: the Canary Island endemic *Drusa glandulosa* is apparently most closely allied to Chilean species of *Bowlesia* and *Homalocarpus*, although no explanation for such a large geographical disjunction has been offered; and the recently discovered *Naufraga balearica* from Majorca finds its closest affinities with South American genera.

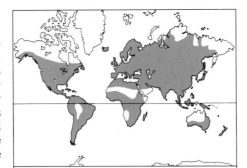

**Number of genera:** about 300
**Number of species:** 2,500–3,000
**Distribution:** near-cosmopolitan, largely temperate uplands.
**Economic uses:** important foods, herbs, spices or flavoring plants (eg carrot, parsnip, celery, parsley, fennel, dill, anise, angelica); source of gum resins, medicines and perfumes; a few ornamentals.

**Diagnostic features.** Most of the Umbelliferae are herbaceous, annuals, biennials or perennials, with hollow internodes; sometimes they are creepers (*Hydrocotyle*), stoloniferous (*Schizeilema*), rosette plants (*Gingidia*) or cushion plants (*Azorella*). Several of the herbaceous species develop some degree of woodiness but genuinely woody tree-like or shrubby species also occur, examples being *Eryngium bupleuroides*, *E. sarcophyllum* and *E. inaccessum* of the Juan Fernández Islands which develop a woody trunk, *Myrrhidendron* species from mountain summits above 3,000m (about 10,000ft) in Central and South America, and several shrubby species of *Bupleurum* (eg *Bupleurum fruticosum*). Several species are spiny, such as the thistle-like *Eryngium* species and the New Zealand species of *Aciphylla* with rigid leaf- and bract-segments tipped by needle-sharp spines.

The leaves are alternate, without stipules, and the leaves usually dissected (ternate or variously pinnate). Entire leaves are found in *Hydrocotyle* and *Bupleurum*, the latter often with parallel venation and frequently resembling monocotyledons.

The main type of inflorescence found in the Umbelliferae is a simple or compound umbel, although sometimes much modified and reduced to a single flower as in some species of *Hydrocotyle* and *Azorella*. In *Eryngium* the flowers are stalkless and crowded into a dense head surrounded by spiny bracts. Dichasia are found in *Petagnia*. The characteristic umbel is a flat-topped inflorescence in which the individual flower stalks (pedicels) arise from the same point on the rays (peduncles) and are of different lengths so as to raise all the flowers to the same height. A compound umbel is one in which the ultimate umbels (termed umbellets or umbellules) are themselves arranged in umbels. Bracts are frequently present at the base of the rays of a compound umbel

forming an involucre, and bracteoles are present at the base of the umbellets where they form an involucel. The bracts and bracteoles vary in number and size.

The flowers of an umbel and of the component umbellules open in sequence from the outer whorls to the center. Most umbels are protandrous (with the stamens maturing before the pistil), although a few genera are protogynous (with the pistil maturing before the stamens), such as *Hydrocotyle* and *Sanicula*. Sexual differentiation in the umbels is quite marked in some cases and varies from genus to genus according to the degree of protandry, ranging from a few male (staminate) flowers per umbel to umbels which are composed of nothing but male (staminate) flowers. Moreover, the percentage of perfect (bisexual) flowers in the latter cases is higher in the primary umbels and progressively lower in the successive umbels until the last umbels produced are almost entirely composed of male flowers. The degree of organization of the umbel is highly developed in some cases and comparable to that found in the capitula of the Compositae. In some umbels whole flowers which are functionally unisexual take on the role of stamens or pistils (eg *Astrantia*, *Petagnia*, *Sanicula*) or the umbels themselves take on the role of stamens or pistils in more complex inflorescences.

The marginal flowers in the umbel are sometimes irregular as in *Daucus carota* (carrot), *Turgenia latifolia* and *Artedia squamata*, thus serving as an attraction to insect pollinators. The visual impact is also enhanced by an increase in the number and size of the umbellules and a closer spacing of the individual flowers. The bracts forming the involucre may also become enlarged, colored and showy as in various *Eryngium* and *Bupleurum* species. One of the most remarkable examples is the Mexican species *Mathiasella bupleuroides*, which has a showy involucre and involucels which are reminiscent of a malvaceous corolla, surrounding umbellules of staminate flowers with petals and naked pistillate flowers.

Most Umbelliferae are "promiscuous" plants in that they are pollinated by a wide array of insects, mostly flies, mosquitoes or gnats, or some of the unspecialized bees, butterflies and moths. Self-fertility is the normal situation and self-sterile plants are very rare. Pollination is often by geitonogamy, that is the pistils may be pollinated by the anthers of adjacent flowers in the same umbel. A curious feature of the family is the almost complete lack of hybridization – hardly any attested records of interspecific hybrids are known. This leads to serious problems in breeding programs.

The flower of the umbellifers is basically uniform, almost monotonously so, consisting of five petals, five free stamens, a greatly reduced calyx, an inferior ovary with two carpels and two locules, and a stylopodium

1

4b

4a

7

6a

6b

9

5

8

10b

10a

2

11

12a

3

12b

supporting two styles. There is a single pendulous, anatropous ovule in each locule. Variations on this basic theme are limited: irregular corollas, the outer petals being sometimes larger and radiate, and unisexuality. A feature showing considerable variation that has been largely overlooked until recently is the stylopodium, the swollen, often colorful nectar-secreting base of the styles which is characteristic of the family. This organ varies widely in shape, size, color and nectar secretion.

The fruit shows a quite remarkable range of variation. Basically it is a dry schizocarp which splits down a septum (commisure) into two one-seeded mericarps which normally remain for some time suspended from a common forked stalk, the carpophore, finally separating at maturity. The outer surface of the mericarp normally has five primary ridges, one dorsal, two lateral and two commisural, and between them four secondary vallecular ridges, all of which run longitudinally from the base to the stylar end of the fruit. In the furrows between the primary ridges, in the ridges themselves or all over the fruit, oil cavities or resin canals (vittae) are often found. Crystals of calcium oxalate may be present in the pericarp. The fruit surface may bear spines, hooks, hairs or tubercles of various kinds; in some fruits the lateral ridges are extended into wings. All these features are related to their dispersal strategy: variations in shape, size, color, wings and spines are numerous; some fruits are remarkable constructions and scarcely bear any resemblance to the basic umbelliferous type, such as those of *Petagnia*, *Scandix* and *Thecocarpus*. The seed has oily endosperm and a small embryo.

**Classification.** The Umbelliferae is usually divided into three subfamilies and several tribes following the system proposed (1897–98) by Drude. Other classifications

Umbelliferae. 1 *Eryngium biscuspidatum* shoot with spiny leaves and bracts, and flowers in a dense head ( $\times\frac{2}{3}$ ). 2 *Centella asiatica* creeping leafy stem bearing axillary flowers ( $\times\frac{2}{3}$ ). 3 *Sanicula europaea* leafy shoot and inflorescences (compound umbels) ( $\times\frac{2}{3}$ ). 4 *Heracleum sphondylium* (a) leaf shoot bearing large inflorescences—note the outer flowers are irregular and have deeply cut petals ( $\times\frac{2}{3}$ ); (b) regular flower from center of inflorescence ( $\times6$ ). 5 *Eryngium maritimum* barbed fruit—a schizocarp ( $\times6$ ). 6 *Petroselinum crispum* (a) schizocarp—comprising two mericarps ( $\times8$ ); (b) cross section of a single mericarp with a central seed and canals (vittae) in the fruit wall ( $\times12$ ). 7 *Psammogeton canescens* schizocarp ( $\times8$ ). 8 *Daucus carota* schizocarp with spines on the ridges ( $\times6$ ). 9 *Artedia squamata* winged schizocarp ( $\times3$ ). 10 *Hydrocotyle vulgaris* (a) schizocarp ( $\times10$ ); (b) cross section of schizocarp showing flattened appearance, and a narrow wall (commisure) between the two single seeded mericarps each with prominent ridges ( $\times10$ ). 11 *Sanicula europaea* schizocarp ( $\times6$ ). 12 *Peucedanum ostruthium* (a) winged schizocarp ( $\times4$ ); (b) cross section of schizocarp ( $\times4$ ).

have been produced by later authors, notably that of the Russian Kozo-Poljansky in 1915 which relied heavily on the anatomical features of the fruit, but they have not gained much acceptance. The three subfamilies are clearly natural but their division into tribes, especially in the subfamily Apioideae, is not entirely satisfactory and is evidently artificial in places. An outline of Drude's system, indicating the main genera, is as follows.

SUBFAMILY HYDROCOTYLOIDEAE
Fruit with a woody endocarp, without a free carpophore; no secretory canals (vittae) or only in the primary ribs; stipules present.
HYDROCOTYLEAE. Mainly Southern Hemisphere. Fruits with a narrow commisure, flattened laterally. *Hydrocotyle*.
MULINEAE. Southern Hemisphere. Fruits with a flattened or rounded back. *Azorella*.

SUBFAMILY SANICULOIDEAE
Fruit with a soft parenchymatous endocarp; base of style surrounded by a ring-like disk; secretory canals various.
SANICULEAE. Ovary with two locules; fruit two-seeded, with broad commissure; vittae distinct. *Eryngium, Astrantia, Sanicula*.
LAGOECIEAE. Ovary with one locule; fruit one-seeded; vittae indistinct. *Lagoecia, Petagnia* (both Mediterranean).

SUBFAMILY APIOIDEAE
Fruit with a soft endocarp, sometimes hardened by woody subepidermal layers; style on apex of disk; stipules absent.
ECHINOPHOREAE. Fruit enclosed by hardened stalks of male flowers. *Echinophora*.
SCANDICEAE. Parenchyma around carpophore with crystal layer. *Scandix, Chaerophyllum, Anthriscus, Myrrhis*.
CORIANDREAE. Parenchyma without crystal layer; fruits usually ovoid-spherical, nut-like, with woody subepidermal layer. *Coriandrum*.
SMYRNIEAE. Mericarps rounded outward. *Smyrnium, Conium, Cachrys, Scaligeria*.
APIEAE (Ammieae). Primary ridges of mericarps all similar; seeds semicircular in sections. *Bupleurum, Pimpinella, Apium, Seseli, Oenanthe, Ligusticum, Foeniculum*.
PEUCEDANEAE. Lateral ridges much broader, forming wings; seeds narrow in section. *Angelica, Ferula, Heracleum, Pastinaca*.
LASERPITEAE. Vallecular ridges on mericarps very distinct, often extended into wings. *Laserpitium, Thapsia*.
DAUCEAE (Caucalideae). Mericarps with spines on ridges. *Daucus, Torilis, Caucalis*.

The Umbelliferae has frequently been associated with the mainly tropical family Araliaceae in the order Umbellales or sometimes united with it into a single family. No clear dividing line can be drawn between the Umbelliferae and Araliaceae and nearly every vegetative or floral feature that characterizes the Umbelliferae can be found in the Araliaceae; there are also similarities in chemistry and pollen characters between them. They probably both arose from a common ancestral stock and have evolved

separately and to some degree in parallel. The Cornaceae in the broad sense is regarded by some authorities as related to the Umbellales and included within it or treated as a separate but parallel order. The whole complex has been held to have arisen from the Hamamelidales, Rosales, Myrtales or Rhamnales and recent studies suggest a relationship between the Cornales and the Rosales on the one hand, and the Umbelliferae with the Sapindales on the other.

**Economic uses.** One of the remarkable features of umbellifers is the wide range of uses made of different species, ranging from food and fodder to spices, poisons and perfumery. However, only the carrot (*Daucus carota*) is a near staple food and also used as animal feed. Carrots and parsnips (*Pastinaca sativa*) are the only umbellifers of international repute as root crops, but other members of the family have been so used, whether cultivated or not, such as the tubers of the great earthnut (*Bunium bulbocastanum*) and the pignut (*Conopodium majus*). Species of *Lomatium*, the largest genus of umbellifers in the United States of America, have been true staple foods for several groups of Indians in the northwest of the country and in western Canada. Stems, petioles and leaves may be used for food as in angelica (*Angelica* and *Archangelica* species), celery (*Apium graveolens*) and lovage (*Levisticum officinale*).

Herbs used for flavoring include chervil (*Anthriscus cerefolium*), fennel (*Foeniculum vulgare*, also used as a salad vegetable) and parsley (*Petroselinum crispum*). Spices derived from fruits or seeds (which contain essential oils) are numerous in the Umbelliferae, examples being dill (*Anethum graveolens*), coriander (*Coriandrum sativum*), cumin (*Cuminum cyminum*), caraway (*Carum carvi*) and anise (*Pimpinella anisum*). Several of these are used for flavoring alcoholic beverages, especially anise.

Many umbellifers have medicinal uses, for gastrointestinal complaints, cardiovascular ailments and as stimulants, sedatives, antispasmodics, etc. They are also a source of gum resins and resins, such as asafetida, derived from *Ferula asafoetida* and other species, the exudate being collected from cuts made at the base of the stem or at the top of the root; another is galbanum, an oleo-gum resin obtained from *Ferula galbaniflua*. There are many poisonous species, the most celebrated being hemlock (*Conium maculatum*), which was responsible for the death of Socrates but is also used as a medicine.

Only a few umbellifers are grown in gardens for ornament. Examples are *Eryngium giganteum* and various cultivars, *Astrantia* (masterwort), *Bupleurum fruticosum*, *Ferula communis* and *F. tingitana*, a variegated form of *Aegopodium podagraria*, and *Heracleum* species (cow parsley), especially the spectacular *Heracleum mantegazzianum* (giant hogweed). V.H.H.

# ASTERIDAE

## GENTIANALES

### LOGANIACEAE
*Buddleias and Strychnine*

The Loganiaceae is a diverse family of trees, shrubs and climbers. They are important to Man, being a source of timber, of well-known ornamentals and of some lethal poisons, notably strychnine.

**Distribution.** Members of the family are found in the tropics, subtropics and temperate zones of the world. They grow in dry lowland habitats, rarely above 3,000m (10,000ft). Though widespread they are seldom abundant, never forming dense stands but usually found singly or in small groups of minor ecological importance.

**Diagnostic features.** The leaves are opposite, entire, pinnately-nerved, often with reduced stipules. The wood sometimes has phloem tubes scattered throughout the xylem (intraxylary phloem). The flowers are regular, bisexual and are borne in terminal cymes or rarely solitary; the four or five lobes of the

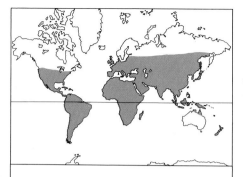

**Number of genera:** about 30
**Number of species:** about 600
**Distribution:** tropics, subtropics and temperate regions.
**Economic uses:** strychnine and curare from *Strychnos* species and several ornamentals, including buddleias.

calyx are always imbricate; the corolla is tubular, four- or five-lobed (occasionally up to 16-lobed) and usually imbricate to varying degrees but sometimes valvate. The corolla-

lobe characters are useful in the classification of the family. There are four or five stamens (rarely 16), attached in a ring to the petals. The ovary is superior, formed of two fused carpels, and in some genera is sunk in a disk. There are two to five locules, each with one to numerous axile ovules. The style is two-lobed. The fruit is a capsule or berry. The seeds have fleshy endosperm and a straight embryo, and are often winged.

**Classification.** The family can be classified conveniently into seven tribes: POTALIEAE (sometimes separated as a family, the Potaliaceae), BUDDLEIEAE (sometimes separated as the Buddleiaceae), ANTONIEAE (sometimes separated as the Antoniaceae), GELSEMIEAE, STRYCHNEAE (sometimes separated as the Strychnaceae), LOGANIEAE and SPIGELIEAE (sometimes separated as the Spigeliaceae). Some authors have used anatomical characters as a basis for the division of the family into two subfamilies, the LOGANIOIDEAE (intraxylary phloem present, superficial cork development, hairs simple) and the BUDDLE-IOIDEAE, comprising the single tribe Bud-

Loganiaceae. 1 *Fagraea lanceolata* (a) flowering shoot ($\times\frac{2}{3}$); (b) five-lobed corolla opened and showing epipetalous stamens ($\times\frac{1}{2}$); (c) calyx tube and style with a capitate stigma ($\times\frac{1}{2}$); (d) fruit—a berry ($\times\frac{2}{3}$). 2 *Buddleia crispa* (a) leafy shoot and terminal inflorescence ($\times\frac{2}{3}$); (b) flower ($\times2\frac{2}{3}$); (c) half flower ($\times3$); (d) stamen ($\times8$); (e) gynoecium ($\times5$). 3 *Strychnos tieute* leafy shoot with axillary inflorescences and coiled spines ($\times\frac{2}{3}$). 4 *Spigelia marilandica* (a) cross section of ovary ($\times4$); (b) fruit ($\times2$). 5 *Logania campanulata* flower ($\times1\frac{1}{3}$).

dleieae (intraxylary phloem absent, cork developing from the pericycle, hairs stellate or glandular).

Chemistry and anatomy have provided some clues to the relationships of the Loganiaceae with other families. It is a very mixed family, probably containing several distinct groups which are no more related to each other than to parts of other families. The Buddleioideae seems to be related to the Scrophulariaceae whereas the Loganioideae resembles more closely the Apocynaceae and Rubiaceae.

**Economic uses.** Many members of the Loganiaceae are extremely poisonous, causing death by convulsions. Poisonous properties are largely due to indo-derivative alkaloids such as those found in *Strychnos*, *Gelsemium* and *Mostuea*. Glycosides in the form of pseudo-indicans are also present, as loganin in *Strychnos*, and the related substance aucubin in *Buddleia*.

Two large genera have more pleasant associations, for they are well-known ornamentals, *Buddleia* and *Fagraea*. The garden and cool greenhouse buddleias are of tropical or subtropical origin, mostly from China. They are shrubs or trees with clusters of fragrant flowers in shades of lilac, orange

or white. *Buddleia davidii* is widely cultivated, its flowers being particularly attractive to butterflies. *Fagraea* is an Asiatic genus with some large trees which yield good timber, eg *Fagraea fragrans*, *F. elliptica* and *F. crenulata*. *F. fragrans* is often planted as an ornamental tree in tropical climates on account of its large showy flowers. Some of the shrubby species have exceptionally large flowers up to 30cm (12in) across (eg *F. auriculata*), which are probably pollinated by bats. S.W.J.

# GENTIANACEAE
## Gentians

The Gentianaceae is a family of annual and perennial herbs (and a few shrubs), including the gentians and other plants with medicinal "bitter principles."

**Distribution.** The family is cosmopolitan. Many species are arctic and mountain rosette herbs, many are found in salty or marshy areas, and some live on decayed vegetation, for example *Voyria* in tropical America and West Africa.

**Diagnostic features.** The plants are usually rhizomatous, and have opposite leaves (alternate in *Swertia*), no stipules, and regular, bisexual flowers borne in cymose

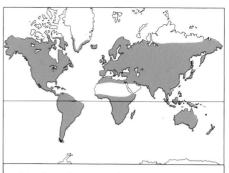

**Number of genera:** about 80
**Number of species:** about 900
**Distribution:** cosmopolitan.
**Economic uses:** many cultivated ornamentals, and several species yield bitter principles used medicinally and in flavorings.

inflorescences. There are four or five (rarely more) united sepals and usually four or five united petals forming a bell- or salver-shaped corolla. The stamens are attached to the corolla and are equal in number to and alternate with the lobes. Occasionally the lobes have long thread-like projections (as in the East African *Urogentias*). The ovary is superior and has two united carpels with a

Gentianaceae. 1 *Chironia purpurascens* (a) flowering shoot ($\times\frac{2}{3}$); (b) corolla opened out showing epipetalous stamens with coiled anthers ($\times 1$); (c) cross section of ovary with two parietal placentas ($\times 3$). 2 *Voyria primuloides* (a) habit ($\times\frac{2}{3}$); (b) flower from above ($\times\frac{2}{3}$); (c) half flower with epipetalous stamens and globose stigma ($\times 3$). 3 *Gentiana depressa* (a) habit ($\times\frac{2}{3}$); (b) ovary with part of wall cut away ($\times 3$); (c) corolla opened out ($\times 1$). 4 *Sabatia campestris* (a) flowering shoot ($\times\frac{2}{3}$); (b) section of ovary ($\times 2$); (c) corolla opened out ($\times 1$).

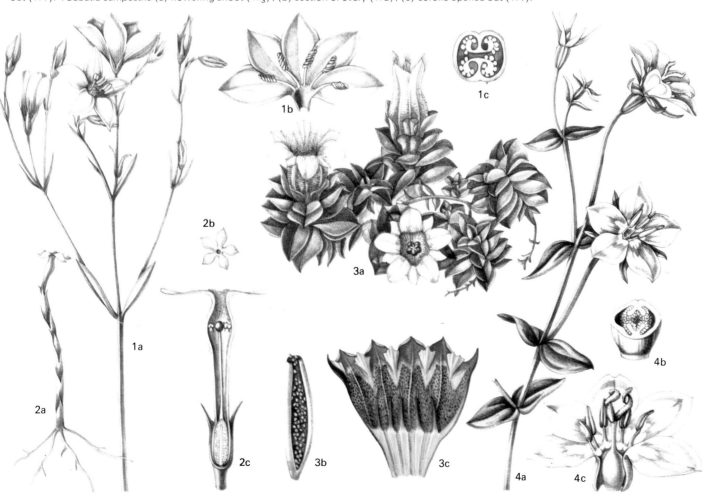

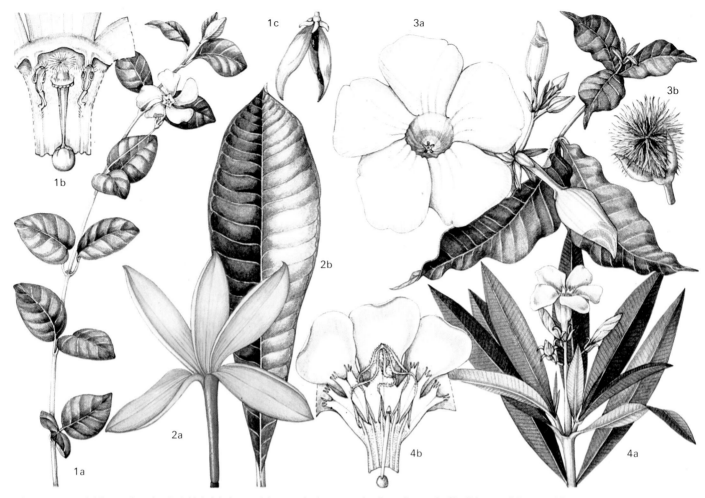

Apocynaceae. 1 *Vinca minor* (periwinkle) (a) shoot with opposite leaves and solitary flower ($\times\frac{2}{3}$); (b) part of dissected flower showing epipetalous stamens and thickened, hairy stigma ($\times$3); (c) paired fruits ($\times$1). 2 *Plumeria rubra* (a) flower ($\times$1); (b) leaf ($\times$1). 3 *Allamanda cathartica* var *grandiflora* (a) flowering shoot ($\times$1); (b) dehiscing fruit ($\times$1). 4 *Nerium oleander* (a) flowering shoot ($\times\frac{2}{3}$); (b) dissected flower with two petals removed, showing the sagittate (arrow-shaped) anthers prolonged into spines which are united at their tips ($\times1\frac{1}{2}$).

glandular disk at the base; there is usually one locule with two parietal placentas, sometimes two locules each with an axile placenta; the ovules are anatropous and usually numerous. The style is simple, and the stigma simple or two-lobed. The fruit is usually a dehiscent capsule, rarely a berry, with small endospermous seeds. The vascular bundles are bicollateral, ie they have phloem strands lying inside the stele as well as outside. Mycorrhiza are very common.

**Classification.** The generic limits in this family are still a matter of debate, but at present the large genera recognized include *Gentiana* (about 400 species, cosmopolitan, except Africa), *Gentianella* (125 species, north and south temperate zones except South Africa), *Sebaea* (100 species, Old World tropics to New Zealand) and *Swertia* (100 species, chiefly north temperate). *Menyanthes*, *Nymphoides* and *Villarsia*, once included in this family, are now referred to Menyanthaceae.

Some authorities regard the Gentianaceae as being allied to the Menyanthaceae. Others see it as being allied to the Loganiaceae, and perhaps Asclepiadaceae and Apocynaceae or even Melastomataceae.

**Economic uses.** Many species of *Gentiana* and *Sabatia* (rose pinks) are cultivated for ornament. The bitter principles of the rhizomes are of medicinal value as in gentian root (*Gentiana*), chiretta (*Swertia*) and, more locally, *Centaurium*. In France a popular aperitif drink, Suze, is made from gentian. In southern Africa, *Chironia baccifera* is fried in butter and applied to sores; it is also a purgative and is said to clear the complexion. Portland powder, formerly a popular remedy for gout, was made up from equal parts gentian, centaury (*Centaurium*), *Teucrium germander*, *Aristolochia clematitis* and *Ajuga chamaepitys*. A yellow dye has been extracted from the seeds of the European *Blackstonia perfoliata*.        D.J.M.

## APOCYNACEAE
### Periwinkles and Oleanders

The Apocynaceae is a large, tropical family of tall rain forest trees, many smaller trees, shrubs and lianas, all usually of rather isolated occurrence, and also a few temperate perennial herbs.

**Distribution.** The family is pantropical, with a few temperate representatives such as *Vinca*. The tropical rain forests and swamps of India and Malaya have small to very tall, evergreen, often buttressed trees such as

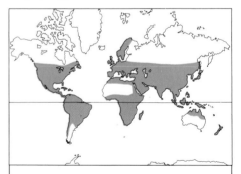

**Number of genera:** about 180
**Number of species:** about 1,500
**Distribution:** throughout tropics, particularly in rain forest regions.
**Economic uses:** drugs, alkaloids, latex (rubbers) and ornamentals such as oleanders and periwinkles.

*Alstonia* and *Dyera*. *Cerbera* and *Ochrosia* are smaller, evergreen trees growing along coasts from Madagascar to northern Australia. Smaller, deciduous trees, such as *Carissa*, *Wrightia* and *Holarrhena* occur sporadically in the deciduous forests of Africa and India. Evergreen trees and shrubs are more common, eg *Rauvolfia* and *Tabernaemontana* (tropical America, India,

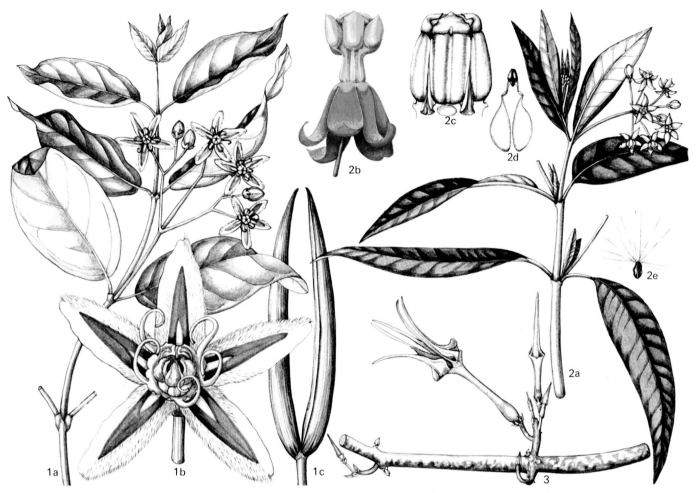

Asclepiadaceae. 1 *Periploca graeca* (a) shoot with opposite leaves and axillary inflorescence ( ×⅔) ; (b) flower with coiled corona filaments ( ×3) ; (c) fruit comprising two follicles ( ×⅔). 2 *Asclepias curassavia* (a) leafy shoot and umbellate inflorescence ( ×⅔) ; (b) flower with down-curved petals and anthers united around the style—a gynostegium ( ×3) ; (c) gynostegium ( ×10) ; (d) pair of pollinia united by thread-like caudicles ( ×14) ; (e) seed with terminal cluster of hairs ( ×10). 3 *Ceropegia stapeliiformis* shoot with flowers in which the corolla tube widens at the base ( ×⅔).

Burma and Malaya) and *Acokanthera* (Kaffir plum). The widely cultivated frangipani (*Plumeria*) originates from Central America. Forests of South America, Africa and Madagascar have many lianas, such as *Landolphia*. The oleanders (*Nerium*) are native to waterside habitats in the Mediterranean region.

**Diagnostic features.** The leaves are simple, opposite or in whorls and rarely have stipules. The sap of all parts of the plant is a milky latex. The inflorescence is a cyme. The flowers are bisexual, regular and are often large, showy and fragrant. There are usually five sepals fused into a tube and five petals also united below into a tube and with free lobes. There are five stamens with the anthers joined to each other. The ovary is superior or half-inferior, formed of two united or free carpels. There are one or two locules, each with two to numerous pendulous, anatropous ovules. The fruits are paired, either fleshy and not splitting or dry and splitting. The seeds may or may not have endosperm and the embryo is straight.

**Classification.** The Apocynaceae is divided into two subfamilies: the PLUMERIOIDEAE (stamens free from the stylar head, anthers full of pollen, seeds usually hairless) and the

APOCYNOIDEAE (stamens firmly joined to stylar head, anthers empty at base, seeds hairy). The family is most closely related to the Asclepiadaceae.

**Economic uses.** Cardiac glycosides are obtained from *Cerbera, Thevetia, Apocynum, Nerium, Strophanthus* and *Acokanthera*. *Strophanthus* seeds yield ouabain and cymarin. *Rauvolfia* produces the alkaloids reserpine and rescinnamine. Latex (rubber) is of commercial importance from some species of *Landolphia, Carpodinus, Hancornia, Funtumia* and *Mascarenhasia*. Ornamentals include *Amsonia, Nerium* (oleanders), *Vinca* (periwinkles), *Carissa* (Natal plum), *Allamanda, Plumeria* (frangipani), *Thevetia* (yellow oleander) and *Mandevilla*.

H.P.W.

## ASCLEPIADACEAE
*Milkweeds and Wax Plant*

The Asclepiadaceae is a fairly large family best known for ornamentals of the genera *Asclepias* (milkweeds and butterfly flowers) and *Hoya* (wax plant).

**Distribution.** This family is principally tropical and subtropical with many representatives in South America. There are several large genera in southern Africa but temperate

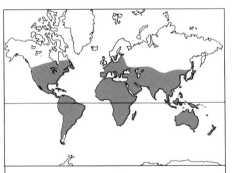

**Number of genera:** about 250
**Number of species:** 1,800 to 2,000
**Distribution:** mainly tropical and subtropical centers in S America and southern Africa.
**Economic uses:** Many ornamentals (eg milkweeds and butterfly flowers), low quality "down" from seeds, and local medicinal uses.

regions generally have very few.

**Diagnostic features.** The Asclepiadaceae is a family of perennial herbs, shrubs, woody climbers or trees, sometimes succulent and usually with milky sap. The leaves are usually opposite or whorled, rarely alternate, simple and generally entire; in some

Oleaceae. 1 *Forsythia viridissima* (a) shoot with flowers borne on previous year's side shoots ($\times\frac{2}{3}$); (b) vertical section of ovary ($\times 4$); (c) part of corolla opened out to show epipetalous stamens ($\times 1$). 2 *Fraxinus platypoda* (a) winged fruit—samaras ($\times\frac{2}{3}$); (b) vertical section of samara base ($\times 2$); (c) vertical section of seed ($\times 2$). 3 *Phyllyrea vilmoriniana* fleshy fruits ($\times\frac{2}{3}$). 4 *Syringa vulgaris* (a) leaves and inflorescence ($\times\frac{2}{3}$); (b) half flower ($\times 1\frac{1}{2}$); (c) cross section of ovary ($\times 6$); (d) corolla opened out to show epipetalous stamens ($\times\frac{2}{3}$); (e) dehisced fruits—bilocular capsules ($\times\frac{2}{3}$).

succulent taxa they are caducous or vestigial. Minute stipules are normally present. The inflorescence is most often a cyme or may be racemose or umbelliform. The flowers are regular and bisexual. The calyx, comprising five partly fused sepals, is deeply divided with the odd sepal in a posterior position. The corolla is formed of five fused petals with the lobes contorted or valvate; both or either the corolla and stamens may bear secondary appendages of various types which form a double or single corona. The filaments are short or may be completely absent and the anthers are usually fused to the gynoecium to form a gynostegium. The anthers have two locules and the pollen of each half-anther is usually united into a waxy pollinium; each pollinium bears a translator or arm for pollen transfer and the two pollinia are united by means of a corpusculum or gland. In the subfamily Periplocoideae the pollen is granular, free and united in tetrads and is transferred by means of the spoon-shaped translator which ends in a sticky disk. The translators and pollinia adhere to the heads and legs of insect visitors by means of the sticky base or the corpusculum and the retiring visitor thus carries away the whole

pollen mass and on visiting another flower may deposit pollen on the stigma surface.

The ovary is semi-inferior and consists of two almost separate carpels, each with a style but a common, five-lobed, large stigma. Each carpel contains numerous ovules in several rows on a single placenta. The fruit consists of a pair of follicles (often only one develops fully). The seeds are usually flattened, ovate to oblong and bear a coma of long, silky hairs; endosperm is present and the embryo is straight.

**Classification.** There are two subfamilies:
PERIPLOCOIDEAE: pollen granular and in tetrads.
CYANCHOIDEAE: pollen massed in pollinia; four tribes: ASCLEPIADEAE, SECAMONEAE, TYLOPHOREAE (which contains succulent members such as *Ceropegia* and *Stapelia*) and GONOLOBEAE.

The Asclepiadaceae is closely related to the Apocynaceae from which it differs by the very specialized androecium and pollen-transfer-system and the presence of a gynostegium.

**Economic uses.** Several genera are grown as ornamentals in warmer areas, eg the milkweeds and butterfly flowers of the genus *As-*

*clepias*, wax plant (*Hoya carnosa*), *Stephanotis floribunda* and many succulents such as *Stapelia, Huernia, Caralluma* and *Ceropegia*.

In some regions, such as the southern United States of America, the coma hairs are used as a low-quality "down" and several species are traditionally considered by the local Indian populations to have medicinal properties (such as emetics and purgatives).

D.B.

## OLEACEAE
*Olives, Ashes and Lilacs*

The Oleaceae is a medium-sized family of trees and shrubs, widely distributed in temperate and tropical regions and containing several genera of economic or horticultural value, eg *Olea* (olive), *Fraxinus* (ash), *Jasminum* (jasmine) and *Syringa* (lilac).

**Distribution.** The family is almost cosmopolitan in range with the 29 genera and 600 species showing diverse distribution patterns but with concentrations in Southeast Asia and Australasia. Genera of restricted distribution include *Abeliophyllum* (Korea), *Amarolea* (eastern North America), *Haenianthus* (West Indies), *Hesperelaea* (northwest Mexico, Guadalupe Islands),

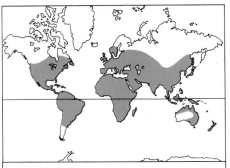

**Number of genera:** 29
**Number of species:** about 600
**Distribution:** almost cosmopolitan, centered in SE Asia and Australasia.
**Economic uses:** olives and olive oil, timber from *Fraxinus* species, and lilac, forsythia, jasmine and other cultivated ornamentals.

*Noronhia* (Madagascar, Mauritius, Comoro islands), *Notelaea* (eastern Australia), *Picconia* (Canary Islands, Madeira, Azores) and *Tessarandra* (Brazil). Others of wide distribution include *Fraxinus* (Eurasia and North America), *Jasminum* (Eurasia, Africa, Australia, Oceania and tropical America) and *Ligustrum* (Europe to northern Iran, Asia, Indomalaysia and New Hebrides).

**Diagnostic features.** All are deciduous or evergreen shrubs, the latter sometimes woody climbers or scramblers. The indumentum characteristically consists of scale-like peltate hairs, in addition to normal hairs, often giving a grayish or silvery appearance to the twigs and leaves. The leaves are usually opposite, without stipules, simple, trifoliolate or pinnate, often entire or lobed. The inflorescence is basically a dichasial cyme but usually modified so as to give the appearance of a raceme, panicle or fascicle. The flowers are bisexual, rarely unisexual (the male and female then being borne on separate plants, either alone or with bisexual flowers). The petals are sometimes absent, usually four but ranging from two to six or twelve, free or united, often with a long or short tube. There are usually four sepals. The stamens are two or four, attached to the petals, alternating with the carpels, with short filaments. The ovary is superior, of two fused carpels, containing two locules usually with two (sometimes one, or four or numerous) anatropous ovules per locule, attached to the apex, side or base of the partition. The style is simple with an entire, bilobed or bifid stigma. The fruits are various – capsule, berry, nut, drupe or samara, dry or fleshy, dehiscent or indehiscent, with one to four seeds.

**Classification.** The most recent classification divides the family into two subfamilies, and seven tribes, as follows:

SUBFAMILY OLEOIDEAE
Ovules two in each locule (rarely more), pendulous. Petals usually in fours, occasionally fives or sixes, sometimes absent.

FRAXINEAE. Fruit dry, indehiscent (samara). Leaves deciduous, imparipinnate, rarely trifoliolate or unifoliolate. Only genus *Fraxinus*.

OLEEAE. Fruit either fleshy (drupe or berry) or a bilocular capsule. Leaves evergreen or deciduous, simple, rarely lobed or pinnatisect. *Syringa, Ligustrum, Olea, Tetrapilus, Linociera, Haenianthus, Tessarandra, Noronhia, Notelaea, Gymnelaea, Amarolea, Osmanthus, Siphonosmanthus, Phillyrea, Picconia, Hesperelaea.*

SUBFAMILY JASMINOIDEAE
Ovules one, four or numerous in each locule, or if two then ascending. Petals four to twelve, never absent.

JASMINEAE. Low shrubs with simple or one- to three-pinnatisect leaves, or erect or scrambling shrubs or woody climbers with imparipinnate trifoliolate or unifoliolate leaves. Corolla large with well-developed tube and four to twelve lobes. Fruit a capsule or berry. *Menodora, Jasminum.*

FONTANESIEAE. Shrubs with deciduous, simple leaves. Corolla with free petals or united at the base in pairs. Fruit indehiscent, compressed, with a surrounding wing. Only genus *Fontanesia.*

FORSYTHIEAE. Shrubs with deciduous simple trilobed or trifoliolate leaves. Corolla gamopetalous, four-lobed. Fruit a leathery or hard capsule or indehiscent compressed with a surrounding wing. *Abeliophyllum, Forsythia.*

SCHREBEREAE. Shrubs or trees with evergreen imparipinnate or simple leaves. Corolla gamopetalous with four lobes. Fruit a woody capsule. *Comoranthus, Schrebera, Noldeanthus.*

MYXOPYREAE. Scrambling climbers. Leaves simple. Corolla fused, with a short tube and four lobes. Fruit fleshy. Only genus *Myxopyrum.*

The relationships of this family are uncertain. It shows affinity with the other families of the Gentianales in which it is placed here but differs, for example, in its placentation and structure of the androecium. It has also been placed along with the Loganiaceae in an order Loganiales or in an order on its own, the Ligustrales. It shows clear signs of derivation from some of the polypetalous families.

**Economic uses.** The main species of economic importance in the family is the olive (*Olea europaea*). Several genera provide valuable trees or shrubs grown in parks or gardens. *Fraxinus* includes the European ash (*Fraxinus excelsior*) which yields a valuable timber, the American or white ash (*F. americana*) which is also commercially important, and the flowering or manna ash (*F. ornus*) which is cultivated in Sicily for its sweet exudate (manna). *Forsythia* comprises several species of popular early-flowering shrubs, the yellow flowers appearing before the leaves, especially *Forsythia suspensa* and *F. viridissima* (both natives of China) and hybrids between them (*F.* × *intermedia*).

*Jasminum*, a tropical and subtropical genus of some 200–300 species, contains several popular widely cultivated species, including the common jasmine, *Jasminum officinale* and winter jasmine, *J. nudiflorum*; some species are cultivated for perfume such as *J. sambac* (Arabian jasmine). *Ligustrum* contains several ornamental trees or shrubs, sometimes used for hedging, including the common privet, *Ligustrum vulgare* (Europe and Mediterranean). *Osmanthus* and *Siphonosmanthus* species are also cultivated for ornament, and the fragrant flowers and leaves of *Osmanthus fragrans* are used to perfume tea in China. *Syringa vulgaris* is the common lilac which is one of the most popular of cultivated shrubs. Other species are grown as garden ornamentals.     V.H.H.

# POLEMONIALES

## NOLANACEAE

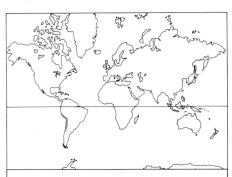

**Number of genera:** 2
**Number of species:** 83
**Distribution:** Chile and Peru.
**Economic uses:** some ornamentals.

The Nolanaceae is a small distinctive family of herbs or low shrubs many species of which are seashore plants with fleshy leaves.

**Distribution.** The family is confined to temperate and subtropical Chile and Peru.

**Diagnostic features.** The leaves are alternate (opposite when in the region of the flowers), simple, often fleshy and glandular-hairy, and are without stipules. The flowers are bisexual, solitary, axillary and regular. There are five connate and persistent sepals. In bud the corolla is pleated like a fan; the five-lobed, blue to pink or white flowers unfold to the shape of a bell or a funnel. The five stamens are unequal, alternating with the corolla lobes; the anthers have two locules, with longitudinal dehiscence. The disk is well developed. The style is single, and the stigma two-to five-lobed. The ovary is superior and has five usually free carpels, which are often divided longitudinally or transversely to give 10–30 segments; placentation is axile. The fruits are one- to seven-seeded nutlets which correspond in number to the divisions of the carpels. The seeds have endosperm and a curved or spiral embryo.

**Classification.** The six species of *Alona* are

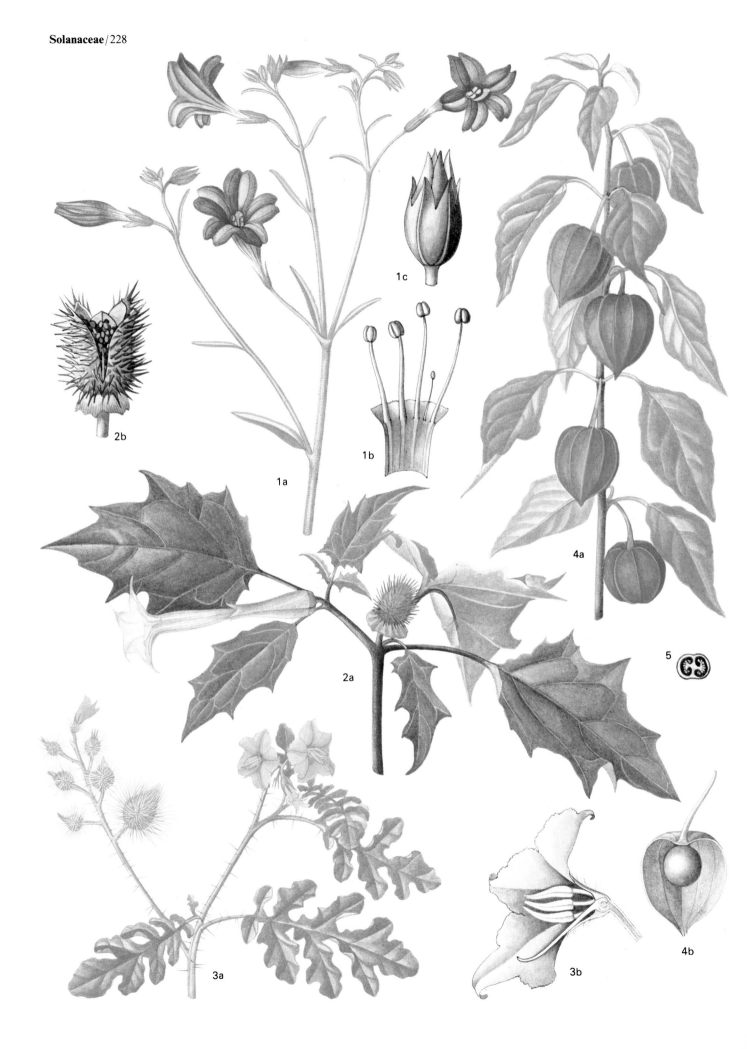

1a

1b

1c

2a

2b

3a

3b

4a

4b

5

low shrubs with the nutlets of the fruit broadly fused to each other. The 77 species of *Nolana* are either herbs or low shrubs, with the nutlets only slightly fused or not fused at all and often further divided longitudinally and transversely.

The family has resemblances to the Convolvulaceae and the Boraginaceae, but is usually considered closest to the Solanaceae.
**Economic uses.** A few species of *Nolana*, but principally the Chilean *Nolana paradoxa*, are grown as garden annuals.　　　　D.M.M.

## SOLANACEAE
*The Potato Family*
The Solanaceae, a cosmopolitan family of herbs and a few shrubs and trees, is one of the most important serving mankind. It contains not only many essential vegetables and fruits such as potatoes, tomatoes, aubergines, paprika, chillies, green and red peppers and Cape gooseberries, but also garden ornamentals such as the petunia. Many species are poisonous (eg deadly nightshade).
**Distribution.** The family is widely distributed throughout tropical and temperate regions. Solanaceous species occur on every continent, but are particularly concentrated in Australia and in Central and South America, where approximately 40 genera are endemic. Its great concentration in South America has led to the hypothesis that the family may have originated in that subcontinent.
**Diagnostic features.** Most Solanaceae are erect or climbing, annual or perennial herbs; some are shrubs (eg *Lycium*, *Cestrum*), and a few are small trees (eg some *Cyphomandra* species and *Dunalia*). The leaves vary greatly in size and shape, and are entire or variously dissected; they are always without stipules and usually alternate. The inflorescence is typically an axillary cyme or combination of cymes, though in some cases it is reduced to a solitary flower (eg *Datura*, *Nierembergia*, *Mandragora*). The flowers are bisexual, usually regular and composed of five (rarely three to ten) sepals and five (rarely up to ten) petals. The sepals are partly fused, usually persistent and often enlarged around the fruit (eg *Physalis*, *Nicandra*). The petals are variously fused, making the corolla round and flat (eg *Solanum*, *Lycopersicon*), bell-shaped (eg *Nicandra*, *Withania*, *Mandragora*) or tubular (eg *Cestrum*, *Nicotiana*); rarely the corolla is two-lipped as in *Schiz-*

Solanaceae. 1 *Salpiglossis atropurpurea* (a) flowering shoot ($\times\frac{2}{3}$); (b) part of flower showing two pairs of unequal stamens and a single infertile reduced stamen ($\times 1\frac{1}{3}$); (c) fruit ($\times 2$). 2 *Datura stramonium* var *tatula* (a) flowering shoot ($\times\frac{2}{3}$); (b) fruit—a capsule ($\times\frac{2}{3}$). 3 *Solanum rostratum* (a) flowering shoot ($\times\frac{2}{3}$); (b) flower with two petals and two stamens removed ($\times 2$). 4 *Physalis alkekengi* (a) shoot showing fruits enclosed in a persistent orange-red calyx ($\times\frac{2}{3}$); (b) calyx removed to show the fruit ($\times\frac{2}{3}$). 5 *Nicotiana tabacum* cross section of ovary showing two locules and axile placentas ($\times 6$).

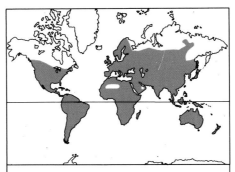

**Number of genera:** about 90
**Number of species:** 2,000–3,000
**Distribution:** cosmopolitan with centers in Australia, Central and S America.
**Economic uses:** food plants (potato, aubergine, tomato, peppers), ornamentals, alkaloids used in medicine and tobacco.

*anthus*. There are five (rarely four to eight) stamens attached to the corolla tube and alternating with the petal lobes. The anthers are usually touching but not fused, equal, though occasionally unequal (eg *Salpiglossis*), bilocular, and dehisce inwards or by terminal pores. The ovary is superior, of two fused carpels with a single style, and usually contains two locules, sometimes more, generally with numerous axile ovules. The fruit, containing many seeds, is either an indehiscent berry (eg *Solanum*, *Atropa*, *Capsicum*, *Saracha*, *Nicandra*), or less commonly a capsule (eg *Datura*, *Hyoscyamus*, *Scopolia*, *Salpiglossis*). The seeds contain copious endosperm.
**Classification.** More than 90 genera, containing between 2,000 and 3,000 species, have been variously included in the Solanaceae. The family can be divided into five tribes, the first three having curved embryos, the last two having straight or only slightly curved embryos.
NICANDREAE. Ovary with three to five locules, the walls dividing the placentas irregularly. Only genus *Nicandra* (one species).
SOLANEAE. Ovary with two locules. Includes *Lycium* (80–90 species), *Atropa* (four species), *Hyoscyamus* (20 species), *Physalis* (100 species), *Capsicum* (50 species), *Solanum* (about 1,700 species), *Lycopersicon* (seven species), *Mandragora* (six species).
DATUREAE. Ovary with four locules, the walls dividing the placentas equally. Only genera *Datura* (10 species), *Solandra* (10 species).
CESTREAE. Five fertile stamens. Includes *Cestrum* (150 species), *Nicotiana* (21 species), *Petunia* (40 species).
SALPIGLOSSIDEAE. Two or four fertile stamens. Includes *Salpiglossis* (18 species), *Schizanthus* (15 species).

The family is closely related to the Scrophulariaceae, from which it is most easily distinguished by the presence of phloem on both the inner and outer sides of the xylem in the vascular bundles, and by the

development of oblique ovaries. The flowers may also be used as a distinguishing feature, those of the Scrophulariaceae being typically irregular, while those of the Solanaceae are usually regular. *Schizanthus* with its highly irregular flowers forms a borderline genus. Although sometimes placed in the Scrophulariaceae, its anatomical features comply with those associated with the Solanaceae. The family also has affinities with the Nolanaceae, Convolvulaceae, Boraginaceae and Gesneriaceae.
**Economic uses.** Various species of *Browallia*, *Brunfelsia*, *Datura*, *Nicotiana*, *Nierembergia*, *Petunia*, *Salpiglossis*, *Schizanthus*, *Solanum* and *Solandra* are cultivated for their showy flowers. Some *Capsicum* and *Solanum* species are widely grown for their colorful fruits, while certain *Cestrum*, *Lycium*, *Solanum* and *Streptosolen* species are popular shrubs. *Physalis* provides the Chinese lantern plant, *Physalis alkekengi*, which is extensively used in dried floral arrangements.

Among the most familiar solanaceous food plants are the potato (*Solanum tuberosum*), aubergine (*S. melongena*), tomato (*Lycopersicon esculentum*) and the peppers (various *Capsicum* species which include paprika, chillies, one kind of pimenta, cayenne pepper, green peppers, red peppers and sweet peppers). Others, popular in tropical America but relatively little known outside this area, include the husk tomato (*Physalis pubescens*), the tomatillo (*P. ixocarpa*), the Cape gooseberry (*P. peruviana*), the tree tomato (*Cyphomandra betacea*), the pepino (*Solanum muricatum*), the cocona (*S. topiro*), the lulita (*S. hirsutissimum*) and the naranjilla or lulo (*S. quitoense*).

Tobacco (*Nicotiana tabacum*), grown extensively for use for smoking, chewing and snuff manufacture, is one of the most popular and yet harmful plants in the world. Many *Nicotiana* species contain the highly toxic alkaloid nicotine, which is used to advantage as a powerful insecticide.

Plants that are both poisonous and of medicinal use are found in most solanaceous genera. Lesser-known examples include *Cestrum*, *Nicandra* and *Physalis*. The notoriously poisonous members, however, are belladonna or the deadly nightshade (*Atropa belladonna*), jimson weed or stramonium (*Datura stramonium*), the mandrake (*Mandragora officinarum*) and black henbane (*Hyoscyamus niger*). These plants have all been used medicinally since earliest times. They contain alkaloids of the tropane group. Steroid alkaloids are characteristic of many *Solanum* and some *Capsicum* and *Lycopersicon* species.　　　　J.M.E.

## CONVOLVULACEAE
*Bindweeds, Morning Glory and Sweet Potato*
The Convolvulaceae is a family of herbaceous and woody, often climbing plants, containing a few members that are important food sources, weeds or ornamentals.

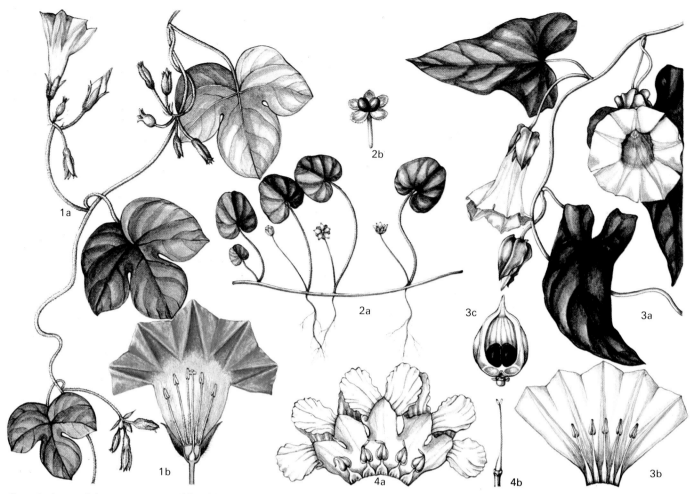

Convolvulaceae. 1 *Ipomoea purpurea* (a) twining stem with axillary flowers ($\times\frac{2}{3}$); (b) flower with corolla opened out showing stamens inserted at its base with a superior ovary surmounted by a thin style and lobed stigma ($\times 1$). 2 *Dichondra repens* (a) habit ($\times\frac{2}{3}$); (b) fruit comprising two mericarps ($\times 2$). 3 *Calystegia sepium* (a) twining stem with solitary axillary flowers ($\times\frac{2}{3}$); (b) corolla opened out to show epipetalous stamens ($\times 1$); (c) fruit with part of wall removed ($\times 1$). 4 *Erycibe paniculata* (a) corolla opened out showing each lobe with two divisions at the apex and stamen filaments with broad bases ($\times\frac{2}{3}$); (b) gynoecium ($\times 3$).

**Distribution.** The family is represented throughout temperate and tropical regions of the world, in a wide range of habitats.

Many representatives have long trailing and twining stems (eg *Ipomoea* and *Calystegia*), and are especially characteristic of rich bushy vegetation, or open, drier places (including sand dunes). In dry, Mediterranean or semidesert climates woody shrubs are more common, many of these possessing trailing or climbing young branches. The woody species are particularly characteristic of tropical or subtropical areas, and in scrub or open woodland large shrubs or even trees over 10m (33ft) high occur. A few species have exploited such habitats as salt marshes, mountain tops and freshwater. *Cuscuta* (dodder) is more or less parasitic on host plants.

**Diagnostic features.** The leaves are alternate, simple, rarely with stipules. The flowers are bisexual, regular, often with an involucre of bracts. They comprise five free sepals (sometimes united), five fused petals and five stamens fused to the base of the corolla tube. The ovary is superior, of two (rarely three to five) fused carpels forming two locules each with two (rarely one or four) axile ovules.

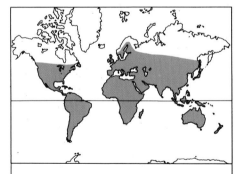

**Number of genera:** about 50
**Number of species:** about 1,800
**Distribution:** cosmopolitan.
**Economic uses:** ornamentals, eg *Ipomoea* (morning glory), food plants (sweet potatoes and yams of the genus *Ipomoea*), and many minor uses.

The style may be simple or forked and the stigmas simple, lobed or capitate. The fruit is a capsule, often dehiscent. The seeds are sometimes hairy, with little endosperm and a curved embryo often with folded cotyledons.

**Classification.** The characters of the ovary, style(s) and stigma(s) are considered important in delimiting three to about ten tribes according to opinion. The most distinct of these, the DICHONDREAE (two genera *Dichondra* and *Falkia*), the HUMBERTIEAE (one genus, *Humbertia*), and the CUSCUTEAE (one genus, *Cuscuta*) are sometimes separated as distinct families, the Dichondraceae, Humbertiaceae and Cuscutaceae, respectively.

The Convolvulaceae is most closely related to the Solanaceae, Boraginaceae and Polemoniaceae.

**Economic uses.** The most important crop plant is *Ipomoea batatas* (sweet potato), hundreds of varieties of which, some known (incorrectly) as yams, are cultivated throughout tropical regions, Japan being a major producer. Many species of this family also have local uses as foods (particularly at times of famine) and as medicines. The roots of *Convolvulus scammonia* (scammony) and of *Ipomoea purga* (jalap) yield a drug used medicinally as a cathartic. A number of species, particularly of the genera *Ipomoea* and *Convolvulus* (some of the bindweeds), are grown as garden ornamentals, notably the morning glory (*Ipomoea purpurea*).     C.A.S.

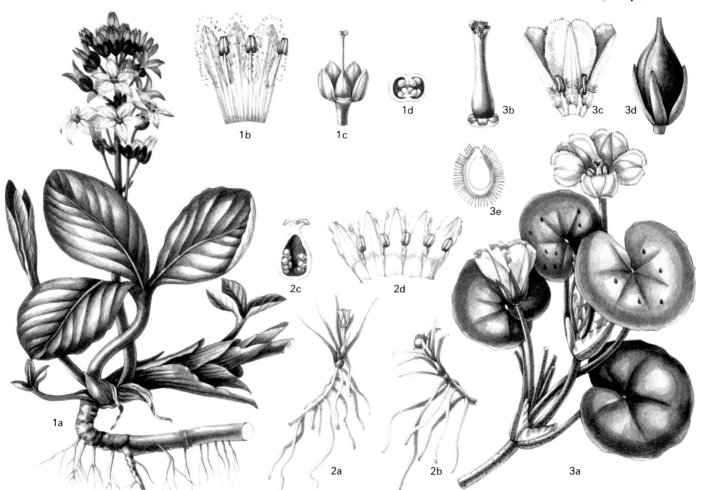

Menyanthaceae. 1 *Menyanthes trifoliata* (a) habit showing trifoliolate leaves and hairy flowers in an erect raceme ($\times \frac{2}{3}$); (b) part of hairy corolla showing epipetalous stamens ($\times 2$); (c) gynoecium and calyx with one sepal removed ($\times 2$); (d) cross section of unilocular ovary with ovules on two parietal placentas ($\times 4$). 2 *Liparophyllum gunnii*, entire plant (a) with flower and (b) with fruit ($\times \frac{2}{3}$); (c) vertical section of gynoecium ($\times 4$); (d) corolla opened out to show epipetalous stamens ($\times 3$). 3 *Nymphoides peltata* (a) shoot with peltate leaves and solitary flowers ($\times \frac{2}{3}$); (b) gynoecium ($\times 2$); (c) part of corolla with basal crests of hairs and epipetalous stamens ($\times 1$); (d) fruit—a capsule ($\times 1$); (e) seed ($\times 3$).

# MENYANTHACEAE

*Bogbean*

The Menyanthaceae is a small family of aquatic or wetland herbs which includes some graceful ornamentals.

**Distribution.** The genus *Nymphoides* is almost cosmopolitan. *Menyanthes* and *Nephrophyllidium* are found in the Northern Hemisphere, while *Liparophyllum* and *Villarsia* are found in the Southern Hemisphere.

**Diagnostic features.** Most species are perennial with tufted rootstocks or horizontal creeping rhizomes, but some species of *Nymphoides* and *Villarsia* are annual. The leaves are alternate, with sheathing petioles, simple, linear to orbicular, or trifoliolate, and without stipules. The flowers are unisexual or bisexual, regular and often heterostylous; they are borne in simple or branched cymes or racemes, or in dense heads or clusters. There are five sepals, united at the base, persistent in fruit and sometimes joined to the ovary below. The petals, also five and united at the base, are yellow, white or pink, and usually have hairs or crests on the inside. The five stamens are fused to the base of the corolla tube and alternate with the corolla lobes; the anthers are versatile. The ovary is

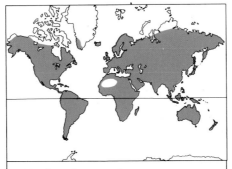

**Number of genera:** 5
**Number of species:** about 40
**Distribution:** almost cosmopolitan.
**Economic uses:** limited local uses as food and in medicine, and several aquatic weeds.

superior, of two fused carpels, with one locule containing two parietal placentas bearing few or numerous ovules. The style is simple and bifid. The fruit is a capsule and is regularly or irregularly dehiscent, or somewhat fleshy and indehiscent. The seeds may be winged, and have copious endosperm and a small embryo.

**Classification.** *Menyanthes* and *Nephrophyl-*

*lidium* both have one species only. They are tough, rhizomatous herbs of wet places. *Menyanthes* has trifoliolate leaves and is widespread in the northern boreal zone, while *Nephrophyllidium* has simple, kidney-shaped leaves and is confined to the north Pacific from Honshu in Japan to Washington in northwest North America. *Liparophyllum* also has one species and it is a small, creeping herb with tufted linear leaves, occuring in Tasmania and New Zealand in mountain bogs. *Nymphoides*, with about 20 species, has the habit of the water lilies (*Nymphaea*) but in many species the flowers appear to arise directly from the petiole; the majority of the species are tropical. *Villarsia*, with about 16 species, is usually found as a terrestrial plant in wet places; the majority of the species are found in Australia and Southeast Asia but one species is found in South Africa. *Nymphoides* and *Villarsia* are distinguished largely by their habit and type of inflorescence, but in Australia, at least, these distinctions are somewhat unsatisfactory and a case could be made for a revision of the genera.

The Menyanthaceae is related to the Gentianaceae, differing in habit, alternate

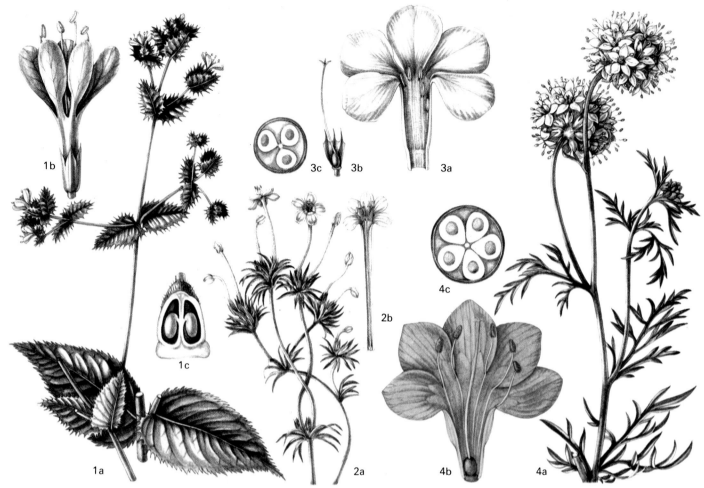

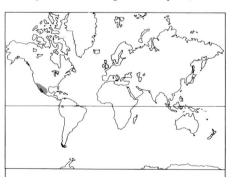

Polemoniaceae. 1 *Loeselia cordifolia* (a) shoot with toothed, opposite leaves and small cymose clusters of flowers ($\times\frac{2}{3}$); (b) flower ($\times 4$); (c) vertical section of ovary showing basal disk and ovules on axile placentas ($\times 12$). 2 *Linanthus androsaceus* (a) flowering shoot ($\times\frac{2}{3}$); (b) flower opened to show anthers inserted at apex of long corolla tube ($\times 1$). 3 *Phlox paniculata* (a) corolla opened to show irregular insertion of stamens ($\times 1\frac{1}{2}$); (b) part of calyx and entire gynoecium ($\times 1\frac{1}{2}$); (c) cross section of ovary ($\times 14$). 4 *Gilia achilleifolia* (a) flowering shoot ($\times\frac{2}{3}$); (b) flower opened out showing insertion of stamens between corolla lobes ($\times 4$); (c) cross section of fruit ($\times 6$).

leaves and valvate arrangement of the petals.
**Economic uses.** *Menyanthes* contains the bitter glucoside menyanthin which is used in medicine as a tonic and to bring down fever. In Scandinavia the leaves are occasionally added to beer or used as a tea substitute. In Russia they are eaten as an emergency food. Several species of *Nymphoides* are found as weeds in ricefields and irrigation channels and some are cultivated as ornamentals, eg *N. cordata*.                                 C.D.C.

## LENNOACEAE
The Lennoaceae is a small family of fleshy herbs parasitic on the roots of shrubs.
**Distribution.** The family is found in deserts, desert scrub and sometimes coastal sand dunes, in southwestern North America; one species of *Lennoa* is recorded from Colombia.
**Diagnostic features.** The Lennoaceae are simple or branched herbs, lacking chlorophyll and brownish when dry. The leaves are reduced to bract-like scales. The flowers are bisexual, regular (rarely irregular), borne in heads or a compact thyrse. The calyx is deeply five- to ten-lobed, the corolla is salver-shaped and has a five- to ten-lobed

limb. There are as many stamens as corolla lobes, inserted in one or two series in the throat of the corolla. The anthers have two locules dehiscing by longitudinal slits, and are usually essentially sessile. The ovary is superior, comprising six to fourteen fused carpels and six to ten locules. Each locule is divided into two parts by a false septum, each part with a single, anatropous, axile

**Number of genera:** 3
**Number of species:** 4–5
**Distribution:** SW N America, Colombia.
**Economic uses:** none.

ovule. The solitary style is simple and the stigma is peltate, crenulate or obscurely lobed. The fruit is a capsule enclosed in the persistent perianth, finally breaking up into 12–28 one-seeded nutlets. The seeds have endosperm and a rounded, rather undifferentiated embryo.
**Classification.** In *Ammobroma* (one species) the inflorescence is a laterally extended head and the sepals have simple, moderately stiff hairs. The other two genera, however, have the flowers borne in axillary cymes to form a compact, spike-like thyrse and the sepals have capitate-glandular hairs: *Lennoa* (two to three species) has eight stamens with partially free filaments borne in two series, while *Pholisma* (one species) has five to seven anthers in one series and with the filaments joined by their whole length to the corolla.

This family was formerly included in the Ericales but on embryological and pollen data is now considered to be close to the Hydrophyllaceae and Boraginaceae.
**Economic uses.** Lennoaceae has no current economic uses, but the underground stems of *Ammobroma* were formerly an important food for the Indians of the southwestern North American deserts.               D.M.M.

# POLEMONIACEAE

*Phlox*

The Polemoniaceae, which includes the showy and familiar *Phlox* widely grown in gardens, is a smallish family with a great diversity of habit, ranging from trees and lianas to small leafless annuals. Of particular interest is the large variety of pollination mechanisms employed by different members.

**Distribution.** The family ranges from the tropics to high latitudes in both hemispheres but is predominantly New World, with most species occurring in North America, particularly the west.

**Diagnostic features.** Polemoniaceae are perennial or annual herbs, less commonly shrubs, lianas or small trees. The leaves are alternate or opposite, simple or compound, and are without stipules. The flowers are bisexual, usually regular and axillary or terminal, solitary or in small cymose clusters of dense heads. There are five sepals, fused into a tube, rarely only at the base. The petals, also five, are fused to form a flat and round, or bell-shaped or funnel-shaped corolla. The five stamens are fused to the corolla tube, alternating with the corolla lobes; the anthers have two locules, with longitudinal dehiscence. The ovary is superior, inserted on the basal disk, comprising three (rarely two or four) fused carpels and the same number of locules, each containing one to numerous anatropous ovules, on axile placentas. The simple style has a stigma with three (rarely two or four) lobes. The fruit is a capsule, usually splitting open along the midribs of the carpels and having one to numerous seeds, usually endospermic. The embryo is straight or slightly curved; the seed-coat is often sticky when wetted.

The family exhibits a notable array of pollination mechanisms. Pollination by bees is most widespread in North American Polemoniaceae, from which pollination by humming birds (eg *Gilia, Ipomopsis, Loeselia, Polemonium*), flies (*Gilia, Linanthus, Polemonium*) and beetles (*Ipomopsis, Linanthus*) seems to have arisen independently

in various genera. Butterfly and moth pollination is known as an advanced condition in several genera, while in the tropical tribes pollination by bats (*Cobaea*), hummingbirds (*Cantua, Loeselia, Huthia*) and hawkmoths (*Cantua, Cobaea*), as well as bees (*Bonplandia, Loeselia, Huthia*), is known.

**Classification.** The family is divided into five tribes:

COBAEEAE. Lianas with pinnate leaves, flowers solitary and calyx regular. Tropical America. Only genus *Cobaea*. This tribe is sometimes regarded as a separate family, the Cobaeaceae.

CANTUEAE. Small trees and shrubs with simple or pinnatifid leaves, flowers in axillary or terminal clusters and calyx regular. Confined to the northern Andes. *Cantua, Huthia*.

BONPLANDIEAE. Shrubs or herbs with simple to pinnatifid leaves, and flowers which may be solitary, in pairs or clustered; calyx slightly irregular. Tropical America. *Bonplandia, Loeselia*.

POLEMONIEAE. Perennial (rarely annual) herbs, with simple to pinnate leaves and calyx usually herbaceous; corolla regular and stamens inserted (often very irregularly) in the corolla tube or throat. Temperate America and Eurasia. *Polemonium, Allophyllum, Collomia, Gymnosteris, Phlox* and *Microsteris*.

GILIEAE. Annual or rarely perennial herbs, with simple, entire or much-dissected leaves; calyx with membranous sinuses which rupture in age; corolla regular or irregular and stamens inserted, usually regularly, in the throat of the corolla-tube or between lobes. Arid southwestern North America, extending to moister areas and into warm temperate South America. *Gilia, Ipomopsis, Eriastrum, Langloisia, Navarretia, Leptodactylon* and *Linanthus*.

The five tropical genera (tribes Cobaeeae, Cantueae, Bonplandieae) are mostly woody plants with large to medium corollas, winged seeds with little or no endosperm and large fleshy cotyledons. The 13 temperate genera (tribes Polemonieae, Gilieae), on the other hand, are mostly herbaceous with medium to small corollas, wingless, endospermic seeds and smaller cotyledons.

The Polemoniaceae is most closely related to the Hydrophyllaceae, the two families together constituting a group within the Polemoniales which in turn has affinities with a group comprising the Nolanaceae, Solanaceae and Convolvulaceae.

**Economic uses.** Many species of *Polemonium, Phlox* and *Gilia* are cultivated as garden ornamentals for their colorful flowers.                D.M.M.

# EHRETIACEAE

The Ehretiaceae is a medium-sized family of trees and shrubs (and a few herbs) chiefly important economically for a number of trees that provide strong timber.

**Distribution.** Species are distributed through-

out the tropics and subtropics. The main centers of distribution are in Central and South America.

**Diagnostic features.** The leaves are simple, alternate, entire or with toothed margins and without stipules. The inflorescence is usually cymose and the flowers are regular and bisexual. The calyx consists of five sepals which are fused into a long or deeply divided tube. They are usually green and leaf-like, but sometimes membranous and enlarged, enclosing the fruit. In most species they are persistent. The petals are fused to form a corolla tube with five (occasionally four or six) lobes. The five stamens are attached to the corolla tube and alternate with the lobes. The ovary is superior, consisting of two to four fused carpels and two to four locules, each containing paired ovules generally attached basally to the axis. It is usually surmounted by a single terminal style with two or four branches or lobes at the tip; some species have two free styles each with a simple stigmatic tip. The fruit is a drupe often enclosed by the persistent calyx. The seeds may be endospermic or non-endospermic.

**Classification.** The genera can be separated on various vegetative and floral features, including the form of the calyx, style and fruit. For example, both *Cordia* and *Patagonula* have two-lobed stylar arms but the calyx of the former is only slightly enlarged in the fruit, while that of the latter is much enlarged. Both *Rochefortia* (spiny shrubs) and *Coldenia* (herbs or shrubs with hairy leaves) have two free styles. *Ehretia, Cortesia* and *Saccellium* all have flowers with a single style but only in the latter genus does the calyx become enlarged to envelop the fruit. *Halgania* has the anthers united into a cone around the style and in *Auxemma* the winged calyx surrounds the fruit.

There is no doubt about the affinity of this family to the Boraginaceae, and in fact some botanists include it within the latter.

**Economic uses.** A number of species of

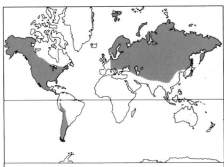

**Number of genera:** 18
**Number of species:** about 300
**Distribution:** mainly N America, but also Chile, Peru, Europe and N Asia.
**Economic uses:** many ornamentals, eg *Phlox, Polemonium* and *Gilia*.

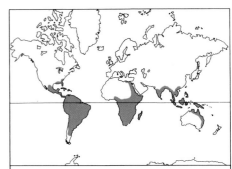

**Number of genera:** 13
**Number of species:** about 400
**Distribution:** tropics and subtropics with centers in C and S America.
**Economic uses:** valuable timber, some edible fruits, and a few medicinal decoctions of limited local value.

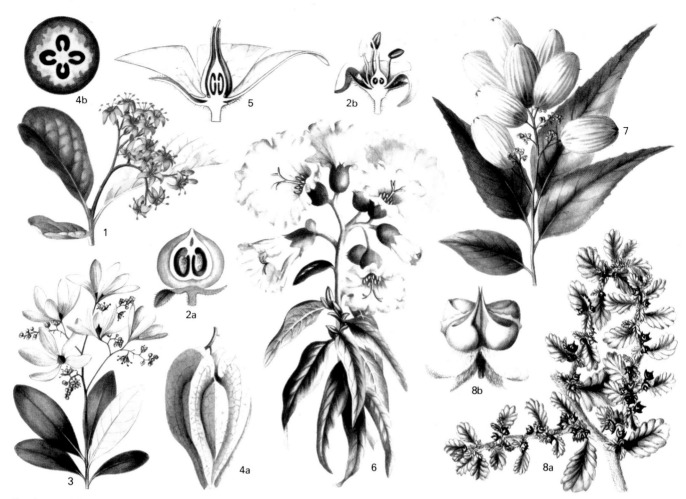

Ehretiaceae. 1 *Ehretia mossambicensis* leafy shoot and inflorescence ($\times\frac{2}{3}$). 2 *Ehretia buxifolia* (a) vertical section of fruit ($\times4$); (b) half flower ($\times4$). 3 *Patagonula americana* leafy shoot with flower and fruit having a persistent star-like calyx ($\times\frac{2}{3}$). 4 *Auxemma oncocalyx* (a) fruit surrounded by five-winged calyx ($\times\frac{2}{3}$); (b) cross section of fruit ($\times1\frac{1}{3}$). 5 *Halgania littoralis* half flower showing anthers united around the ovary ($\times2$). 6 *Cordia decandra* leafy shoot and inflorescence ($\times\frac{2}{3}$). 7 *Saccellium lanceolatum* leafy shoot with flowers and fruits that are surrounded by an inflated calyx ($\times\frac{2}{3}$). 8 *Coldenia procumbens* (a) leafy shoot and fruit ($\times\frac{2}{3}$); (b) fruit that separates into four nuts ($\times6$).

*Cordia* (*Cordia gerascanthus*, *C. alba*, *C. dodecandra* and *C. alliodora*, native to tropical Central and South America) are valued for their timber which is used for making furniture and the beams and doors in houses. The fruit of some species (*C. gharaf* and *C. rothii* in Africa and India and *C. sebestina* in Mexico and the Caribbean) is edible. A decoction from the leaves and fruits of some species (eg *C. boisseri*) is used as a treatment for cold symptoms. The wood of *Ehretia elliptica* is used in Mexico for making the handles of agricultural implements, and the leaves of *E. philippensis* are used in the Philippines as a curative for intestinal disorders including dysentery. The South American genus *Patagonula* contains at least two species (*Patagonula americana* and *P. batensis*) valued for their timber which is used both for building and furniture.

S.R.C.

## HYDROPHYLLACEAE

The Hydrophyllaceae is a smallish but widely distributed family of herbs and small shrubs containing several attractive temperate garden plants such as species of *Nemophila*, *Wigandia* and *Phacelia*.

**Distribution.** The family is almost cosmopolitan.

**Diagnostic features.** The members of the family are annual or perennial herbs or undershrubs. The leaves are alternate (rarely opposite), usually hairy or glandular, simple or compound, and without stipules. The usually blue to purple flowers are often borne in cymes or cincinni, and are bisexual and regular, with five free sepals with or without auricles between the lobes. The five-lobed corolla is fused at the base and is wheel- or bell-shaped or funnel-like, with five stamens inserted at the base of the corolla tube. The stamens often have a pair of small appendages at the base of each filament, and these protect nectaries which serve to attract pollinating bees. The ovary is superior, comprising two fused carpels and one or two locules, each containing two to many sessile or pendulous ovules, on parietal placentas when unilocular, or on axile placentas when bilocular. There are one or two styles. The fruit is a loculicidal capsule, and the many seeds have a small straight embryo and fleshy endosperm.

**Classification.** The family is a homogeneous one, being split into genera chiefly on the basis of technical fruit and seed characters. Its closest relation is the Polemoniaceae.

**Economic uses.** The Hydrophyllaceae contains a number of ornamental genera such as *Nemophila* and *Phacelia*, grown in gardens. Some of the more attractive phacelias are *Phacelia sericea* with silvery, silky foliage and bluish purple or white corollas; *P. campanularia* with deep blue corollas and *P. tanacetifolia* with hairy parts, fern-like

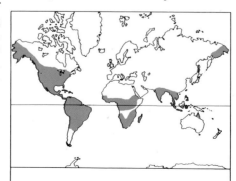

**Number of genera:** about 18
**Number of species:** about 250
**Distribution:** cosmopolitan.
**Economic uses:** several ornamentals.

Hydrophyllaceae. 1 *Phacelia minor* shoot with leaves and flowers ( ×⅔). 2 *P. tanacetifolia* half flower showing appendages on bases of filaments
( ×4). 3 *P. franklinii* (a) cross section of ovary with ovules on parietal placentas ( ×18) ; (b) dehiscing capsule with many seeds ( ×1½).4 *Hydrolea
floribunda* stamen before (left) and after (right) dehiscence ( ×6). 5 *H. spinosa* (a) half flower ( ×4) ; (b) cross section of ovary showing ovules
on placentas adnate to the septa ( ×10). 6 *Hydrophyllum virginianum* (a) shoot with pinnate leaf and flowers in a head-like cyme ( ×⅔); (b)
dehiscing capsule with two seeds ( ×4).

compound leaves, and compact cymes of blue flowers; all are North American.

*Hydrophyllum*, from the Greek, *hydor*: water and *phyllon*: leaf – a reference to the watery appearance of the foliage of some species – is the type-genus of the family. They are chiefly perennials with mainly basal leaves, and greenish, white or violet flowers, with protruding stamens, borne in open or head-like cymes. *Wigandia* contains six species from the American tropics; several are planted for their bold foliage in bedding displays, eg *Wigandia caracasana*, with long leaves and violet and white flowers. *Nama* species include the perennial *Nama rothrockii*, a small attractive purplish-headed species with dentate leaves from dry sandy habitats in western Nevada and California. *Draperia systyla* from California is a softly hairy, diffuse perennial herb, with heads of pale violet flowers, and *Lemmonia californica*, also from California, makes an attractive annual with small white flowers and hairy leaves.                                    B.M.

# BORAGINACEAE
*Forget-me-not and Alkanet*
The Boraginaceae is a relatively large family

of annual to perennial herbs, shrubs, trees and a few lianas. About 30 genera are used as ornamentals and several species are of medicinal value or used as dyes or herbs.
**Distribution.** The family is found throughout temperate and subtropical areas of the world with a major center of distribution in the

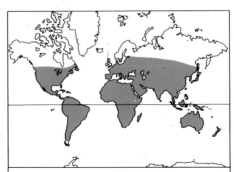

**Number of genera:** about 100
**Number of species:** 2,000
**Distribution:** throughout temperate and subtropical regions, but centered in the Mediterranean region.
**Economic uses:** ornamentals (eg heliotrope, forget-me-not) pot-herbs (alkanet, comfrey) and red dye.

Mediterranean region. It is less frequent in cool temperate and tropical regions.
**Diagnostic features.** The stems are usually covered in rough hairs which also occur on the leaves and in the inflorescences. The leaves are generally alternate, simple, usually entire, without stipules and often bearing cystoliths. The inflorescence is very characteristic of the family and usually consists of one or more determinate scorpioid or helicoid cymes (cincinni) which uncoil progressively as the flowers open. The flowers are regular (irregular in *Echium* and some related genera) and usually bisexual although female flowers borne on separate plants are quite frequent. There are five sepals, free or connate at the base, sometimes unequal in size. The corolla has five lobes and is salver-shaped to bell-shaped, the tube often with scales at the base or mouth. There are five stamens inserted on the corolla, sometimes unequal, often with basal appendages. The anthers have two locules and the filaments often have a nectariferous disk at the base. The ovary is superior, of two fused carpels, with two locules often becoming four by means of false septa. There is one erect, ascending or

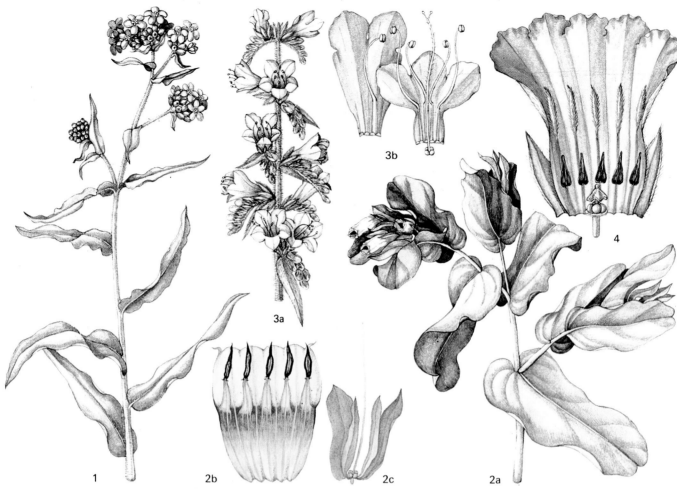

Boraginaceae. 1 *Anchusa officinalis* leafy shoot and inflorescence of regular flowers ($\times\frac{2}{3}$). 2 *Cerinthe major* (a) leafy shoot and inflorescence ($\times\frac{2}{3}$); (b) corolla opened out to show five epipetalous stamens ($\times 2$); (c) calyx and four-lobed gynoecium with a gynobasic style, ie arising from the base of the ovary between the lobes ($\times 2$). 3 *Echium vulgare* (a) inflorescence of irregular flowers ($\times\frac{2}{3}$); (b) flower dissected to show epipetalous stamens and four-lobed ovary with a thin, gynobasic style which is forked at its tip ($\times 2$). 4 *Heliotropium* sp. flower opened out with five arrow-shaped stamens and the ovary with a terminal style and an umbrella-shaped expansion below the stigma ($\times 2$).

horizontal, anatropous ovule in each locule. The single style is gynobasic or terminal, usually simple or capitate but two- or four-lobed in some genera. The fruit consists of four (rarely two) nutlets or is a drupe. The seeds are with or without endosperm and with a curved or straight embryo.

The flowers are predominantly insect-pollinated, with blue, white, pink or yellow flowers. Many have pendulous, bee-pollinated flowers, such as those of *Borago* and *Symphytum*, and several outbreeding mechanisms are to be found, such as hetero-styly in *Pulmonaria* and separate female flowers in *Echium*; some species are self-incompatible.

**Classification.** The family can be divided into two subfamilies:

HELIOTROPIOIDEAE. Style terminal, simple or bilobed, with a ring of hairs near the tip; fruit a drupe; seed with endosperm (*Heliotropium*, *Tournefortia*).

BORAGINOIDEAE. Style gynobasic; fruit of two or four separate nutlets. This subfamily can be further subdivided into five tribes, mainly on style and fruit characters: CYNO-GLOSSEAE (eg *Omphalodes*, *Cynoglossum*, *Rindera*); ERITRICHIEAE (eg *Echinospermum*, *Eritrichium*, *Cryptanthe*); BORAGINEAE (eg *Symphytum*, *Borago*, *Anchusa*); LITHO-SPERMEAE (eg *Myosotis*, *Lithospermum*, *Arnebia*) and ECHIEAE (*Echium*).

The family is included in the order Polemiales with other tubiflorous families such as the Hydrophyllaceae, Polemon-iaceae and Convolvulaceae. The Ehretiaceae is often regarded as comprising two sub-families of the Boraginaceae (Cordioideae and Ehretioideae). Other authorities treat all the subfamilies of the Boraginaceae as separate families.

**Economic uses.** Several genera are cultivated for ornament, for example *Heliotropium* (heliotrope), *Mertensia* (Virginia bluebells), *Myosotis* (forget-me-not), *Pulmonaria*, *Echium*. *Symphytum officinale* is commonly used as a pot-herb (comfrey) and *Alkanna tinctoria* (alkanet) is a source of red dye used to stain wood and marble and to color med-icines, wines and cosmetics. The borage (*Borago officinalis*) is a traditional garden herb used since the Middle Ages for its reputed medicinal and culinary value, and to flavor drinks, but today is grown more for its attractive, bright blue flowers and as a source of nectar for bee-feeding.                D.B.

# LAMIALES

## VERBENACEAE
*Teaks and Verbenas*

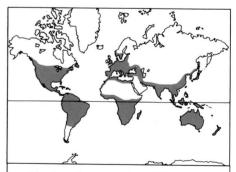

**Number of genera:** about 75
**Number of species:** over 3,000
**Distribution:** tropical and subtropical, a few temperate.
**Economic uses:** timbers (teaks), essen-tial oils, teas, herbal medicines, fruits, gums, tannins and ornamentals.

This large family contains a number of useful plants (notably teak) and ornamentals. The Verbenaceae includes herbs, shrubs, trees

Verbenaceae. 1 *Clerodendrum thomsoniae* (a) shoot with cymose inflorescence of flowers which have an inflated, winged calyx ( ×$\frac{2}{3}$) ; (b) fruit—a four-lobed drupe ( ×1). 2 *Verbena chamaedrifolia* (a) shoot bearing opposite leaves and a terminal raceme ( ×$\frac{2}{3}$) ; (b) flower with irregular corolla ( ×2) ; (c) corolla opened out to show epipetalous stamens ( ×6) ; (d) gynoecium showing lobed ovary surmounted by a single style with a lobed stigma ( ×3). 3 *Vitex agnus-castus* (a) shoot bearing digitate leaves and cymose inflorescences ( ×$\frac{2}{3}$) ; (b) flower ( ×2) ; (c) corolla opened out to show stamens of two lengths (didynamous) ( ×3) ; (d) fruit ( ×4) ; (e) cross section of fruit showing four locules ( ×4).

and many lianas; several members are thorny xerophytes.

**Distribution.** The family has an almost entirely tropical and subtropical distribution, with a few genera and species native to temperate zones.

**Diagnostic features.** The leaves are usually opposite, rarely whorled or alternate, entire or divided and without stipules. The flowers are usually irregular and bisexual, arranged in racemose or cymose inflorescences. The calyx is four- or five-lobed or toothed, the corolla is tubular and also four- to five-lobed. The stamens are four (rarely two or five), alternating with the corolla lobes, with bilocular anthers opening lengthwise. The ovary is superior, consisting of two (rarely four or five) fused carpels normally divided early into four (or more) locules by formation of false septa. Placentation is axile, with two erect, rarely pendulous, ovules per carpel, leading to one ovule in each locule after formation of the false septa. The style is terminal, occasionally arising from between the ovary lobes. The fruit is a drupe, less commonly a capsule or schizocarp. The seeds have a straight embryo and little or no endosperm is present.

**Classification.** The family is basically classified according to the type of inflorescence. The VERBENOIDEAE, with a racemose spicate inflorescence includes the genera *Verbena*, *Lantana*, *Lippia*, *Priva* and *Citharexylum*. The group with cymose inflorescences, which frequently occur as panicles and corymbs, or may even be reduced to a single axillary flower, is divided according to fruit structure into the VITICOIDEAE, the NYCTANTHOIDEAE and the CARYOPTERIDOIDEAE. The fruit of the Viticoideae is a drupe, and genera include *Tectona*, *Vitex*, *Callicarpa* and *Clerodendrum*. Those genera whose fruit is a capsule with two locules, valves and seeds, such as *Nyctanthes*, are included in the Nyctanthoideae. The Caryopteridoideae have a four-valved, capsule-like fruit, eg *Caryopteris*.

The woody Verbenaceae is generally considered to be closely related to the herbaceous Labiatae, although the Labiatae shows a constancy in pollen-type, while the Verbenaceae shows great variation.

**Economic uses.** The family includes many genera of economic value. The most important is *Tectona grandis* (Southeast Asia), the source of teak, a durable and water-resistant timber much used in shipbuilding. *Citharexylum* (Mexico and South America), commonly called zither wood, is used to make musical instruments. *Vitex celebica* produces a fine timber and *V. agnus-castus* (chaste tree or monk's pepper tree) is the source of a valuable oil; other species yield edible fruits, gum and tannin. *Petitia* is another timber-producing genus, as is *Premna* (Malaya), which produces a beautifully veined wood used by the Japanese for knife handles. The South American shrub *Lippia citriodora*, lemon verbena, bears densely glandular, scented leaves which are used in herbal teas. Other species yield valuable essential oils.

A number of genera are cultivated for their ornamental value, including *Lippia citriodora*, with lemon-scented foliage; *Lantana camara*, a half-climbing shrub, the flowers of which open as pink and yellow and change to red and orange; *Holmskioldia sanguinea*, known as the Chinese hat plant for its large spreading calyx; *Verbena* species for their showy blooms; and species of *Petrea*, *Clerodendrum*, *Vitex*, *Caryopteris* and *Callicarpa*. *Verbena officinalis* (vervain) is used for a number of herbal remedies, eg to treat skin diseases.                 S.A.H.

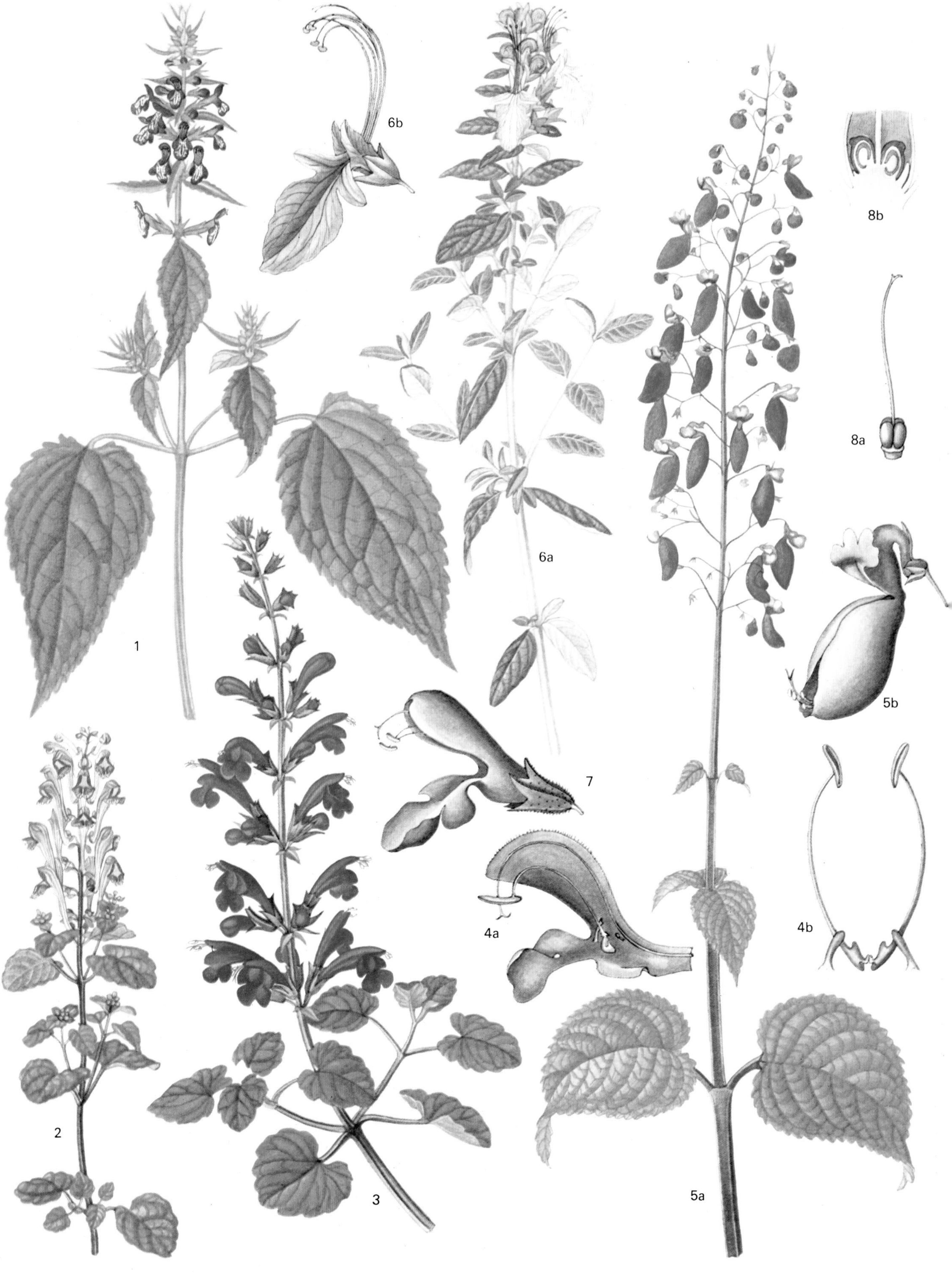
6b
8b
8a
1
6a
5b
7
4a
4b
2
3
5a

# LABIATAE

*The Mint Family*

The Labiatae (or Lamiaceae) is a large and natural family of mostly herbs and undershrubs containing many useful plants such as sage (*Salvia*) and mint (*Mentha*).

**Distribution.** Few regions of the world lack labiates; they grow in almost all types of habitat and at all altitudes, from the Arctic to the Himalayas, Southeast Asia to Hawaii and Australasia, throughout Africa and in the New World from north to south; a few genera such as *Salvia*, *Scutellaria* and *Stachys* are almost cosmopolitan. One of the regions of greatest concentration of species is the Mediterranean basin, where such genera as *Micromeria*, *Phlomis*, *Rosmarinus*, *Sideritis* and *Thymus* are characteristic components of the maquis and the garrigue. In general, labiates are plants of open ground; only a few genera are found in tropical rain forest (eg *Gomphostemma*).

**Diagnostic features.** Most species are shrubby or herbaceous; trees are extremely rare but do occur in the huge South American genus *Hyptis*, where some species reach 12m (40ft). The stems often have a characteristic square shape. The leaves are mostly simple, opposite and decussate (each pair at right angles to the next) and are without stipules. The plants are often covered in hairs and glands that emit an aromatic fragrance.

The flowers of all labiates are essentially bisexual, but in many species of *Mentha*, *Nepeta* or *Ziziphora*, for instance, up to 50% of the plants may have flowers in which the male organs are reduced and sterile and the flower is functionally female. In these flowers, the corollas are often smaller and paler colored. The flowers are irregular and basically comprise: five fused sepals forming a funnel- or bell-shape, sometimes two-lipped; five fused petals; four or two epipetalous stamens, either of two lengths or all nearly equal; and a superior ovary of two fused carpels which form four distinct locules each with one basal ovule. A very characteristic feature of the family is the gynobasic style, arising from the base of and between the lobes of the ovary. The fruits consist of four one-seeded indehiscent achene-like nutlets. The seeds have little or no endosperm.

Labiatae. 1 *Stachys sylvatica* shoot with opposite leaves and terminal inflorescence (×⅔). 2 *Scutellaria indica* var *parvifolia* flowering shoot (×⅔). 3 *Salvia porphyrantha* flowering shoot (×⅔). 4 *Salvia* sp (a) section of flower showing stamen with much elongated connective (×2); (b) detail of stamens (×3). 5 *Coleus freederici* (a) flowering shoot showing the square stem characteristic of the family (×⅔); (b) detail of flower (×2). 6 *Teucrium fruticans* (a) flowering shoot (×⅔); (b) detail of flower (×2). 7 *Rosmarinus officinalis* flower with stigma and stamens projecting from the corolla (×⅔). 8 *Lamium marulatum* typical labiate four-lobed ovary (a) entire (×2) and (b) in vertical section (×9) showing style attached to base of ovary (ie gynobasic).

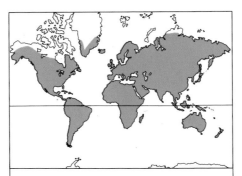

**Number of genera:** about 200
**Number of species:** about 3,000
**Distribution:** cosmopolitan, mostly on open ground.
**Economic uses:** ornamentals (salvia, lavender, coleus), herbs (mint, sage, thyme, marjoram) and essential oils.

There is a very wide range of corolla-shape and staminal position in the family. Usually, there is a clear division into an upper and a lower lip. In most of the temperate genera the upper, often hooded, lip consists of two lobes and the lower of three, forming a convenient landing platform for insects seeking nectar; the stamens are protected by, or included within, the upper lip. In most tropical genera (eg *Coleus*) there is a different organization: an upper lip of four lobes and a lower of one lobe; the stamens lie along the lower lip or ascend from it. In yet other genera (eg *Teucrium*) the upper lip is absent and the lower consists of five lobes, the stamens often being completely exposed. The corolla-tube also varies greatly: frequently there are protective devices, such as rings of hair or constrictions or folds in the throat of the corolla, which protect the nectar, secreted at the base of the ovary, from too-easy access by pollinators.

Various types of specialized pollination mechanisms occur; these are usually linked with insects, sometimes moths or butterflies, or with birds. The most advanced type is found in *Salvia*. A visiting insect knocks its head against one end of an elongated curved staminal connective which effectively prevents access to the nectar. As it does so, the other end of the staminal arm comes down by means of an articulating joint, tapping the insect's back with pollen. Although in most labiates pollination is accomplished by insects, there are many scarlet-flowered species in the New World, often with very long corolla-tubes, which are pollinated by long-tongued humming-birds. Explosive types of pollination mechanism are known in *Hyptis* and in *Aeollanthus* (tropical Africa). In these cases the mechanism consists of a tight juxtaposition of stamens and corolla lobes: the stamens are held under tension by the enfolding lobes of the corolla so that when an insect lands on the lower lip the stamens are abruptly released and a cloud of pollen dusts the pollinator.

Terpenes are present in many members of the family. Sometimes they act as growth inhibitors for other species: in the Californian *Salvia leucophylla* the presence in the air of terpenes exuded from its leaves prevents nearby grasses from germinating or growing.

**Classification.** The 200 or so genera are not readily grouped into higher natural units. Although some nine or ten subfamilies are currently recognized, many are unsatisfactory: numerous genera or closely allied groups of genera are very isolated in the family and do not fit happily into the existing hierarchy; the vast majority of temperate genera are in the subfamily STACHYOIDEAE (eg *Stachys*); most tropical genera are in subfamily OCIMOIDEAE (eg *Coleus*); *Scutellaria*, *Lavandula*, *Ajuga* and *Rosmarinus* are placed in independent subfamilies.

Generally regarded as one of the most highly evolved of all dicotyledonous families, the Labiatae is closely related to the Verbenaceae, primarily a woody tropical family, generally without essential oils and usually without a deeply four-lobed ovary. The small aquatic family Callitrichaceae is also considered to be an ally of the Labiatae.

**Economic uses.** A large number of labiates are cultivated either as ornamentals or as kitchen herbs. Upward of 60 genera are grown in temperate regions alone. Some of the best-known are *Mentha* (mint), *Monarda*, *Nepeta* (catmint), *Origanum* (marjoram or oregano), *Phlomis*, *Salvia* (sage), *Stachys*, *Thymus* (thyme) and *Ajuga* (bugle). Many are grown for their combined virtues of attractive flowers and pleasant fragrance: the essential oil from *Lavandula* (lavender) is mostly obtained from wild plants. In the tropics, *Coleus* and *Plectranthus*, better known as houseplants in cooler regions, are widely grown for their colorful and variegated foliage. So too are several showy species of *Salvia* and of *Leonotis*. *Ocimum sanctum*, a holy plant for Hindus, is frequently grown near temples.

Many species are cultivated commercially. Mostly these are the aromatic herbs of Mediterranean origin, such as mint, marjoram and thyme, so commonly used in flavoring food; but others which are important sources of essential oils used in perfumery and pharmacy are not infrequently cultivated in the tropics and subtropics. Various species of *Ocimum* (basil and sweet basil) are much grown; a species of *Pogostemon* is the source of patchouli, much used in perfumery in Southeast Asia; *Perilla* is grown in India for perilla oil, used in printing-inks and paints.

In different parts of the world, native species of Labiatae are much used by the local people: in Turkey and elsewhere, *Sideritis* leaves provide a tea-like drink; in Iran, *Ziziphora* is used to flavor yoghurt; and in India and Southeast Asia, tubers of *Coleus rotundifoluis* (hausa potato) are eaten as a potato substitute.                                    I.H.

Callitrichaceae. 1 *Callitriche verna* (a) habit showing creeping submerged and erect aerial shoots (×⅔); (b) male flower (left) comprising a single stamen and female flower (right) comprising an ovary with two styles (×20); (c) male and female flower in a single leaf axil (a rare occurrence) (×20); (d) cross section of ovary (×34); (e) fruit with winged lobes (×23). 2 *C. deflexa* habit showing solitary female flowers on long stalks (×2⅔). 3 *C. asagraei* (a) habit showing narrow submerged leaves (×1⅓); (b) habit showing spatula-shaped aerial leaves (×1⅓).

## TETRACHONDRACEAE

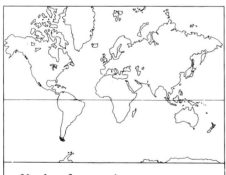

**Number of genera:** 1
**Number of species:** 2
**Distribution:** North Island, New Zealand, and Patagonia.
**Economic uses:** none.

This small family comprises a single genus of small creeping herbs.

**Distribution.** There is one species of *Tetrachondra* in New Zealand and another in Patagonia. This distribution illustrates the botanical affinities between Australasia and South America, seen also in other families such as the Winteraceae and Eucryphiaceae, and in the genus *Laurelia* (Monimiaceae).

**Diagnostic features.** In habit it is similar to *Crassula aquatica*. The leaves are simple, opposite and without stipules. The petioles connect across the stem, and are more or less fleshy with obscure glandular spotting. The stems root at the nodes. The flowers are bisexual, regular, solitary and terminal or axillary, with four fused sepals and four fused petals. There are four stamens which alternate with the petals and are inserted on the corolla. The ovary is superior, composed of four fused carpels, and is lobed to the base, from where the single gynobasic style arises. There is one erect ovule in each of the carpels. The fruits consist of four brown, one-seeded nutlets joined together at the base. The seeds have copious endosperm and a cylindrical embryo.

**Classification.** *Tetrachondra hamiltonii* occurs on North Island, New Zealand, in the Ruahine Range in wet open habitats such as damp meadow and carpeting river bottoms; *T. patagonica* occurs in South America.

The family is probably related to the Labiatae although *Tetrachondra* has been placed in the genus *Veronica* (Scrophulariaceae), in *Mentha* (Labiatae) and in the Boraginaceae by some authorities.

**Economic uses.** None are known.          B.M.

## CALLITRICHACEAE

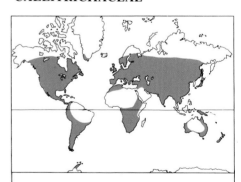

**Number of genera:** 1
**Number of species:** about 17
**Distribution:** cosmopolitan, centered in temperate zones.
**Economic uses:** none.

Most species in this family are submerged aquatics comprising one genus, *Callitriche*.

**Distribution.** *Callitriche* is almost cosmopolitan in occurrence, with most species found in the temperate zones of both hemispheres. Its distribution in the tropics is erratic: it is perhaps dispersed by migrating birds.

**Diagnostic features.** Some species are entirely submerged aquatics, with underwater pollination; others are amphibious; and a few are terrestrial with aerial pollination. All the species are small and delicate, annual or perennial herbs. The stems are usually erect above and creeping and rooting below. When submerged they are elongate and when emersed contracted. The leaves are decussate, but often rosette-forming at the tips of floating stems. The submerged leaves are usually linear and often have a forked apex. The floating or aerial leaves are linear, elliptic, oblong or spathulate. Stipules are absent.

The flowers are minute, unisexual and usually solitary or, rarely, one male and one female in the same leaf axil. There are no sepals or petals. The male flower consists of one stamen with a slender filament and an anther with two locules opening lengthwise, the slits joining at the top. The female flower consists of a single naked ovary with two elongate styles. The ovary has two carpels, each longitudinally divided into two locules. There is a single pendulous, anatropous ovule in each locule. The fruit is four-lobed, with each lobe winged or keeled. At maturity the fruit splits into four one-seeded nutlets. The seeds have a fleshy endosperm.

**Classification.** The relationships of the Callitrichaceae, like those of other reduced dicotyledonous aquatics, are disputed. The presence of four nutlets has often been cited as indicating a relationship to the Labiatae or Boraginaceae. In the latter families the septum is median but in the Callitrichaceae it is transverse. Most reduced dicotyledonous aquatics have single-seeded fruits so perhaps it is unwise to put too much stress on this character. In many ways *Callitriche* resembles the Tetrachondraceae.

Most species of *Callitriche* are very plastic in form, adapting to different conditions, and their identification is often very difficult.

**Economic uses.** *Callitriche* has no known economic use. Several species are very sensitive to specific pollutants. In southern Germany it is possible to predict the abundance of certain pollutants by the species composition and state of the *Callitriche* plants. C.D.C.

## PHRYMACEAE

The Phrymaceae is a family of one genus of erect perennial herbs.

**Distribution.** The family is restricted to northeastern Asia and eastern North America.

**Diagnostic features.** The branches are quadrangular and bear simple, opposite, ovate leaves with coarsely toothed margins and without stipules.

The flowers are bisexual, irregular and borne in axillary or terminal spike-like racemes. The calyx is two-lipped, the upper with two and the lower with three lobes. There are five fused petals which also form a bilabiate structure, the lower lip of which is three-lobed and much larger than the upper

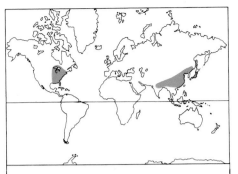

**Number of genera:** 1
**Number of species:** 1–3
**Distribution:** E N America and NE Asia
**Economic uses:** none.

two-lobed portion. There are four stamens (two longer than the rest) inserted above the middle of the corolla tube and projecting slightly beyond it. The ovary is superior, comprising a single carpel, and bears a terminal style with a forked stigma. There is a single erect ovule inserted at the base of the single locule. The fruit is enclosed by the persistent reflexed, ribbed calyx, and is a one-seeded nutlet with a membranous pericarp. The seed has no endosperm and the embryo is oblong with broad, folded cotyledons.

**Classification.** This family consists of a single genus (*Phryma*) and one, two or three species, depending on interpretation.

The family is closely related to the Verbenaceae. The main feature of distinction is the unilocular ovary containing a single ovule in *Phryma*. Otherwise, the two families are similar.

**Economic uses.** None are known. S.R.C.

# PLANTAGINALES

## PLANTAGINACEAE
*Plantains*

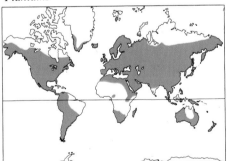

**Number of genera:** 3
**Number of species:** about 253
**Distribution:** temperate, and mountains in the tropics.
**Economic uses:** none.

The Plantaginaceae is a family of annual to perennial herbs. Most species belong to the genus *Plantago*, which has about 250 species.

**Distribution.** The family is widely distributed in temperate regions and on mountains in

the tropics. The bipolar temperate genus *Littorella* has two aquatic or shore species, while *Bougueria* has a single species (*Bougueria nubigena*) native to the Andes of southern Peru, Bolivia and northern Argentina.

**Diagnostic features.** The plants are herbaceous, but some have persistent woody stems. The entire or variously divided leaves are alternate, in basal rosettes or, rarely, opposite, with a basal sheath. They lack stipules. The flowers are small, usually bisexual and wind-pollinated, few or many in axillary spikes, rarely solitary. The calyx has four lobes which are usually imbricate. The corolla is scarious and regular, with a short tube and four lobes. There are four (rarely two or three) stamens alternating with the corolla lobes. The anthers are versatile, dehiscing inwards. The ovary is superior, of two fused carpels and two locules (rarely one), with one to several ovules on axile placentas in each locule. The style is solitary and the stigma simple. The fruit is a capsule with transverse (circumscissile) dehiscence, rarely indehiscent. The seed has a fleshy endosperm, and is mucilaginous when wet; the embryo is erect or curved.

**Classification.** *Plantago* has a dehiscent capsule with two to numerous seeds, while *Littorella* and *Bougueria* have indehiscent, single-seeded fruits, the former being aquatic or littoral herbs with separate male and female flowers, the latter small perennial herbs with bisexual flowers either alone or together with male and female flowers on the same plant.

The relationships of the family are not clear; it possesses affinities with the Polemoniaceae and allied families on the one hand and with the Scrophulariaceae and its relatives on the other.

**Economic uses.** The family has no economic value, except perhaps in a negative sense, since several species of *Plantago* are troublesome weeds. D.M.M.

# SCROPHULARIALES

## COLUMELLIACEAE

The Columelliaceae is a small family comprising a single genus, *Columellia*, with four species of evergreen trees and shrubs confined to the Andes of South America.

The leaves are opposite, simple and without stipules. The flowers are bisexual, slightly irregular and borne in terminal cymes. There are five sepals forming a five-lobed tube that is fused to the ovary, and the petals form a five-lobed corolla which is fused into a short tube. There are two stamens inserted near the base of the corolla and these alternate with the adaxial and lateral corolla lobes. The pollen sacs are contorted. The ovary is inferior, of two fused carpels and is imperfectly two-locular. There are numerous ovules on parietal placentas.

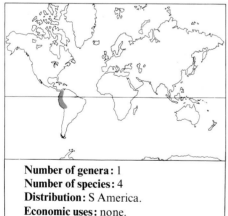

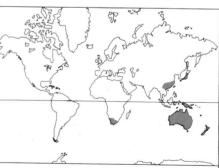

Myoporaceae. 1 *Myoporum viscosum* (a) shoot with flowers in leaf axils ($\times\frac{2}{3}$); (b) flower comprising a five-lobed fused calyx and corolla, four stamens and a simple style ($\times 1\frac{1}{3}$); (c) corolla opened out showing epipetalous stamens alternating with the corolla lobes ($\times 2$); (d) calyx and gynoecium ($\times 2\frac{2}{3}$); (e) fruit—a drupe ($\times 2$); (f) cross section of fruit ($\times 2$). 2 *Eremophila bignoniiflora* (a) shoot bearing linear leaves, flower with irregular two-lipped corolla, and fruits ($\times\frac{2}{3}$); (b) corolla opened out showing stamens of two lengths ($\times\frac{2}{3}$); (c) stamen with divergent anthers and longitudinal dehiscence ($\times 2$); (d) fruit ($\times\frac{2}{3}$); (e) vertical section of fruit ($\times\frac{2}{3}$). 3 *Stenochilus glaber* leafy shoot with flowers and fruits ($\times\frac{2}{3}$).

**Number of genera:** 1
**Number of species:** 4
**Distribution:** S America.
**Economic uses:** none.

The style is terminal with a two-lobed stigma. The fruits are capsular and contain numerous small seeds with fleshy endosperm and a tiny straight embryo. Perhaps the best-known species is *Columellia oblonga*, a small tree with silky downy shoots, oblong leaves and yellow flowers in terminal leafy clusters.

The affinities of this family are uncertain. It has been associated with the Gentianales and Rosales, but is here placed in the Scrophulariales. No economic uses for the family are known.                    B.M.

# MYOPORACEAE

## Emu Bushes

The Myoporaceae is a small family of chiefly trees and shrubs.

**Distribution.** The family is found mainly in Australia and the South Pacific area, with a few species scattered in South Africa, Mauritius, eastern Asia and the West Indies.

**Diagnostic features.** The leaves are alternate, rarely opposite, entire or toothed, without stipules and often glandular, scaly or woolly. The flowers are bisexual, usually irregular, axillary and either solitary or in cymose clusters. The calyx consists of five fused sepals, and the corolla of five fused petals forming a tube with five lobes. There are four stamens fused to the corolla tube and alternating with the lobes. The anther locules run into one another. The ovary is superior, of two fused carpels, forming two locules each with one to eight ovules, or three to ten locules by segmentation, each with only one ovule, these being pendulous. Placentation is axile. The style is simple. The fruit is a drupe, the seeds having little endosperm and a straight or slightly curved embryo.

**Classification.** *Myoporum* with about 32 species is the most widespread genus and the

**Number of genera:** 4
**Number of species:** about 150
**Distribution:** mainly Australia and S Pacific area.
**Economic uses:** ornamentals and a few are useful timber trees.

type-genus of the family. In Australia, the genus consists of small trees or prostrate shrubs with alternate leaves, small creamy white to lavender flowers with a short corolla tube and a four- or five-lobed limb, and cream, yellow or purple drupes.

The emu bushes, *Eremophila*, are restricted to Australia. There are about 105 species. They differ from *Myoporum* in having more colorful flowers and tubular, often two-lipped, corollas. The 15 species of the closely related genus *Stenochilus* (here included in *Eremophila*) are also restricted to Australia.

Both species of *Oftia* are endemic to South Africa, the leaves of *Oftia africana* being flat, dentate and pubescent and those of *O. revoluta* having curled margins, but still dentate and pubescent. Both species have solitary, axillary, white flowers and globose drupes, and are placed in the Scrophulariaceae by some botanists.

The family is represented in the West Indies by *Bontia daphnoides*, the only species in its genus. It is a shrub or small tree with fleshy elliptical or lanceolate leaves, and solitary, rarely clustered, brownish-green flowers. The drupes are yellowish when mature. It is planted in places such as Barbados and Trinidad and is perhaps not native there but in northern South America.

The family is most closely related to the Scrophulariaceae and Gesneriaceae.

**Economic uses.** Creeping boobialla, *Myoporum parvifolium*, is a useful ground-cover plant in Australian gardens on account of its spreading prostrate habit. *Myoporum insulare*, a small tree with dense, bright green foliage which is fire- and wind-resistant, and *M. floribundum*, an aromatic shrub with narrow drooping leaves. *Eremophila* species have great horticultural potential in areas with hot, dry soils.

Several species, notably *Eremophila mitchelli* and *Myoporum sandwicense*, provide useful timber.                              B.M.

# SCROPHULARIACEAE
*The Foxglove Family*

The Scrophulariaceae is a large family consisting mainly of north temperate herbs and a few shrubs and lianas. *Paulownia* is the sole tree genus. Some of the herbaceous genera are semiparasitic, taking part of their nourishment from the roots of their host plants, most frequently members of the Gramineae. Several genera provide attractive garden ornamentals.

**Distribution.** The Scrophulariaceae is a cosmopolitan family, most of the larger genera, eg *Pedicularis* (500 species), *Penstemon* (250 species), *Verbascum* (360 species), *Linaria* (150 species), *Mimulus* (100 species), *Veronica* (300 species), *Castilleja* (200 species), being mainly north temperate while *Hebe* (130 species) and *Calceolaria* (350 species) are southern genera from Australasia and South America respectively. As there are no large trees in the family, it is relatively poorly represented in densely forested regions of the world.

**Diagnostic features.** The leaves are usually alternate or opposite, and are evergreen in *Hebe*; they are without stipules, and both simple and pinnately lobed or incised shapes are represented. The inflorescence is racemose or cymose, and great variation can

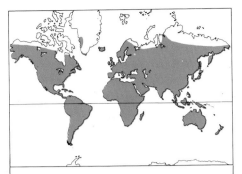

**Number of genera:** about 220
**Number of species:** about 3,000
**Distribution:** cosmopolitan, mainly north temperate.
**Economic uses:** the cardiac drugs digitalin and digoxin from the common foxglove; many ornamental garden genera, including *Antirrhinum*, *Veronica*, *Calceolaria*, *Penstemon*, *Mimulus* and *Verbascum*.

occur even within genera, as, indeed, can the size and shape of the bracts and bracteoles which reach their most spectacular in *Castilleja* where they are often brightly colored. The flowers are bisexual and usually irregular, sometimes markedly so as in the spurred *Linaria*; they are almost regular in some species of *Verbascum*. The most usual pattern is exemplified by *Antirrhinum*: the calyx is five-lobed, the corolla is five-lobed and two-lipped. The four stamens, two of which are longer than the others, are attached to the petals. The two-locular, introrse anthers split longitudinally by slits. There is a nectar-secreting disk at the base of the ovary, which is superior and consists of two united carpels and two locules aligned in the median plane. Numerous anatropous ovules occur on axile placentas in each locule. The style is single, often bilobed.

Numerous modifications to floral structure have occurred, some reflected in "reduction series" of floral parts, whereby organs exhibit several lower numbers than the assumed fundamental number – in this case five, at least for sepals, petals and stamens. *Verbascum* and *Capraria* have five fertile stamens, *Scrophularia* and *Penstemon* have four fertile stamens with a sterile staminode replacing the fifth posterior one, *Castilleja* and many species of *Linaria* have four with no staminode, *Stemotria* has three and *Calceolaria* and *Veronica* have two. Greatest reduction in combined floral parts is exhibited by *Veronica* and *Hebe* with four sepals, four petals and two stamens.

Further variation in floral structure, particularly corolla shape, reflects the evolution of specialized insect-pollination mechanisms. The commonest situation appears to be dichogamy, as in *Scrophularia*, where the ovule comes to maturity before the anthers. Cross-pollination is also encouraged in many species by the stigma extending beyond the anthers, so that the visiting insect

carrying pollen from another plant touches it first. In *Scrophularia* the female flowers have a relatively short style and are visited by wasps; an open flower with a relatively short tube in *Verbascum* and *Veronica* is suited to flies and bees, while a long tube with anthers and stigma arranged so as to touch the back of the visiting bee is exemplified by *Digitalis* and *Linaria*. *Pedicularis* and *Euphrasia* have "loose pollen" and spiny anthers which are protected by the upper lip of the corolla and shake the pollen on to the pollinator as it lands on the lower lip.

The fruit is usually a dry capsule with various types of dehiscence, depending on the position and plane of the split, and is rarely indehiscent, in which case it can be dry or succulent; the seeds may also be dispersed through pores. The seeds have endosperm and are smooth or have a variously and often intricately sculptured surface, and are sometimes angled or winged. The embryo is straight or slightly curved.

**Classification.** The family is usually divided into three subfamilies, based on the vernation of the corolla lobes and on the arrangement of leaves.

VERBASCOIDEAE (Pseudosolaneae). Two tribes and about 10 genera, which have the two posterior corolla lobes (equivalent to the upper lip) overlapping the lateral lobes in bud, and all the leaves alternate. There are often five stamens. The best-known genus is *Verbascum* (mullein).

SCROPHULARIOIDEAE (Antirrhinoideae). Seven tribes and over 100 genera, which have the lobes similarly overlapping, but at least the lower leaves are opposite, and the fifth stamen is staminodial or absent. This subfamily contains such well-known genera as *Calceolaria*, *Linaria*, *Antirrhinum*, *Scrophularia*, *Penstemon* and *Mimulus*. The tribe Selagineae, with up to eight genera, and with a single seed per locule, is sometimes considered as a separate subfamily or even family, the Selaginaceae, and has strong affinities with the Globulariaceae.

RHINANTHOIDEAE. Three tribes and over 100 genera, which differ from the other subfamilies in their vernation: the two posterior corolla lobes are overlapped by one or both of the lateral lobes in bud. This group includes *Veronica* and *Hebe* (always classified together because of their similarity in floral structure), *Digitalis* and the many semiparasitic genera of the family (such as *Castilleja*, *Euphrasia*, *Melampyrum*, *Odontites*, *Pedicularis*, *Rhinanthus*, and *Striga*).

The fused corolla and superior ovary of two fused carpels have led to the Scrophulariaceae being classified with a group of families of superficial resemblance often termed the Tubiflorae. The best-known of these are the Orobanchaceae, Gesneriaceae, Bignoniaceae and Acanthaceae. Some members of the Solanaceae are readily separable from the Scrophulariaceae only on the alignment of the ovary.

2

4a

3

1

7a

5

9

7b

8

4b

6

Globulariaceae. 1 *Globularia trichosantha* (a) leafy shoot with an erect capitulate inflorescence ( ×⅔); (b) lower deeply three-lobed portion of corolla with four epipetalous stamens ( ×6); (c) upper petal ( ×6); (d) calyx opened out and gynoecium ( ×6). 2 *G. salicina* flowering shoot ( ×⅔). 3 *Poskea socotrana* (a) flowering shoot ( ×⅔); (b) corolla opened out showing five petal lobes, four epipetalous stamens and ovary with a single style crowned by a forked stigma ( ×16).

**Economic uses.** This large family is of limited economic use. Perhaps the best-known application are the drugs digitalin and digoxin extracted from certain species of *Digitalis*. Many genera are well known as garden ornamentals, such as species of *Antirrhinum* (snapdragons), *Veronica* (speedwells), *Hebe*, *Calceolaria* (slipper flowers), *Penstemon* (beard tongues), *Mimulus* (monkey flower), *Digitalis* (foxglove) and

Scrophulariaceae. 1 *Erinus alpinus* habit showing rosette of leaves and terminal inflorescence of irregular flowers ( ×⅔). 2 *Verbascum betonicefolium* shoot with alternate leaves and inflorescence of irregular flowers ( ×⅔). 3 *Rhinanthus minor* shoot with opposite leaves and inflorescence ( ×⅔). 4 *Linaria vulgaris* (a) shoot with linear leaves and inflorescence ( ×⅔); (b) half flower with spurred corolla and stamens of two lengths ( ×3). 5 *Digitalis obscura* leafy shoot and inflorescence ( ×⅔). 6 *Veronica fruticans* leafy shoot and inflorescence ( ×⅔). 7 *Scrophularia macrantha* (a) lower lip of corolla opened out showing four stamens with anthers linked in pairs and a central, small staminode ( ×4⅔); (b) cross section of ovary showing two locules and axile placentas ( ×6). 8 *Sibthorpia europaea* dehiscing fruit—a capsule ( ×10). 9 *Penstemon lyallii* leafy shoot with irregular flowers and young fruits ( ×⅔).

*Nemesia*. Other cultivated genera include *Collinsia*, *Cymbalaria*, *Nierembergia*, *Torenia*, *Verbascum* and *Wulfenia*.

Some species, particularly semiparasites in subfamily Rhinanthoideae, are serious weeds, principally of cereal crops: *Rhinanthus minor* (yellow rattle) and *Pedicularis palustris* (lousewort) are common in north temperate regions, while *Centranthera humifusa* parasitizes grasses and sedges in India. The most destructive members of the family occur in the tropical genus *Striga* (witchweeds); they are root parasites with no root hairs and thus rely entirely on the host plant for water and mineral nutrition; *S. orobanchoides* can flower and seed entirely below ground.                          I.B.K.R.

## GLOBULARIACEAE

The Globulariaceae is a small family of two genera, *Globularia* with about 28 species and *Poskea* with two.

**Distribution.** The family is endemic to the Mediterranean region in a wide sense, with species in Macaronesia, Socotra, Somalia, northern Europe and the Alps as well as the Mediterranean basin. *Poskea* is endemic to Socotra and Somalia.

**Diagnostic features.** All members of the

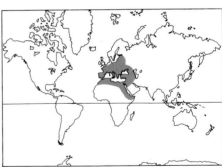

**Number of genera:** 2
**Number of species:** about 30
**Distribution:** Mediterranean, extending to Macaronesia, Socotra, Somalia, N Europe and the Alps.
**Economic uses:** rock garden ornamentals.

Globulariaceae are herbaceous or shrubby perennials with alternate, entire, smooth leaves without stipules. The flower heads are capitulate and surrounded by an involucre. They are arranged in paniculate (*Globularia*) or spicate (*Poskea*) inflorescences. The flowers are bisexual and somewhat irregular. The calyx is tubular and five-lobed, and the corolla of five fused petals is two-lipped with the upper lip two-lobed and the lower three-lobed. There are four stamens inserted on the upper part of the corolla tube. The anthers have a single locule and open by means of a single longitudinal slit. The ovary is superior, of one free carpel and one locule, with a single anatropous, pendulous ovule. There is a single style. The fruit is a one-seeded achene enclosed in the calyx. The seed has endosperm and a straight embryo.

**Classification.** The family Globulariaceae has been considered by some authorities to be related to the Scrophulariaceae, particularly to the southern African tribe Selagineae with which it has sometimes been united into a single family. Hutchinson, however, chooses to include the Globulariaceae within his order Lamiales, comprising the Labiatae, Myoporaceae, Globulariaceae and Selaginaceae (ie the Selagineae detached from the Scrophulariaceae). Recent pollen evidence, however, generally supports a close relationship between the Scrophulariaceae and the Globulariaceae and suggests that the latter family should be kept within the Scrophulariales. Globulariaceous pollen grains are isopolar and three-colporate with compound apertures; similar grains are found in some genera of Scrophulariaceae, but not in the Lamiales.

**Economic uses.** The family has no major economic use. A few species of the genus *Globularia* are occasionally cultivated as ornamental rock-garden plants.

D.B.

# GESNERIACEAE
## African Violets and Gloxinias

The Gesneriaceae is a large family comprising mostly tropical herbs and shrubs. It includes many popular cultivated ornamentals, the best-known being the gloxinias and African violets.

**Distribution.** The 125 genera and 2,000 or so species are mostly pantropical, but some are temperate, in the Americas from Mexico to Chile, East, West and South Africa, Madagascar, Southeast Asia, Polynesia, Australasia, China, Japan and southern Europe.

**Diagnostic features.** Gesneriads are often regarded as tropical counterparts of the essentially temperate family Scrophulariaceae, and are herbs and shrubs, rarely trees, with opposite or alternate, sometimes basal leaves (rarely a single leaf), which are simple, entire or toothed (rarely pinnatisect), and without stipules. The underground parts may be fibrous woody tubers, scaly rhizomes or aerial stolons. The flowers are bisexual, irregular and borne in racemes, cymes or singly. There are five sepals, usually tubular at the base, with five petals also fused into a basal tube, the free ends being oblique, two-lipped or rarely rotate. The two or four stamens often cohere in pairs and release pollen by longitudinal slits. The ovary is superior or inferior and has a single locule containing numerous ovules, usually on two

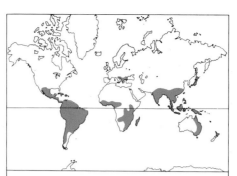

**Number of genera:** about 125
**Number of species:** about 2,000
**Distribution:** pantropical with some temperate species.
**Economic uses:** many popular cultivated ornamentals, including African violets and gloxinias.

parietal or intrusive placentas. The style is single, crowned with a two-lobed or mouth-shaped stigma. An annular, lobed or one-sided nectary lies between the ovary and petals. The fruits are rounded or elongated capsules or rarely berries, and contain many small seeds, with or without endosperm, and with straight embryos.

Evolution of about half of the New World gesneriads has been partly by co-adaptation with bird pollinators, notably the hummingbird family which is restricted to the Americas. Typical hummingbird flowers are two-lipped, often red as in *Columnea*, *Asteranthera* and some *Sinningia* species. Other pollinators such as bees, bats, butterflies, moths and flies have also been active in gesneriad evolution. In *Hypocyrta*, *Besleria* and *Alloplectus* some species have pouched corollas with constricted throats, the significance of which is still not clear. The Old World genus *Aeschyanthus* is considered a parallel development with *Columnea* in being bird-pollinated. Flowers are an important part of the pollination system in gesneriads but extra-floral attraction also exists in some species, such as strikingly colored leaf- and sepal-hairs, or leaf pigmentation with stained-glass-like optical properties when viewed against the light.

**Classification.** Notable New World genera include *Columnea*, 250 species of shrubs and climbers, often epiphytic; *Sinningia*, 60 species of herbs, some popularly known as gloxinias; *Achimenes*, 20 species of often hairy herbs with red to blue flowers; *Episcia*, 10 species of small trailing evergreens; *Gesneria* (46 species) and *Rhytidophyllum* (20 species), two related genera with yellow-green, white or red flowers; *Gloxinia* (not to be confused with the popular gloxinias), 15 species of herbs with lilac bell-flowers or cinnabar-red pouch flowers; *Smithiantha*, four Mexican species with green or purple-brown velvety leaves and pyramids of orange-red or yellowish tubular flowers; *Phinaea*, 10 species with whitish flowers; *Kohleria*, 20 species often with racemes of orange-red flowers

patterned inside with contrasting spots and with brown-green velvety hairy leaves.

Notable Old World genera include *Ramonda*, three species of stemless, hairy herbs from southern Europe with showy flowers on leafless scapes; *Saintpaulia*, 12 East African species mostly of rosette herbs; *Aeschyanthus*, 70 species of trailing or climbing shrubs from the Far East; *Streptocarpus*, 130 African species of evergreen herbs often with foxglove-like flowers; *Cyrtandra*, 350 species from Southeast Asia and Oceania; *Jankaea* (one species) and *Haberlea* (one species), rosette alpines with lilac or violet flowers native to southern Europe; *Chirita*, 80 species of tropical Asian herbs with fleshy, often transparent parts and large whitish, blue, purplish or yellow clustered flowers; *Titanotrichum*, with a single species from China and Taiwan, with tubular flowers bright yellow outside, blotched red-brown with a narrow yellow margin inside; *Conandron*, three Japanese species of alpine rosette herbs regarded as the counterpart of *Ramonda*; *Petrocosmea*, 15 species from Southeast Asia similar to *Saintpaulia*.

About half the genera are placed in the Old World subfamily Cyrtandroideae, with cotyledons of unequal length, unlike the New World Gesnerioideae with equal cotyledons. This division is supported by data from pigment chemistry and chromosome number patterns. Each subfamily is divided into tribes as follows, with representative genera given in parentheses.

CYRTANDROIDEAE: CYRTANDREAE (*Cyrtandra*), TRICHOSPOREAE (*Trichosporum* = *Aeschyanthus*), KLUGIEAE (*Rhynchoglossum*), LOXONIEAE (*Loxonia*), DIDYMOCARPEAE (*Ramonda, Chirita, Streptocarpus*).
GESNERIOIDEAE: GESNERIEAE (*Gesneria*), GLOXINIEAE (*Achimenes, Sinningia*), EPISCIEAE (*Episcia, Columnea*), BESLERIEAE (*Besleria*), NAPEANTHEAE (*Napeanthus*), CORONANTHEREAE (*Asteranthera, Mitraria, Sarmienta*).

The temperate Andean genera *Asteranthera*, *Mitraria* and *Sarmienta*, all climbers with red flowers, and *Rhabdothamnus*, a shrubby New Zealand genus with attractive red-striped yellow flowers, do not easily fit

Gesneriaceae. 1 *Chrysothemis pulchella* shoot with opposite leaves and inflorescences ($\times\frac{2}{3}$). 2 *Aeschynanthus microtrichus*, fruit—an elongate capsule ($\times\frac{2}{3}$). 3 *Columnea crassifolia* shoot with alternate leaves and solitary, two-lipped flower ($\times\frac{2}{3}$). 4 *Ramonda myconi* basal rosette of leaves and inflorescences ($\times\frac{2}{3}$). 5 *Gesneria cuneifolia* (a) basal rosette of leaves and solitary flowers ($\times\frac{2}{3}$); (b) flower with part of calyx and corolla cut away to show two stamens with anthers cohering together ($\times 2$). 6 *Aeschynanthus pulcher* flowering shoot ($\times\frac{2}{3}$). 7 *Streptocarpus caulescens* leafy shoot and inflorescences ($\times\frac{2}{3}$). 8 *Aeschynanthus pulcher* half flower showing corolla tube constricted at the base, four-lobed ovary crowned by a long style and stamens with curved filaments ($\times 1$).

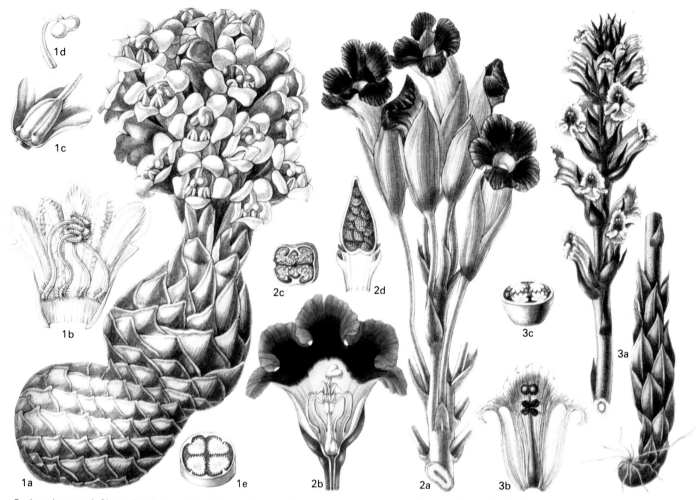

Orobanchaceae. 1 *Cistanche violacea* (a) habit showing swollen underground organ and flower spike ($\times\frac{2}{3}$); (b) flower opened out showing linked anthers ($\times1$); (c) ovary and sepals ($\times1$); (d) bilobed stigma ($\times1$); (e) cross section of unilocular ovary ($\times2$). 2 *Aeginetia pedunculata* (a) inflorescence ($\times\frac{2}{3}$); (b) flower opened out ($\times1\frac{1}{2}$); (c) cross section of ovary ($\times4$); (d) vertical section of ovary ($\times4$). 3 *Orobanche major* (a) habit showing single, erect, stem with scale leaves, and terminating in a flower spike ($\times\frac{2}{3}$); (b) flower opened out showing stamens of two lengths with anthers linked in pairs ($\times1\frac{1}{2}$); (c) section of ovary ($\times5$).

into either Old or New World subfamilies, and their own subfamily MITRARIOIDEAE has been proposed by certain botanists.

Herbaceous families related to the Gesneriaceae are the Scrophulariaceae, Orobanchaceae, and Lentibulariaceae; the chiefly woody Bignoniaceae is also florally similar but has woody fruits that often have two locules, and winged seeds, as well as divided leaves. The gesneriad ovary may be superior, as in Scrophulariaceae, or inferior, but in contrast to the Scrophulariaceae the ovary usually has a single locule not two. The usually parietal placentas of gesneriads with superior ovaries differ from the basal placentas of butterworts (Lentibulariaceae). Orobanches are parasites and lack chlorophyll, so also differ from gesneriads.

**Economic uses.** Some species have been reported as being used in rural medicine, but the importance of the family lies in its cultivated ornamentals. Popular garden and house-plant genera include *Achimenes*, *Columnea*, *Episcia*, *Gesneria*, *Haberlea*, *Hypocyrta*, *Kohleria*, *Mitraria*, *Ramonda*, *Saintpaulia* (African violet), *Sinningia* (gloxinias), *Smithiantha*, *Streptocarpus* (Cape primrose) and *Aeschynanthus*.          B.M.

# OROBANCHACEAE
## Broomrapes and Toothworts

The Orobanchaceae is a family of total parasites almost completely lacking green coloration.

**Distribution.** The family is chiefly north temperate Eurasian. The largest genus, *Orobanche*, the broomrapes, with about 140 species, is quite common throughout temperate Eurasia. Elsewhere in the world the family is not at all well represented; there are a few tropical and American species.

**Diagnostic features.** Almost all the species are rooted in the soil, but few have any extensive rooting system. Instead there is either a congested mass of short, thick roots or a large, single or complex swollen organ. At one or more points on this underground structure there are connections, via swollen, clamp-like haustoria, with the root of the host plant, from which virtually all the nourishment of the parasite is obtained. Above the ground there is often a single erect stem bearing brownish scale leaves, terminating in the flower spike. The plants are mostly annuals, but some are perennial and in *Lathraea squamaria* there is a rhizome bearing fleshy, whitish scale leaves resembl-

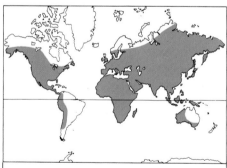

**Number of genera:** about 14
**Number of species:** about 180
**Distribution:** mainly temperate Eurasia.
**Economic uses:** none.

ing a set of dentures and giving the plant its common name of toothwort.

The flowers are irregular, bisexual and arranged in racemes or spikes. The calyx comprises two to five fused sepals forming a two- to five-toothed tube. The corolla is two-lipped, comprising five fused petals. There are four stamens (two of which are longer than the others) inserted below the middle of the corolla tube and alternating with the lobes.

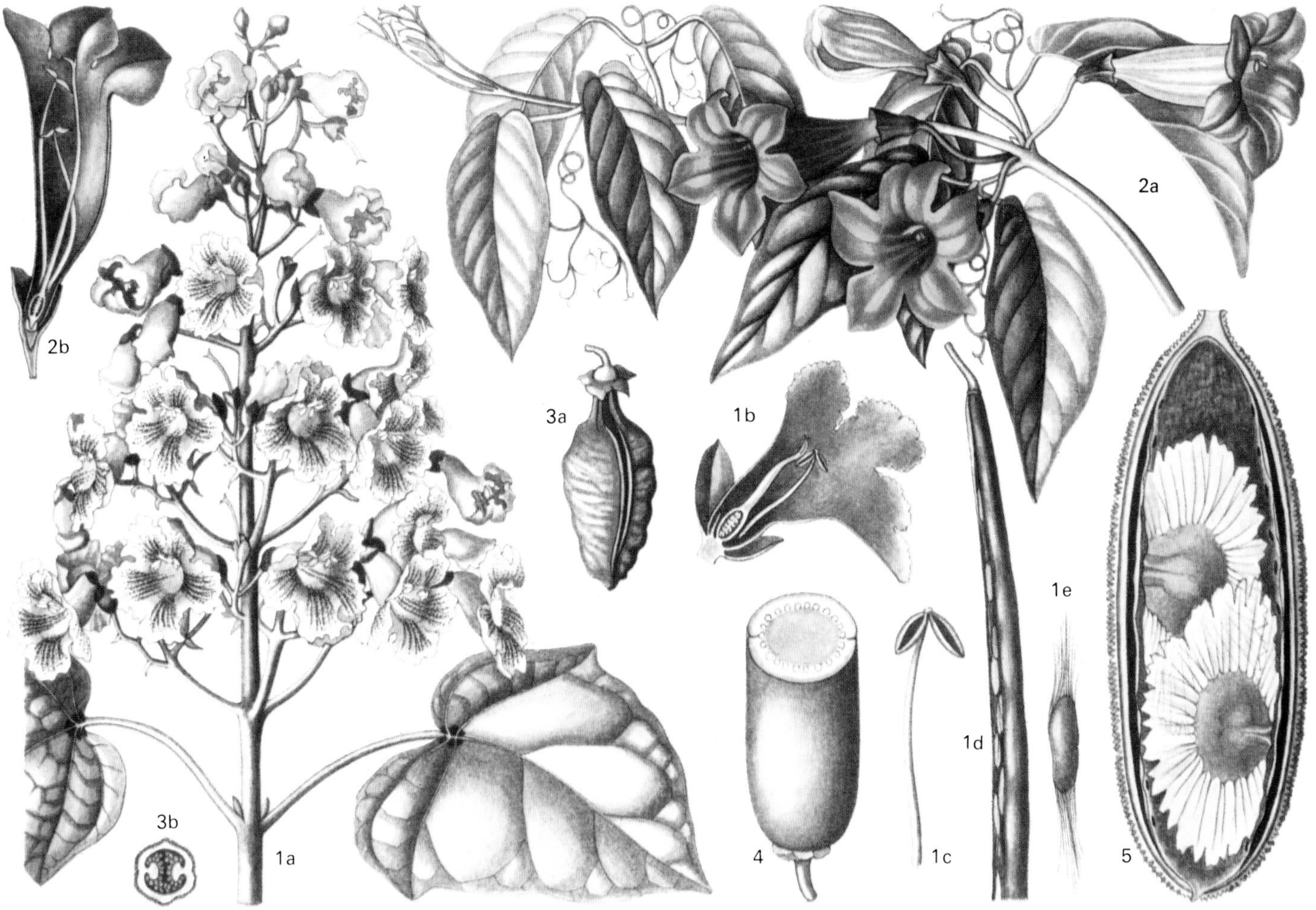


Bignoniaceae. 1 *Catalpa ovata* (a) shoot bearing flowers in a terminal panicle and simple opposite leaves ($\times\frac{2}{3}$); (b) half flower with fertile and infertile stamens ($\times 1\frac{1}{3}$); (c) stamen with divergent anthers ($\times 2\frac{2}{3}$); (d) part of fruit ($\times\frac{2}{3}$); (e) seed bearing tufts of hair ($\times 1\frac{1}{3}$). 2 *Bignonia capreolata* (a) flowering shoot with compound leaves in which the terminal leaflet is tendril-like ($\times\frac{2}{3}$); (b) half flower with all stamens fertile ($\times 1$). 3 *Eccremocarpus scaber* (a) dehiscing fruit ($\times\frac{2}{3}$); (b) cross section of ovary ($\times 5$). 4 *Parmentiera cereifera* base of fleshy indehiscent fruit in cross section showing numerous seeds ($\times\frac{2}{3}$). 5 *Pithecoctenium aubletii* vertical section of fruit ($\times\frac{2}{3}$).

A fifth stamen is either a staminode or absent altogether. The anthers open lengthwise. The ovary is superior, comprising two or rarely three fused carpels and a single locule with four parietal placentas bearing numerous ovules. The style is single. The fruit is a capsule opening by two valves, and the small, very numerous seeds have a fleshy endosperm and a minute embryo.

Some species are confined to a particular host, some to a related range (eg a family) of hosts, while others are more catholic. It is possible that in the last case there are different strains within the species which are adapted for parasitizing different species of host. This has not been substantiated, owing both to the difficulty often experienced in germinating the seeds experimentally and to the difficulty in tracing subterranean connections between the parasite and host roots in mixed vegetation.

The seeds of broomrapes (*Orobanche*) and most other genera are very small and light, but those of the toothworts are substantially larger. Since many of the broomrapes are annuals, the development of small, light seeds appears to be an adaptation allowing for large numbers to be produced, thus increasing the chances of some finding a suitable host plant. These seeds are very effectively spread by wind.

**Classification.** The family is very closely related to the Scrophulariaceae, many members of which are semiparasitic and with which it shares almost all its floral features. There is, in fact, no clear-cut difference between the two families, and there seems to be much in favor of uniting them. Indeed, some genera, including the toothworts (*Lathraea*), appear variously in one family or the other according to the judgment of different authors.

**Economic uses.** The family is of no economic importance except that in warm temperate climates some species may be serious pests of crop plants. In the Mediterranean region, for example, *Orobanche crenata* often infects fields of beans and peas and considerably decreases the yield. C.A.S.

## BIGNONIACEAE

*Catalpa*

The Bignoniaceae is a family of trees and shrubs, the majority of which are lianas.

**Distribution.** The family is mainly tropical, and primarily centered in northern South America, with relatively few genera elsewhere. *Catalpa* and *Campsis*, from Southeast Asia, are also present in the New World.

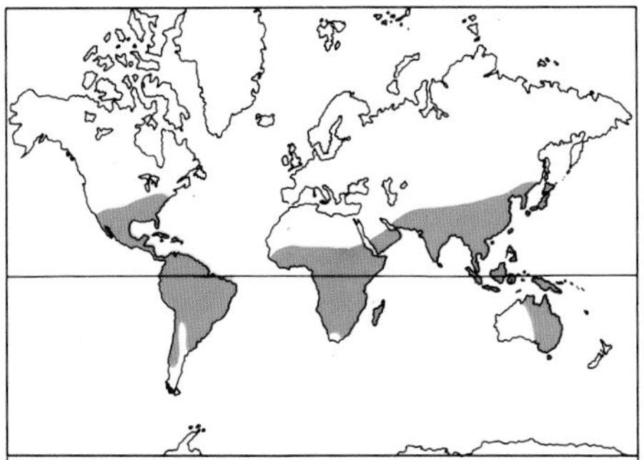

**Number of genera:** about 120
**Number of species:** about 650
**Distribution:** mainly tropical, centered on S America.
**Economic uses:** timber and many cultivated ornamentals.

**Diagnostic features.** Except for a few genera such as *Incarvillea*, the species of this family are nearly all woody, usually adapted in one way or another to climbing in the humid forests where they abound; often they have

Acanthaceae. 1 *Acanthus longifolius* (a) leaf, stem and terminal inflorescence ($\times\frac{2}{3}$); (b) flower with part of corolla cut away to show four stamens ($\times$1); (c) ovary entire and in cross section ($\times$1). 2 *Thunbergia grandiflora* flowering shoot ($\times$1). 3 *Reullia dipterocanthus* (a) corolla opened out to show epipetalous stamens ($\times$1); (b) calyx and gynoecium ($\times$1). 4 *Beloperone guttata* leaf and flower opened out showing two stamens with a broad connective ($\times$2). 5 *Justicia* sp flower opened out ($\times$3). 6 *Justicia patentiflora* vertical section of ovary with ovules on axile placentas ($\times$9).

twining stems, and frequently the terminal leaflet of pinnately-leaved species is modified into a tendril. The leaves are usually opposite, decussate, without stipules, and are usually compound; glands are often present at the base of the petiole. The showy flowers are clustered usually in a cymose arrangement, with relatively inconspicuous bracts and bracteoles; the calyx tube usually bears five lobes and is sometimes two-lipped; the larger, bell- or funnel-shaped corolla shows the same basic structure, and has four epipetalous stamens arched under the upper lip, holding the anthers in readiness for a suitable pollinating agent. Sometimes the number of stamens is reduced to two (eg *Catalpa*); often the missing stamens (assuming five to be the basic number, to agree with the number of calyx- and corolla-lobes) are represented as staminodes, as in the Scrophulariaceae. The two anther locules of each stamen are characteristically divergent, and dehisce by longitudinal slits. The single, superior ovary, with a longish style and two-lobed stigma, bears a nectariferous disk, and is composed of two fused carpels, usually bilocular with axile placentation, occasionally (eg *Eccremocarpus*) unilocular with two forked parietal placentas. The

ovules are numerous, anatropous, and usually develop into flat, winged seeds in a septicidal or loculicidal capsule. A few genera (eg *Kigelia*) have an indehiscent, fleshy fruit with unwinged seeds. The seeds have no endosperm and a straight embryo.

**Classification.** The family is usually divided into about five tribes, mostly on the basis of ovary structure, along with fruit and seed characters. It is closely related to the Scrophulariaceae.

**Economic uses.** The family is of some importance both for timber and for its ornamental species. *Tabebuia* (West Indian boxwood) and *Catalpa* (useful as a fence-post material) are the most commonly exploited timbers. Many genera provide often spectacular ornamental trees, particularly in the tropics: examples are *Spathodea*, *Kigelia* (the sausage tree), *Tabebuia* (the poui, gold tree or araguaney – the Venezuelan national tree), *Crescentia* (calabash tree) and *Jacaranda*. Climbers such as *Campsis* (trumpet vine), *Bignonia* and *Eccremocarpus* species are popular, as are the tender vines *Doxantha unguiscati* (cat's-claw), *Tecomaria* (Cape honeysuckle), *Pandorea* (Australian bower plant) and *Pyrostegia*.                                I.B.K.R.

# ACANTHACEAE
*Black-eyed Susan and Sea Holly*

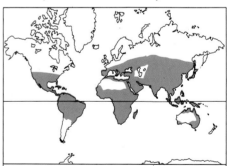

**Number of genera:** about 250
**Number of species:** about 2,500
**Distribution:** cosmopolitan, but centered in tropics.
**Economic uses:** mainly cultivated as ornamentals (*Aphelandra*, black-eyed Susan) with limited local medicinal uses.

The Acanthaceae is a family consisting mainly of tropical shrubs, but including some temperate species of which the best known is *Acanthus*, whose leaves are said to have provided the motif of the Corinthian capitals in classical Greek temples.

**Distribution.** There are about 250 genera and over 2,000 species mainly native to the tropics, although some extend into temperate regions. The main centers of distribution are Indomalaysia, Africa, Brazil and Central America.

**Diagnostic features.** Most species are shrubby or herbaceous, but there are a few trees. Drought-resisting, semi-aquatic and climbing habits are also found in this family. The leaves are opposite and decussate, simple, without stipules, and often have cystoliths which show up as streaks on the lamina.

The flowers are bisexual, usually irregular or two-lipped, solitary or arranged in cymes or racemes. The bracts and bracteoles subtending the individual flowers are often large and petaloid, enclosing the flower. There are four or five sepals and the same number of fused petals. There are two or four stamens attached to the petals, and sometimes one or more staminodes, Sometimes, one of the anther locules on the stamen is much smaller than the other. The ovary is superior, of two fused carpels forming two locules each with two to numerous ovules on axile placentas. The style is simple, usually long with two stigmas. The ovary ripens into a capsule in which the seeds are usually borne on small hook-like outgrowths. The seeds have no endosperm and usually large embryos. The testa of some genera (eg *Crossandra* and *Blepharis*) is covered with hairs or scales which become sticky or slimy on wetting.

**Classification.** The genera can be separated on a range of characters, including the size and nature of the bracts, form of the corolla, number and form of the stamens and staminodes, and the number of ovules in the ovary. For example, *Acanthus*, *Aphelandra*, *Crossandra* and *Thunbergia* all possess flowers with four stamens, while those of *Sanchezia*, *Eranthemum*, *Mackaya*, and *Odontonema* contain only two stamens. The latter two genera have inflorescences with small and inconspicuous bracts. *Filtonia* and *Graptophyllum* both have flowers with markedly two-lipped corollas, the latter possessing staminodes in addition to the stamens.

This family is closely related to the Scrophulariaceae. The irregular, five-part flowers, with reductions in stamen numbers and a bicarpellate superior ovary are common features to both families.

**Economic uses.** A number of the Acanthaceae are cultivated as ornamentals. For example *Aphelandra squarrosa*, native to Brazil, is a popular plant with its tubular yellow flowers and conspicuous bracts. *Crossandra*, a genus of about 50 species, includes a few suitable for greenhouse cultivation. *Crossandra nilotica* produces large showy red flowers over a period of about six weeks. *Thunbergia* produces several popular climbing plants, notably *Thunbergia alata* (black-eyed Susan). Some spec-

ies of *Barleria*, *Beloperone*, *Eranthemum* and *Justicia* are also cultivated.

Some species of *Acanthus* are used medicinally. An extract of the boiled leaves of *Acanthus ebracteatus* (sea holly) is used as a cough medicine in parts of Malaya, while the roots of *A. mollis* (bear's breech) are used to treat diarrhoea in some parts of Europe. The leaves and flowers of *Blechum pyramidatum* are used as a diuretic and for the treatment of coughs and fevers in some parts of Central and South America.                    S.R.C.

# PEDALIACEAE
*Sesame*

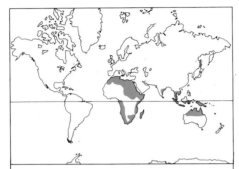

**Number of genera:** 12
**Number of species:** about 50
**Distribution:** dry and shore areas of Africa, Madagascar, Indomalaysia and Australia.
**Economic uses:** sesame seeds and oil, and occasional uses as vegetables.

The Pedaliaceae is a family of annual or perennial herbs and some shrubs.

**Distribution.** Members of the family are found in Africa, Madagascar, Indomalaysia and Australia, chiefly in desert or seashore habitats.

**Diagnostic features.** The leaves are opposite, or the uppermost may be alternate. They are simple, entire or lobed, without stipules, and often possess glandular hairs. The flowers are usually solitary, or borne in few-flowered axillary cymose clusters, with glands at the base of the stalks. They are bisexual and irregular with a calyx of five connate sepals and a tubular corolla of five fused petals. Apart from *Trapella*, which has two stamens, the androecium consists of four fertile epipetalous stamens two of which are longer than the rest; the fifth (posterior) stamen is replaced by a small staminode. The anthers are often contiguous in pairs and have two locules with longitudinal dehiscence. The ovary is superior and normally consists of two fused carpels bearing a long style with two stigmas. The ovary has two to four locules with one to numerous ovules borne on axile placentas. The fruit is a capsule or nut, often with hooks as in the South African *Harpagophytum procumbens* (the grapple plant). The seeds have a straight embryo and thin endosperm.

**Classification.** The chief genera are *Pedalium*

one species), *Sesamum* (30 species), *Ceratotheca* (15 species), *Harpagophytum* (eight species) and *Uncarina* (five species). The family is related to the Martyniaceae and the Bignoniaceae.

**Economic uses.** The most important species economically is sesame (*Sesamum indicum*), an annual herb native to tropical Asia but widely cultivated, especially in India, for its seeds, which are used to coat bread and confections and from which sesame oil is expressed. The oil is used for cooking and in the manufacture of soap and margarine, and the residue is used for cattle feed. The seeds of *S. angustifolium* are used for similar purposes. The leaves of a number of African species, notably *Ceratotheca sesamoides* and *Pedalium murex*, are used as vegetables.
                    S.R.C.

# HYDROSTACHYDACEAE

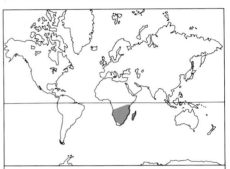

**Number of genera:** 1
**Number of species:** 22
**Distribution:** southern Africa and Madagascar, but most species in Madagascar.
**Economic uses:** none.

The Hydrostachydaceae is a small family of submerged freshwater aquatic herbs consisting of a single genus, *Hydrostachys*, with about 22 species.

**Distribution.** Most of the species are endemic to Madagascar. A few species are found in southern Africa from Tanzania and Zaire southwards to South Africa.

**Diagnostic features.** The plants are found attached to rocks and stones in flowing water. They come into flower as the water level drops. The roots are closely attached to rocks. The stems are flat and disk-shaped or thick and tuber-like. The leaves develop in tufts and have scales at the base. These scales probably represent intrapetiolar stipules. The leaves are very variable in shape, not only within the family but often within species. They are elongate and vary from simple to three times pinnate. The ultimate leaf segments are usually linear but occasionally scale-like. The compound leaves have been frequently misinterpreted as branched stems, particularly when the leaflets bear stipule-like scales at their bases. The flowers are unisexual (male and female on separate plants), and are borne in dense spikes which emerge above the surface of the

Martyniaceae. 1 *Proboscidea fragrans* leafy shoot covered with hairs and bearing irregular flowers in a terminal raceme ($\times\frac{2}{3}$). 2 *Proboscidea louisianica* half flower showing stamens of two lengths ($\times 1$). 3 *Martynia annua* (a) flowering shoot ($\times\frac{2}{3}$); (b) part of corolla with two fertile stamens with united anthers and two staminodes ($\times 2$); (c) young fruit ($\times\frac{2}{3}$). 4 *Martynia lutea* (a) cross section of ovary showing T-shaped parietal placentas ($\times 4$); (b) gynoecium ($\times 1$); (c) fruit (a capsule) showing horns that aid dispersal by animals ($\times\frac{2}{3}$).

water, each flower being borne in the axil of a bract. Sepals and petals are lacking. The male flower is reduced to a single stamen, while the female flower is reduced to a single ovary with two divergent styles. The ovary is superior, of two fused carpels and one locule containing numerous ovules borne on two parietal placentas. The fruit is a capsule opening by two equal valves. The seeds have no endosperm.

**Classification.** In habit the Hydrostachydaceae closely resembles the Podostemaceae and in many works these two families are considered to be related. Recent morphological and embryological work has shown that the Hydrostachydaceae should be assigned to the group of orders Polemoniales – Lamiales – Plantaginales – Scrophulariales (the Tubiflorae), and that they are probably closely allied to the Solanaceae and Plantaginaceae.

**Economic uses.** None are known.　　C.D.C.

## MARTYNIACEAE
### Unicorn Plant

The Martyniaceae is a small family of herbs containing only three genera and about 13 species.

**Distribution:** Members of the family are re-

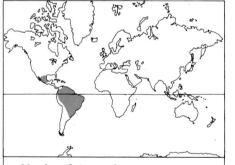

**Number of genera:** 3
**Number of species:** about 13
**Distribution:** chiefly drier parts of tropical and subtropical S America and Mexico.
**Economic uses:** limited use for fruits.

stricted to the New World. Two genera, *Proboscidea* (9 species) and *Craniolaria* (3 species) are found in tropical and subtropical South America, while the monotypic *Martynia* is confined to Mexico. They prefer dry and coastal regions.

**Diagnostic features.** All species are herbs, some annuals, the perennials often having tuberous roots. They are characteristically covered with sticky hairs, and the leaves are opposite or alternate, and without stipules.

The terminal inflorescence, which is sometimes subtended by bracts, bears racemes of usually showy bisexual flowers which are more or less two-lipped. The calyx is either spathe-like or composed of five free sepals. The corolla has a cylindrical, bell-shaped or funnel-shaped tube which has five lobes and is often curved. The stamens are attached to the petals; in *Proboscidea* and *Craniolaria* there are four, two longer than the rest, and the fifth is represented by a posterior staminode; *Martynia* has two fertile stamens and three staminodes. The anthers are bilocular, coherent in pairs and opening by slits. The ovary is superior and surrounded at the base by a nectariferous disk; there are two fused carpels, but a single locule with few to many anatropous ovules on parietal placentas. The single style is slender and the forked stigma is sensitive so that when the insect pollinator touches it its two flat lobes close up. The fruit is a loculicidal capsule with the persistent style forming a usually hooked projection at the end. Animal dispersal is further aided by the sticky outer wall which later splits from the apex, falling off to reveal the woody inner wall of the ovary which becomes more or less four-loculed by coherence of the two winged

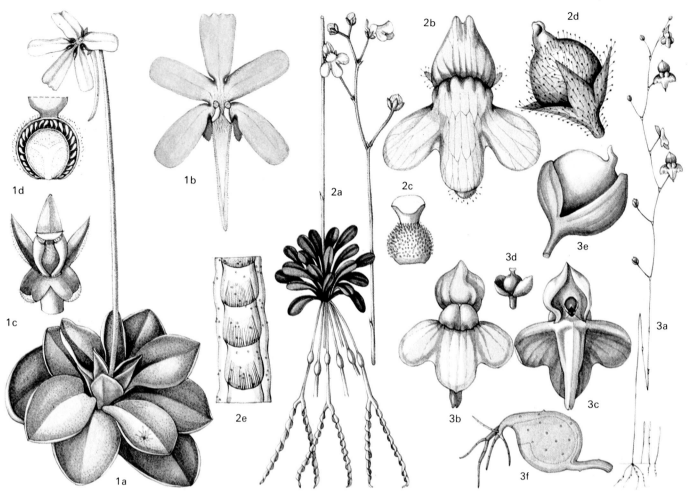

Lentibulariaceae. 1 *Pinguicula moranenis* (a) habit (×⅔); (b) spurred flower (×1); (c) calyx, ovary and stamens (×3); (d) vertical section of ovary showing the large free-central placenta (×8). 2 *Genlisea africana* (a) habit (×1); (b) flower (×4½), (c) gynoecium with sessile stigma (×12); (d) fruit—a capsule (×8); (e) section of pitcher (×14). 3 *Utricularia subulata* (a) habit (×1); (b) flower, front view (×6); (c) flower, back view (×6); (d) gynoecium and calyx (×6); (e) fruit (×12); (f) trap with projecting bristles around the entrance (×40).

placentas. The seeds are sculptured, somewhat compressed but not winged, with a straight embryo and no endosperm.

**Classification.** The Martyniaceae is closely related to the Bignoniaceae, but has its greatest affinities with another smallish family, the Pedaliaceae, in which it is sometimes included. However, the parietal placentation in the unilocular ovary distinguishes it. The Gesneriaceae, a large family often considered to be part of this complex of families, has a similar placentation, but in a usually inferior ovary; this and other differences in fruit structure separate it from the others.

**Economic importance.** Certain species are cultivated as "unicorn plants" on account of their horned fruits. Young fruits can also be pickled and eaten.                    I.B.K.R.

## LENTIBULARIACEAE
### Bladderworts and Butterworts

The Lentibulariaceae is a small family of carnivorous and often rootless, occasionally epiphytic, herbs found in water and other moist habitats. It contains the bladderworts (*Utricularia*) and butterworts (*Pinguicula*).

**Distribution.** The tropical and temperate, aquatic and terrestrial bladderworts are

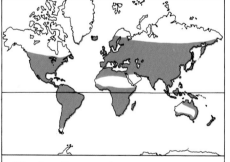

**Number of genera:** 4
**Number of species:** about 180
**Distribution:** cosmopolitan.
**Economic uses:** ornamental curiosities (bladderworts and butterworts).

represented by about 120 species and the 46 temperate species of terrestrial butterworts are found in Eurasia and the Americas. *Polypompholyx*, native to south and southwestern Australia, contains two species, while *Genlisea*, native to South and Central America and Africa, contains 15 species.

**Diagnostic features.** The leaves are simple, entire, alternate, sometimes arranged in rosettes and covered with glandular hairs (*Pinguicula*) or bearing bladders (*Utricu-*

*laria*). Digestive processes occur in contact with the glandular hairs or in the bladders. The distinction between stem and leaf is blurred in *Utricularia* where the morphology of the bladders has been variously interpreted. The traps consist of a hollow bag borne at the end of a stalk, with a small entrance near to or opposite the stalk. Around the entrance are usually some projecting bristles, so arranged that an insect or crustacean passing the bladder will tend to be guided towards its mouth. The entrance itself is closed by a hermetically sealed semicircular valve which bears four hairs. If these are touched the valve is triggered and the rush of water drags the animal inside.

The inflorescence is a raceme or a spike of irregular flowers with two to five calyx lobes, and a five-lobed, two-lipped corolla, the lower lip more or less spurred. Two stamens are borne on the petals. The ovary is superior, comprising two united carpels and a single locule, with one to numerous ovules on a free central placenta. The stigma is usually sessile, with two lobes. The fruit is a capsule which is indehiscent, or opens by two to four valves. The seeds lack endosperm.

**Classification.** *Utricularia* and *Polypompholyx* species bear bladders, but the latter has four calyx lobes, whereas *Utricularia* has two. *Genlisea* has rosettes of leaves and bottle-like pitchers containing bands of hairs and digestive glands, whilst *Pinguicula* is covered in glandular hairs and has no bladders or pitchers. *Pinguicula* and *Genlisea* have a five-lobed calyx. The Lentibulariaceae is allied to the Scrophulariaceae, but differs by the placentation and the carnivorous habit.

**Economic uses.** Some *Utricularia* species can become weeds in ricefields. The family is of great interest to biologists on account of the carnivorous habit. Several species of *Pinguicula* are in cultivation.                    D.J.M.

# CAMPANULALES

## CAMPANULACEAE

*The Bellflower Family*

The Campanulaceae is a family that contains mostly herbs (annual, biennial or more often perennial), but rarely shrubs or undershrubs, which often produce large showy flowers that are predominantly blue in color. Species from genera such as *Campanula* (bellflowers), *Symphyandra* (pendulous bellflowers), *Phyteuma* (horned rampions), *Edraianthus* and *Jasione* are popular as garden ornamentals.

**Distribution.** The vast majority of the Campanulaceae have their native haunts in the north temperate zone. The Southern Hemisphere is exceedingly poor in the Campanulaceae, except for South Africa where seven small endemic genera are found. South America can show only certain species of *Wahlenbergia*, *Legousia* and *Cephalostigma*. The latter is the only genus of the family confined to the tropics; it is also represented in Africa and Asia. In Australia and New Zealand, only several species of *Wahlenbergia* occur.

**Diagnostic features.** The leaves are alternate, sometimes opposite or whorled, simple or rarely pinnate, and without stipules.

The flowers are regular and bisexual, with floral parts normally in fives. They are borne singly or more often in inflorescences (either racemes or cymes). The calyx tube is united with the ovary. There are small appendages present in the sinuses between the calyx teeth in *Michauxia* and certain species of *Campanula*, *Edraianthus* and *Symphyandra*. The petals are nearly always partially or wholly united (sympetalous) and the corolla is inserted at the line where the calyx becomes free from the ovary. The petals are not united in *Jasione*, *Asyneuma*, *Michauxia*, *Cephalostigma* and *Lightfootia*, and in *Phyteuma* when fully developed. There are as many stamens as corolla lobes. The anthers are free but tend to be united in certain genera and species; they show introrse dehiscence. The ovary is inferior to semi-inferior but in *Cyananthus*, *Codonopsis* and *Campanumoea* it is superior. It comprises five, three or two

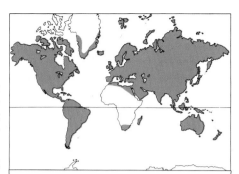

**Number of genera:** about 35
**Number of species:** about 600
**Distribution:** mainly found in the N Temperate Zone with a few in the S Hemisphere.
**Economic uses:** many species used as ornamentals, eg the bellflowers (*Campanula*).

fused carpels, with five, three or two locules (rarely one or six or 10), each containing a large number of ovules on axile placentas. The style is simple with as many stigmas as carpels.

The fruit is a capsule. It opens in a variety of ways but in some genera, such as *Peracarpa* of the Himalayas and *Merciera* of South Africa it is indehiscent. Only two genera, *Canarina* and *Campanumoea*, are known to produce berries. The seeds are numerous and small, with a straight embryo and fleshy endosperm. Anatomically, the members of the family are distinguished by the constant presence of latex-containing vessels in their tissues. The reserve material is not starch as in the majority of plants, but a polysaccharide called inulin.

The flowers are generally large and showy. The predominantly blue color is particularly attractive to bees but there are many other pollen carriers from various groups of insects. A glandular disk (in a few cases a cup) at the base of the style secretes honey. It is usually covered by the dilated bases of the stamens which allow the insertion of an insect's proboscis between them.

An interesting floral mechanism has been developed in the family, facilitating cross-pollination and hindering self-pollination. The flowers are distinctly protandrous, and pollen is shed upon the style (with the stigma lobes closed up against one another) in the bud. The style is either sticky (as in *Wahlenbergia*) or hairy, to hold the pollen. Technically, this is the male stage in the development of the flower and stamens wilt quickly as the corolla opens. After a time (usually several days) when the bulk of the pollen has been carried away by insects, the stigma lobes unfold and the female stage sets in. In some species the stigma lobes coil spirally and may touch the remnants of pollen deposited on the style, thus effecting self-pollination.

Species of *Legousia* produce, in addition to normal flowers, small, abortive ones

which fail to open and are self-pollinated – a case of cleistogamy (pollination and fertilization before the flower has opened). *Campanula* has several cleistogamous species.

**Classification.** The family Campanulaceae, in its restricted sense (that is, excluding the aberrant genera *Cyphia*, *Pentaphragma* and *Sphenoclea*) is rather natural and homogeneous but its subdivision presents serious problems because there seems to be little correlation between the various diagnostic features. Based on the morphology of ovary and capsule, the following three subtribes may be distinguished which, however, do not necessarily represent separate phyletic (evolutionary) lines:

CAMPANULINAE. The ovary is inferior and carpels superposed on the calyx teeth. The fruit is a capsule opening laterally, sometimes indehiscent, in one genus a berry. Genera include *Campanula*, *Symphyandra*, *Adenophora*, *Legousia*, *Michauxia*, *Ostrowskia* and *Canarina*.

WAHLENBERGINAE. The ovary may be inferior or semi-inferior to superior. The carpels are superposed on the calyx teeth. The fruit is a capsule opening on the top, in one genus a berry. Genera include *Wahlenbergia*, *Campanumoea*, *Codonopsis*, *Cyananthus*, *Roella*, *Githopsis*, *Lightfootia*, *Edraianthus* and *Jasione*.

PLATYCODINAE. The ovary is inferior to semi-inferior. The carpels alternate with the calyx teeth, and the fruit is a capsule opening terminally or laterally. The type-genus is *Platycodon*.

The Cyphiaceae, Lobeliaceae, Pentaphragmataceae and Sphenocleaceae, often regarded as belonging to the Campanulaceae, are here treated as separate families.

The Cyphiaceae is a link between bellflowers and lobelias, having regular flowers, stamens fused into a tube and free anthers. There are four genera and about 70 species in South Africa. The Pentaphragmataceae is distinguished by the asymmetric leaves, regular flowers arranged in scorpioid cymes and baccate fruits. The only genus *Pent-*

Campanulaceae. 1 *Canarina eminii* shoot with opposite leaves and axillary, regular flowers ($\times\frac{2}{3}$). 2 *Trachelium rumelianum* shoot with alternate leaves and terminal inflorescences ($\times\frac{2}{3}$). 3 *Campanula rapunculoides* (a) leafy shoot and racemose inflorescence ($\times\frac{2}{3}$); (b) half flower showing free anthers to stamens, and inferior ovary surmounted by a single style with a lobed stigma ($\times 1$); (c) cross section of trilocular ovary with ovules on axile placentas ($\times 3$). 4 *C. rapunculus* dehisced fruit—a capsule ($\times 1\frac{1}{3}$). 5 *Phyteuma orbiculare* inflorescence ($\times\frac{2}{3}$). Lobeliaceae. 6 *Lobelia cardinalis* cv "Red Flush" showing irregular flowers with stamens united around the style—features which differ from the Campanulaceae; note that in the upper flowers of the inflorescence the stigma is not protruding between the anthers, but it is in the lower flowers; these are the male and female stages of the flowering cycle ($\times\frac{2}{3}$). 7 *Pratia arenaria* (a) creeping stem ($\times\frac{2}{3}$); (b) flower ($\times 4$).

*aphragma* has about 30 species (mostly succulent) confined to Southeast Asia and Malaysia. Some authors relate it to the Begoniaceae. The Sphenocleaceae contains one species, *Sphenoclea zeylanica*, an annual herb that occurs in wet habitats in Central and Southeast Asia. The stem is hollow and flowers are regular, small, borne in dense spikes. Certain characters suggest affinity with the Lythraceae.

It seems possible that the Campanulaceae (or, rather, their evolutionary ancestors) are the basic stock from which the huge family Compositae evolved. This contention is supported by the morphological evidence (head-like inflorescences in *Jasione* and *Phyteuma*, connate anthers in certain species and genera), protandry, presence of latex and, last but not least, of inulin. This peculiar polysaccharide occurs in both the orders Campanulales and Asterales.

**Economic uses.** Many, if not all, species of *Campanula* (bellflowers), *Edraianthus*, *Symphyandra* (pendulous bellflowers), *Phyteuma* (horned rampions) and *Jasione* are valued ornamentals, being highly attractive and easy to grow; a number of them are extremely useful in rock gardens. *Adenophora*, *Michauxia*, *Ostrowskia* (giant bellflower), *Trachelium* (blue throatwort), *Codonopsis* and *Platycodon* (balloon flower) are also popular with gardeners. *Campanula rapunculus* or rampion is one of the very few species of (minor) nutritive value, the roots and leaves being used occasionally in salads.

M.K.

## LOBELIACEAE
*Lobelias*

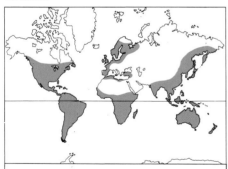

**Number of genera:** about 30
**Number of species:** about 1,200
**Distribution:** cosmopolitan, especially common in the Americas and tropics.
**Economic uses:** many ornamentals and *Lobelia inflata* yields medicinally important alkaloids.

The Lobeliaceae is a medium-sized family with life forms ranging from tiny annuals to trees, often of weird habit. Species of the Hawaiian genus *Cyanea* resemble palms. Desert species of *Lobelia* and *Monopsis* have developed needle-like leaves and the Australian *Lobelia gibbosa* is succulent. Certain representatives of *Clermontia* are epiphytes and several *Lobelia* species are aquatic.

**Distribution.** The family is worldwide, being particularly rich in the Americas and containing many tropical species and genera.

**Diagnostic features.** The leaves are alternate, simple and without stipules. The flowers are irregular, mostly blue, red or violet, normally bisexual (but several Australian species of *Lobelia* and *Pratia* are known to have male and female flowers on separate plants), and are inverted on their axis through 180°. They are arranged in racemes or panicles lacking the terminal flower; single-flowered species are rare. The calyx is five-lobed and joined to the ovary. The corolla has five fused petals (rarely free), which are two-lipped, with the lips either equal or unequal. The corolla tube is curved in most species. The five stamens are free from or joined to the corolla, and the anthers are fused into a tube. The upper three stamens are longer than the lower two. The ovary, of fused carpels, is inferior or rarely superior and has two or three locules (rarely one) with numerous ovules on axile placentas. The style is simple with two or three stigmas. The fruit is either indehiscent and pulpy or dry, or more often dehiscent, opening in various ways.

The flowers achieve cross-fertilization by being protandrous. The style pushes through the tube formed by the fused anthers and drives the pollen out at the top, where it is collected by insects. The style having emerged fully, the stigmas separate and the female stage of the flower begins. Exceptions are the African genus *Monopsis*, with anthers and stigmas ripening simultaneously, and *Lobelia dortmanna* (water lobelia), where the stigmas unfold within the anther tube and are self-fertilized by the pollen deposited there. Pollen carriers include a variety of aphids, bees and butterflies; some large-flowered species are pollinated by birds.

Many members of the family are strongly poisonous. For instance, *Laurentia longiflora* from South America supplies isotomin, a heart poison, and the mere smell of the Chilean *Lobelia tupa* may cause poisoning.

**Classification.** Two tribes of genera, based primarily on the morphology of the ovary and fruit, are currently recognized.

LOBELIEAE. Ovary conical at the top; fruit dehiscing; includes *Lobelia*, *Siphocampylus*, *Laurentia* and *Monopsis*.

DELISSEAE. Ovary flat at the top; fruit indehiscent, pulpy or dry; includes *Centropogon*, *Burmeistera*, *Hypsela*, *Pratia*, *Clermontia*, *Cyanea* and *Delissea*.)

The Lobeliaceae is an advanced type of the Campanulaceae from which it differs in its irregular flowers, connate anthers and the presence of different alkaloids. However, the relationships between the families are so close that some authorities include the Lobeliaceae as a subfamily of the Campanulaceae.

**Economic uses.** The genus *Lobelia* has many ornamental representatives which are much valued for their long flowering-period.

Perhaps the most familiar is *Lobelia erinus*, a native of South Africa, which has been in cultivation since the 17th century and is commonly used for bedding. *Lobelia cardinalis* (cardinal flower), *L. splendens*, *L. amoena* and *L. fulgens* are effective in the herbaceous border. Species of *Pratia*, *Centropogon*, *Downingia*, *Monopsis*, *Laurentia* and *Hypsela* are also grown for ornament.

*Lobelia inflata*, in addition to being the source of important alkaloids, yields a remedy for asthma and whooping cough. A decoction of the roots of *Lobelia syphilitica* was used by the American Indians to cure venereal diseases. The berries of *Centropogon* and *Clermontia* are edible.

M.K.

## STYLIDIACEAE
*Trigger Plant*

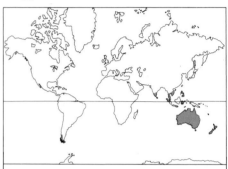

**Number of genera:** 6
**Number of species:** about 150
**Distribution:** centered in Australia.
**Economic uses:** some ornamentals.

The Stylidiaceae is a small family of subtropical and temperate annual or perennial herbs and a few shrubs. The largest genus, *Stylidium* (103 species) is sometimes called trigger plant or springback, in reference to the movable column in the flower, characteristic of the family.

**Distribution.** The family is confined to Australia except for a few species in New Zealand, South America, eastern Asia and India. They are usually somewhat adapted to drought conditions, but *Donatia* and *Phyllachne* are important in the wet bogs of New Zealand and southernmost South America.

**Diagnostic features.** The leaves are alternate or in a basal rosette, simple, and usually without stipules. The flowers are irregular and bisexual or unisexual. There are five persistent sepals, usually more or less united in two lips, and the corolla is usually deeply five-lobed, with one smaller downward-directed lobe (labellum) and the other four ascending, in pairs. There are two (rarely three) stamens, usually fused with the style to form a column; the anthers open outwards, often concealing the entire or two-lobed stigma. The column protrudes and curves in such a way that when an insect lands on the lower lip of the corolla, the column moves elastically up and down, and thus assists the transfer of pollen to and from the insect's back. The ovary is inferior, of two fused carpels, and has one or two locules, with few or numerous anatropous ovules attached to the septum or a central placenta. The fruit is a two-valved capsule with small seeds containing endosperm and a minute embryo.

**Classification.** The family is usually subdivided into two subfamilies:

DONATIOIDEAE. Petals free and stamens two or three, also free. The only genus is *Donatia*, which is placed by some authorities in its own family, the Donatiaceae.

STYLIDIOIDEAE. Petals united at the base and stamens two, fused with the style as far as the apex. The tribe PHYLLACHNEAE has anther locules which are connate at the apex to form two curved anthers, each with a single cell (*Phyllachne* and *Forstera*). The tribe STYLIDEAE has anthers with two locules (*Oreostylidium*, *Levenhookia*, *Stylidium*).

The affinities of the Stylidiaceae are still open to debate, but the pollen of most of them is like that of the Campanulaceae and other characters do not contradict this.

**Economic uses.** A few species are cultivated for ornament. The most important are the western Australian evergreen shrubs belonging to *Stylidium* which are often referred to as *Candollea* by horticulturists.     D.M.M.

## BRUNONIACEAE

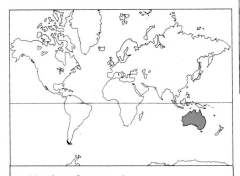

**Number of genera:** 1
**Number of species:** 1
**Distribution:** Australia and Tasmania.
**Economic uses:** none.

The Brunoniaceae is an Australian family containing a single genus of silky-haired perennial herbs, *Brunonia*.

**Distribution.** The single species of the genus, *Brunonia australis*, whose colloquial name, blue pincushion, is an apt description of the flowers, is restricted to Australia and Tasmania.

**Diagnostic features.** All leaves arise from the basal part of the plant, and are entire, without stipules, and up to about 10cm (4in) long. The flowers are bisexual, regular and arranged in dense hemispherical heads subtended by a series of bracts. There are five fused sepals, forming a tube which is not fused to the ovary, but the sepal lobes persist on the fruit. The corolla consists of five blue lobes, fused together at the base but free and spreading above. There are five stamens inserted at the base of the corolla tube and

the anthers cohere together around the style within the corolla tube. The ovary is superior and has a single locule with one erect ovule. There is a single style which is capped by a small stigma surrounded by a collar or indusium. The fruit is an achene with a single seed which lacks endosperm and has a straight embryo.

**Classification.** The family is thought to be related to Goodeniaceae in which it has often been included, both having an indusium around the stigma, but at the same time the family bears some similarity to the Dipsacaceae and Compositae in inflorescence structure.

**Economic uses.** None are known.     B.M.

## GOODENIACEAE
*Leschenaultias and Scaevolas*

**Number of genera:** 14
**Number of species:** about 300
**Distribution:** centered in Australia.
**Economic uses:** several garden and greenhouse ornamentals.

The Goodeniaceae is a smallish family of herbs and some shrubs. It contains some outstanding ornamentals, cultivated in gardens in warmer climates and in greenhouses in temperate regions.

**Distribution.** The family is largely confined to Australia, particularly western Australia, with only a few species of *Scaevola* extending to New Zealand, Africa, eastern Asia, the West Indies, Central and South America and the Pacific Islands, and *Selliara radicans* occurring in Australia, New Zealand and central Chile.

**Diagnostic features.** The plants are herbs, sometimes woody at the base, or rarely shrubs. The leaves are alternate or radical, rarely opposite, and entire to pinnatisect; there are no stipules. The flowers are irregular, bisexual and axillary, either solitary or borne in axillary spikes, racemes or cymes. The calyx tube is usually attached to the ovary, and has five (rarely three) lobes or is reduced to a ring. The corolla tube is slit almost to the base on the upper side, sometimes fused to the ovary, the limb being unequally or equally five-lobed, and usually two-lipped, yellow or white to blue, rarely red to purplish. There are five stamens, free or slightly joined to the corolla, and alternating with the corolla lobes. The anthers are attached at their base to the filaments, and

are free or united in a ring round the style. The ovary is inferior or partly so, rarely superior, formed of two fused carpels and has two locules (rarely one), each with one to numerous erect or ascending, anatropous or rarely campylotropous ovules. The style has a terminal outgrowth (indusium) surrounding the usually two-lobed stigma. The fruit is an indehiscent drupe or nut, or a capsule, the seeds being usually flat and often winged, with endosperm and an erect embryo.

**Classification.** Most species belong to *Goodenia* and *Scaevola* (about 90 species each) and *Dampiera* (about 60 species). Like the related family Brunoniaceae, which is sometimes included in it, the Goodeniaceae is characterized by the cup-shaped or two-lipped indusium, often finely hairy, which contains the stigma at the apex of the style and which seems to be involved in aiding pollen capture from insect visitors.

**Economic uses.** Several species of *Goodenia*, *Dampiera*, *Leschenaultia* and *Scaevola* are grown for ornament in greenhouses or, in warmer climates, in gardens. *Leschenaultia* is considered to contain some of the most beautiful greenhouse shrubs, including *Leschenaultia biloba*, which is prized for its blue flowers. *Selliera radicans*, a hairless creeping perennial, is sometimes grown in moist rock gardens.     D.M.M.

# RUBIALES

## RUBIACEAE
*Gardenias, Coffee and Quinine*

The Rubiaceae is one of the largest flowering plant families. Most tropical species are trees or shrubs while all temperate ones are herbaceous. Coffee is the most important product of this family and is obtained mainly from *Coffea arabica* and *C. canephora*. *Cinchona* species yield the drug quinine. Among the many tropical species that are cultivated as ornamentals is *Gardenia jasminoides* (Cape jasmine), while the best-known European genera are *Galium* (the bedstraws) and *Asperula* (woodruff).

**Distribution.** This very widespread family is concentrated in the tropics and subtropics with some species represented in temperate regions and even the Arctic and Antarctic.

**Diagnostic features.** The leaves are opposite or whorled, simple and usually entire. The presence of stipules is particularly characteristic; these are sometimes fused at each node, leaf-like (as in *Galium* and *Asperula*), or inserted in the axil of the petiole (intrapetiolar) or between the petioles. The flowers are borne in panicles or cymes or are aggregated into congested heads. They are usually bisexual and regular (although one of the sepals is often enlarged), with four or five free sepals, four or five fused petals and four or five stamens, the latter borne on the corolla tube whose mouth is frequently filled with flattened, ribbon-like hairs. The ovary

Rubiaceae. 1 *Ixora chinensis* (a) flowering shoot showing stipules between the petioles ($\times\frac{2}{3}$); (b) tubular corolla opened out to show epipetalous stamens and simple style with a lobed stigma and vertical section of ovary ($\times1\frac{1}{2}$). 2 *Asperula suberosa* flowering shoot ($\times\frac{2}{3}$). 3 *Mussaenda luteola* flower with one calyx lobe much enlarged ($\times1\frac{1}{2}$). 4 *Coffea arabica* (a) fruits ($\times\frac{2}{3}$); (b) cross section of fruit ($\times1$). 5 *Nauclea pobeguinii* vertical section of fruit ($\times\frac{1}{2}$). 6 *Sherbournia calycina* (a) flowering shoot ($\times\frac{2}{3}$); (b) vertical section of ovary ($\times2$); (c) cross section of ovary ($\times3$).

is inferior (very rarely superior), with one to many carpels (normally two) and as many locules, each containing one to many anatropous ovules on axile, apical or basal (rarely parietal) placentas. The style is simple, and the stigma capitate or variously lobed. The fruit is a capsule, berry, drupe or schizocarp. The seeds are sometimes winged; they contain a straight or curved embryo, and endosperm may be present or absent.

Interesting examples of association with ants are found in the genera *Myrmecodia* and *Hydnophytum*, native to tropical Asia and Australia. All members of these genera are epiphytes which cling to branches of trees with their roots. Large swellings, containing a network of cavities, develop on the roots, and are inhabited by ants. Although it has been supposed that the plant and insects have developed a symbiosis of mutual benefit – the ant guarding the plant and providing it with extra nutrients, receiving shelter in return – it has not been proved that they are in fact interdependent.

**Classification.** Although the Rubiaceae forms a very clear-cut group, its intrafamilial classification is controversial and there has been disagreement about which characters are most suitable for delimiting tribes and

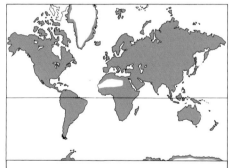

**Number of genera:** about 500
**Number of species:** about 7,000
**Distribution:** mainly in the tropics and subtropics with a few species in temperate and cold regions.
**Economic uses:** coffee and quinine, and some lesser-known drugs such as ipecacuanha, dyes from madder and gambier, and many ornamentals (eg *Gardenia*, woodruffs and bedstraws).

subfamilies. Older systems recognize two subfamilies, characterized, respectively, by one or many ovules per locule. Recently, however, three completely different subfamilies have been proposed, as follows:

RUBIOIDEAE. Calcium oxalate raphides (needle-shaped crystals) present in leaves; hairs on stems and leaves often with cross-walls; heterostyly common; stipules often divided into many slender branches; seeds without endosperm. Predominantly herbaceous – 11 tribes.

CINCHONOIDEAE. Raphides absent; hairs without cross-walls; complete heterostyly absent; stipules rarely divided; seeds with endosperm. Predominantly woody – 17 tribes.

GUETTARDOIDEAE. Raphides absent; hairs without cross-walls; heterostyly absent; seeds lack endosperm. Woody – one tribe.

The affinities of the Rubiaceae seem to lie equally with the Gentianales and Dipsacales, both of which generally have opposite leaves and bicarpellate ovaries. Opinions vary as to whether the family is best placed in one or other, or kept as a distinct order. The Rubiaceae resembles members of the Gentianales (particularly the Loganiaceae) in having well-developed stipules which often bear special glands (colleters), and in possessing certain alkaloids; on the other hand, it differs in the absence of internal phloem and in its inferior ovary. These two features are typical of the Dipsacales, but the latter tend to lack stipules and do not synthesize alkaloids.

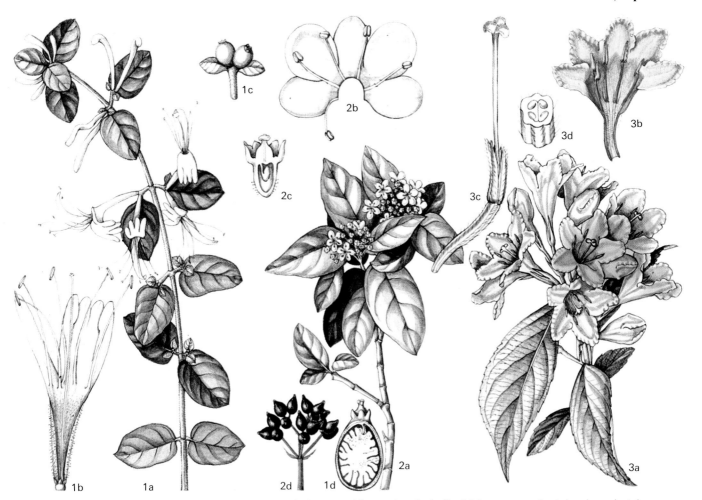

Caprifoliaceae. 1 *Lonicera biflora* (a) twining stem with opposite leaves and flowers in pairs ($\times\frac{2}{3}$); (b) flower opened out showing epipetalous stamens ($\times 2$); (c) paired fruit—berries ($\times 4$); (d) vertical section of fruit ($\times 3$). 2 *Viburnum tinus* (a) leafy shoot and inflorescences ($\times\frac{2}{3}$); (b) corolla and stamens ($\times 3$); (c) vertical section of gynoecium showing capitate stigma and pendulous ovule ($\times 3$); (d) fruits ($\times\frac{2}{3}$). 3 *Weigela amabilis* (a) leafy shoot and inflorescence ($\times\frac{2}{3}$); (b) corolla opened out ($\times 1$); (c) calyx, style and stigma ($\times 3$); (d) section of ovary ($\times 9$).

**Economic uses.** The best-known products of the Rubiaceae are coffee (*Coffea*) and quinine (*Cinchona* species). Members of the family also provide the drug ipecacuanha (*Cephaelis*) and the dyes madder (*Rubia*) and gambier (*Uncaria*). Of the large number of tropical flowering shrubs in the family, *Gardenia* is perhaps the most popular ornamental but others, such as *Bouvardia, Hamelia, Manettia, Randia* and *Rondeletia*, are also cultivated for their blooms. Other garden plants include *Asperula, Galium* and *Houstonia*. F.K.K.

# DIPSACALES

## ADOXACEAE

The Adoxaceae is a family consisting of one genus with one species of perennial, rhizomatous herbs, *Adoxa moschatellina*.

**Distribution.** *Adoxa* is widely distributed in Europe, North America, and northern and central Asia, on mountain rocks, and in hedges and woods.

**Diagnostic features.** The plants bear ternate radical leaves on long petioles, as well as a pair of opposite, shortly petiolate, trisected or ternate leaves on the erect, unbranched

flowering stem. The inflorescence is a condensed cyme with each branch giving rise to two other branches. There are one terminal and four lateral regular, bisexual, greenish flowers. The terminal flower has a two-lobed calyx (or bracts) and a four-lobed greenish corolla, and four stamens alternating with the corolla lobes. However, the stamen filaments are deeply divided so that at first sight there appear to be eight stamens. The lateral flowers have a three-lobed calyx (or

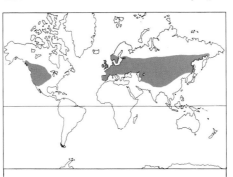

**Number of genera:** 1
**Number of species:** 1
**Distribution:** Europe, N America, northern and central Asia.
**Economic uses:** none.

bracts) and a five-lobed greenish corolla and five (apparently 10) stamens. In both types of flower nectar is secreted by a ring round the base of the stamens. The flowers are visited by various small insects, particularly flies. The ovary is semi-inferior, of three to five fused carpels and three to five locules, each with a single pendulous ovule. The style is three- to five-lobed. The fruit is a drupe and the seed contains a small embryo surrounded by copious endosperm.

**Classification.** The affinities of the Adoxaceae are in some doubt. At one time *Adoxa* was included in the Caprifoliaceae but it differs from the members of this family in the longitudinal splitting of the stamen filaments almost to the base, and the semi-inferior ovary. Relationships have been suggested with the Araliaceae and the Saxifragaceae.

**Economic uses.** None are known. S.R.C.

## CAPRIFOLIACEAE
*Elders and Honeysuckles*

The Caprifoliaceae is a smallish family mainly of small trees and shrubs, several of which are climbers and familiar ornamentals.

**Distribution.** Although generally distributed throughout the world, the family is richest in temperate parts of eastern North America and eastern Asia, and absent from the Sahara, southern and tropical Africa, but with one species in the East African mountains and Madagascar.

**Diagnostic features.** The Caprifoliaceae are mostly small trees or shrubs, with a number of lianas (as in some species of *Lonicera*). *Triosteum* is herbaceous, and *Sambucus* contains both shrubs and herbs. The leaves are usually opposite, simple and without stipules. An exception is *Sambucus* which has stipules and sometimes pinnate leaves, while in *Viburnum* and *Leycesteria* stipules are represented as nectar glands. The flowers are bisexual, and either regular or irregular. The inflorescence is basically cymose, often with pairs of flowers which are sometimes joined towards the bases of the ovaries. The calyx consists of four or five partly fused sepals. The calyx tube is fused to the ovary wall and is surmounted by usually five small teeth. The corolla, which rises above the ovary, usually has five spreading lobes and two lips. The shape of the tube is very variable, being very short in *Viburnum*, while in *Lonicera* it can be long and narrow or shortish and salver-shaped. The lobes are imbricate in bud, except in *Alseuosmia* and related genera and *Sambucus*. The stamens arise from the corolla tube, alternate with the lobes; occasionally one of the five is absent, as in *Linnaea*. The anthers have two locules and open by longitudinal slits, usually inwards. The ovary is inferior and formed usually of three to five united carpels, containing one to five locules. It is surmounted by a single style bearing a capitate or lobed stigma. There is a single pendulous ovule (numerous in *Leycesteria*) on an axile placenta in each locule. The fruit is most often a berry and the seeds typically have a small, straight embryo with copious endosperm. *Diervilla* and *Weigela* have a capsular fruit.

**Classification.** The chief genera are *Abelia* (30 species), *Diervilla* (three species), *Leycesteria* (six species), *Lonicera* (200 species), *Sambucus* (40 species), *Triosteum* (six species), *Viburnum* (200 species) and *Weigela* (12 species).

*Alseuosmia*, a genus of about eight species, from New Zealand, is sometimes separated as the family Alseuosmiaceae. Two other genera usually included in the Caprifoliaceae are transferred to this small family if recognized, namely *Periomphale* (two species) and *Memecylanthus* (one species), both from New Caledonia. They all have alternate leaves and valvate corolla lobes, and the pollen structure suggests that they may belong nearer to the Escallonioideae (a subfamily of the Saxifragaceae) and/or the Loganiaceae.

The other small genera, from Southeast Asia, *Carlemannia* (three species) and *Silvianthus* (two species), have also been split off as a further separate family, the Carlemanniaceae. Affinities with the Hydr-

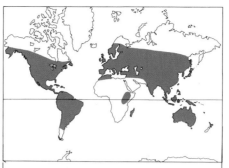

**Number of genera:** about 18
**Number of species:** about 450
**Distribution:** cosmopolitan, centered in E N America and E Asia.
**Economic uses:** many ornamentals (eg honeysuckle, snowberry) and elderberries used in wine-making.

angeoideae (a subfamily of the Saxifragaceae) are suggested by their two stamens with the anthers joined round the style and the numerous seeds in a two-chambered, four- to five-valved capsule.

The Caprifoliaceae as a whole is closely related to the Rubiaceae, from which the majority can be most readily distinguished by the absence of stipules.

**Economic uses.** Although a few species such as *Lonicera japonica* present a problem as weeds, the family is best-known for the numerous hardy ornamental shrubs it contains. Notable among these are *Lonicera* (honeysuckles), *Symphoricarpos* (snowberry, coralberry), *Sambucus* (elder) and species of *Viburnum*, *Abelia*, *Leycesteria* and *Weigela*. Elderberries are often used in wine-making.                              I.B.K.R.

# VALERIANACEAE
## Spikenard and Valerians

The Valerianaceae is a medium-sized family, more than half of whose members are in the genus *Valeriana*. They are mostly herbs, but a few have a shrubby habit, while in parts of South America some are cushion plants.

**Distribution.** The Valerianaceae is primarily

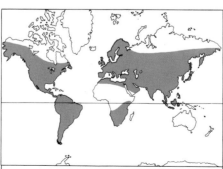

**Number of genera:** 13
**Number of species:** about 400
**Distribution:** mainly N Hemisphere and S America
**Economic uses:** limited uses as medicines, in perfumes, in salads and as ornamentals.

a Northern Hemisphere family, being absent from Australia and much of Africa. However, there is a diversity of genera in South America, particularly in the mountainous areas. The Mediterranean region is also a center of diversity with two genera, *Centranthus* (nine species) and *Fedia* (about three species), being endemic to this region.

**Diagnostic features.** The leaves are opposite, often pinnatisect, without stipules, and usually have clasping bases; when dry the plants give off a strong, characteristic odor, due to the presence of valerianic acid and derivatives, particularly in the roots.

The inflorescence is cymose, usually with numerous, bracteate, often crowded flowers, as in the part-inflorescences of *Valerianella*. The flowers are usually bisexual and irregular. The calyx, situated on the top rim of the ovary, is scarcely visible in flower, but shows great variation in structure in fruit. It is basically five-lobed but is considerably modified by unrolling into a "parachute" of up to 30 plumose segments in the fruit of *Valeriana* and *Centranthus*, while in *Valerianella*, a genus of about 80 species of small, annual weeds, it shows many variations in structure from being virtually absent in some species to being bladder-like or plumose in others. The petals are united, with usually five imbricate lobes and often with a long corolla tube which is particularly strongly developed in *Centranthus*, some species of which also have a long spur at the base containing nectar; pollination of such flowers is by moths. The stamens, which are attached to the upper part of the corolla tube, vary in number from four in *Patrinia* and *Nardostachys*, three in *Valeriana* and two in *Fedia* to one in *Centranthus*.

The ovary, of three united carpels, is inferior and has one locule by unequal development of the carpels, with a single pendulous, anatropous ovule. The style is simple and slender. The fruit is a cypsela, dry and indehiscent, parallel in many ways to the bicarpellate fruit of the Compositae.

**Classification.** The genera are mostly well delimited. The reduction in number of stamens and the highly modified calyx of some genera is an indication of how derived a particular genus is within the family. The Valerianaceae is in the order Dipsacales, generally recognized as an advanced group of families showing adaptations paralleled in the Compositae.

**Economic uses.** Species of *Valeriana* have medicinal properties, and both root and leaf extracts are used locally to treat nerve complaints. Some species of *Valerianella* provide corn salad or lamb's lettuce, used as a salad mainly in Continental Europe. A few species of the Valerianaceae produce perfumes and dyes; best-known is *Nardostachys jatamansi* from the Himalayan region, the spikenard of old. The best-known garden-ornamental is *Centranthus ruber*, red valerian.                              I.B.K.R.

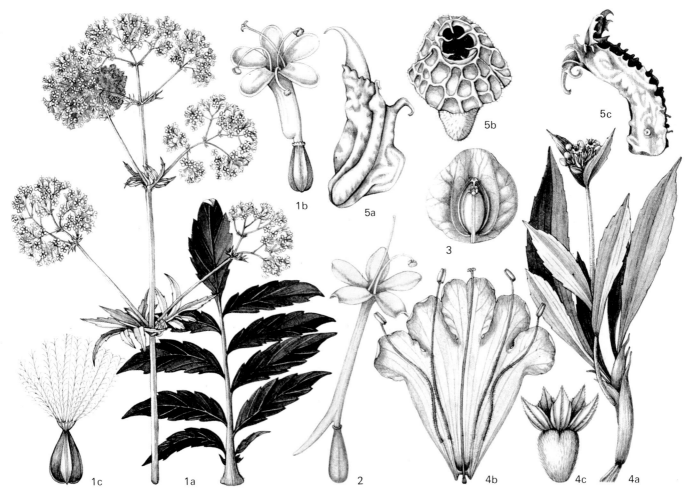

Valerianaceae. 1 *Valeriana officinalis* (a) pinnatisect leaf and cymose inflorescence ( ×$\frac{2}{3}$) ; (b) flower with small corolla spur ( ×6) ; (c) fruit crowned by plumose calyx ( ×6). 2 *Centranthus lecoqii* flower with distinct spur and single stamen ( ×4). 3 *Patrinia villosa* fruit ( ×4). 4 *Nardostachys jatamangi* (a) leafy shoot and inflorescence ( ×$\frac{2}{3}$) ; (b) flower opened out showing four stamens and one style ( ×4) ; (c) fruit ( ×4). 5 *Valerianella* species have very varied fruits due to the growth of the calyx ; shown here are those of (a) *V. echinata*, (b) *V. vesicaria* and (c) *V. tuberculata*.

# DIPSACACEAE
*Teasel and Scabious*

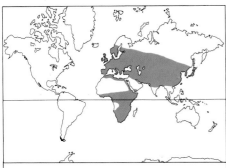

**Number of genera:** 11
**Number of species:** 350
**Distribution:** Europe to E Asia and central to southern Africa.
**Economic uses:** floriculture and teasel used in cloth industry.

The Dipsacaceae is a small family of herbs or subshrubs which provides some fine ornamentals as well as that rare product of the plant kingdom, a natural tool – the teasel.
**Distribution.** The family's center of diversity is in the Mediterranean region and the Near East, but extending to northern Europe, eastern Asia and central to southern Africa.
**Diagnostic features.** The Dipsacaceae are annual to perennial herbs and rarely shrubs. The leaves are opposite or in whorls and are without stipules. The flowers are in dense, cymose heads (capitula) subtended by involucral bracts, often with longer marginal flowers, and rarely in a spike of verticillasters. They are bisexual or female, and irregular, each with a basal epicalyx (involucel) of fused bracteoles which may be expanded into a "corona," often subtended by a receptacular scale.

The calyx is small, and cup-shaped or divided into four, five or numerous teeth or bristles. The corolla has four or five subequal lobes, or is two-lipped. There are two or four stamens, growing on the petals and alternating with the corolla lobes. The ovary is inferior, of two fused carpels and a single locule with a single pendulous ovule. The style is slender and the stigma either simple or two-lobed. The fruit is dry and indehiscent, enclosed in the epicalyx and often surmounted by a persistent calyx. It has a single seed with endosperm; the embryo is straight.

Perennial species occur in all genera but annuals have evolved several times in dry habitats as parallel lines of specialization, especially in *Cephalaria*, *Knautia*, *Pterocephalus* and *Scabiosa*. The capitula are generally conspicuous and insect-pollinated, with outbreeding assured by frequent self-sterility. Only some annual groups in, for example, *Pterocephalus* and *Scabiosa*, have inconspicuous capitula with reduced flowers which are predominantly self-pollinated.
**Classification.** The family is usually divided into two tribes: MORINEAE and DIPSACEAE. The Morineae has flowers in verticillasters in the upper leaf-axils, a leaf-like calyx and either two stamens, or four stamens two of which are longer than the others. The Dipsaceae has flowers in capitula (the outer flowers often enlarged), a scaly or bristly calyx and four (rarely two) similar stamens. Only *Morina* is included in the first tribe, which is sometimes separated as a distinct family, the Morinaceae.

Differentiation in the Dipsacaceae has been most evident in fruit characters. In less advanced species of *Cephalaria* and *Succisa* the involucel and calyx are inconspicuous and undifferentiated, the flowers are subtended by leafy bracts and the fruits are small

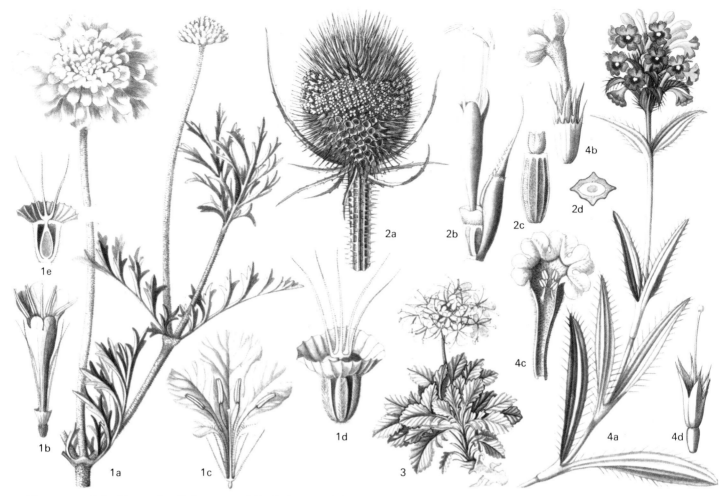

Dipsacaceae. 1 *Scabiosa anthemifolia* var *rosea* (a) leafy shoot and inflorescences ($\times\frac{2}{3}$); (b) inner flower with bristle-like calyx-segments ($\times 3$); (c) larger outer flower opened out ($\times 2$); (d) fruit with epicalyx (involucel) expanded into an umbrella-shaped extension and crowned by spines ($\times 3$); (e) vertical section of fruit ($\times 2$). 2 *Dipsacus fullonum* (a) dense flower head surrounded by spiny bracts ($\times\frac{2}{3}$); (b) flower ($\times 3$); (c) fruit ($\times 3$); (d) cross section of fruit ($\times 5$). 3 *Pterocephalus perennis* flowering shoot ($\times\frac{2}{3}$). 4 *Morina betonicoides* (a) leafy shoot and inflorescence ($\times\frac{2}{3}$); (b) flower ($\times 2$); (c) corolla tube opened to show stamens of two lengths ($\times 2$); (d) gynoecium and calyx ($\times 2$).

hard grains. The subtending bracts are elongated and rigid in *Dipsacus* to provide a catapult mechanism for dispersal as animals brush against the thistle-like fruiting heads. The bracts are reduced in other genera that develop adaptations to wind dispersal, which is achieved in *Scabiosa*, for example, by a thin, translucent umbrella produced by extension of the involucel limb, and in *Pterocephalus* by the numerous long calyx bristles which bear feather-like hairs. Adaptations to animal dispersal are seen in several genera. *Knautia* has a fatty outgrowth at the fruit base that attracts ants which carry off the fruit, while annual species of *Cephalaria* and *Scabiosa* develop, from the involucel and calyx respectively, rough, often branched, spines.

This family bears a superficial resemblance to the Compositae, in that the inflorescence usually forms a head surrounded by a calyx-like involucre of bracts. It is, however, readily distinguished by the stamens, with free anthers protruding from the corolla, and by the fruit being enclosed in an involucel of united bracteoles and crowned by the calyx composed of four, five or more bristles or teeth.

**Economic uses.** Many species of *Cephalaria*, *Morina*, *Pterocephalus* and *Scabiosa* (scabious) are cultivated for ornament, while teasel (*Dipsacus sativus*) is used on a limited scale for raising the nap on cloth.

D.M.M.

# CALYCERACEAE

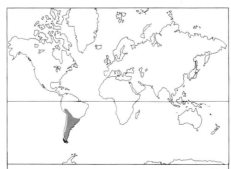

**Number of genera:** 6
**Number of species:** about 52
**Distribution:** S America.
**Economic uses:** none.

The Calyceraceae is a small family of annual and perennial herbs.
**Distribution.** The family is entirely South

American, being most abundant in the Andes south from Bolivia, but extending eastwards through Paraguay to Uruguay and southern Brazil and throughout much of Argentina to southern Patagonia. Most species occur in dry soils supporting open scrub or steppe vegetation.

**Diagnostic features.** The leaves are alternate, entire to pinnatisect, in a basal rosette and without stipules. The flowers are bisexual or rarely unisexual, regular or irregular, clustered in a head with the receptacle surrounded by one to two series of involucral bracts. The heads of flowers may be solitary or in cymes, and are stalked or sessile. The receptacular bracts are free, united into groups of two or three, or absent. There is a calyx of four to six teeth or lobes, persisting in the fruit. The corolla is cylindrical, with a four- to six-lobed limb. The stamens are four to six, alternate with the corolla lobes and joined into a tube around the style, the filaments being partially free and inserted at various levels in the corolla tube. The anthers are free, with inward dehiscence. The ovary is inferior and has one locule with a solitary, anatropous, apical, pendulous ovule. The style is protruding and thread-

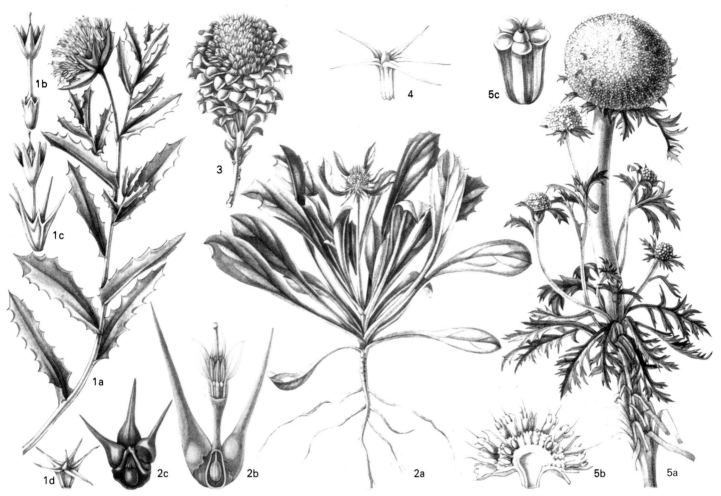

Calyceraceae. 1 *Calycera crassifolia* (a) leafy shoot and flowers in a head surrounded by involucre of bracts ( $\times\frac{2}{3}$ ); (b) floret immediately after opening ( $\times 2$ ); (c) old floret ( $\times 2$ ); (d) fruit—a ribbed achene with a persistent calyx ( $\times\frac{2}{3}$ ). 2 *Acicarpha spathulata* (a) habit ( $\times\frac{2}{3}$ ); (b) fertile marginal floret with spiked calyx-lobes and parts of the corolla tube and ovary wall removed to show stamens in tube around the style and single pendulous ovule ( $\times 4$ ); (c) vertical section of achene ( $\times 4$ ). 3 *Moschopsis rosulata* shoot ( $\times\frac{2}{3}$ ). 4 *Calycera herbacea* var *sinuata* achene ( $\times 2$ ). 5 *Nastanthus patagonicus* (a) habit ( $\times\frac{2}{3}$ ); (b) vertical section of capitulum ( $\times 2$ ); (c) achene ( $\times 6$ ).

like with a rounded stigma. The fruits are achenes with persistent calyx lobes at the apex, and may be free or united with each other. They contain a pendulous seed with fleshy endosperm and a straight embryo.

**Classification.** The six genera can be distinguished as follows: in *Moschopsis* the involucre is not well developed; in *Acicarpha* the heads of flowers have sterile central and fertile marginal florets, the outer achenes being joined in their lower part; *Gamocarpha* has the receptacular bracts united into groups of two to three; *Calycera* has dimorphic achenes, the outer being spiny; *Nastanthus* has winged achenes and a large receptacle, while *Boopis* has ribbed achenes and a small receptacle.

Because the flowers are borne in heads subtended by involucral bracts, this family bears an obvious resemblance to the Compositae and Dipsacaceae, while the pollen form and arrangement of anthers are similar to those found in the Goodeniaceae and related families. The floral morphology is similar to that of the Compositae but the ovule is apical in the Calyceraceae, not basal, so that it is usually treated as an aberrant relative of the Dipsacaceae, which it further

resembles in the attachment of the filaments and in certain embryological features.

**Economic uses.** None are known.

D.M.M.

# ASTERALES

## COMPOSITAE

*The Sunflower Family*

The Compositae or Asteraceae is one of the largest families of flowering plants, with about 1,100 currently accepted genera and 25,000 species. Most of its members are evergreen shrubs or subshrubs or perennial rhizomatous herbs, but tap-rooted or tuberous-rooted perennials, and biennial and annual herbs are also frequent; large trees are infrequent, as are epiphytes, and true aquatics are rare. Some tropical island and montane species are giant tree-like herbs, the so-called cabbage trees; many are scramblers, some are true climbers and not a few are succulents, with fleshy leaves or stems. The family includes lettuces, artichokes and sunflowers, as well as chrysanthemums, dahlias and numerous other popular garden flowers, not to mention

weeds like dandelions, thistles and sow thistle.

**Distribution.** The family is of worldwide distribution, being absent only from the Antarctic mainland, and is particularly well represented in semiarid regions of the tropics and subtropics, such as the Mediterranean region, Mexico, the Cape Province of South Africa, and the woodland, wooded grassland, grassland and bushland formations of Africa, South America and Australia. Compositae are also abundant in arctic, arctic-alpine, temperate and montane floras throughout the world. Only in the tropical rain forests are they poorly represented.

**Diagnostic features.** The leaves are alternate or opposite, rarely whorled, and without stipules; they are simple (rarely compound), pinnately or palmately veined, and sessile or with petioles, often with an expanded, sheathing or auricled base; they are usually lobed or toothed, sometimes succulent, rarely ending in a tendril, sometimes reduced to scales and quickly falling. Anatomically, Compositae are characterized by the presence of resin canals (in all except most members of the tribe Lactuceae) or latex ducts (in all Lactuceae and a few genera of

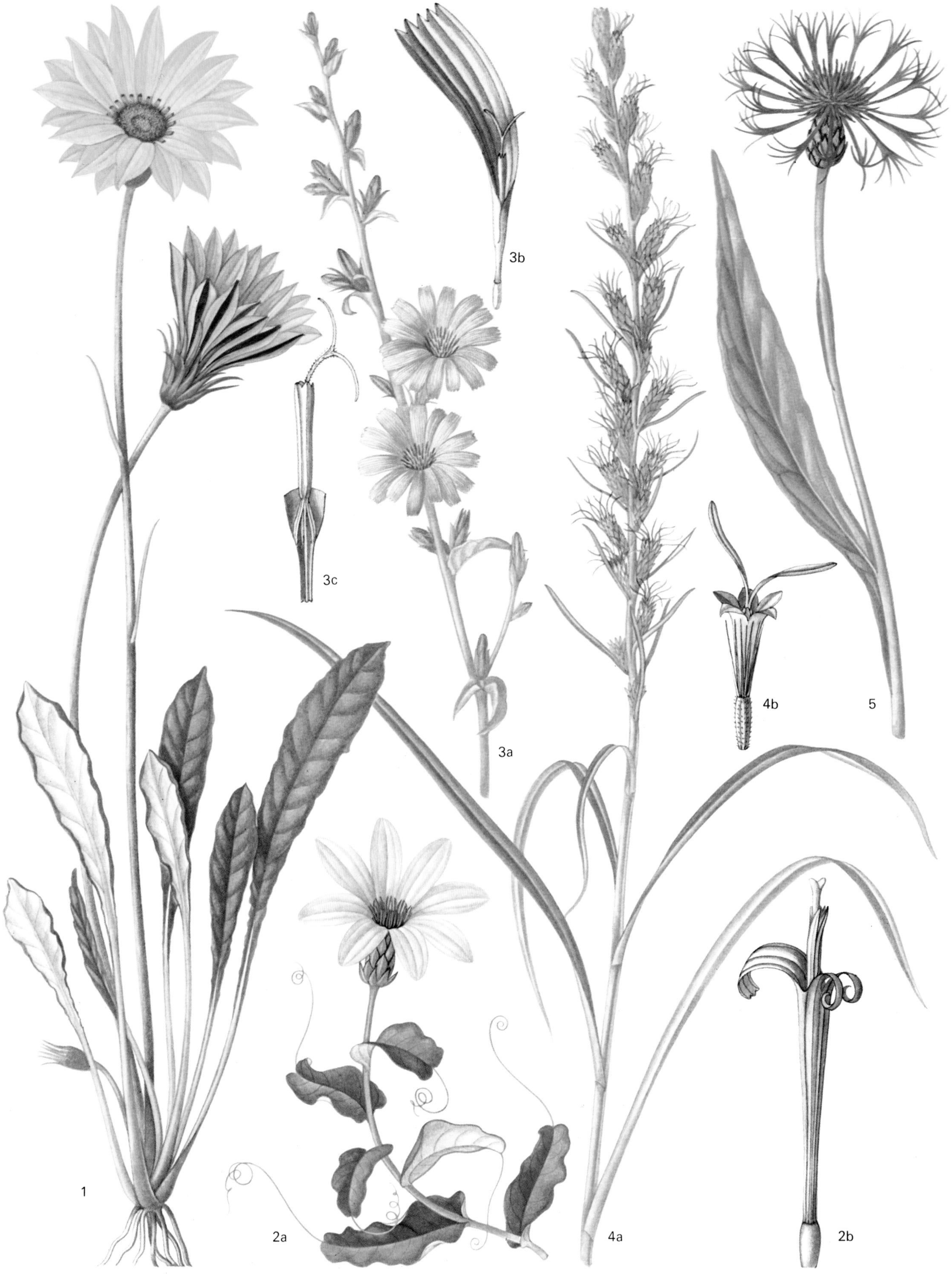
1
2a
2b
3a
3b
3c
4a
4b
5

Cardueae and Arctotideae). Characteristic biochemical features include the presence of the polysaccharide inulin, instead of starch, in the subterranean parts and fatty oils in the seeds.

The familiar daisies, thistles and dandelions exemplify one of the most characteristic features of the Compositae – the head-like inflorescence, known as a capitulum, made up of numerous small individual flowers called florets, and surrounded by an involucre of protective bracts. The whole resembles a single flower, and is usually taken as such by the layman; indeed, biologically it functions as a single flower. This type of inflorescence is constant throughout the family, though sometimes much modified in various ways.

Further characteristic floral features include the inferior ovary with one locule and one basal ovule; the modified calyx, known as a pappus, made up of hairs, scales, bristles or awns (sometimes more or less fused), which acts as an aid to the distribution of the fruit, although it is sometimes completely lacking; the corolla of fused petals, with the five stamens inserted on the corolla tube and united by their anthers into a cylinder surrounding the style; and the two style arms, which bear hairs or papillae on their tips and external surfaces while the receptive stigmatic surfaces are on the inner sides.

This arrangement of anthers and style arms is associated with a particular mode of pollen presentation. The anthers ripen before the stigmas, and discharge their pollen into the tube formed by the cylinder of fused anthers. At this stage, the style is short and the style arms are pressed together. The style then elongates up the anther tube, from which the pollen is swept by the hairs of the style arms and presented at the apex of the anther tube to any visiting pollinator. Only later do the style arms separate to expose the stigmatic surfaces. Each floret thus goes through a staminate (male), then a pistillate (female) stage. Finally, the style arms may

Compositae. 1 *Gazania linearis* a low-growing perennial herb producing a basal rosette of leaves and terminal radiate (ie outer ray and inner disk florets) flower heads (capitula) that are subtended by a series of green bracts ($\times\frac{2}{3}$). 2 *Mutisia oligodon* (a) a scrambling perennial herb showing alternate leaves with tendrils and the terminal capitulum ($\times\frac{2}{3}$); (b) bisexual bilabiate floret with a three-toothed outer lip and a two-toothed inner lip ($\times 3$). 3 *Cichorium intybus* (endive) (a) flowering shoot with capitula of ligulate florets only ($\times\frac{2}{3}$); (b) ligulate floret ($\times 2$); (c) ligulate floret with corolla removed to show stamens inserted in the corolla tube and anthers united in a tube around the style ($\times 4$). 4 *Liatris graminifolia* (a) flowering shoot bearing discoid capitula (ie only with disk florets) ($\times\frac{2}{3}$); (b) disk floret with a regular five-lobed corolla ($\times 4$). 5 *Centaurea montana* leaf and terminal capitulum of disk florets, the outer ones being sterile and enlarged ($\times\frac{2}{3}$).

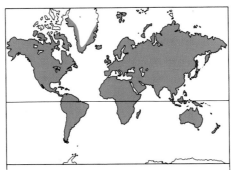

**Number of genera:** about 1,100
**Number of species:** about 25,000
**Distribution:** cosmopolitan.
**Economic uses:** food plants (eg lettuce, artichokes, sunflower), ornamentals (chysanthemums, dahlias), insecticide (pyrethrum) and medicines and drugs.

recurve sufficiently for the stigmas to make contact with pollen from the anthers of their own floret. In this way, in self-compatible species, self-pollination may be effected should cross-pollination for some reason have failed to occur. Some florets, however, are exclusively staminate or pistillate.

Florets in the individual capitula are arranged in a racemose or indeterminate manner, with the outer ones opening first. The capitula themselves, however, are cymosely disposed in the total overall inflorescence, which is very variable in size, shape, construction, number of capitula and disposition on the plant. Commonly, the capitula are arranged in terminal or terminal and upper axillary corymboid or paniculoid cymes or thyrses. In some genera, the capitula are secondarily aggregated into spikes or compound capitula; sometimes, as in *Echinops*, each individual capitulum in such a compound cluster is reduced to the one-flowered condition and the whole is sometimes surrounded, as in *Lagascea*, *Elephantopus* and some *Sphaeranthus* species, by a secondary involucre of its own.

The most common type of floret is the disk floret, which is usually bisexual, sometimes staminate or sterile, and has a tubular corolla with five (rarely four) apical lobes or teeth. Other types are the bilabiate floret, which is two-lipped with a three-toothed (or four-toothed) outer (lower) lip and a two-toothed (or one-toothed) inner (upper) lip; the ray floret, usually pistillate or sterile, and strap-shaped with three or fewer apical teeth; the ligulate floret, which is bisexual and strap-shaped with five apical teeth; the reduced ray floret, which is pistillate and has a short, usually narrow ray; and the filiform floret, which is pistillate, with a narrow corolla tube and oblique or truncate apex. Rarely, the corolla is highly reduced or even absent.

Capitula may be homogamous, with all the florets bisexual, or heterogamous, in which case the inner florets are usually bisexual and the outer florets pistillate or

sterile. Less commonly, other arrangements may be found, and in some cases the capitula are unisexual, either pistillate (with all florets pistillate) or staminate (with all florets staminate). Unisexual capitula may be borne on the same plant (the monoecious condition) or on different plants (the dioecious condition). Homogamous capitula may be discoid (with all florets disk), ligulate (with all florets ligulate) or bilabiate (with all florets bilabiate). Heterogamous capitula are usually radiate (with outer ray florets and inner disk florets), disciform (with outer filiform florets and inner disk florets) or radiant (with inner bisexual disk florets and outer enlarged, sterile disk florets).

In common use, as in *Bellis* or in *Helianthus*, the central regular florets are collectively called the "disk" and the outer irregular florets are termed the "rays".

The involucre may be of one or more rows of overlapping bracts; only rarely is it absent. On the receptacle, between the florets, there may occur chaffy bracts, scales, bristles or hairs; most often, however, the receptacle is naked. The anthers may be dorsifixed (fixed to the filaments by their upper surface), with pollen grains within the anthers below the level of attachment, or basifixed (fixed at their bases), with all the pollen grains above the insertion. At the apex of the anther, there is usually a sterile appendage; the base of each anther lobe may be rounded, acute or lobed, short or long; when long and attenuate, it is referred to as tailed. The style arms vary in shape, length and distribution of the hairs or papillae, configuration of the stigmatic surfaces, and in the shape of the apex, which may be truncate, rounded, tapered or variously appendaged.

The fruit is one-seeded, indehiscent, nearly always dry, and is termed a cypsela. It may be angular, rounded, variously compressed, or curved, ornamented or winged in various ways; rarely it is a drupe, with fleshy endocarp. It is often furnished with an apical pappus (a ring of fine hairs which can act as a parachute for wind-dispersal). The seed has no endosperm and a straight embryo.

**Classification.** The classification of the Compositae is in a state of transition. The arrangement into twelve tribes, which has been generally accepted for the last 20 years, is now seen to be in need of modification, in the light of recent discoveries in biochemistry, pollen analysis, micromorphology, anatomy and cytology. Not a few genera have been shown to be misplaced, while others require segregation into distinct tribes. Below tribal level, the classification into subtribes and genera is likely to be much modified in the light of new knowledge, and the number of accepted species is likely to undergo reduction. The following arrangement recognizes two subfamilies and 17 tribes. The tribes Eupatorieae and Senecioneae are intermediate between the two

8c

8b

8a

6

10

7

9

11

subfamilies in some respects, and might warrant segregation together into a third subfamily.

## Subfamily Lactucoideae

Capitula homogamous, ligulate, bilabiate or discoid, less often heterogamous, radiate or disciform; disk florets usually with long, narrow lobes, purplish, pinkish or whitish, less often yellow; anthers dorsifixed; style arms usually with single stigmatic area on inner surface; pollen ridged, ridged and spiny, or spiny.

1. LACTUCEAE. Capitula ligulate; latex ducts present; resin canals mostly absent; pollen usually ridged and spiny or spiny; mostly herbs; leaves alternate. Nine subtribes, 70 genera, 2,300 species, worldwide, mainly in the Northern Hemisphere. *Catananche* (Mediterranean), *Chondrilla* (Eurasia), *Cichorium* (Mediterranean), *Crepis* (Northern Hemisphere, Africa), *Hieracium* (temperate, except Australasia), *Hypochoeris* (Northern Hemisphere and South America), *Lactuca* (mostly Northern Hemisphere, Africa), *Lapsana* (Eurasia), *Picris* (Eurasia and Mediterranean), *Prenanthes* (temperate Northern Hemisphere), *Scolymus* (Mediterranean), *Scorzonera* (Eurasia), *Sonchus* (Old World), *Taraxacum* (mostly Northern Hemisphere), *Tragopogon* (Eurasia).

2. MUTISIEAE. Capitula bilabiate or ligulate, rarely radiate or discoid; latex ducts absent; anthers usually tailed; pollen and style arms various; woody or herbaceous; leaves alternate, rarely opposite. Three subtribes, 90 genera, 1,000 species, mostly South American. *Barnadesia* (South America), *Gerbera* (Africa and Asia), *Mutisia* (South America), *Perezia* (New World), *Stifftia* (tropical South America).

3. EREMOTHAMNEAE. Capitula radiate; latex ducts absent; anthers shortly tailed; pollen spiny; style arms elongate, acute; shrubs; leaves alternate. One subtribe, one genus, one species, South Africa. *Eremothamnus*.

4. ARCTOTIDEAE. Capitula radiate, rarely discoid; latex ducts usually absent; anthers obtuse to acute, not tailed; pollen spiny; style arms short with swollen papillose zone below the style arms, usually with collar of hairs at its base; herbs or shrubs; leaves alternate. Three subtribes, 15 genera, 200 species, mostly South African. *Arctotis* (including *Venidium* (South Africa), *Gazania* (South Africa).

5. CARDUEAE. Capitula discoid, homogamous or with sterile outer florets; latex ducts usually absent; anthers acute, often

tailed; pollen spiny, not ridged; style arms usually short, often with papillose zone below them as in Arctotideae; herbs or rarely shrubs; leaves alternate. Three well-marked subtribes, 80 genera, 2,600 species, mostly Eurasian. *Arctium* (temperate Eurasia), *Carduus* (Eurasia), *Carlina* (Eurasia), *Carthamus* (Mediterranean), *Centaurea* (mostly Eurasia and Africa), *Cirsium* (temperate Northern Hemisphere), *Cnicus* (Mediterranean), *Cousinia* (Asia), *Cynara* (Mediterranean, southwest Asia), *Echinops* (Eurasia, Africa), *Jurinea* (Eurasia), *Onopordum* (Mediterranean, southwest Asia), *Saussurea* (mostly temperate Asia), *Silybum* (Mediterranean), *Xeranthemum* (Mediterranean).

6. VERNONIEAE. Capitula discoid, rarely ligulate, homogamous; latex ducts absent; anthers obtuse to acute, rarely tailed; pollen spiny, ridged and spiny, or ridged; style arms elongate, gradually attenuate, acute or obtuse, hairy; herbs or shrubs, rarely trees; leaves alternate or opposite. One subtribe, 50 genera, 1,200 species, mostly tropical. *Stokesia* (southeast United States of America), *Vernonia* (pantropical).

7. LIABEAE. Capitula radiate or discoid; latex ducts absent; anthers acute or shortly tailed; pollen spiny; style arms as in Vernonieae; herbs, shrubs or small trees; leaves opposite or whorled. One subtribe, 15 genera, 120 species, New World, mostly tropical. *Liabum* (Central and South America).

8. EUPATORIEAE. Capitula discoid, homogamous; latex ducts absent; anthers obtuse to acute, not tailed; pollen spiny; style arms elongate, club-shaped, papillose; herbs or shrubs; leaves opposite or alternate. Perhaps three subtribes, 120 genera, 1,800 species, mostly New World. *Ageratum* (New World), *Ayapana* (New World), *Chromolaena* (New World), *Eupatorium* (temperate Northern Hemisphere), *Liatris* (North America), *Mikania* (mostly New World tropics), *Piqueria* (Central and South America), *Stevia* (New World).

## Subfamily Asteroideae

Capitula heterogamous, radiate or disciform, less often discoid; disk florets usually with short, broad lobes, usually yellow; anthers basifixed; style arms usually with two distinct stigmatic areas; pollen spiny; latex ducts absent.

9. SENECIONEAE. Involucral bracts usually in one row, often with an outer series of reduced bracts; receptacle naked; anthers rounded to acute, sometimes tailed; style arms usually truncate, apically minutely hairy, less often variously appendaged; herbs or shrubs, sometimes more or less succulent; leaves alternate. Three subtribes, 85 genera, 3,000 species, worldwide. *Cacalia* (temperate Northern Hemisphere), *Cineraria* (Africa), *Crassocephalum* (Africa), *Doronicum* (Eurasia), *Emilia* (Old World tropics), *Euryops* (Africa and Arabia), *Gynura* (Africa and Asia), *Kleinia* (mostly Africa), *Ligularia* (mostly eastern Asia),

*Othonna* (South Africa), *Petasites* (Eurasia), *Senecio* (worldwide), *Tussilago* (Eurasia).

10. TAGETEAE. Involucral bracts in one or two series, free or united, usually with pellucid glands; receptacle naked; anthers obtuse to acute, not tailed; style arms truncate or variously appendaged; herbs or shrubs; leaves opposite or alternate, usually with pellucid glands containing strongly scented essential oils. Two subtribes, 20 genera, 250 species, New World, mostly Mexico and Central America. *Tagetes* (New World), *Pectis* (New World).

11. HELIANTHEAE. Involucral bracts in several series; receptacle naked or scaly; anthers obtuse to acute, not tailed; style arms various, tapered and hairy throughout or truncate with apical hairs or hairy appendages; pappus usually of awns or scales; herbs, less often shrubs; leaves opposite, less often alternate, frequently roughly hairy. Perhaps 26 subtribes, 250 genera, 4,000 species, worldwide but mostly New World. *Acanthospermum* (South America), *Ambrosia* (mostly New World), *Argyroxiphium* (Hawaii), *Arnica* (northern temperate and arctic), *Bidens* (including *Coreopsis* and *Cosmos*) (worldwide), *Dahlia* (Central America), *Echinacea* (North America), *Eriophyllum* (North America), *Espeletia* (South America), *Flaveria* (mostly Central America), *Gaillardia* (New World), *Galinsoga* (Central and South America), *Guizotia* (Africa), *Helenium* (North America), *Helianthus* (mostly North America), *Heliopsis* (North America), *Lagascea* (New World tropics), *Layia* (North America), *Madia* (New World), *Parthenium* (New World), *Polymnia* (New World), *Ratibida* (North and Central America), *Rudbeckia* (North America), *Sanvitalia* (North America), *Silphium* (North America), *Spilanthes* (pantropical), *Tithonia* (Central America), *Tridax* (Central and South America), *Wyethia* (North America), *Xanthium* (New World), *Zinnia* (New World).

12. INULEAE. Involucral bracts in several series; receptacle naked, sometimes scaly; anthers usually tailed; style arms truncate and apically hairy, rounded or variously appendaged; herbs or shrubs; leaves alternate. Perhaps three subtribes. 180 genera, 2,100 species, worldwide. *Anaphalis* (mostly temperate Northern Hemisphere), *Antennaria* (mostly arctic and temperate Northern Hemisphere), *Blumea* (Old World tropics), *Buphthalmum* (Eurasia), *Gnaphalium* (worldwide), *Helichrysum* (Africa, Madagascar, Australasia, Eurasia), *Helipterum* (southern Africa, Australia), *Inula* (Eurasia and Africa), *Leontopodium* (Eurasia), *Raoulia* (Australasia), *Sphaeranthus* (Old World tropics).

13. ANTHEMIDEAE. Involucral bracts in several series, usually with thin, dry, transparent tips or margins; receptacle naked or scaly; anthers obtuse to acute, not tailed; style arms truncate, fringed with short hairs;

Compositae (continued). 6 *Ursinia speciosa* shoot bearing pinnatisect (deeply cut) leaves and radiate capitulum ($\times \frac{2}{3}$). 7 *Bellis perennis* showing basal rosette of leaves and solitary radiate capitula ($\times \frac{2}{3}$). 8 *Helianthus giganteus* (a) female ray floret ($\times 4$); (b) bisexual disk floret ($\times 6$); (c) disk floret opened out ($\times 6$). 9 *H. angustifolius* flowering shoot ($\times \frac{2}{3}$). 10 *Argyranthemum broussonetti* flowering shoot ($\times 2$). 11 *Leontopodium haplophylloides* flowering shoot ($\times 2$).

herbs or less often shrubs; leaves alternate, vary rarely opposite, often much divided, strongly scented. Perhaps four subtribes, 75 genera, 1,200 species, mostly Northern Hemisphere. *Achillea* (temperate Northern Hemisphere), *Anacyclus* (Mediterranean), *Anthemis* (Europe, Mediterranean, southwest Asia), *Argyranthemum* (Macaronesia), *Artemisia* (mostly Northern Hemisphere), *Chrysanthemum* (Europe, Mediterranean), *Dendranthema* (mostly eastern Asia), *Leucanthemum* (Europe, Mediterranean, southwest Asia), *Lonas* (Mediterranean), *Matricaria* (Eurasia), *Santolina* (Mediterranean), *Tanacetum* (Europe, southwest Asia).

14. URSINIEAE. Involucral bracts in several series, often with thin, dry transparent tips or margins; receptacle scaly; anthers obtuse to acute, not tailed; style arms truncate, fringed with short hairs; shrubs or herbs; leaves alternate or opposite. Perhaps three subtribes, eight genera, 120 species, South African. *Lasiospermum* (South Africa), *Ursinia* (mostly South Africa).

15. CALENDULEAE. Involucral bracts in one or two series; receptacle naked; anthers acute, more or less tailed; style arms truncate, with apical hairs; pappus absent; cypselas often curiously shaped; herbs or shrubs; leaves alternate or opposite. One subtribe, seven genera, 100 species, Africa, Europe and southwest Asia. *Calendula* (mostly Mediterranean), *Dimorphotheca* (South Africa), *Osteospermum* (Africa and southwest Asia).

16. COTULEAE. Involucral bracts in one or two series; receptacle naked; anthers obtuse to acute, not tailed; style arms truncate, fringed with short hairs; mostly herbs; leaves alternate. One subtribe, 10 genera, 120 species, mostly Southern Hemisphere. *Cotula* (Australasia, subantarctic islands, South Africa, South America).

17. ASTEREAE. Involucral bracts in two or more series; receptacle naked, or rarely scaly; anthers obtuse, not tailed; style arms with a shortly hairy triangular to lanceolate apical appendage; herbs, shrubs or rarely small trees; leaves alternate or less often opposite. Perhaps three subtribes, 120 genera, 2,500 species. *Aster* (mainly Northern Hemisphere), *Baccharis* (New World), *Bellis* (Eurasia), *Brachylaena* (Africa, tribal position uncertain), *Brachycome* (Australasia), *Callistephus* (eastern Asia), *Conyza* (mostly tropical), *Erigeron* (mostly Northern Hemisphere), *Felicia* (Africa), *Grindelia* (temperate North and South America), *Haplopappus* (New World), *Olearia* (Australasia), *Solidago* (Northern Hemisphere, mostly New World).

The relationships of the Compositae are uncertain, a fact recognized in classification by their allocation as the only family in a distinct order, the Asterales. Certain features exhibited by the Compositae are also known in other families, but their occurrence is insufficiently correlated to permit the selec-tion of any one of these as indicating a close relationship. Some of the basic features of the Compositae probably evolved very rapidly very early in the history of the family, thus setting it apart very clearly from any other. They include the involucrate capitulum, the receptacle without bracts, the inferior ovary with a single locule, the capillary pappus, the pollen-presentation mechanism and the cypsela. The tribe Mutisieae is most heterogeneous in many respects, such as the pollen grain wall structure and corolla form, and gives the appearance of a rather loosely allied assemblage of archaic (though often highly specialized) forms. The bilabiate corolla is most frequent in this tribe, and certain considerations of the floral biology of racemose inflorescences indicate that this is the least specialized of the various types exhibited by the Compositae.

Amongst the families often postulated as allied to the Compositae – Rubiaceae, Caprifoliaceae, Dipsacaceae, Valerianaceae, Stylidiaceae, Goodeniaceae, Brunoniaceae, Calyceraceae, Campanulaceae – the first four have basically cymose, not racemose, inflorescences, and regular, not irregular, flowers; they also differ widely from the Compositae in biochemical features, in which they resemble more such families as the Cornaceae. The Stylidiaceae, Goodeniaceae and Brunoniaceae resemble the Compositae in being mostly racemose and irregular and in possessing inulin; on the other hand, they resemble the previous families, and differ from the Compositae, in some other biochemical features commonly held to be of taxonomic significance. The Calyceraceae is too poorly known biochemically to be assessed, but the Campanulaceae, whilst differing from the Compositae in embryological details, resembles it biochemically in the presence of inulin and the absence of tannins and iridoid compounds. On the other hand, there are many biochemical similarities between the Compositae and the Umbelliferae, Araliaceae and Pittosporaceae, and, more remotely, the Rutaceae and eventually the Magnoliales.

**Economic uses.** The Compositae are of incalculably great indirect economic importance to Man as major contributors to the diversity, and therefore the stability and sustainable productivity, of the drier (wooded grassland, grassland, bushland and semidesert) vegetation types throughout the world, especially in tropical and subtropical areas. In proportion to its size, however, the direct economic importance of the family is comparatively small. It includes food plants, sources of raw materials, medicinal and drug plants, ornamentals and succulents, and, on the debit side, weeds and poisonous plants.

The most commercially important food plant is *Lactuca sativa* (lettuce); others include *Cichorium endivia* (endive), *C. intybus* (chicory), *Scorzonera hispanica* and *Tragopogon porrifolius* (salsify), *Cynara scolymus* (globe artichoke), *Helianthus tuberosus* (Jerusalem artichoke), and the culinary herb tarragon (*Artemisia dracunculus*).

*Helianthus annuus* (sunflower), *Carthamus tinctorius* (safflower) and *Guizotia abyssinica* (niger seed) are grown for their seeds, which are important sources of edible and drying oils. *Tanacetum cinerariifolium* is the main commercial source of natural pyrethrum, used as an insecticide. *Parthenium argentatum* (guayule) and *Taraxacum bicorne* have been utilized as minor sources of rubber. *Brachylaena huillensis* yields a commercial timber (muhugu) of high durability. Many members of this chemically rich family have long been used in folk medicines. Certain *Artemisia* species, such as *Artemisia cina* and *A. maritima*, yield santonin, used as a vermifuge; *A. absinthium* is the source of the essential oil used to flavor absinthe. *Anthemis nobilis* produces chamomile.

The Compositae contribute largely to gardens throughout the world as ornamental plants. They include species and/or hybrids of the following genera: *Gerbera, Mutisia, Arctotis, Gazania, Echinops* (globe thistle), *Stokesia* (Stokes' aster), *Ageratum, Senecio* (*S. × hybridus*, florists' cineraria), *Tagetes* (African and French marigolds), *Bidens* (cosmos), *Dahlia* (*D. × hortensis*, dahlia), *Gaillardia, Helianthus* (sunflower), *Zinnia, Helichrysum* (everlastings), *Dendranthema* (florists' chrysanthemum), *Leucanthemum* (ox-eye daisy), *Ursinia, Calendula* (marigold), *Aster* (Michaelmas daisy), *Callistephus* (florists' aster), *Olearia* (daisy bush) and *Solidago* (golden rod). Of these, *Dahlia, Dendranthema* and *Callistephus* are the most important, each with many thousands of cultivars. Many species of *Kleinia, Senecio* and *Othonna* are grown by succulent plant enthusiasts.

Several Compositae have become widespread, sometimes noxious, weeds, frequently in areas far removed for their original homes. They include *Chondrilla juncea* (skeleton weed), *Sonchus oleraceus* (sow thistle), *Taraxacum* species (dandelions), *Cirsium arvense, C. vulgare* and *Carduus nutans* (thistles), *Ageratum conyzoides, Chromolaena odorata, Mikania micrantha, Crassocephalum crepidioides*, a number of *Senecio* species, *Ambrosia artemisiifolia, Tridax procumbens, Acanthospermum hispidum, Xanthium spinosum* and *X. strumarium* (cocklebur), *Bidens pilosa* (black Jack) and *Helichrysum kraussii*. Poisonous *Senecio* species are especially serious weeds of pasture as they are responsible for more deaths of domestic stock than all other poisonous plants together. The wind-borne pollen of the ragweeds *Ambrosia artemisiifolia* and *A. trifida* is one of the main causes of hay fever in the regions of North America where these species occur.

C.J.

# MONOCOTYLEDONS

# ALISMATIDAE

## ALISMATALES

### BUTOMACEAE
*Flowering Rush*

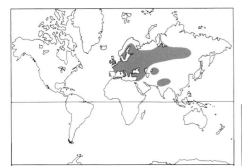

**Number of genera:** 1
**Number of species:** 1
**Distribution:** Europe and temperate Asia, in aquatic and marshy habitats.
**Economic uses:** cultivated as an ornamental and the rhizomes are eaten in parts of Russia.

The Butomaceae is a family consisting of a single herbaceous aquatic species, *Butomus umbellatus*, widely cultivated under its common name, the flowering rush.

**Distribution:** *Butomus* usually occurs in swamps, ditches and along pools, lakes and rivers. It is widespread in Europe and temperate Asia, and has become naturalized in North America.

**Diagnostic features.** *Butomus* is a rhizomatous perennial with linear leaves up to 1m (3ft) long or more. The leaves are triangular in transverse section and arise in two rows along the rhizome. The inflorescence is umbel-like, consisting of a single terminal flower surrounded by three cymes. The flowers are regular and bisexual. There are three petal-like sepals which are pink with darker veins, and persist in the fruit. The three petals are like the sepals, but somewhat larger. There are six to nine stamens. The pollen grains have a single distal aperture. The carpels are superior, six to nine in number and slightly united at the base; when ripe they are obovoid and crowned by a persistent style. The ovules are numerous and found scattered over the inner surface of the carpel wall, except on the midrib and edges. The fruit is a follicle. The seeds have no endosperm and a straight embryo.

**Classification.** In floral anatomy, development and structure, the Butomaceae is in many ways very similar to the dicotyledonous family Nymphaeaceae. However,

*Butomus* undoubtedly belongs to the Monocotyledons and is related to the Alismataceae, Limnocharitaceae and Hydrocharitaceae, and possibly also the Aponogetonaceae.

**Economic uses.** *Butomus* is frequently cultivated as a decorative plant. In parts of Russia the rhizomes are used as food.

C.D.C.

### LIMNOCHARITACEAE
*Water Poppy*

**Number of genera:** 3
**Number of species:** about 12
**Distribution:** tropics and subtropics, in aquatic habitats.
**Economic uses:** *Limnocharis* is cultivated for food and *Hydrocleis* (water poppy) as an ornamental.

The Limnocharitaceae is a small family of annual and perennial aquatic herbs.

**Distribution.** The monotypic genus *Tenagocharis* is found in tropical Africa, India, Malaysia and northern Australia. *Hydrocleis* and *Limnocharis* are found in tropical and subtropical America. *Limnocharis flava* has become naturalized in India and Southeast Asia.

**Diagnostic features.** All species have secretory ducts containing latex. The juvenile leaves are linear and usually submerged, the mature leaves differentiated into petiole and blade. The leaf blades are ovate to cordate with distinct, curved, parallel nerves. The inflorescence is usually umbel-like; occasionally the flowers are solitary. The flowers are showy, regular and bisexual. There are three sepals which are green and persistent in fruit, and possess latex tubes. There are three petals alternating with the sepals and these are white or yellow, delicate and not persistent. There are six to nine or numerous stamens; staminodes are often present. The pollen has four or more pores. The carpels are superior, three to numerous, free, in one or rarely two whorls. The ovules

are numerous and scattered over the inner surface of the carpel wall. The fruit is a follicle which opens by the adaxial (ventral) suture to release numerous seeds that have no endosperm and a curved or folded embryo.

**Classification.** *Tenagocharis* has relatively small white petals, lacks staminodes and in some respects resembles *Damasonium* in the Alismataceae. The New World genera have large, showy, yellow petals and usually some staminodes.

On the basis of the diffuse parietal placentation of the ovules the Limnocharitaceae is often considered to be allied to the Butomaceae. In older works both families are combined. However, recent phytochemical, embryological and anatomical studies indicate a close relationship with the Alismataceae and not with the Butomaceae.

**Economic uses.** *Limnocharis flava* is cultivated for food in India and Southeast Asia, the leaves being eaten as an alternative to spinach or endive or used as fodder for pigs. *Hydrocleis nymphoides* (water poppy) has large decorative blossoms which are shining yellow with a reddish-brown center; it has been grown in heated greenhouses in Europe since 1830.

C.D.C.

### ALISMATACEAE
*Water Plantains*

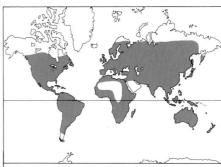

**Number of genera:** 11
**Number of species:** about 100
**Distribution:** cosmopolitan, centered in New World, in aquatic and marshy habitats.
**Economic uses:** important food for wildlife, some species decorative and *Sagittaria* cultivated for its edible tubers.

The Alismataceae is a small family of aquatic or amphibious plants widely distributed throughout the world. Very few are cultivated, but many provide food for wildlife.

**Distribution.** The family as a whole is cosmopolitan. The majority of species are found in the New World.

**Diagnostic features.** Most species are robust perennials, but some may be annual or perennial depending on the water regime;

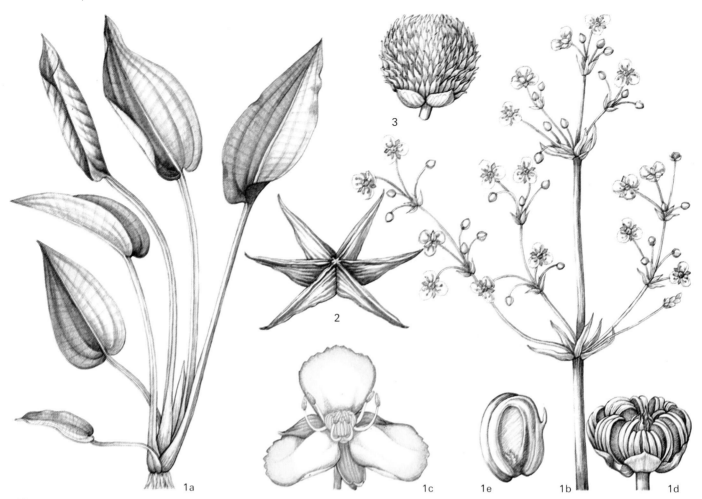

Alismataceae. 1 *Alisma plantago-aquatica* (a) habit showing leaves with sheathing bases, long petioles and leaf blade with main veins parallel to the edges (×⅔); (b) inflorescence with flowers arranged in whorls (×1); (c) flower with three green sepals, three pink petals, six stamens and gynoecium of numerous free carpels (×8); (d) fruit—a head of nutlets, each with longitudinal ribs (×6); (e) single nutlet with a persistent style (×8). 2 *Damasonium alisma* star-shaped whorl of dehiscent fruits which are united at their bases (×4). 3 *Sagittaria sagittifolia* fruiting heads composed of numerous nutlets (×3).

generally speaking, in permanent water they are perennial and in seasonal water they behave as annuals. The stems are corm-like or stoloniferous. Most species have two forms of leaf: the juvenile leaves are linear and usually submerged while the mature leaves vary from linear to ovate or occasionally sagittate, and are usually emergent. Most species have a distinct petiole. with an expanded, sheathing base. Secretory ducts are present. The inflorescence is usually compound with whorls of branches, but some species have umbel-like inflorescences and others have solitary flowers. The flowers are regular, bisexual or unisexual (male and female on separate plants in *Burnatia*). There are three sepals which usually persist in fruit. The petals are three, usually conspicuous, white, pink, purple, occasionally with yellow or purple spots, but they rarely last for more than one day. In *Burnatia* and *Wisneria* the petals are minute and occasionally absent in female flowers. The stamens are three, six, nine or numerous. The ovary is superior, comprising three to numerous free carpels in one whorl or in a clustered head; each carpel contains one (rarely two) anatropous ovules. The fruit is a

head of nutlets (except in *Damasonium* which has six to ten dehiscent or semi-dehiscent, several-seeded fruits in one whorl which are more or less united at the base or adnate to the elongated receptacle and spread star-like in fruit). The seeds have no endosperm and a curved or folded embryo.

**Classification.** The generic delimitation in the Alismataceae is somewhat unsatisfactory. The Old World genera *Baldellia*, *Caldesia* and *Ranalisma* are somewhat artificially separated from each other and the New World genus *Echinodorus*. *Limnophyton* and *Wisneria* are found in tropical Africa and Asia. *Wisneria* shows a remarkable superficial resemblance to *Aponogeton* (Aponogetonaceae). *Burnatia* is monotypic and found in tropical and southern Africa. *Luronium* is also monotypic but is endemic to Europe; it is probably related to *Caldesia*. *Damasonium*, as noted above, has an unusual fruit and a somewhat disjunct distribution: about four species in Europe, North Africa and the Orient, one in southern Australia and one in western North America. It has been suggested that it should be recognized as a distinct family. *Alisma* has nine species, and is almost cosmopolitan

through introductions, although probably indigenous in the Northern Hemisphere. *Sagittaria* has 20 or perhaps more species, the majority being found in the New World.

The Alismataceae, at least superficially, resembles the dicotyledonous family Ranunculaceae but there are considerable anatomical and embryological differences. There is no doubt that the Alismataceae belongs to the monocotyledons and that it possesses many primitive characteristics. It is probably more closely related to the Limnocharitaceae than the Butomaceae. The genera *Tenagocharis* (Limnocharitaceae) and *Damasonium* (Alismataceae) are very similar.

**Economic uses.** *Sagittaria sagittifolia* is cultivated in China and Japan for its edible corms. The roots of *Sagittaria latifolia* were used as food by North American Indians, and today are eaten by the Chinese in North America. Most genera are important for providing food for wildlife. Several species of *Sagittaria*, *Echinodorus* and *Alisma* are cultivated as decorative poolside plants, eg *Sagittaria sagittifolia*, *S. lancifolia*, *S. latifolia* and *S. montevidensis*. *Sagittaria subulata* and *S. latifolia* are sometimes used as aquarium plants.                     C.D.C.

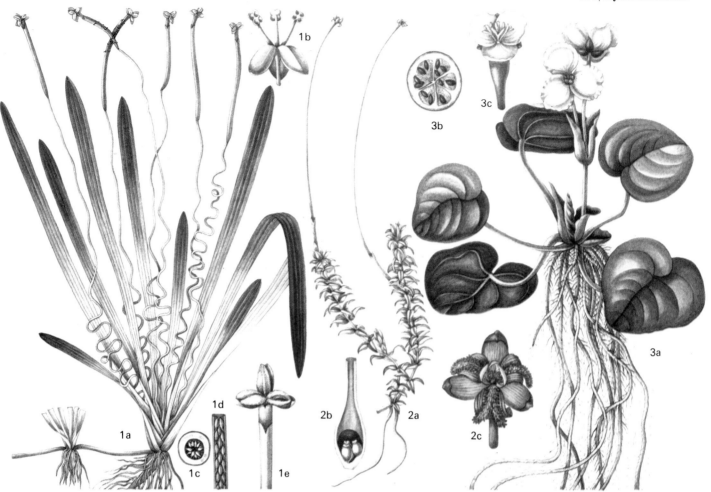

Hydrocharitaceae. 1 *Vallisneria spiralis* (a) habit showing stolons bearing new plants, ribbon-shaped leaves and long-stalked female flowers that reach the surface of the water ($\times\frac{2}{3}$); (b) male flower which separates from the parent and floats to the water surface ($\times 12$); ovary (c) in cross section ($\times 8$) and (d) in vertical section ($\times 4$); (e) female flower ($\times 4$). 2 *Elodea canadensis* (a) habit showing female flowers on long stalks ($\times\frac{2}{3}$); (b) vertical section of ovary ($\times 5$); (c) female flower with three, forked styles ($\times 5$). 3 *Hydrocharis morsus-ranae* (a) general habit of this free-floating plant. shown here with male flowers ($\times\frac{2}{3}$); (b) cross section of fruit ($\times 2$); (c) female flower ($\times\frac{2}{3}$).

# HYDROCHARITALES

## HYDROCHARITACEAE
*Canadian Waterweed and Frog's Bit*

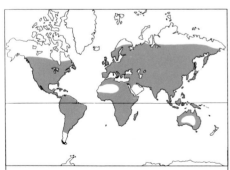

**Number of genera:** 15
**Number of species:** about 106
**Distribution:** cosmopolitan but mainly tropical, in marine and freshwater habitats.
**Economic uses:** pernicious weeds and some aquarium plants.

The Hydrocharitaceae is a family of marine and freshwater aquatics which produces some fine aquarium ornamentals and has specialized pollination mechanisms.

**Distribution.** This family is found in a wide variety of aquatic habitats throughout the world, though the majority of species are tropical.

**Diagnostic features.** The family consists of annual or perennial aquatic herbs, having either a creeping monopodial rhizome with leaves arranged in two vertical rows, or an erect main shoot with roots at the base and spirally arranged or whorled leaves. The leaves are simple, usually submerged, or sometimes floating or partly emergent. They are very variable in shape: from linear to orbicular, with or without a petiole and with or without sheathing bases. The flowers are arranged in a forked, spathe-like bract or between two opposite bracts. They are usually regular though occasionally slightly irregular (eg *Vallisneria*), and either bisexual or unisexual (male and female then being borne on separate plants). The perianth segments are in one or two series of three (rarely two) free segments; the inner series when present are usually showy and petal-like. The stamens are one to numerous, in one or more series; the inner ones are sometimes sterile (in *Lagarosiphon* the staminodes function as sails for the free-floating male flowers). The pollen is globular and free but in the marine genera *Thalassia* and *Halophila* the pollen grains are liberated in chains like strings of beads. The ovary is inferior with two to fifteen united carpels containing a single locule with numerous ovules on parietal placentas which either protrude nearly to the center of the ovary or are incompletely developed. There are as many styles as carpels. The fruits are globular to linear, dry or pulpy, dehiscent or more usually indehiscent and opening by decay of the pericarp. The seeds are usually numerous with straight embryos and no endosperm.

The family has a remarkable variety of specialized pollination mechanisms. Some genera such as *Egeria*, *Hydrocharis* (frog's bit) and *Stratiotes* have relatively large showy flowers that are insect-pollinated but some species are reported to be self-pollinating before the flower has opened (cleistogamous). In *Hydrilla*, *Lagarosiphon*, *Maidenia*, *Nechamandra* and *Vallisneria* the male flowers become detached from the mother plant and rise to the surface where they

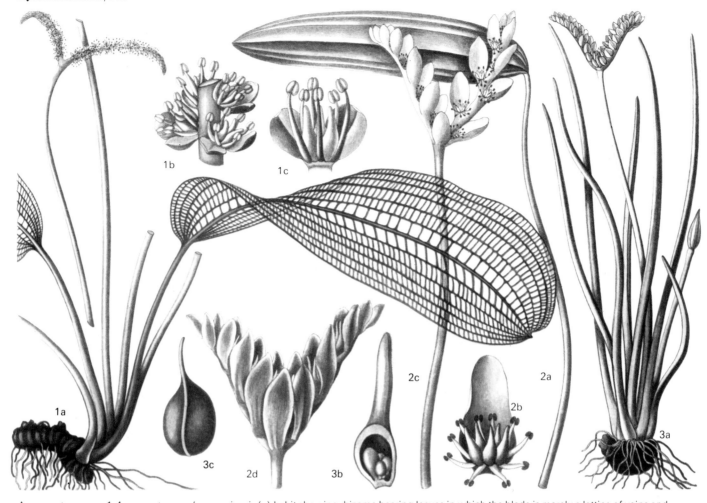

Aponogetonaceae. 1 *Aponogeton madagascariensis* (a) habit showing rhizome bearing leaves in which the blade is merely a lattice of veins and nerves, and a forked inflorescence born on a long stalk ( ×⅔) ; (b) portion of inflorescence ( ×4) ; (c) flower consisting of two perianth segments, six stamens and three sessile carpels ( ×6). 2 *Aponogeton distachyos* (a) aerial leaf ( ×⅔) ; (b) flower ( ×2) ; (c) inflorescence ( ×⅔) ; (d) infructescence with persistent perianth segments ( ×⅔). 3 *Aponogeton spathaceus* (a) habit showing tuber-like corms, tufts of strap-like leaves and forked inflorescence ( ×⅔) ; (b) vertical section of carpel showing sessile ovules ( ×16) ; (c) fruit—a leathery follicle ( ×6).

expand and then drift or sail to the female flowers. In *Elodea* the male flowers may become detached or remain on the mother plant but in either case the anthers explode and scatter pollen grains over the surface of the water.

**Classification.** The Hydrocharitaceae is usually divided into three subfamilies: HYDROCHARITOIDEAE, THALASSOIDEAE and HALOPHILOIDEAE. The latter two subfamilies are monogeneric (*Thalassia, Halophila*) and grow in tropical seas; the male spathe contains one flower, pollination takes place underwater and the pollen is liberated in chains. The genera of the Hydrocharitoideae are pollinated at or above the water surface with globular pollen grains. They occur in fresh water except for *Enhalus*.

The Hydrocharitaceae does not possess secretory ducts and frequently contains anthocyanin pigments. These characters, along with the pollen morphology and numerous embryological features, indicate a relationship with the Butomaceae and more remotely with the Nymphaeaceae. In most systems the Hydrocharitaceae is given a central position in the group known as the Helobiae, which includes the orders Alis-

matales, Hydrocharitales and Najadales, but today it looks as if they may represent a very primitive group within the monocotyledons. **Economic uses.** Many species are attractive or interesting aquarium plants. Several introductions, however, have become serious weeds in their new habitats, eg *Hydrilla* in the United States of America, *Elodea canadensis* (Canadian waterweed) in Europe and *Lagarosiphon* in New Zealand.   C.D.C.

# NAJADALES

## APONOGETONACEAE
*Water Hawthorn*

The Aponogetonaceae comprises a single genus of freshwater aquatic or amphibious perennial herbs with corms or rhizomes.
**Distribution.** The family is found in the warm and tropical regions of the Old World and in northern Australia; most species are found in Africa and Madagascar. *Aponogeton distachyos* (the water hawthorn or Cape pondweed), a native of South Africa, has become naturalized in southern Australia, western South America and Western Europe.
**Diagnostic features.** The stems are reduced to

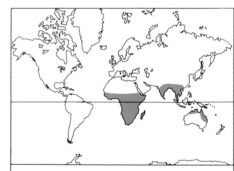

**Number of genera:** 1
**Number of species:** about 45
**Distribution:** warmer and tropical regions of the Old World, in aquatic or marshy habitats.
**Economic uses:** tubers eaten by Man and livestock.

tuber-like corms or elongated and branched rhizomes. The roots are fibrous. Many species grow in temporarily still or flowing water and live through the dry period as a dormant tuber. The leaves are borne in tufts at the tips of the stems. The majority of *Aponogeton* species have two forms of leaf: the juvenile leaves are usually strap-like, while the mature leaves are usually stalked with an expanded, linear, elliptic or oblong blade mostly with a distinct midrib and one or more pairs of parallel main nerves connected by numerous cross veins. In the Madagascan lace plant, *A. madagascariensis* (*A. fenestralis*) the whole blade often consists of no more than a lattice of veins and nerves. Submerged leaves are usually thin and are often undulate or contorted. Several species develop thick, leathery floating leaves. In some species, such as *A. junceus*, the leaves are reduced to elongated midribs and resemble rushes.

The flowers are usually bisexual or occasionally unisexual; some species are agamospermous (forming seeds without fusion of gametes). The inflorescences are spike-like and are borne on long stalks that emerge above the water surface. In bud, each spike is enveloped in a spathe. In all the Asian and most of the Australian species the spikes are single; in the majority of the African species they are paired and in some species from Madagascar there are up to ten spikes on a single stalk. The number and form of the floral parts is very variable. The perianth segments may be absent or up to six, they may be petal-like or bract-like and they may be persistent or caducous. The stamens are in two or more whorls, usually six or more in number. The ovary is superior with two to nine free, mostly sessile carpels each with two to numerous anatropous ovules on a basal placenta. The top of each carpel narrows into a style with an adaxial stigmatic ridge. Each carpel matures into a free, leathery follicle containing one to numerous seeds. The seeds have a straight embryo and no endosperm; the testa is usually single but sometimes is split into two, an inner one closely fitting the embryo and a loose transparent outer one.

**Classification.** The Aponogetonaceae is considered to be allied to the Potamogetonaceae-Najadaceae complex of families.

**Economic uses.** The tubers of several species are eaten by humans and their livestock. Many species make decorative aquarium plants, notably *A. distachyos*. As a result of large-scale trade the Madagascan lace plant (*A. madagascariensis*) has become extinct in many localities in Madagascar.

C.D.C.

# SCHEUCHZERIACEAE
*Arrowgrass*

The Scheuchzeriaceae consists of a single genus (*Scheuchzeria*) of marsh plants, with inconspicuous flowers and a graceful, grass-like appearance, hence their common name, arrowgrass.

**Distribution.** *Scheuchzeria* is restricted to cold north temperate zones, being especially common in cold sphagnum bogs.

**Diagnostic features.** The plants are slender perennial herbs with alternate, linear leaves, with a sheath at the base which embraces the

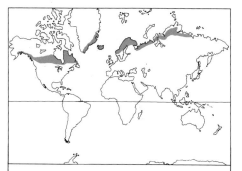

**Number of genera:** 1
**Number of species:** 2
**Distribution:** cold north temperate zone, especially sphagnum bogs.
**Economic uses:** none.

stem and terminates in a ligule. The flowers are regular, bisexual and borne in terminal racemes with bracts. The perianth is of two similar whorls each of three free segments. There are two whorls of three free stamens with basifixed anthers. The ovary is superior, comprising three to six carpels, united only at their bases, with three to six locules, each containing two or few basal, erect, anatropous ovules. Each carpel bears a sessile stigma. The fruit is composed of a number of one- or two-seeded follicles, the seeds having no endosperm. The embryo has a rounded cotyledon and a small plumule.

**Classification.** Some botanists consider that the family should also include *Triglochin*, which here is placed in a separate family, the Juncaginaceae. However, there is no doubt about the close relationship between *Scheuchzeria* and *Triglochin* as they have many features in common, both vegetatively and floristically. The six carpels in two whorls and the free perianth segments are regarded as primitive and may indicate a relationship with the Alismataceae.

**Economic uses.** None are known. S.R.C.

# JUNCAGINACEAE

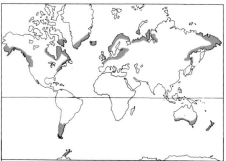

**Number of genera:** 3
**Number of species:** 14
**Distribution:** temperate and cold regions of the N and S Hemispheres, in marshy habitats.
**Economic uses:** limited, local uses as food.

The Juncaginaceae is a small family of annual and perennial marsh herbs.

**Distribution.** The family is found in temperate and cold regions of the Northern and Southern Hemispheres, mainly around the coasts.

**Diagnostic features.** The rhizome produces fibrous or tuberous roots and flat linear leaves, which are sheathing at the base, and usually radical (sometimes floating).

The inflorescence is a raceme or spike. The flowers are regular, without bracts, bisexual or unisexual (male and female on separate plants or male, female and bisexual on the same plant). They have a green or red perianth of two series of three free segments. There are four or six stamens with subsessile anthers. The ovary is superior and consists of four or six free or partly united carpels, each containing a single basal, anatropous ovule (rarely apical and orthotropous). The styles are short or absent, and the stigmas often feathery. The fruit is a follicle and the seeds have a straight embryo and no endosperm. The flowers are wind-pollinated, and in one species of *Triglochin* there are boat-shaped pockets under the anthers which first collect the pollen before it is dispersed by the wind.

**Classification.** The genera *Triglochin* (12 species), *Maundia* (one species) and *Tetroncium* (one species) are closely related to each other and to the Lilaeaceae. They show similarities with other monocotyledonous aquatic and marsh plants of the families Aponogetonaceae and Najadaceae. The possession of a double whorl of carpels is regarded as a primitive feature and possibly indicates evolutionary links with the Alismataceae and the Butomaceae.

**Economic uses.** The leaves of *Triglochin maritima* of the north temperate zone are edible, and the Australian aborigines use the rhizome of *T. procerum* as food.

S.R.C.

# LILAEACEAE

The Lilaeaceae is a family of aquatic and marsh herbs having only a single genus with a single species. The plants are notable for their strikingly unusual inflorescences.

**Distribution.** *Lilaea* is found in permanent or temporary shallow water in western North America and western South America southwards to Chile and Argentina. It is also found in Victoria, Australia, but is probably introduced.

**Diagnostic features.** *Lilaea* is a tufted, grass-like annual. The leaves are basal, simple, linear and cylindrical with membranous, sheathing bases that converge across the top to form a short ligule. The inflorescence is very complex; each leaf axil bears one or two female flowers and a stalked spike bearing bisexual and male flowers. The female flowers are enclosed in the sheathing leaf base and consist of a completely naked carpel with a thread-like style up to 30cm (12in) long that arises laterally at the top of the carpel but at length ascends. The fruit is an

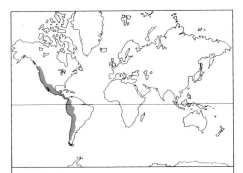

**Number of genera:** 1
**Number of species:** 1
**Distribution:** temperate western N and S America, in aquatic and marshy habitats.
**Economic uses:** none.

occasionally auriculate base. The flowers are unisexual, small, axillary, solitary or in small clusters; male and female are borne on the same plant in some species, on different plants in others. The male flowers are enclosed in a perianth-like, membranous sheath (spathe) which terminates in two thickened lips; the stamen is solitary and the anther subsessile. Pollination takes place under water, the pollen grains being ellipsoidal and without apertures. The female flowers are either naked or surrounded by a spathe. The ovary is superior and consists of a single carpel tapering into a short style with two to four linear stigmas. There is a single basal, anatropous ovule. The fruit is a single-seeded nutlet. The seeds have a straight embryo and no endosperm.

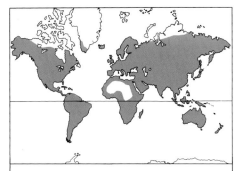

**Number of genera:** 2
**Number of species:** about 100
**Distribution:** cosmopolitan, in aquatic habitats.
**Economic uses:** important food plants for many animals.

angled nutlet, often with hooks or horns at the apex. The bisexual flowers have a single, bract-like perianth segment arising from the base of the solitary and sessile anther and one naked carpel with a single style that is very variable in length from flower to flower. Female and bisexual flowers have a single, basal, erect, anatropous ovule in each carpel. The fruits of the bisexual flowers are flattened nutlets with a dorsal ridge and wavy lateral wings. The male flowers consist of a single, bract-like perianth segment arising from the base of a solitary, stalkless anther. The seeds have a straight embryo and no endosperm.

The single species, *Lilaea scilloides*, is found in shallow water and usually starts to flower as the water level drops. It passes through the dry period as a seed and is a characteristic plant of vernal pools.

**Classification.** The structure of the flower and inflorescence of *Lilaea* is disputed. Some authors consider that the bisexual flowers are partial inflorescences consisting of one male and one female flower and that the perianth is an outgrowth of the connective. In spite of the confusion in the interpretation of the flower there is little doubt that the Lilaeaceae are helobian monocotyledons. They can be considered to represent a bridge between the Zannichelliaceae and the Najadaceae.

**Economic uses.** None are known.

C.D.C.

# NAJADACEAE
The Najadaceae is a family of small submerged aquatic annual or perennial plants containing one genus, *Najas*.

**Distribution.** *Najas* species occur throughout temperate and warm regions of the world.

**Diagnostic features.** The stems are slender, either sparsely branched and diffuse or much-branched and condensed. The leaves appear to be opposite but are usually crowded in leaf axils and thus may be described as pseudo-whorled or in bunches. Each leaf is simple, linear and usually toothed at the margin with a sheathing and

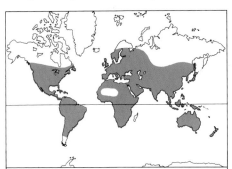

**Number of genera:** 1
**Number of species:** 50
**Distribution:** cosmopolitan, in aquatic habitats.
**Economic uses:** food for fish, green fertilizer and as a packing material.

**Classification.** The affinities of the Najadaceae are obscure. The family is usually placed in the superorder Alismatidae, which includes the orders Alismatales, Hydrocharitales as well as the Najadales. Its nearest relatives are probably the Aponogetonaceae and Potamogetonaceae. About 50 species are recognized but identification is difficult.

**Economic uses.** In warmer regions *Najas* species can be problematical weeds in ricefields and irrigation ditches. However, *Najas* provides a valuable fish food, and being relatively easily gathered, is also frequently used as a green fertilizer and as packing material.

C.D.C.

# POTAMOGETONACEAE
*Pondweeds*
The Potamogetonaceae is a family of aquatic herbs which are a familiar sight all over the world, often inhabiting ditches and ponds.

**Distribution.** All species grow in fresh or somewhat brackish water. *Potamogeton* is cosmopolitan and occurs in a wide variety of aquatic habitats. *Groenlandia* is found in Western Europe and northern Africa to southwestern Asia; it shows a preference for water which is rich in nutrients but is sensitive to organic pollution.

**Diagnostic features.** The plants are usually perennial but some are annual. The stems are elongate, flexible, erect, creeping or floating. Some species develop specialized winter buds (turions) on the creeping stems. The leaves are alternate in *Potamogeton* and opposite or whorled in *Groenlandia*. They are simple, entire and often have two forms: the floating leaves are broad and the submerged ones narrower and often linear or capillary. Sheathing bases are usually present in *Potamogeton* but except for the leaves subtending the inflorescence absent in *Groenlandia*. The inflorescence is a stalked spike. The flowers are bisexual, regular and somewhat inconspicuous. The perianth consists of four free, bract-like, clawed scales inserted opposite each stamen (the perianth segments are often considered to be outgrowths of the connective). There are four stamens, each of which is joined to a perianth segment; the anthers are sessile and have two locules. The ovary is superior and consists of four (rarely fewer) free or partly united carpels, each containing a single campylotropous ovule. The stigmas are sessile or on short styles. The fruit is a drupe with a bony endocarp and a fleshy exocarp in *Potamogeton* or a nutlet with a thin pericarp in *Groenlandia*. The seeds have a well-developed hypocotyl and no endosperm.

**Classification.** *Groenlandia* differs from *Potamogeton* in having opposite or whorled leaves without sheathing bases, and a thin pericarped nutlet as fruit. There is only one species (*Groenlandia densa*). *Potamogeton*, with about 100 species, is the largest exclusively aquatic genus of flowering plants. Most of the species have aerial pollination but the subgenus *Coleogeton* is pollinated underwater.

The family Potamogetonaceae takes a somewhat central position in the superorder Alismatidae, the helobian group. It is usually positioned between the more or less terrestrial families Juncaginaceae and Scheuchzeriaceae and the very reduced aquatic families Ruppiaceae and Zannichelliaceae. The Potamogetonaceae is probably very

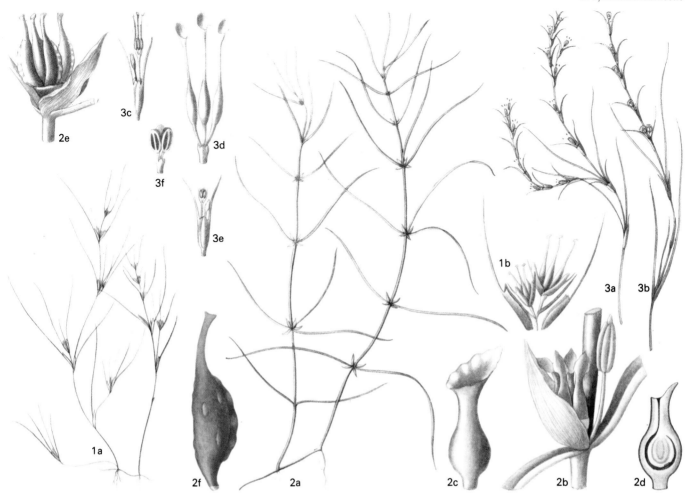

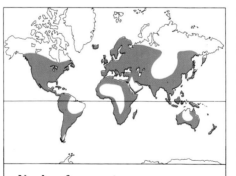

Zannichelliaceae. 1 *Althenia filiformis* (a) habit showing narrow leaves with sheaths joined to their bases ( ×⅔) ; (b) female flowers ( ×4). 2 *Zannichellia palustris* (a) habit showing leaf sheaths free from the leaves ( ×⅔) ; (b) male flower of one stamen and female flower of five carpels ( ×6) ; (c) carpel ( ×16) ; (d) vertical section of carpel ( ×16) ; (e) fruits ( ×4) ; (f) fruit ( ×8). 3 *Lepilaena preissii* (a) shoot with female flowers ( ×⅔) ; (b) shoot with male flowers ( ×⅔) ; (c) female flowers ( ×2) ; (d) gynoecium of three free carpels with spoon-shaped stigmas ( ×6) ; (e) male flowers ( ×2) ; (f) stamens ( ×6).

closely allied to the marine families Posidoniaceae and Zosteraceae.

**Economic uses.** Several species of *Potamogeton* have been reported to be a nuisance in canals and ditches. However, they are very important food plants for many animals. The starchy rootstock of *Potamogeton natans* and the turions have been used as a source of human food.  C.D.C.

## ZANNICHELLIACEAE
### Horned Pondweed
The Zannichelliaceae is a small family of submerged aquatic plants usually with graceful, slender leaves and stems, and inconspicuous flowers.

**Distribution.** The family is found throughout the world in fresh and brackish water.

**Diagnostic features.** The stems are thin and flexible, the upper parts usually floating above and the lower parts creeping and rhizomatous below. The leaves are alternate, opposite or in bunches; they are simple, entire and linear or reduced to a sheath. The leaf base is sheathing, with the sheath free from or partly joined to the leaf. These sheaths can be morphologically interpreted as sheath-like stipules, free or partly joined to the leaf. The

**Number of genera:** 4
**Number of species:** about 7
**Distribution:** cosmopolitan, in freshwater or brackish habitats.
**Economic uses:** some species stabilize mud and purify polluted water.

flowers are unisexual, solitary or clustered in the leaf axils; male and female are either on the same (monoecious) or on separate plants (dioecious). The perianth is either a small cup-like sheath, or a few scales, or absent. The stamens are solitary, or two or three united. The pollen is more or less spherical and pollination takes place under water. The ovary is superior and consists of one to nine

free carpels, each with a single, pendulous ovule. The styles are simple and usually persist in fruit. The stigma is conspicuous, usually irregularly flattened or spoon-shaped, and the margin of the stigma is either entire or fringed. The fruit is a nutlet and the seeds are without endosperm.

**Classification.** The cosmopolitan *Zannichellia* has one species (*Z. palustris* – the horned pondweed) which has the leaf sheath free from the leaf and the leaves usually opposite or in false whorls. The plants are monoecious. *Althenia* and *Lepilaena* have the leaf sheaths joined to the leaves for most of their length, the leaves being alternate and the plants dioecious. *Althenia* is found in the Mediterranean, Persia and South Africa, usually in brackish water; there are one or two species. *Lepilaena* has four species in Australia and New Zealand. *Althenia* and *Lepilaena*, although geographically isolated should perhaps be totally or partially united. A new genus *Vleisia* has recently been described from South Africa.

The Zannichelliaceae belongs in the superorder Alismatidae, the helobian group of monocotyledons. There are convincing arguments to support the theory that it has evolved from the marine family Cymodoceaceae, and certainly it has close affinities with this family. The two families are often united and are, in turn, probably allied to the Potamogetonaceae.

**Economic uses.** None are known, but *Zannichellia* can be considered a beneficial plant as it stabilizes mud and contributes to the purification of the water.                    C.D.C.

## RUPPIACEAE
*Ditch Grasses*

The Ruppiaceae consists of a single genus (*Ruppia*) of submerged aquatic herbs commonly known as ditch grasses.

**Distribution.** *Ruppia* is found throughout the world. It usually grows in brackish water in coastal areas but a few species are found inland in freshwater in South America and New Zealand. Plants have been collected at 4,000m (about 13,000ft) in the Andes.

**Diagnostic features.** The stems are slender, the upper parts floating and the lower parts creeping, with alternate or opposite leaves which are simple, hair-like and somewhat toothed at the apex. The leaf bases are enlarged and sheathing, and the inflorescence is a short terminal raceme which appears umbel-like. The flowers are bisexual, small and borne in pairs on slender axillary stalks. These axillary stalks are at first short and enveloped in a spathe-like leaf base but at the opening of the flower bud or following pollination they elongate and sometimes become coiled. The perianth is lacking or vestigial. There are two stamens with sessile bilocular anthers. The ovary is superior with four, or rarely more, free carpels, each with a single, pendulous, campylotropous ovule. The carpels are at first sessile but become stalked in fruit. The style is lacking and the stigma is a small disk on top of the carpel. The fruit is usually a nutlet but occasionally a drupe with a somewhat spongy exocarp. The seeds have no endosperm.

**Classification.** The Ruppiaceae probably represents a family that has evolved from the Potamogetonaceae by reduction. Pollination takes place under water in some races, while in others the pollen floats on the water

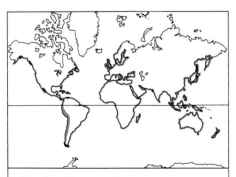

**Number of genera:** 1
**Number of species:** about 7
**Distribution:** cosmopolitan, in brackish water.
**Economic uses:** none.

surface and contacts floating stigmas. As with many other aquatics, the flowers are reduced and the plants are very plastic in form so there is little agreement as to how many species exist. It seems likely that there are about seven species but some authorities recognize only one or perhaps two polymorphic species.

**Economic uses.** None are known.       C.D.C.

## ZOSTERACEAE
*Eel Grasses*

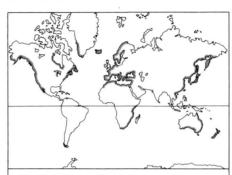

**Number of genera:** 3
**Number of species:** 18
**Distribution:** mainly in temperate seas of N and S Hemispheres.
**Economic uses:** dried leaves and stems used as packing material, also mixed with plaster or cement to add strength.

The Zosteraceae is a marine family of grass-like herbs which live entirely submerged.

**Distribution.** These plants are found in salt water only, distributed in the temperate seas of the Northern and Southern Hemispheres. A few species extend into tropical seas.

**Diagnostic features.** The stem is a creeping rhizome, monopodial in *Phyllospadix* and *Zostera* (eel grass) or sympodial in *Heterozostera*. The leaves are linear and grass-like with distinctly sheathing bases. The inflorescence is a flattened spadix, enclosed at flowering time in the spathe (the sheath of the uppermost leaf). The flowers are unisexual, the male and female being borne either on separate plants or on the same plant; in the latter case male and female are arranged alternately along the spadix. Each male consists of one stamen with two free, bilocular anthers joined by a ridge-like connective. The pollen grains are long threads of the same specific gravity as sea water so when discharged they float freely and have a good chance of adhering to the stigmas. In most species a curious hook-like process (retinaculum) is found beside each stamen or in the female spadixes of *Phyllospadix* where they alternate with the female flowers. The female flower consists of a single, naked carpel with a short style and two relatively long stigmas. The ovule is solitary, straight and pendulous. The fruit is ovoid or ellipsoidal with a dry, thin pericarp or it is crescent-shaped with the pericarp differentiated into a soft exocarp and a hard,

fibrous endocarp. The seed has no endosperm.

**Classification.** *Zostera* (12 species) is widely distributed, but *Heterozostera* (one species) occurs on coasts of Australia and Chile and *Phyllospadix* (five species) occurs on the coasts of Japan and Pacific North America.

The family is probably related to the Posidoniaceae and Potamogetonaceae. In several treatments they are all included within the Potamogetonaceae.

**Economic uses.** The dried leaves and stems of *Zostera* are used as a packing material, particularly for Venetian glass. They are also mixed with plaster or cement as a strengthening material. *Zostera* also has considerable indirect economic importance by supporting flora and fauna which themselves provide food for many bird and fish species.   C.D.C.

## POSIDONIACEAE

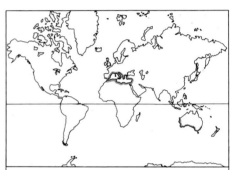

**Number of genera:** 1
**Number of species:** 3
**Distribution:** Mediterranean sea and S Australian coast.
**Economic uses:** the fibers are used in textiles and packing materials.

The Posidoniaceae is a family of submerged marine perennials with a single genus, *Posidonia*.

**Distribution.** *Posidonia* shows a wide disjunction. Two species are found in the extratropical waters of Australia and the third occurs in the Mediterranean; there are some doubtful records from the western Atlantic coast of Europe.

**Diagnostic features.** The stem is a creeping monopodial rhizome. The leaves are linear with a distinct auriculate and ligulate leaf sheath which surrounds the stem. The leaf blade and sheath have numerous dark spots and stripes resulting from local accumulations of tannin.

The inflorescence is a stalked cyme. The flowers are bisexual or male and the perianth is lacking. The three stamens are sessile; the anthers are bilocular and have a broad connective that is somewhat produced beyond the locules; the anthers are extrorsely dehiscent. The pollen is thread-like and pollination takes place under water. The carpel is superior, solitary and naked, containing a single, campylotropous, ventrally attached ovule. The stigma is sessile and irregularly lobed. The fruit has a fleshy pericarp, and the seed is filled completely by the embryo.

**Classification.** The Mediterranean species *Posidonia oceanica* differs from the Australian ones in that the leaf sheaths are incompletely wrapped around the stem, the ligule is very short, the bracts of the inflorescence are larger than its sheaths and the seed coat is without a wing. The Australian species are relatively simple to distinguish; *P. australis* has membranous leaves, 6mm to 14mm (0.2in to 0.5in) wide with about 11 nerves and two to seven terminal flowering spikes, while *P. ostenfeldii* has leathery leaves, 1mm to 4mm (0.04in to 0.2in) wide with about seven nerves and 6–14 terminal flowering spikes.

The Posidoniaceae certainly belongs to the helobian group (ie the superorder Alismatidae), and is usually considered to be allied to the Potamogetonaceae and Zosteraceae.

**Economic uses.** *Posidonia australis* is the source of posidonia fiber, cellonia or lanmar. Either alone or mixed with wool it is used for making sacks and coarse fabrics. It is also used for packing and stuffing. C.D.C.

## CYMODOCEACEAE

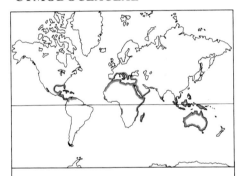

**Number of genera:** 5
**Number of species:** 16
**Distribution:** mainly tropical and subtropical seas.
**Economic uses:** none.

The Cymodoceaceae is a family of sea-grasses, which grow submerged in seawater, where they provice both food and shelter for fish.

**Distribution.** The Cymodoceaceae are marine plants found in tropical and subtropical seas with a few species in warm temperate waters. The genus *Amphibolis* is limited to the temperate seas of Australia.

**Diagnostic features.** The stem is creeping, either herbaceous and monopodial or woody and sympodial. The leaves are arranged in two vertical rows, linear or awl-shaped with a sheathing base and a leaf tip that is very variable in outline. The flowers are unisexual, naked, and either terminal on a short branch or, in *Syringodium*, arranged in a cymose inflorescence. Male and female are borne on separate plants. The male flower consists of two dorsally united stamens (some authorities consider the male flower to be an inflorescence of two one-stamened flowers). The pollen is thread-like and pollination takes place in water. The female flower consists of two free carpels with a long (*Halodule*) or short style and two or three stigmas. Each carpel contains a solitary pendulous ovule. The fruit is indehiscent and one-seeded, either with a hard or fleshy exocarp (eg *Amphibolis*) or consisting of a fleshy bract enclosing the fertilized ovaries (eg *Thalassodendron*). The seeds have no endosperm.

**Classification.** The genera are *Cymodocea* (four species), *Syringodium* (two species), *Amphibolis* (two species), *Halodule* (six species) and *Thalassodendron* (two species). The family is allied to the Zannichelliaceae.

**Economic uses.** The beds of sea-grasses are of no direct use to Man, but many fish use them both as food and as spawning grounds. C.D.C.

# TRIURIDALES

### TRIURIDACEAE
The Triuridaceae is a small family of small tropical herbs which live on dead or decaying matter.

**Distribution.** The family is native to tropical America, Africa and Asia.

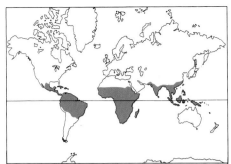

**Number of genera:** 7
**Number of species:** 80
**Distribution:** tropical America, Africa and Asia.
**Economic uses:** none.

**Diagnostic features.** The plants are small, colorless or reddish-purple saprophytes with scale leaves. The small flowers are regular and bisexual, or unisexual with male and female flowers on the same or separate plants, and are borne in racemes. The perianth consists of one series of three to ten segments which may be equal or unequal in size, and which become reflexed after flowering. In a number of species the segments are terminated by a tuft of hairs. There are two to six free stamens with short filaments and anthers with two or four locules, which sometimes dehisce transversely. The connective is sometimes extended into a long appendage (as in *Andruris*). In the male flowers there may be three fertile stamens and three staminodes; female flowers may possess staminodes. The gynoecium consists of numerous, superior, free carpels immersed in the receptacle, each carpel containing a single erect basal ovule with one integument. Each carpel has its own style which may be inserted terminally, laterally or basally. The fruit is a crowded mass and the erect seeds contain white oily endosperm.

**Classification.** The chief genera are *Triuris* (one species), *Sciaphila* (50 species), *Seychellaria* (three species) and *Andruris* (16 species). The relationships of the Triuridaceae are obscure, its closest relative being the Alismataceae.

**Economic uses.** None are known. S.R.C.

# COMMELINIDAE

## COMMELINALES

### XYRIDACEAE
*Yellow-eyed Grasses*
The Xyridaceae is a small family of herbaceous marsh plants.

**Distribution.** The family is mainly tropical and subtropical in distribution in marshy areas, especially in the southeastern United States, tropical America, and southern Africa.

**Diagnostic features.** The plants are some-times annual, but mostly perennial, rush-like herbs with leaves normally arising in tufts from the top of the rootstock. The leaves are linear, either flat and slender or circular in cross-section, and distinctly sheathing at the base.

The flowers are bisexual, slightly irregular and arranged in globose or cylindrical heads. Each flower is in the axil of an overlapping stiff or leathery bract. There are three sepals, of which the inner initially forms a hood-shaped structure over the petals, while the two lateral sepals are small and keel-shaped.

The corolla consists of three (normally yellow) petals fused to form a short or long tube, opening out into three lobes at the top. There are three fertile stamens opposite the corolla lobes and in some species there are three small staminodes alternating in position with them. The stamens are attached to the corolla tube by their short, flat filaments, and the anthers have two locules, opening lengthwise. The ovary is superior, made up of three fused carpels forming a single locule with numerous ovules inserted basally or on three parietal placentas. It is surmounted by

Rapateaceae. 1 *Rapatea pandanoides* habit showing oblong leaves with large sheathing leaf bases ( ×⅓). 2 *R. paludosa* (a) inflorescence with involucre of two bracts ( ×⅔); (b) half flower showing stiff outer perianth-segments, inner petaloid perianth-segments, epipetalous stamens with basifixed anthers, and superior ovary containing a single ovule in each locule and crowned by a simple style ( ×5⅔); (c) capsule dehiscing by three valves to reveal one seed in each locule ( ×5⅔). 3 *Schoenocephalium arthrophyllum* (a) habit showing linear leaves and inflorescences (×⅓); (b) dehiscing capsule with two seeds in each locule ( ×4).

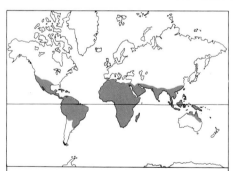

**Number of genera:** 2
**Number of species:** about 240
**Distribution:** chiefly in SE US, tropical America and southern Africa, in marshy habitats.
**Economic uses:** aquarium ornamentals and local medicinal uses.

a simple or three-lobed style. The corolla tube is persistent on the fruit, which is a capsule containing numerous seeds each with a small apical embryo surrounded by copious endosperm.

**Classification.** The genus *Xyris* (yellow-eyed grass—about 240 species) has a much wider distribution than *Achlyphila* (one species

tropical South America) and can be separated largely on the basis of floral characteristics. For example, *Xyris* has no appendages at the style base, and the inner sepal is distinctly hooded.

The family is related to the Commelinaceae and the Eriocaulaceae.

**Economic uses.** The leaves and roots of two North American species, *Xyris ambigua* and *X. caroliniana*, have been used as domestic remedies in treating colds and skin diseases respectively. A few species of *Xyris* are used as aquarium plants.                    S.R.C.

## RAPATEACEAE

The Rapateaceae is a small tropical family of perennial herbs.

**Distribution.** The family is native to tropical South America, with one genus (*Maschalocephalus*) in West Africa, often in swampy habitats.

**Diagnostic features.** The leaves arise from a thick rhizome or fleshy rootstock. They are narrow or linear, usually twisted and sheathing at the base, and may attain a length of 1.5m (5ft), as in *Rapatea paludosa*. The flowers are regular and bisexual, borne in heads of spikelets enclosed in two large

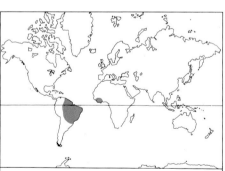

**Number of genera:** 16
**Number of species:** about 80
**Distribution:** swamps, habitats in tropical S America with one genus in Liberia.
**Economic uses:** none.

spathes. The perianth consists of two whorls, the outer of three rigid lobes, the inner of three petal-like segments fused at the base and spreading out into three broad ovate lobes. The six stamens are adnate to the corolla tube with anthers basifixed and opening by one, two or four apical pores, or by terminal slits. The ovary of three fused carpels is superior, with one to three locules, each containing one or several ovules on basal or axile placentas. It is surmounted by a simple style. The fruit is a capsule containing several seeds, each with copious, mealy endosperm.

**Classification.** One system of classification groups the genera into two subfamilies according to features of the carpels.

SAXOFRIDERICIOIDEAE. Carpels with several ovules which are either axile or on the septa; seeds prismatic or pyramidal; includes *Saxofridericia*, *Stegolepis* and *Schoenocephalium*.

RAPATEOIDEAE. Carpels with one ovule inserted at or near the base; seeds oval or oblong; includes *Rapatea*, *Cephalostemon* and *Monotrema*.

Alternatively the genera can be classified on a number of other characters including the nature of the inflorescence and details of the anthers. For example, *Rapatea* (flowers with pedicels), *Cephalostemon* (flowers sessile, anthers dehiscing by a terminal slit), *Schoenocephalium* (flowers sessile, anthers dehiscing by two terminal pores) and *Amphiphyllum* (flowers sessile, anthers dehiscing by one terminal pore) all have clusters of flowers subtended by a common involucre of bracts. On the other hand, *Windsorina* (flowers with pedicels), *Stegolepis* (flowers sessile) and *Maschalocephalus* (flowers subsessile) all have clusters of flowers without a common involucre.

The distinct calyx and corolla, lack of nectaries or nectar, superior ovary and capsular fruit place the Rapateaceae close to the other families in the Commelinales (Commelinaceae, Xyridaceae, Mayacaceae). The Rapateaceae is most closely related to the Xyridaceae.

**Economic uses.** None are known.

S.R.C.

# MAYACACEAE

The Mayacaceae is a monogeneric family of small, mat-forming aquatic or amphibious herbs, some of which are used as ornamental aquarium plants.

**Distribution.** The family is primarily American, extending from the southwestern United States to Paraguay. One species, *Mayaca baumii*, is found in Angola. The plants grow on mud or in shallow water.

**Diagnostic features.** The stems are usually branched, the lower parts creeping and rooting and the upper parts erect or floating, often densely packed together in mats. The leaves are spirally arranged and clothe the stem; they are simple, linear, up to about 3cm (1.2in) long, and notched at the apex. The flowers are regular, bisexual, borne singly on long stalks and subtended by membranous bracts which usually become reflexed after flowering. The perianth is in two whorls, the outer sepal-like with three elongate segments persistent in fruit, the inner petal-like with three broad white, pink or violet segments. Three stamens alternate

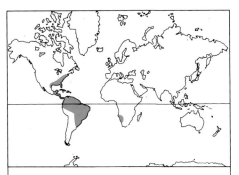

**Number of genera:** 1
**Number of species:** about 10
**Distribution:** chiefly America, from SW USA to Paraguay, with one species in Angola, in aquatic and marshy habitats.
**Economic uses:** some species used as decorative aquarium ornamentals.

with the petal-like segments; the anthers open by an apical pore or pore-like slit. The ovary is superior, of three fused carpels forming a single locule with numerous orthotropous ovules attached in two rows to three parietal placentas. There is a single style, simple or slightly three-lobed. The fruit is a three-valved capsule. The seeds have a pitted or net-like surface; they contain endosperm and the small embryo is situated under an embryostega, as in the Commelinaceae.

**Classification.** *Mayaca*, the only genus, contains 10 species. The Mayacaceae is allied to the Commelinaceae, from which it differs in having parietal placentation, anthers opening by pores or pore-like slits and non-sheathing leaves.

**Economic uses.** Some species are cultivated as decorative aquarium plants.

C.D.C.

# COMMELINACEAE
*The Spiderwort Family*

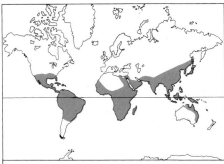

**Number of genera:** 38
**Number of species:** about 600
**Distribution:** tropical, subtropical and warm temperate.
**Economic uses:** garden ornamentals, greenhouse and house plants: *Commelina*, *Tradescantia* (spiderwort, wandering Jew), *Rhoeo*, *Cyanotis* and *Zebrina*

The Commelinaceae is a medium-sized family of succulent annual or perennial herbs, many of which are popular as garden, greenhouse and house ornamentals.

**Distribution.** The Commelinaceae in general prefer damp conditions and are mostly found in tropical, subtropical and warm temperate regions. A few are found in the southern United States of America, China, Japan and Australia.

**Diagnostic features.** The plants are either stemless or have jointed, succulent aerial stems with alternate leaves. The leaves are flat, entire, and with a closed basal sheath. The adventitious roots on the underground stem are fibrous or occasionally swollen and tuberous.

The inflorescence is essentially a cyme and is borne either at the end of the stem or in the axil of a leaf. The flowers are bisexual, usually regular but occasionally irregular. The perianth consists of two series of three segments, the outer series of three generally free, imbricate green sepals, the inner series of three free, usually equal petals. In a few species the petals are fused into a tube while rarely one of the three petals is reduced in size. The stamens are in two series of three with the filaments generally free. In some genera only three of the stamens bear anthers, the other three being sterile and modified as staminodes. One genus (*Callisia*) possesses only one functional stamen and no staminodes. In several genera the filaments are adorned by brightly colored hairs. An unusual feature in a few species is that the anthers liberate their pollen through an apical pore instead of through lengthwise slits. The ovary is superior, of three fused carpels with three locules (rarely two), each containing one to several orthotropous ovules on axile placentas. It is surmounted by a single style terminating either in a flattened stigmatic head or in three stigmatic branches. The fruit is a thin-walled dehiscent capsule, but rarely may be fleshy and indehiscent. The seeds usually have a rough or ridged surface covered by an aril. They contain copious, mealy endosperm, the embryo being situated under a disk-like structure (embryostega) on the seed coat.

**Classification.** There are various systems used in classifying the genera of this family. One system divides the family into two subfamilies — the TRADESCANTOIDEAE (regular flowers) and the COMMELINOIDEAE (irregular flowers). Further divisions are made by splitting the former group into subgroups dependent on the presence of either three or six fertile stamens and the latter into subgroups dependent on the orientation of the flower buds toward the axis (stem).

In the other primary division of the family, the genera can be grouped according to whether or not the inflorescence perforates and emerges near the base of the leaf sheath which subtends it. Those in which it does not include *Forrestia* (petals free) and *Coleotrype*

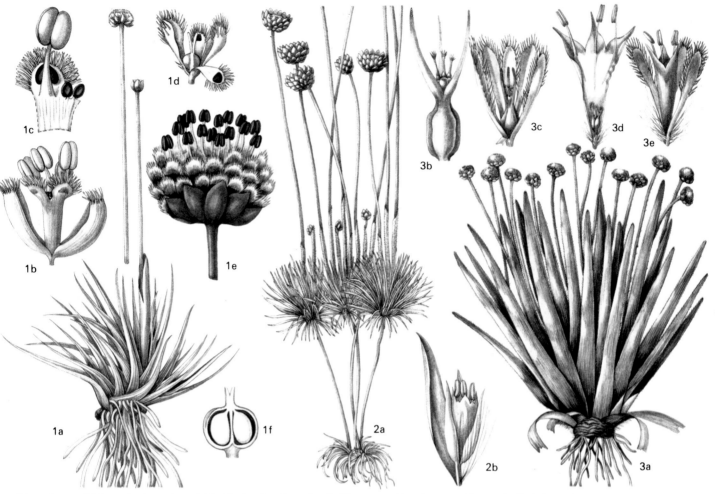

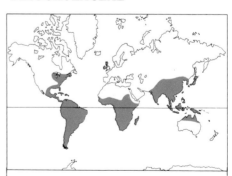

Eriocaulaceae. 1 *Eriocaulon aquaticum* (a) habit showing dense head of flowers and basal rosette of leaves ($\times\frac{2}{3}$); (b) male flower with free outer and fused inner perianth segments ($\times 8$); (c) inner perianth segment from male flower showing vestigial stamen and fertile stamen with gland behind ($\times 12$); (d) female flower ($\times 8$); (e) head of male and female flowers ($\times 3$); (f) vertical section of fruit (a capsule) showing pendulous seeds ($\times 12$). 2 *Syngonanthus laricifolius* (a) habit ($\times\frac{2}{3}$); (b) male flower ($\times 15$). 3 *Paepalanthus riedelianus* (a) habit ($\times\frac{2}{3}$); (b) gynoecium ($\times 16$); (c) female flower ($\times 8$); (d) inner perianth of male flower opened out to show three stamens and vestige of ovary ($\times 10$); (e) male flower ($\times 8$).

(petals fused), while those which have the inflorescence emerging from the top of the leaf sheath include *Callisia* and *Cochliostema* (one to three stamens), *Cyanotis* (six stamens and sepals fused at the base) and *Rhoeo* (six stamens and sepals free).

The family shows a number of similarities to the Flagellariaceae and the Mayacaceae with respect to such vegetative characters as leaves bearing a closed sheath, regular or irregular flowers with biseriate perianth differentiated into sepals and petals, and stamens usually three or six in number, a

Commelinaceae. 1 *Commelina erecta* shoot showing sheathing leaf bases, flowers with three petals, three stamens and three staminodes ($\times\frac{2}{3}$). 2 *Gibasis graminifolia* (a) leafy shoot and inflorescence, each flower with six stamens ($\times\frac{2}{3}$); (b) fruit ($\times 3$). 3 *Zebrina pendula* leafy stem and solitary flower ($\times\frac{2}{3}$). 4 *Tradescantia sillamontana* (a) leafy shoot with inflorescence subtended by boat-shaped, leafy bracts ($\times\frac{2}{3}$); (b) flower with six stamens ($\times 2$); (c) cross section of trilocular ovary ($\times 10$). 5 *Rhoeo spathacea* shoot showing rosette of bromeliad-like leaves and inflorescence with boat-shaped bracts ($\times\frac{1}{2}$). 6 *Tradescantia navicularis* juvenile plant (left) and adult shoot (right) ($\times\frac{2}{3}$).

superior fused gynoecium and endospermic seeds with an embryostega. The Commelinaceae is probably derived from the Alismataceae and Butomaceae but shows evolutionary advance in having fused carpels.

**Economic uses.** The family has no agricultural importance. However, the genera *Commelina* (about 180 species), *Tradescantia* (about 35 species), *Zebrina* (about four species), *Cyanotis* (about five species), *Dichorisandra* (about 30 species) and *Rhoeo* (one species), are well known as house pot plants or as garden ornamentals. Most species of *Tradescantia* are sold under the popular names tradescantia, wandering Jew or wandering sailor and *Tradescantia virginiana* is known as the spiderwort.

An extract of leaves and stems of the tropical African perennial herb *Aneilema beninense* is used as a laxative, while the leaf sap of *Floscopa scandens* is used in tropical Asia to treat inflammation of the eyes. The young shoots and leaves of *T. virginiana* are edible and can be used in salads while the rhizomes of some species of *Commelina* are edible as are the leaves of *Commelina clavata*. Some species are weeds.                    S.R.C.

# ERIOCAULALES

## ERIOCAULACEAE

**Number of genera:** 13
**Number of species:** about 1,200
**Distribution:** mainly tropical and subtropical, centered in the New World.
**Economic uses:** limited use as "everlastings."

The Eriocaulaceae is a largish family of perennial or occasionally annual herbaceous plants often with grass-like leaves.

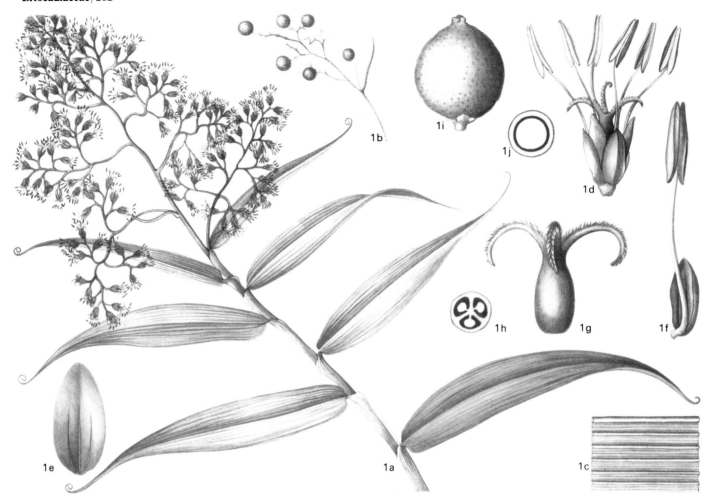

Flagellariaceae. 1 *Flagellaria guineensis* (a) shoot bearing a terminal branching racemose inflorescence and leaves with tightly sheathing bases and tips formed into a coiled tendril ($\times\frac{2}{3}$); (b) tip of shoot bearing fruit ($\times\frac{2}{3}$); (c) lower surface of leaf showing parallel veins ($\times 2$); (d) bisexual flower ($\times 3\frac{1}{3}$); (e) perianth segment ($\times 7\frac{1}{2}$); (f) stamen attached to perianth segment ($\times 5$); (g) gynoecium with broad style and three hairy stigmatic surfaces ($\times 5$); (h) cross section of ovary ($\times 5$); (i) fleshy indehiscent fruit ($\times 2\frac{2}{3}$); (j) cross section of fruit ($\times 1\frac{1}{3}$).

**Distribution.** The family is found throughout the tropics and subtropics, and a few species grow in temperate regions. The majority are found in the New World. Most species are found in swampy places or seasonally inundated regions, some are truly aquatic and others grow in dry areas.

**Diagnostic features.** The stems are corm-like or elongate. The leaves are in basal rosettes or growing from the stem; they are usually linear and somewhat grass-like. The flowers are regular, unisexual and borne in dense heads subtended by an involucre of bracts. The heads are solitary or in umbels. The peduncles usually extend beyond the leaves, which may sheath them at the base. Within each head the male and female flowers are mixed or the male flowers are in the center surrounded by female flowers; occasionally the male and female flowers are on separate plants. The perianth is in two series but not clearly differentiated into sepals and petals. The segments of the outer perianth are two or three in number and free, united or partly united; the inner series has two or three united or partly united segments or is lacking. The stamens are as many or twice as many as the outer perianth segments and inserted on the inner perianth segments

when these are present. The pollen grains are spiraperturate. The ovary is superior, with two or three fused carpels and two or three locules and a single terminal style bearing two or three elongate stigmas. There is a solitary, orthotropous, pendulous ovule in each locule. The fruit is a membranous, loculicidally dehiscent capsule. The seeds contain copious floury endosperm and a small embryo.

**Classification.** The family is relatively uniform in habit and the genera are distinguished on almost microscopic characters of the flowers. The largest genera are *Eriocaulon* (400 species), *Paepalanthus* (485 species), *Sygonanthus* (195 species) and *Leiothrix* (65 species).

The family has no near relatives and occupies a somewhat isolated position in the monocotyledons. It is usually placed near Xyridaceae and Rapateaceae as they also have heads of flowers, but these families show affinities with the Commelinaceae.

**Economic uses.** There are no reported economic uses for the family apart from their widespread sale when dried and stained as ornamental "everlastings." Some species of *Eriocaulon* are found as weeds in ricefields, but are not troublesome.                    C.D.C.

# RESTIONALES

## FLAGELLARIACEAE

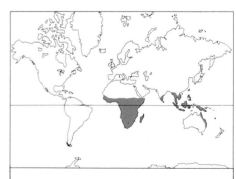

**Number of genera:** 1–3
**Number of species:** 3–7
**Distribution:** tropical and subtropical.
**Economic uses:** the stems of one species used in basket-making.

The Flagellariaceae is a small family of tropical herbs, often climbers.

**Distribution.** The family is tropical and subtropical from Africa, through Sri Lanka and Malaysia to the Pacific.

**Diagnostic features.** The erect or climbing stems are divided or undivided and arise from a sympodial creeping or floating rhizome. The leaves are long, sometimes ending in a tendril, and have at the base a leaf sheath embracing the stem. In *Hanguana* the leaves have petioles and are mostly radical.

The flowers are regular, bisexual or unisexual, and arranged in terminal branching racemes. The perianth is composed of six free or slightly fused segments in two whorls of three. They may be petaloid as in *Flagellaria* and *Hanguana* or scale-like as in *Joinvillea*. There are two whorls each of three stamens with free filaments, which may be attached to the base of the perianth segments. The ovary is superior, of three fused carpels and three locules, with a single ovule in each locule, attached to the central axis. The ovary is surmounted by a single style which terminates in three stigmatic lobes. The fruit is fleshy or drupe-like, containing a seed with a small embryo surrounded by copious endosperm.

**Classification.** Each of the three genera has distinctive characteristics which may be sufficiently disparate to suggest that this family is not a natural grouping.

*Flagellaria*, with three species native to tropical Africa, Indomalaysia, Australia and the Pacific, is a genus of climbers with dichotomously branching stems, solid internodes, leaves with closed sheathing bases and tips modified as tendrils. There is no covering of hairs or bristles, and no silica present in the leaf, but there are secretory cells. The flowers are bisexual and the perianth segments are free and petaloid.

*Hanguana* is a genus native to Sri Lanka, Indochina and Malaysia, comprising one or two species of robust, erect-stemmed herbs, with petiolate leaves and unisexual flowers (male and female on separate plants), the male possessing a rudimentary ovary and the female six sterile stamens. The perianth segments are greenish or yellow and slightly fused at the base.

*Joinvillea* is another genus of two species native to Malaysia and the Pacific. They are erect herbs but the stems are unbranched, the internodes hollow, and the leaves are long and narrow, with open sheathing bases. They are covered with a dense mat of branched hairs or bristles and contain abundant silica but no secretory cells. The flowers are bisexual and the six, free perianth segments are scale-like or bract-like.

Some botanists claim that these differences as well as others (eg the nature of the pollen) are sufficient to establish a separate family for each genus. Under this scheme of classification *Flagellaria* is the sole genus of the Flagellariaceae; *Hanguana* the sole genus of the Hanguanaceae, and *Joinvillea* the sole genus of the Joinvilleaceae.

*Flagellaria* has some features of anatomy and pollen that indicate a relationship with the Gramineae. *Hanguana* has some features

which suggest relationships with *Lomandra* (Xanthorrhoeaceae) and possibly with the Palmae. All three genera may also have affinities with the Commelinaceae and Mayacaceae, with which they share such features as a biseriate perianth, usually six stamens and a superior ovary with a single style.

**Economic uses.** *Flagellaria indica* is the only species known to be used by Man. Its tough stems are used in Thailand and Malaysia for basket-making. S.R.C.

# CENTROLEPIDACEAE

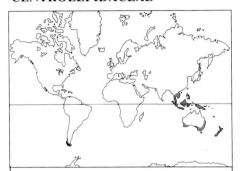

**Number of genera:** 5
**Number of species:** about 30
**Distribution:** Australia and New Zealand to SE Asia, and southern S America.
**Economic uses:** none.

The Centrolepidaceae is a family of small grass-like, rush-like or even moss-like annual or perennial herbs.

**Distribution.** The family is mainly found in Australia and New Zealand but extends to Southeast Asia and southernmost South America. It has obvious Antarctic affinities, with its major development in the Australian-Neozealandic region, from which it had radiated northwards and eastwards.

**Diagnostic features.** The plants are tufted or cushion-like herbs with linear, bristle-like leaves in dense basal rosettes in annuals, and imbricate and crowded along stems in perennials. The structure of the inflorescence and flower has been much discussed. The "flowers" of *Centrolepis* and *Gaimardia*, once thought to be bisexual, are now considered to be reduced inflorescences or "pseudanthia," cymose structures consisting of one (sometimes two) male flowers and two to many female flowers, all flowers being reduced to either one stamen or one carpel. The inflorescence is terminal and spike-like, with one to several pseudanthia within each of two or more glume-like floral bracts. Each pseudanthium may have none to two male flowers and none to many female flowers. The flowers are often fused and sometimes have thin, translucent subtending bractlets; perianth segments are absent. The male flowers have a single stamen with a thread-like filament and a unilocular (rarely bilocular), versatile anther with longitudinal de-

hiscence. The female flowers have a unilocular ovary, with one pendulous, orthotropous ovule, and one style (rarely three to ten, sometimes fused at the base). The fruit has a membranous pericarp, dehiscing by a longitudinal slit (or rarely indehiscent as in *Hydatella*). There is one seed per locule or carpel, with copious endosperm.

**Classification.** The family is usually divided into two tribes: TRITHURIEAE and CENTROLEPIDEAE. The Trithurieae with two-celled anthers comprise the genera *Trithuria* and *Hydatella*. The Centrolepideae with one-celled anthers comprise the genera *Brizula*, *Centrolepis* and *Gaimardia*.

The Centrolepidaceae is most closely related to the Restionaceae.
**Economic uses.** None are known.
D.M.M.

# RESTIONACEAE

**Number of genera:** about 30
**Number of species:** about 320
**Distribution:** mainly S Hemisphere, with centers in S Africa, Australia, Tasmania and New Zealand.
**Economic uses:** limited uses for thatching and broom-making.

The Restionaceae is a small family of rush-like herbs with small, inconspicuous flowers.
**Distribution.** Except for one species in northern Vietnam, all are found in the Southern Hemisphere. The main concentrations of species are in South Africa (in the coastal or subcoastal lowland and mountain areas) and Australia, Tasmania and New Zealand. Others occur in Malawi, the Malay Peninsula, the Chatham Islands, Chile, Patagonia and Madagascar. There are no genera in common between Australasia and South Africa, but interesting links exist between species of *Leptocarpus* in Australasia, Malaysia and South America. This distribution is of significance in theories of continental drift.

The plants grow in a wide range of conditions, but favor seasonally wet habitats which dry out each year. Some species can tolerate very dry conditions, and others grow in standing water. It is of interest that whereas the aerial parts have obvious adaptations to drought, the roots frequently have cortical air cavities which are associated with wet soil conditions.

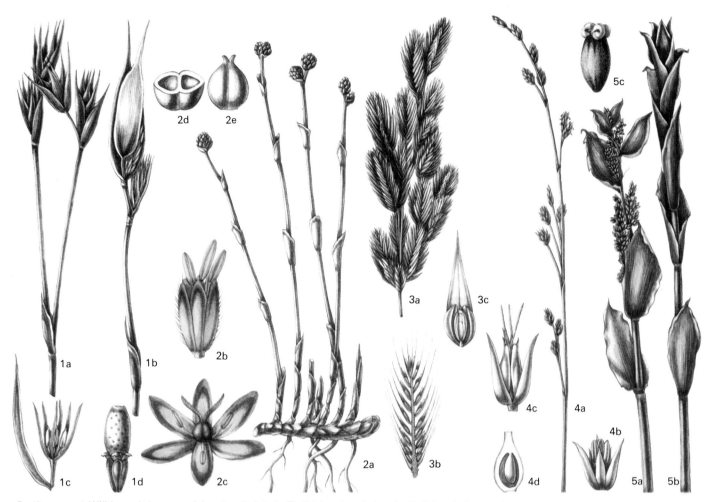

Restionaceae 1 *Willdenowia lucoeana* (a) male spikelets (×⅔); (b) female spikelets (×⅔); (c) male flower with sheath-like spathe (×2); (d) fruit (×2).
2 *Restio monocephalus* (a) habit (×⅔); (b) male flower with three stamens (×4); (c) female flower with two styles and three staminodes (×4); (d)
cross section of fruit (×6); (e) fruit entire (×4). 3 *Thamnochortus insignis* (a) female spikelets (×⅔); (b) vertical section of female spikelet (×1½);
(c) female flower (×3). 4 *Leptocarpus simplex* (a) shoot with male spikelets (×⅔); (b) male flower (×4); (c) female flower (×4); (d) vertical
section of ovary (×4). 5 *Elegia juncea* (a) shoot with male spikelets (×⅔); (b) shoot with hidden female spikelets (×⅔); (c) fruit (×6).

**Diagnostic features.** Species range in height from about 10cm (4in) to 2m (6.5ft). Most have aerial parts composed of several- to many-noded tough, wiry shoots which may be fertile or sterile. The shoots are usually simple or slightly branched, but some species are characteristically much branched. Functional leaf blades are rarely developed but a small, dry blade may be present; usually only a sheathing leaf base which is split to the base can be seen at each node. Leaf bases, sometimes deciduous, may be prominent and well developed although non-photosynthetic. Very few species have ligules. The shoots are modified to carry out photosynthesis. They bear abundant stomata and can have from one to four layers of palisade-like cells. In cross-section the stems may be circular, semi-circular, more or less square or variously ribbed. They are solid or hollow. The rhizomes are creeping or tufted, below 1cm (less than ½in) or up to 2cm (less than 1in) in diameter. In most species they are closely covered with dry brown scales. The roots are fleshy or wiry.

Flowers are small, regular and unisexual, with male and female on separate plants (very rarely on the same plant); they are usually arranged in spikelets in loose inflorescences. The spikelets are one- to many-flowered and are commonly subtended by a sheath-like spathe. The perianth is composed of three to six thin, dry segments in two series. In a few species, no perianth is developed. Opposite the inner three perianth segments are three stamens in the male flowers; a rudimentary ovary may be also present. The female flowers may have a variable number of staminodes. The ovary is superior, consisting of one to three carpels and one to three locules, each containing a single pendulous, orthotropous ovule, attached to the apex. There are one to three styles, free or variously connate.

The fruits are dry and nut-like or occur as three-sided capsules. The seeds have a small embryo and abundant endosperm.

**Classification.** The family is regarded as being taxonomically difficult. There is often a high degree of similarity between male plants of the same genus, making it hard to match male and female plants of the same species using only characters of floral and gross morphology. The varied anatomy of the stems has been found to be of diagnostic and taxonomic significance and can be used as a guide to generic and even specific identity in some instances.

There are probably more than the 30 genera currently recognized. The principal genera include *Restio, Leptocarpus, Elegia, Chondropetalum, Thamnochortus* and *Willdenowia*. There are several monospecific genera. *Anarthria* and *Ecdeiocolea* are now given individual family status as the Anarthriaceae and Ecdeiocoleaceae respectively.

The Restionaceae appears to be closely related to the Centrolepidaceae and to the Juncaceae and Thurniaceae. Although it is difficult to compare a more or less "leafless" family such as the Restionaceae with other leafy ones, this association is supported by a range of characters such as anatomy and seed structure. Recent treatments incline to regard the Restionaceae as more closely related to the Centrolepidaceae than to Juncaceae.

**Economic uses.** Matting, thatching and brooms are variously provided by a few species of this family.                    D.C.

# POALES

## GRAMINEAE
*The Grass Family*

The grasses or Gramineae (Poaceae is a permitted alternative name) comprise some 9,000 species grouped into about 650 genera. Although not the largest, the family is ecologically the most dominant and economically by far the most important in the world. It provides all the cereal crops (including rice), most of the world's sugar and grazing for domestic and wild animals, as well as bamboos, canes and reeds. The grasses also make a major contribution to much of the world's landscape.

**Distribution.** The family is cosmopolitan, ranging from the polar circle to the equator, and from mountain summits to the sea itself; it has been estimated that it is the principal component in some 20% of the earth's cover of vegetation. Few ecological formations lack grasses and many, such as steppe, prairie and savanna, are dominated by them. The great grasslands occupy a climatic zone between forest and desert, but are difficult to equate with any simple climatic parameter owing to the extent to which their distribution has been influenced by other plants and by animals. Indeed the spread of the grasses is a story of reciprocal adaptation, first with the herbivorous mammals and latterly with Man.

**Diagnostic features.** In a typical grass the root system is fibrous and often supplemented by adventitious roots from the lower nodes of the stem. Branching ("tillering") occurs mainly at ground level to form a rosette or tussock, often extended laterally by underground rhizomes or surface stolons to form a close sward. The upright stems are cylindrical, usually hollow but sometimes pithy, and mostly herbaceous although sometimes more or less cane-like or even woody. The leaves are borne in two rows at intervals along the stem, their point of origin being termed a node, and they are composed of two parts, sheath and blade. The sheath, a distinctive feature of grass morphology, tightly invests the stem and gives mechanical support to the soft meristematic zone situated just above each node. Differential growth at this meristem enables the stem to bend upright again after being flattened (lodging) by rain or trampling. At its upper end, the sheath passes into a parallel-veined blade, which also has a meristematic zone at its base, permitting the blade to continue growth despite the removal of its distal parts by grazing or cutting. The blade is typically long and narrow, but may be lanceolate to ovate in tropical shade-loving species; in a few genera it is deciduous from the sheath and occasionally it is narrowed at the base into a false petiole (the sheath being considered homologous with a true petiole). In size, leaves can vary from the bladeless

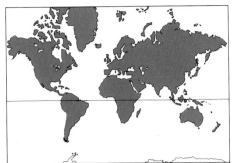

**Number of genera:** about 650
**Number of species:** about 9,000
**Distribution:** cosmopolitan.
**Economic uses:** cereal crops (including wheat, rice, maize, oats, barley, millet, sorghum), fodder for animals, sugar cane, bamboos and canes, lawns and some decorative ornamentals.

sheaths of the Australian *Spartochloa scirpoidea*, to the enormous blades up to 5m (16ft) long of the South American bamboo *Neurolepis nobilis*. At the junction of sheath and blade is a short membranous or ciliate rim (ligule), whose function is uncertain but may be to prevent rain entering the sheath. The combination of basal tillering and meristems at the node and base of the leaf blade enables the grasses to tolerate a degree of burning, trampling or grazing which effectively eliminates most competing families and which must contribute largely to the success of the family.

The inflorescence is a specialized leafless branch system, which usually surmounts the stem. Its basic units are termed spikelets, analogous to the flowers of petaloid plants. They are arranged in various ways, ranging from a single spike or raceme (in the latter case the pedicel is usually so short that the distinction is not very important), through an intermediate stage of several spikes arranged digitately or along an axis, to a many-branched panicle, this sometimes so condensed that it appears spike-like. The inflorescence develops inside the uppermost leaf-sheath, emerging only when almost mature. Extremes of size range from the spectacular 2m (6.5ft) plume of *Gynerium sagittatum*, down to the solitary single-flowered spikelet of *Aciachne pulvinata*.

An understanding of spikelet structure is essential for the identification of grasses. Visible from the outside are two opposite rows of scales arranged alternately along an axis (rhachilla). The two lowest scales (glumes) are empty, but the remainder (lemmas) each form part of a floret, whose floral parts are enclosed by the lemma on the outside and a delicate membranous scale (palea) on the inside. Glumes or lemmas are often produced into one or more, long, stiff bristles termed awns. The floral parts consist of two (rarely none or three) tiny scales called lodicules, three (rarely one to six or

more) stamens, and two (rarely one or three) feathery stigmas surmounting the ovary, which is superior and has a single locule containing one ovule, usually adnate to the adaxial side of the carpel. In origin, the spikelet can be related to a branch system in which the glumes and lemmas represent modified leaf-sheaths and the palea a prophyll (a special bract replacing the first leaf on the inner side of an axillary branch). The lodicules are taken to represent vestigial perianth members, there being intermediates between lodicules and stamens in a few bamboos (eg *Ochlandra*); their function is uncertain, but their turgor may help to open the floret. The basic pattern of two glumes and several florets is remarkably uniform through the family, apart from a few exceptions among the bambusoid grasses. However, an extraordinary number of variations has been developed from this simple basic theme by changes in the size, shape, ornamentation or sex of the parts or by reduction in their number.

Bisexual spikelets are the rule, although some of their florets are often unisexual or barren; occasionally male and female spikelets are borne on the same plant, rarely on separate plants. The florets open for only a few hours to permit the exposure of the sexual organs to wind-pollination, cross-pollination usually being ensured by protandry. The pollen itself is viable for less than a day, the shortest-lived of all angiosperm pollens. It is of a distinctive type, monoporate operculate, with a very finely granular surface.

Variants of normal flowering behavior are sometimes encountered. Apomixis (development of the ovule without fusion with a male gamete) occurs frequently in the family, though always interspersed with spells of sexuality. Cleistogamy (self-pollination within the closed floret) is not uncommon and may be associated with special spikelets hidden in the axils of the leaf sheath or even beneath the soil surface. Proliferation (the modification of spikelets into little leafy shoots) is commonly the result of hormone imbalance brought about by short days at the end of the season in genetically predisposed races, but in some mountain or arctic species the transformation of spikelets into bulbils has become a regular means of propagation. Somewhat similar abnormal growth effects are sometimes produced by hormone weedkillers. Vivipary (the germination of the seed while still attached to the parent) has been recorded in *Melocanna*.

Cytogenetically the grasses engage in a diversity of behavior which raises many problems for those attempting to divide them into discrete species. Some 80% of them have a polyploid chromosome number, and the occurrence of polyhaploidy (halving the chromosome number) has also been demonstrated; apomictic swarms are not unusual; and over 2,000 hybrids have been recorded, 200 of them fertile. In certain tribes a tendency to wide crossing makes it difficult to establish satisfactory generic boundaries (but the cross between *Saccharum* and *Bambusa*, often quoted as an example, is bogus). Nor should the role of purely vegetative clones in the life history of many perennial species be ignored. It has been estimated that a single plant of *Festuca rubra*, which spreads by rhizomes, may be some 250m (over 800ft) in diameter and up to 400 years old, and that a large tussock of *F. ovina* 8m (26ft) across could be 1,000 years old.

The fruit is a caryopsis, although some bamboos have a rather fleshy pericarp (amounting to a berry several centimeters long in *Melocanna*) and a few genera (notably *Sporobolus*) have a loose seed within a free pericarp. Nearly always, however, the disseminule (including the grass "seed" of commerce) incorporates part of the inflorescence, ranging from an adherent palea to the complete inflorescence functioning as a tumbleweed. Dispersal is aided by wind-borne plumes of various kinds, by a great variety of hooks and barbs, occasionally by glue, and sometimes by a dark coloration inviting consumption by birds; *Streptogyna* and *Streptochaeta* are notable for their elegant hair-traps. Awns often play a part in dispersal, but in many species a more important function seems to be their ability to tuck the fruit beneath the soil surface by hygroscopic movements. The embryo is peculiar with no exact homologue among the angiosperms. Its most distinctive feature is the transformation of the cotyledon into a haustorial organ (scutellum) attached to the starchy (rarely liquid) endosperm. In many species the fruit is viable for five years or so; while a few exceed 30 years.

**Classification.** Grasses are rather uniform in their vegetative parts, so that recognition has relied heavily upon their spikelet structure. The older classifications were, in effect, arrangements of spikelet types in a logical sequence of increasing complexity. However, it has gradually emerged that a more natural grouping can be achieved with the aid of such features as internal leaf anatomy, photosynthetic pathway, basic chromosome number and embryo structure. The laborious task of cataloguing these cryptic characters is still incomplete and the significance of differences is not always easy to assess. Consequently grass classification is still in a state of flux and there is no wholly satisfactory overall account available. Nevertheless the basic features of an improved classification are becoming clearer. There are six subfamilies and over 50 tribes, only the larger or more interesting of which are listed below. Quite unrelated grasses may look very similar to the naked eye and even an expert can be misled unless the spikelets are carefully examined with the aid of a lens.

## SUBFAMILY BAMBUSOIDEAE

A group of tribes with certain features of the internal leaf anatomy in common, the most distinctive being known technically as arm cells and fusoid cells. Many genera have unusual features, these being of two types: those forming links with the main stream of monocotyledon morphology, and those standing in isolation. The former are interpreted as primitive, and the latter as relics of discontinued evolutionary trends, together suggesting an ancestral role for the subfamily.

BAMBUSEAE. A few herbs, but mostly shrub or tree-like bamboos up to 40m (130ft) high. A number of genera display apparently primitive features. Thus some have three lodicules, six stamens and three stigmas, recalling the usual three-part monocotyledon type of flower; the number of glumes may be indefinite, some of them furnished with axillary buds which develop into spikelets, these also having buds in the lower axils, and so on; and the leaf- and spikelet-bearing branch systems may be imperfectly separated. Against this must be set the highly developed woody stem, and considerable variation in spikelet structure, including types with single florets or with sexual dimorphism between the florets. Thus they are "primitive" only in a rather restricted sense. A curious aspect of their biology is that many species flower only periodically, the interval being 10–120 years according to species, with surprisingly exact synchronization throughout the geographical range. Mainly tropical forest plants, reaching into warm temperate regions. Chief genera: *Arundinaria, Bambusa, Chusquea, Dendrocalamus, Phyllostachys, Sasa.*

STREPTOCHAETEAE. The only genus has single-flowered spikelets, and is remarkable for the three enormous lodicules, up to 2cm (almost 1in) long, palea cleft to the base, and lemma with long coiled awn; a supposed homology with six perianth members has led some authorities to regard it as exceptionally primitive. It undoubtedly has primitive features (4–5 glumes, 6 stamens, 3 stigmas), but it also has a sophisticated dispersal mechanism. The spikelets dangle from the awns like fish-hooks, and the lodicules deflect animals' hairs into the palea, where they are gripped by the converging sides of the cleft. Rain forest of tropical South America. Only genus: *Streptochaeta.*

STREPTOGYNEAE. The only genus has several-flowered spikelets equipped with another ingenious fish-hook mechanism. The florets have long tangled stigmas, the hair-trap being formed by a springy curved rhachilla segment pressing against the back of the palea. Rain forest in tropical Africa and America. Only genus: *Streptogyna.*

OLYREAE. Little herbs carpeting the ground in deep shade, or rambling canes at the edge of clearings. In common with many forest-dwelling grasses the leaves are often broad, and scarcely recognizable as belonging to grasses. The single-flowered spikelets are unisexual, a phenomenon encountered here and there in the family, the sexes being mixed or in different parts of the inflorescence. A small tribe, mostly from South America. Chief genus: *Olyra.*

PHAREAE. Similar to Olyreae, but the veins of the leaf-blade slant obliquely from the midrib, instead of running parallel to it as in most other grasses. A small tribe of tropical rain forest. Chief genera: *Leptaspis, Pharus.*

PARIANEAE. The single genus is remarkable for the large number of stamens, commonly about 30, in each floret. It is insect-pollinated, though this appears to be a secondary adaptation. Rain forest of South America. Only genus: *Pariana.*

ORYZEAE. A small but important tribe with single-flowered spikelets whose glumes are almost completely suppressed, though in *Oryza* two sterile lemmas simulate glumes. Tropical and warm temperate regions, mainly in swamps; rice is widely cultivated. Chief genera: *Oryza, Zizania.*

## SUBFAMILY CENTOSTECOIDEAE

An enigmatic group with similarities to both Bambusoideae and Panicoideae; probably nearer the former, but occupying an isolated position.

CENTOSTECEAE. Broad-leaved herbs with panicles of single- to several-flowered spikelets. *Lophatherum* is unusual in possessing root tubers. A small tribe of worldwide distribution, mainly in rain forest. Chief genera: *Centosteca, Lophatherum, Orthoclada.*

## SUBFAMILY ARUNDINOIDEAE

An unspecialized subfamily, defined more by the absence of distinctive features than by their presence. For this reason it is commonly regarded as close to the ancestral line from which the first non-bambusoid grasses have been derived.

DANTHONIEAE. Panicles of spikelets with 2–10 flowers; lemma usually bilobed, with an awn from between the lobes; ligule a line of hairs. Subtropics and warm temperate regions, particularly in the Southern Hemisphere; a few, such as moor grass (*Molinia*), in cool temperate regions. Chief genera: *Chionochloa, Danthonia, Molinia, Pentaschistis.*

ARUNDINEAE. Robust grasses with handsome panicles of plumose spikelets. In other respects its genera are not particularly close,

Gramineae. 1 *Arundinaria japonica* (a) tip of shoot showing leaves and inflorescence ($\times\frac{2}{3}$). 2 *Phleum pratense* inflorescence ($\times\frac{2}{3}$). 3 *Stipa capillata* fruit—a caryopsis crowned by a long feathery awn ($\times\frac{2}{3}$). 4 *Aristida kerstingii* fruit with a three-branched awn ($\times\frac{2}{3}$). 5 *Tristachya decora* inflorescence ($\times\frac{2}{3}$). 6 *Avena sativa* (a) inflorescence ($\times\frac{2}{3}$); (b) fruit ($\times6$). 7 *Lolium perenne* habit showing adventitious roots, leaves with sheathing bases and inflorescence ($\times\frac{2}{3}$) 8 *Poa annua* part of leaf showing (from the bottom) the sheathing base expanding into the blade with a ligule inserted where the blade and base meet ($\times6$).

and their traditional association in the same tribe is open to question. Subtropics to cool temperate regions; includes reed grass (*Phragmites*) of marshes throughout the world, and pampas grass from South America (*Cortaderia selloana*) which can be seen as the centerpiece of many a temperate garden. Chief genera: *Arundo, Cortaderia, Phragmites.*

LYGEAE. Spikelets two-flowered, the glumes suppressed; the two lemmas are fused along their margins to form a rigid tube, divided longitudinally by the two paleas which are also fused together. A weird tribe from the Mediterranean; its only species is one of the esparto grasses. Only genus: *Lygeum.*

MICRAIREAE. The only genus is a moss-like plant, unique among the family for its spiral arrangement of the leaves. Australia. Only genus: *Micraira.*

### SUBFAMILY CHLORIDOIDEAE

Characterized by the "Kranz syndrome," an assemblage of anatomical characters associated with an extra loop in the photosynthetic pathway, apparently making it more efficient in high light intensities. The spikelets break up at maturity.

ERAGROSTIDEAE. Panicles or racemes with several-flowered spikelets and three-nerved lemmas. A large tropical tribe, with many pioneer species of bare ground and disturbed places; tef and finger millet are grown as cereals. Chief genera: *Dactyloctenium, Eleusine, Eragrostis, Leptochloa.*

CHLORIDEAE. Racemes of spikelets containing only one fertile floret, with or without extra sterile ones; perhaps rather arbitrarily separated from Eragrostideae. A large tribe, mainly of tropical and subtropical savannas. It includes Bermuda grass (*Cynodon dactylon*), the commonest tropical lawn species; grama (*Bouteloua*) and the dioecious buffalo grass (*Buchloe*) from the North American plains; and *Spartina* from intertidal mudflats on either side of the Atlantic. Chief genera: *Bouteloua, Chloris, Cynodon, Lepturus, Spartina.*

SPOROBOLEAE. Panicles of little single-flowered spikelets, but otherwise scarcely separable from Eragrostideae. Tropics and subtropics in open places. Chief genera: *Muhlenbergia, Sporobolus.*

ZOYSIEAE. Like Chlorideae, but with the spikelets falling entire at maturity and often of bizarre shape. Old World tropics. Chief genera: *Perotis, Tragus, Zoysia.*

ARISTIDEAE. Panicles of single-flowered needle-like spikelets, the lemma with a three-branched awn. Tropics, particularly in hot, dry places. Chief genus: *Aristida.*

### SUBFAMILY PANICOIDEAE

The Kranz syndrome is usually present, though not always well marked. The spikelets are strictly two-flowered, the lower floret male or barren, the upper bisexual and usually different in appearance.

ARUNDINELLEAE. Relatively unspecialized, the spikelets disarticulating between the florets at maturity in the usual way. In many species the emergent panicle bears tiny juvenile spikelets, unlike most grasses whose spikelets are almost fully grown before emergence from the uppermost sheath. Tropical savannas, mainly in the Old World. Chief genera: *Arundinella, Loudetia.*

PANICEAE. Spikelets fall entire at maturity, the grain encapsulated by the hard, often bony, lemma and palea of the upper floret; inflorescence a panicle, or composed of racemes in various arrangements. In some genera (notably *Setaria* and *Pennisetum*) the spikelets are surrounded by an involucre of bristles, derived from modified panicle branches. A large and important pantropical tribe; includes the crops fundi, proso, foxtail and bulrush millets. Chief genera: *Axonopus, Brachiaria, Digitaria, Echinochloa, Panicum, Paspalum, Pennisetum, Setaria.*

ANDROPOGONEAE. Spikelets fall entire at maturity, protected by the tough glumes. They are borne on racemes in pairs, one member of each pair sessile, the other raised on a pedicel. Sometimes both members of a pair are alike, but usually they are quite different, the pediceled often sterile; the disseminule may then be a most complex organ in which the adjacent rachis internode, the pedicel and the modified pediceled spikelet all participate in the protection of the fertile sessile spikelet, sometimes supplemented by special barren spikelet pairs forming a kind of involucre at the base of the raceme. The racemes may be borne in a terminal panicle, but are commonly single or paired, in which case there is often copious axillary branching. The whole system of axillary inflorescences and subtending leaves, themselves highly modified, may then be raised to the top of the stem in a manner that simulates a panicle. A large and important pantropical tribe, often of great morphological complexity; includes the crops sorghum, maize, sugar cane and lemon grass. Chief genera: *Andropogon, Coix, Cymbopogon, Erianthus, Euchlaena, Hyparrhenia, Miscanthus, Saccharum, Sorghum, Themeda, Vetiveria, Zea.*

### SUBFAMILY POOIDEAE

Differs from the other, primarily tropical, subfamilies in many cryptic characters concerning anatomy (notably the absence of micro-hairs on the leaf epidermis), cytology and physiology. The subfamily in fact constitutes a major evolutionary departure, though this is not apparent from the external morphology. Indeed much taxonomic research during the last two or three decades has been directed towards the gradual elimination of misplaced genera from other tribes.

POEAE. Several-flowered spikelets, mostly in panicles, with five- to seven-nerved lemmas longer than the glumes and with or without a straight awn from the tip (compare with the three-nerved lemmas of Eragrostideae). A large pantemperate tribe; includes rye grass (*Lolium perenne*), the foremost species of high-grade temperate pastures; *Festuca* used in fine lawns; and *Poa annua*, possibly the most ubiquitous of all grasses. Chief genera: *Briza, Cynosurus, Dactylis, Festuca, Lolium, Poa.* (*Glyceria* (Glycerieae) and *Melica* (Meliceae) belong to tribes which differ only in detail from Poeae.)

AVENEAE. Distinguished from Poeae by the long papery glumes enclosing the lemmas, which often have a kneed awn from the back; ligule membranous (compare with Danthonieae, formerly included in Aveneae). A large pantemperate tribe; includes the cereal oats. Chief genera: *Anthoxanthum, Avena, Deschampsia, Helictotrichon, Holcus, Phalaris.*

AGROSTIDEAE. A single-flowered variant of Aveneae, and included in that tribe by some authorities. A large pantemperate tribe; includes *Agrostis*, a typical genus of second-rate European pastures, as well as a number of other common hay and forage species. Chief genera: *Agrostis, Alopecurus, Ammophila, Lagurus, Milium, Phleum.*

BROMEAE. Outwardly resembling Poeae, but with the unusual starch grains of Triticeae, thus forming a link between the two tribes. Pantemperate. Chief genus. *Bromus.*

TRITICEAE. One- to several-flowered spikelets arranged in spikes. Further distinguished by the unusual round starch grains in the seed. A peculiarity of the tribe is a tendency to wide hybridization, making the delimitation of genera unusually difficult. A pantemperate tribe notable for its cereal genera wheat, barley and rye. Chief genera: *Aegilops, Agropyron, Elymus, Hordeum, Secale, Triticum.*

STIPEAE. Panicles of single-flowered, needle-like spikelets, often with conspicuous awns; leaves typically harsh and narrow. A small tribe whose largest genus, *Stipa*, is characteristic of dry steppes throughout the world. Chief genus: *Stipa.*

The relationship of grasses to other families is obscure. Their spikelets serve to distinguish them from all other families except the Cyperaceae, and their close superficial resemblance to this family is not maintained in matters of detail, probably indicating no more than parallel evolution from a rather remote common ancestry. The Cyperaceae may be distinguished by the presence of one or more of the following

Gramineae (continued). 9 *Andropogon fastigatus* inflorescences ($\times\frac{2}{3}$). 10 *Imperata cylindrica* inflorescence ($\times\frac{2}{3}$). 11 *Cynodon dactylon* habit showing creeping stem bearing adventitious roots and branching (tillering) at ground level ($\times\frac{2}{3}$). 12 *Brachiaria brizantha* inflorescence ($\times\frac{2}{3}$). 13 *Bromus commutatus* (a) exploded view of a floret showing two scales (the awned lemma and hairy palea) and flower with three stamens and two feathery styles ($\times 4$); (b) spikelet showing lower scales without awns (glumes) and upper scales with awns (lemmas). 14 *Olyra ciliatifolia* leaves with parallel veins and inflorescence ($\times\frac{2}{3}$).

9

10

11

12

13a

13b

14

features: solid triangular stems; leaves without ligules; umbellate inflorescences subtended by leaf-like bracts; spirally arranged spikelet scales.

Some evidence of derivation from a general commelinaceous type is found in the Bambusoideae, most authorities tentatively suggesting the Flagellariaceae as a possible living relative. The tenuous link between grasses and other monocotyledons is thus strongest among inhabitants of the tropical forest zone, pointing to the forest environment as the cradle of the grasses, but leaving their early adoption of wind-borne pollination unexplained. The first grasses to spread from the forest fringes into the dry savanna were probably akin to the present-day Arundinoideae. From them eventually diverged the two major tropical groups, Chloridoideae and Panicoideae; They were also, it is assumed, akin to the Pooideae, which adapted to cool climates and invaded the temperate zone.

The time-scale for the evolution of the family is difficult to establish, there being little help from fossil evidence. This amounts to scarcely more than grass-type pollen at the end of the Cretaceous; florets of an extant genus, *Stipa*, in the Oligocene; and herbivorous mammals with grass-eating teeth in the Miocene, indicating that by this time grassland had become an ecological formation. Geographical distribution casts some light on the subject: whereas the major tribes occupy worldwide climatic belts, their constituent genera tend to be confined to single continents. The implication is that the tribes were in existence before the continents drifted beyond the reach of seed dispersal at the end of the Cretaceous (the timing of this event is itself contentious), generic evolution then continuing upon the separate continents.

**Economic uses.** The adoption of the grasses as a principal source of food was a milestone in human development, many, if not most, of the great civilizations being founded on the cultivation of grass crops. Opportunistic gathering ("ramassage") of wild grain has always been a common practice among primitive people, but domestication entailed the selection of strains whose inflorescence does not shatter to disperse the seed before it can be harvested. This first happened 8,000–10,000 years ago in southwestern Asia and the Middle East, where wild species of *Triticum* and *Hordeum* yielded the cereals wheat and barley. With the spread of agriculture through temperate Europe and Asia, various grasses adapted to life as arable weeds, some of which were themselves domesticated as oats (*Avena sativa*) and rye (*Secale cereale*). Elsewhere archeological evidence is more sketchy, but domestication seems to have occurred somewhat later. Rice (*Oryza sativa*) became the principal cereal of tropical Asia, supported by foxtail millet (*Setaria italica*) and proso (*Panicum mili-*

*aceum*). In Africa the main indigenous cereals are sorghum (*Sorghum bicolor*) and bulrush or pearl millet (*Pennisetum glaucum*), supplemented by a number of rather local minor grains, including finger millet (*Eleusine coracana*), fundi (*Digitaria exilis*), tef (*Eragrostis tef*) and an independently domesticated species of rice (*Oryza glaberrima*). Maize (*Zea mays*) is the indigenous American cereal. Although not a cereal, sugar cane (*Saccharum officinarum*), from Southeast Asia, may be mentioned here.

The second facet of Man's dependence on the grasses springs from the domestication of animals, which was roughly contemporaneous with the beginnings of agriculture. Until recent times livestock rearing was based upon the exploitation of natural grasslands, although the preservation of fodder as hay had been introduced by the Roman era. Sown pastures, based on rye grass, date from the late 12th century in north Italy and from the late 16th century in northern Europe.

Bamboos provide an ideal building material in many parts of the world, and grass is also employed for building construction in the form of thatch and matting. To the engineer grasses are invaluable for stabilizing sand dunes, road verges and other raw soil surfaces. Many species have been used for papermaking, the best known being esparto, a name variously applied to *Stipa tenacissima*, *Ampelodesma tenax* and *Lygeum spartum*.

An aromatic oil is distilled from the leaves of lemon grass (several species of *Cymbopogon*), imparting a citronella scent to soaps and other perfumery. Among a host of minor uses may be mentioned necklace beads (*Coix* involucres), brush bristles (*Sorghum* inflorescence branches), pipe bowls (*Zea* cobs), edible bamboo shoots, clarinet reeds (*Arundo donax* stems), fishing rods (bamboo species) and corn dollies or various garishly dyed inflorescences sold as house decorations. In the form of lawns grasses have an honored place in horticulture, though few of them have been admitted voluntarily to the herbaceous border. A well-known exception is the variegated form of *Phalaris arundinacea*, often called gardener's garters.

The obnoxious properties of grasses lie mainly in their success as weeds of cultivation; those whose spikelets are equipped with pungent spikes or barbs may also become a serious nuisance to domestic animals when present in quantity. Some tropical forage species are known to develop a lethal content of hydrocyanic acid under conditions associated with the partial wilting of lush growth, but still largely unpredictable. Poisonous properties may also arise from fungal infections, St Anthony's fire, an affliction caused by eating cereals infected by ergot (*Claviceps*, especially *C. purpurea*), being the most notorious.                    W.D.C.

# JUNCALES

## JUNCACEAE
*Rushes*

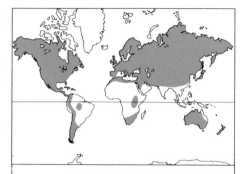

**Number of genera:** 9
**Number of species:** about 400
**Distribution:** worldwide, but chiefly cold temperate or montane regions, in damp habitats.
**Economic uses:** stems used in basketwork, mat-making and chair bottoms, with some waterside ornamentals.

The Juncaceae is a smallish family of tufted perennial or annual herbs (rushes) and, rarely, woody shrubs (*Prionium*).

**Distribution.** The family occurs worldwide but chiefly in cold temperate or montane regions, in wet or damp habitats.

**Diagnostic features.** The perennial species have erect or horizontal rhizomes. The stems are erect, cylindrical and normally only leafy at the base. The leaves are cylindrical or flat and grass-like, sheathing at the base or reduced entirely to sheaths and mostly arranged in basal tufts when present. The flowers are regular, bisexual or unisexual (with male and female on the same or on separate plants), wind-pollinated and consequently small and insignificant, sometimes solitary but more often in open panicles, corymbs or clustered heads. The perianth segments consist of two whorls of three (or rarely one whorl of three) scales which are leathery or thin and papery, sometimes with a dried-up appearance at the tips and edges. They are dull in color, often green, brown or black but sometimes also white or yellowish. There are six or three stamens usually opposite the perianth segments. The ovary is superior, of three fused carpels comprising one or three locules, with few to numerous ovules on axile or parietal placentas. There may be one style or three, but there are always three stigmas. The fruit is a dry capsule which dehisces loculicidally, with one to many small globose, angled or compressed seeds which have starchy endosperm and a straight embryo.

**Classification.** The nine genera include the antarctic rhizomatous perennials *Rostkovia* (monoecious, perianth segments equal in length, obovoid seeds) and *Marsip-*

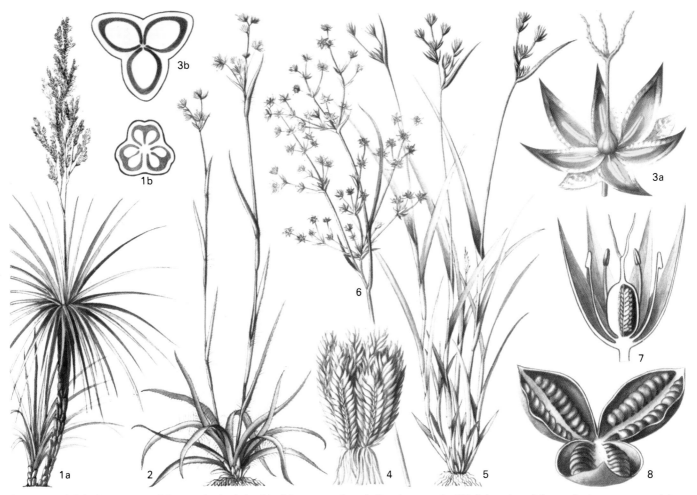

Juncaceae. 1 *Prionium serratum* (a) a woody shrub ($\times\frac{1}{15}$); (b) cross section of trilocular ovary ($\times 20$). 2 *Luzula nodulosa* an herbaceous perennial with sheathing leaf bases ($\times\frac{1}{4}$). 3. *L. spadicea* (a) flower ($\times 20$); (b) cross section of trilocular ovary ($\times 100$). 4 *Distichia muscoides* a low-growing cushion plant with leaves in two rows and terminal capsules ($\times\frac{2}{3}$). 5 *Juncus bufonius* an herbaceous perennial with linear erect leaves with loosely sheathing bases and flowers in dense cymose heads subtended by leaf-like bracts ($\times\frac{2}{3}$). 6 *J. acutiflorus* inflorescence ($\times\frac{2}{3}$). 7 *J. bulbosus* half flower ($\times 14$). 8 *J. capitatus* dehiscing capsule consisting of three valves ($\times 18$).

*pospermum* (similar, but with shorter inner perianth segments, and pointed, spindle-shaped seeds). The Andean rhizomatous perennials of *Andesia* are thick cushion plants with equal perianth segments and ovoid seeds. Other South American groups include the three closely related dioecious genera from the Andes: *Oxychloe* (irregular leaves with spreading blades) and the regular-leaved, cushion plants of *Distichia* and *Voladeria* from Ecuador. Perhaps the most unusual genus of the family is *Prionium*, a group of subarborescent shrubs from South Africa with a terminal tuft of toothed leaves, closed leaf sheaths and flowers in large panicles. *Juncus* is a cosmopolitan genus of herbaceous rushes characterized by their entire leaves with open leaf sheaths and small inflorescences of bisexual flowers producing capsules with many seeds. *Luzula*, another fairly widespread genus, is most common in the Northern Hemisphere, very similar in many ways to *Juncus* but differing in its closed leaf sheaths, hairy leaf blades and three-seeded capsules.

The Juncaceae is currently considered to be related either to the Liliaceae through the putative primitive genus of the family,

*Prionium*, or to the Restionaceae, which it generally resembles in many more respects. The Juncaceae and Restionaceae are both florally reduced families, a feature they have in common with the Cyperaceae and Gramineae; all four are most probably individually derived evolutionary lines from a common stock resembling the modern Commelinales order.

**Economic uses.** The family is not generally of much commercial value. However, among the well-known products are juncio, used in binding, derived from the sea rush, *Juncus maritimus*; palmite, a strong fiber made from the serrate leaves of the palmiet, *Prionium serratum* (*P. palmitum*); and the split rushes used in basket making and chair-bottom manufacture, taken from the stems of the soft rush, *J. effusus*, and the heath rush, *J. squarrosus*.                                C.J.H.

## THURNIACEAE

The Thurniaceae is a small family of perennial sedge-like herbs endemic to Guyana and certain parts of the Amazon valley.

The leaves are elongated and leathery, with sheathing bases and leaf margins either

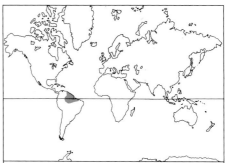

**Number of genera:** 1
**Number of species:** 3
**Distribution:** Guyana and parts of Amazon valley.
**Economic uses:** none.

serrulate, as in *Thurnia sphaerocephala* and *T. polycephala*, or entire, as in *T. jenmanii*. The stems are three- or four-angled and bear one or several heads of flowers subtended by several basal, leafy bracts. The flowers are small, irregular, pendulous and bisexual, with six persistent narrow perianth segments arranged irregularly below the ovary. The six stamens extend far beyond the perianth parts. The ovary of three fused carpels has

Cyperaceae. 1 *Carex decurtata* (a) entire plant showing habit (×⅔); (b) spikelet containing both male and female flowers (×3); female flower (c) entire showing three-branched stigma (×6) and (d) opened out to show the superior ovary (×8); (e) male flower with three stamens (×6). 2 *Cladium tetraquetrum* (a) habit showing sheathing leaf bases (×⅔); (b) inflorescence (×6); (c) spikelet (×6); (d) flower and subtending bract (glume) (×6). 3 *Cyperus compressus* (a) habit (×⅔); (b) flower showing trifid style (×18); (c) spikelet of bisexual flowers (×3); (d) flower with three stamens and a superior, lobed ovary crowned by three styles (×12).

three locules each with one to numerous axile ovules. There are three filiform stigmas. The fruit is a three-angled capsule. The seeds contain endosperm. The family is most closely related to the Juncaceae and has at times been included within it, but anatomical evidence (such as the presence of silica bodies in the leaf epidermis) clearly indicates that the Thurniaceae should be maintained as a distinct family. Indeed, some authorities believe that such anatomical features as the curious inverted vascular bundles observed in the leaves of *T. sphaerocephala* and *T. jenmanii* indicate that the family has no close affinities with the Juncaceae. No economic uses are known.                    S.A.H.

# CYPERALES

## CYPERACEAE
*Reeds and Sedges*

The Cyperaceae is a large family of mainly perennial, and a few annual, grass-like herbs.
**Distribution.** The family is distributed in all parts of the world but more especially in damp, wet or marshy regions of the temperate and subarctic zones. The genus *Carex*

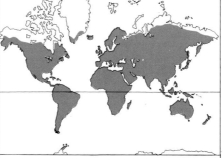

**Number of genera:** about 90
**Number of species:** about 4,000
**Distribution:** cosmopolitan, especially damp temperate and subarctic habitats.
**Economic uses:** stems and leaves used in mat-, hat-, basket- and paper-making and as fodder for animals; some edible tubers (tigernut, Chinese water chestnut), locally-used medicines and several pot plants and water garden ornamentals.

(true sedges), in particular, is of considerable ecological importance.
**Diagnostic features.** The plants have a creeping underground rhizome from which

arise solid aerial stems (culms) which are often three-angled in cross-section, generally unbranched below the inflorescence and often leafless. The leaves, usually arranged in a tuft around the base of the stem, are normally disposed in three ranks; they have a grass-like blade, a closed (or rarely open) sheath around the stem, and usually no ligule.

The small, inconspicuous flowers are bisexual or unisexual (with male and female usually on the same plant), and arranged in spikelets. Each flower is in the axil of a glume (bract). The perianth is represented by scales, bristles or hairs and in some species is entirely absent. There are one to six stamens, but usually three; the filaments are free. The ovary is superior, of two or three fused carpels forming a single locule with a single ovule inserted at the base. The style is divided into two or three teeth or branches and is sometimes persistent on the ripe fruit, which is an achene. The seed contains a small embryo surrounded by copious mealy or fleshy endosperm.
**Classification.** The two major systems for classifying the genera of this family have depended on an initial division into either

three subfamilies or seven tribes. In the former system (after Engler) the subfamilies are SCIRPOIDEAE (including *Cyperus*, *Eriophorum*, *Scirpus*, *Eleocharis*); RHYNCHOSPOROIDEAE (including *Rhynchospora*, *Scirpodendron*, *Cladium* and *Scleria*); and CARICOIDEAE (including *Carex* and *Uncinia*). Both this system and the other (after Hutchinson) are based primarily on features of the inflorescence, flowers and fruit. For example, *Cyperus* and *Scirpus* have several- to many-flowered spikelets while those of *Rynchospora* and *Cladium* are mostly one- or two-flowered. All these genera have bisexual flowers while those of *Scleria* and *Carex* are unisexual.

The Cyperaceae is perhaps closest to the grasses (Gramineae), but the relationship is not close despite the superficial similarities in habit and appearance. The sedges are generally distinguished by the often solid and three-angled stems, the generally closed leaf sheath and the absence of a ligule. Another distinctive feature is that each individual flower is usually subtended by a single glume.

**Economic uses.** The family contains a large number of useful species which are employed for a variety of purposes. *Cyperus* includes the papyrus or paper reed (*Cyperus papyrus*), whose stems provide the papyrus paper much used in ancient times. The stems of a number of other *Cyperus* species, such as *C. malacopsis* and *C. tegetiformis* (Chinese mat grass), are used for mat-making while a few species (eg *C. esculentus*, the chufa, tigernut, earth almond or rush nut) have edible storage organs. Some species, such as *C. longus* (galingale) and *C. articulatus*, have sweet-scented rhizomes and/or roots which can be used in perfumery.

*Carex* includes *Carex atherodes* which is used as a hay grass in the United States of America. The stems and leaves of *C. brizoides* are used as a packing material in some central European countries. Both *C. paniculata* and *C. riparia* are used in stables in place of straw, and *C. dispalatha* is cultivated in Japan for its leaves, which are used to make hats.

Stems of *Cladium mariscus* are used for thatching houses in Europe and parts of North Africa. The stems of the tropical and subtropical American *C. effusum* (saw grass) are a source of cheap paper. The stems of the Pacific species *Eleocharis austro-caledonica* are used for basket-making while *E. tuberosa* (matai, Chinese water chestnut) is cultivated in China and Japan for its edible tubers. The stems and leaves of *Lepironia mucronata* are used for packing and for basket work. *Mariscus umbellatus* has edible rhizomes, while *M. sieberianus* is used as a vermifuge in Sumatra.

*Scirpus* contains a large number of useful species, some of which have medicinal uses; the roots of *Scirpus grossus* and *S. articulatus* are used in Hindu medicine as a curative for

diarrhoea and as a purgative respectively. The tubers of *S. tuberosus* are eaten as a vegetable in Japan and China. The stems of *S. totara* (tropical South America) are used for making canoes and rafts and those of *S. lacustris*, the bulrush of North and Central America, in basketwork, mats and chair seats.

Some species of *Carex*, *Cyperus*, *Leiophyllum* and *Scirpus* are cultivated as pot plants and water garden ornamentals.     S.R.C.

# TYPHALES

## TYPHACEAE
*Reedmace Bulrush and Cattails*

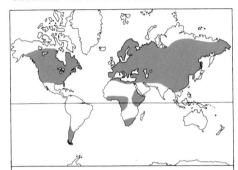

**Number of genera:** 1
**Number of species:** about 15
**Distribution:** almost cosmopolitan in freshwater habitats.
**Economic uses:** leaves used for basketwork, especially from the reedmace bulrush (*Typha latifolia*).

The Typhaceae is a small family of gregarious aquatic herbs consisting of the single genus *Typha* with about 15 species, which includes the reedmace bulrush and cattails.

**Distribution.** The Typhaceae occur in habitats of shallow freshwater such as reed swamps, lakes, rivers and ponds in many temperate and tropical localities from the Arctic Circle to southern South America.

**Diagnostic features.** Most species are very tall plants, growing to 2m (6ft) or more, with long simple stems that are usually submerged at the base. The leaves arise mostly from the subterranean part of the stem and invariably consist of linear, aerial, elongated blades which are rather thick and spongy. The flowers are unisexual and pollinated by the wind; they are characteristically crowded into a long, dense, club-shaped terminal spadix, with the female flowers occurring in the lower half of the inflorescence and the males arranged above them. The two sexes are either indistinguishable and contiguous forming a single "club," or quite separate with an articulate segment of the stem between them to form a double club. The perianth segments are somewhat indeterminate in terms of petals or sepals and uniformly consist of numerous slender threads or elongated spoon-shaped scales

mixed with ovaries or stamens. The stamens have two to five slender, free or fused filaments with linear anthers attached at the base. The crowded female florets each consist of an ovary with a single locule, tapered at the apex into a narrow ribbon-shaped stigma. The fruits are achenes with a long stipe covered in hairs which aid dispersal by the wind. The seeds have mealy endosperm and a long, narrow embryo.

**Classification.** The Typhaceae together with the Sparganiaceae form a very distinct order, but the wider relationships of these two families are somewhat obscure; they probably represent an independent evolutionary line from an ancestor somewhat resembling the Commelinaceae of today.

**Economic uses.** The leaves of the reedmace bulrush (*Typha latifolia*) are used as a weaving material for chair bottoms, mats and baskets, and the plant is also cultivated for the ornamental value of its large brown cylindrical fruiting spears.     C.J.H.

## SPARGANIACEAE
*Bur-reeds*

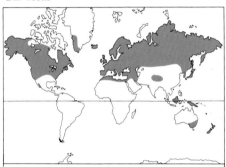

**Number of genera:** 1
**Number of species:** about 15
**Distribution:** mainly temperate zone of the N Hemisphere and in the Arctic, in aquatic and marshy habitats.
**Economic uses:** food and cover for waterfowl.

The Sparganiaceae is a family of perennial aquatic or amphibious herbs, with a single genus *Sparganium* (bur-reeds), which provides cover and food for numerous aquatic organisms, especially waterfowl.

**Distribution.** Most species are found in the temperate and Arctic regions of the Northern Hemisphere, growing gregariously in water or in marshes. One or two species occur in Southeast Asia and Australasia.

**Diagnostic features.** The stems are corm-like with elongate rhizomes. The leaves are linear and arranged in two vertical rows with sheathing leaf bases. The juvenile leaves are thin, strap-shaped and submerged while the mature leaves are floating or erect and emergent. The flowers are unisexual and crowded into separate globose heads; the female heads are towards the base in each inflorescence. The perianth has from three to six elongate scales. The male flower has from one to eight stamens. The female flower has one, occasionally two, or rarely three, superior, united carpels comprising one to three locules with one pendulous ovule in each. Whatever the number of carpels, the style is single and persists as a beak in fruit. The fruit is a drupe with a dry, spongy exocarp and a hard endocarp. The seed has mealy endosperm and a central embryo.

**Classification.** Recent anatomical and embryological work has shown that the Sparganiaceae is closely allied to the Typhaceae, although the affinities of these two families with the rest of the monocotyledons are disputed. There is also some evidence to suggest a relationship to the group known as the Helobiae, which includes the orders Alismatales, Hydrocharitales and Najadales, and it is interesting to note that the Sparganiaceae shares some parasitic fungi with the Araceae. Fossil remains of *Sparganium* indicate that the earlier species were larger and structurally more complex than those of today.

**Economic uses.** *Sparganium* stands provide sheltered nesting and roosting places for wildfowl and in late autumn the fruits form an important part of their diet.

C.D.C.

# BROMELIALES

## BROMELIACEAE
*The Pineapple Family*

**Number of genera:** about 50
**Number of species:** about 2,000
**Distribution:** centered in tropical and warm temperate America.
**Economic uses:** fruits (pineapple), fibers (Spanish moss, pineapple, caroa) and several ornamentals (eg *Billbergia, Cryptanthus, Pitcairnia*).

The Bromeliaceae is a large and distinctive family, including the pineapple, Spanish moss (*Tillandsia*), and various greenhouse ornamentals and house plants.

**Distribution.** The family is tropical or warm temperate, native to the New World with the exception of one species, *Pitcairnia feliciana*, which occurs in West Africa. The family extends from the southern United States of America south to central Argentina and Chile. One species, *Tillandsia usneoides*, has

a distribution as wide as that of the entire family. The family as a whole is adapted to xerophytic conditions and many genera have become epiphytes.

**Diagnostic features.** Most bromeliads are short-stemmed herbaceous plants with basal rosettes of stiff, often spiny, leaves which frequently have colored bases. A few are subshrubs, and some species of the Andean genus *Puya* reach a height of 3m (10ft) and have a similar habit to the giant lobelias of the mountains of tropical Africa. The least specialized members of the family (eg *Pitcairnia, Puya*) are terrestrial plants with fully developed root systems. They have unexpanded leaf bases and leaf hairs which serve only to reduce transpiration. Tank-root types (eg pineapple) have few true roots but the overlapping leaf bases act as reservoirs for water and humus, which are utilized by adventitious roots growing up between the leaf bases. More specialized tank types, which include the majority of genera, have still larger leaf-base tanks, and absorption from the tanks is carried out mainly by specialized leaf hairs (trichomes), not by the roots. The tanks may hold up to 5 liters (over a gallon) of water and contain a considerable flora and fauna, including species of *Utricularia* (bladderwort), tree frogs and various insects. The most specialized genera (eg *Tillandsia*) are entirely epiphytic, have roots only as young seedlings, lack leaf-base tanks and absorb water from the atmosphere by means of multicellular scale-like trichomes. These expand when wetted, so water is drawn into the dead cells of the scale and thence osmotically through the living cells of the stalk of the trichome into the leaf. The scales collapse when dry, permitting gas exchange through the stomata but reducing water loss from the surface of the plant. The plants can thus survive in very dry habitats, but not in very humid habitats such as rain forests.

The inflorescence is terminal, produced out of the center of the tank in tank types, and may be a spike, raceme or panicle. Many bromeliads die after flowering, including some of the genera cultivated for their inflorescence, but these produce suckers and can be readily propagated. The flowers are bisexual (rarely unisexual), regular (somewhat irregular in *Pitcairnia*) and borne in the axils of bracts, which are often brightly colored. The perianth is usually clearly differentiated into greenish calyx and showy petaloid corolla, each with three segments. The six stamens are often attached to the base of the perianth. The ovary, of three fused carpels, may be superior or inferior and contains three locules, each with numerous ovules on axile placentas. Three stigmas are borne on a slender style. The fruit is a berry or capsule. In the pineapple (*Ananas*) and the related *Pseudananas*, the individual fruits fuse and the inflorescence swells to form a multiple fruit. The seeds

contain a small embryo and a mealy endosperm. In several genera the seeds have wings or long feathery or tailed appendages, presumably to assist in dispersal.

In most genera the showy inflorescences and the nectaries on the septa of the ovaries are adaptations to pollination by insects or birds. The flowers are protandrous: when the anthers open, the stigmas are spirally twisted into a head and their receptive surfaces are exposed later. This favors cross-pollination. A few genera (eg *Navia*) appear to be wind-pollinated. Versatile, freely moving anthers are characteristic of the family.

**Classification.** The Bromeliaceae is divided into three subfamilies:

PITCAIRNIOIDEAE. About 13 genera and one-third of the species, containing mainly terrestrial xerophytes; the ovary is superior, the fruit is a capsule, and the seeds are winged or tailed; chief genera *Pitcairnia, Puya, Dyckia*.

BROMELIOIDEAE. About 30 genera, including both terrestrial and epiphytic forms; the ovary is inferior, the fruit is a berry and the seeds have neither wings nor appendages; chief genera *Bromelia, Ananas, Billbergia, Aechmea*.

TILLANDSIOIDEAE. About 6 to 12 genera, all of which are entirely epiphytic; the ovary is superior, the fruit is a capsule and the seeds have feathery appendages formed by the splitting of the elongated outer integument and part of the funicle; chief genera *Tillandsia, Vriesea*.

The family is not closely related to any of the other families of monocotyledons, but shows some affinities with the Commelinaceae and the Zingiberaceae.

**Economic uses.** The pineapple (*Ananas comosus*) is an important edible fruit of the tropics and subtropics. Annual world production exceeds $3\frac{1}{2}$ million tonnes, and about two-thirds of this is consumed in the areas of production. Most commercially grown pineapples are canned or made into juice, which is a good source of vitamins A and B.

Various species produce fibers used locally for making cloth and cordage, notably the pineapple in the Philippines, pita (*Aechmea magdalenae*) in Colombia and caroa (*Neoglaziovia variegata*) in Brazil.

Bromeliaceae. 1 *Aechmea nudicaulis* var *nudicaulis* inflorescence and leaf ($\times\frac{2}{3}$). 2 *Pitcairnia integrifolia* (a) leaf ($\times\frac{2}{3}$); (b) inflorescence ($\times\frac{2}{3}$); (c) vertical section of ovary showing numerous ovules on axile placentas ($\times 4$). 3 *Billbergia pyramidalis* (a) leaf with spiny margins and inflorescence with large red bracts ($\times\frac{2}{3}$); (b) half flower with an inferior ovary crowned by a single style with a lobed stigma and stamens inserted at the corolla base ($\times 1\frac{1}{2}$); (c) cross section of the trilocular ovary showing axile placentas ($\times 4$). 4 *Vriesia carinata* habit ($\times\frac{2}{3}$). 5 *Ananas comosus* (the pineapple) a multiple fruit (produced from the entire inflorescence) and "crown" of leaves produced by continued growth of the axis ($\times\frac{1}{3}$).

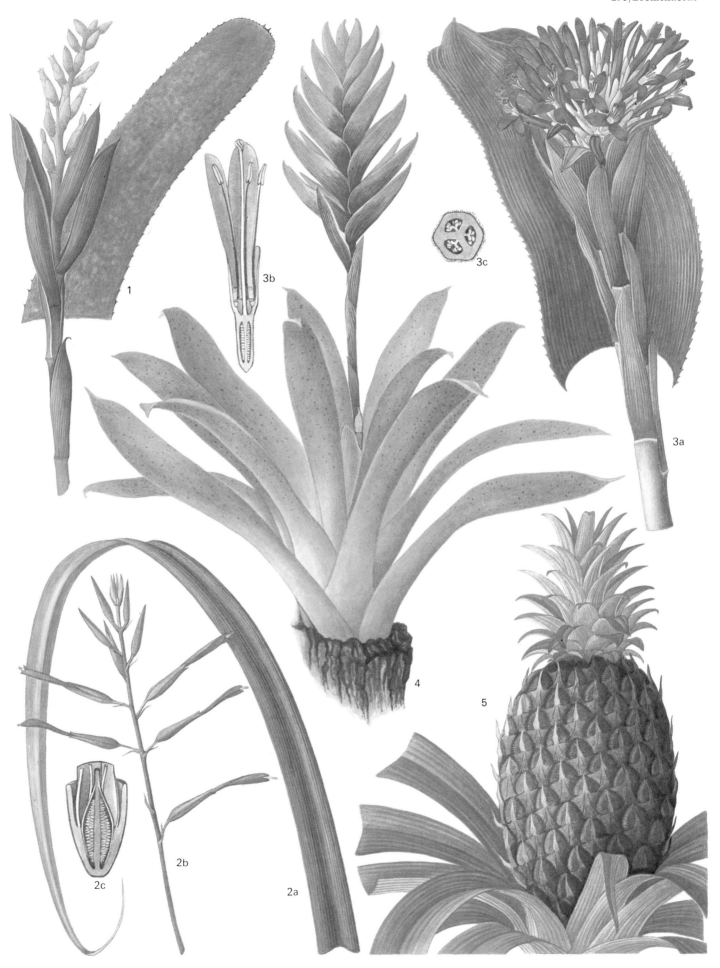

Pineapple leaf fibers have also been used experimentally to produce paper. Pineapple stems and fruits are a possible commercial source of a proteolytic enzyme, bromelain. Spanish moss (*Tillandsia usneoides*) is used as a substitute for horsehair in upholstery. Various genera are grown as ornamentals, in the open in frost-free regions, under glass or as house plants in temperate regions. The foliage alone may be attractive, as in variegated forms of the pineapple, the striped leaves of certain species of *Billbergia*, *Cryptanthus* and *Guzmania*, and the dense rosettes of *Dyckia*, *Nidularium* and *Aregelia*. Other genera also produce showy inflorescences, for example *Pitcairnia*, *Billbergia*, *Aechmea* and *Vriesea*. *Bromelia* and *Neoregelia* are also cultivated.

In parts of the dry tropics, the water retained in the leaf-base tanks of certain native bromeliads may serve as a breeding ground for malaria-carrying anopheline mosquitos. While stretches of open water can easily be sprayed for mosquito control, the tanks of epiphytic and terrestrial bromeliads cannot, and presence of bromeliads has been hampering elimination of malaria in some parts of the tropics.                B.P.

# ZINGIBERALES

## MUSACEAE
*Bananas and Manila Hemp*

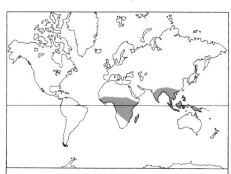

**Number of genera:** 2
**Number of species:** about 40
**Distribution:** chiefly wet tropical lowlands from W Africa to the Pacific.
**Economic uses:** major food crop (banana) and cordage fibers (abaca, Manila hemp).

The Musaceae is a small family mainly of gigantic, tender, evergreen, perennial herbs, the most important cultivated member being the banana.

**Distribution.** The Musaceae is an Old World family spread from West Africa to the Pacific (southern Japan to Queensland). *Musa* is Asian, with greatest diversity in the area Burma–New Guinea, and most species live in the wet tropical lowlands, with outliers in cooler hilly country from southern Japan to the Himalayas; they are essentially jungle weeds of disturbed habitats. A very few outlying occurrences of *Musa* species in the central Pacific and coastal East Africa (Pemba) may well be introduced by Man. The genus *Ensete* is predominantly African but with some representations in Asia as far as Southeast Asia and southern China.

**Diagnostic features.** All species are laticiferous herbs, large to gigantic in size, with pseudostems formed from the leaf sheaths. *Musa* is rhizomatous-stooling and perennial; *Ensete* is unbranched and monocarpic. The leaves are very large, sheathing and spirally arranged; they are oval and have a thick midrib with parallel veins running from it to the margin.

The inflorescences are heavily bracteate and terminal, derived from the growing

Musaceae. 1 *Musa rubra* (a) spike with female flowers below and upper male flowers subtended by large bracts ($\times\frac{2}{3}$); (b) female flower ($\times 1$); (c) male flower ($\times 1$); (d) fruit—a fleshy berry ($\times\frac{2}{3}$); (e) toothed apex to upper perianth ($\times 1$); (f) single lower perianth segment ($\times 1$). 2 *Ensete edule* (a) large herb (up to 10m high) with "stem" formed of sheathing leaf bases; (b) bract subtending numerous flowers ($\times\frac{1}{6}$); (c) male flower ($\times\frac{2}{3}$); (d) bisexual flower ($\times\frac{2}{3}$); (e) cross section of ovary ($\times 1$); (f) seed with window cut out to show embryo ($\times 1$). Strelitziaceae. 3 *Strelitzia reginae* (a) inflorescence—a cincinnus in the axil of a boat-shaped bract in which the flowers unfold in succession ($\times\frac{1}{2}$); (b) half flower ($\times\frac{1}{2}$).

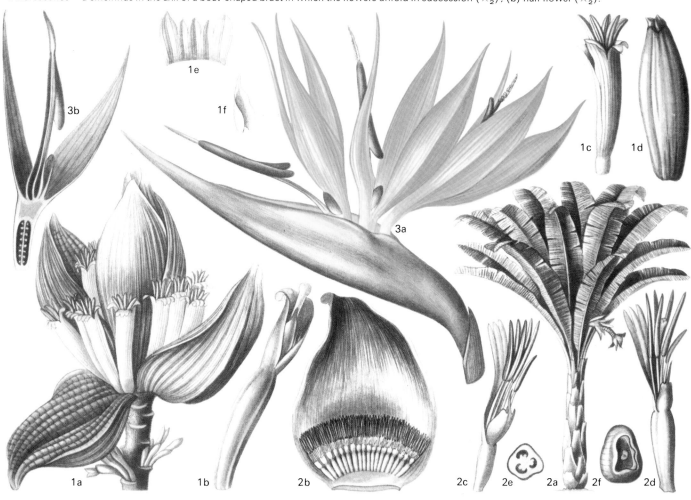

points of the basal corms. The flowers are irregular and usually unisexual, with females in basal clusters and males terminal on the same plant. There are two whorls of three petal-like perianth segments, and five stamens plus a small staminode. The pollen is sticky and pollination is commonly by bats. The ovary is inferior, of three fused carpels and three locules, each containing numerous ovules on an axile placenta. The style is filiform and the stigma lobed. The fruit is a fleshy berry containing numerous stony seeds, and the fruits finally form compact infructescences ("bunches"). The seed has copious endosperm and a straight embryo.

**Classification.** The two genera are *Musa* (30–40 species) and *Ensete* (about six species). The family is placed in the Zingiberales, which includes the Strelitziaceae, Lowiaceae, Zingiberaceae, Marantaceae and Cannaceae. These families are certainly its nearest relatives but no close relationships can be detected. The genus *Heliconia* used to be included in the Musaceae, but is here placed in the Strelitziaceae.

**Economic uses.** The family provides a major food crop, the banana. Cultivated bananas evolved in Southeast Asia from two wild species, *Musa acuminata* and *M. balbisiana*. The plants are grown from suckers and banana varieties are therefore clones. The fruits develop without forming seeds and become filled with the characteristic sweet-acid-aromatic parenchymatous pulp that is eaten. Bananas are of enormous importance as food (locally staple) in the tropics and an important item in international trade.

Abaca or Manila hemp, a declining product of the Philippines used to make ropes and cordage, derives from *M. textilis*. *Ensete ventricosa* (*Musa ensete*, the inset, or Abyssinian banana) is cultivated for its fiber and for food: the stem pulp and young shoots are eaten cooked. Like some dwarf cultivars of *Musa* (eg *M. acuminata* 'Dwarf Cavendish'), it is an occasional greenhouse plant in temperate countries.       N.W.S.

## STRELITZIACEAE
*Bird of Paradise Flower*

The Strelitziaceae is a small and economically unimportant, but varied, attractive and interesting group of tropical banana-like herbs and trees.

**Distribution.** The four genera have a disjunct distribution in the tropics: *Ravenala* (one species, Madagascar), *Phenakospermum* (one species, Guiana), *Strelitzia* (four species, South Africa) and *Heliconia* (about 50 species, tropical America).

**Diagnostic features.** *Heliconia* is herbaceous, but the other three genera tend to woodiness and indeed *Ravenala madagascariensis* can grow into a large tree. The stems are formed by the sheathing leaf bases. The leaves are alternate, in two ranks, medium to very large, with long petioles. They have a thick midrib and numerous pinnately parallel

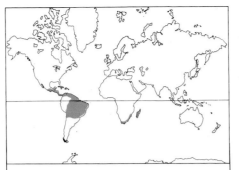

**Number of genera:** 4
**Number of species:** about 55
**Distribution:** tropical America, S Africa, Madagascar.
**Economic uses:** ornamentals, eg bird of paradise flower (*Strelitzia*), traveller's palm (*Ravenala*).

veins extending to the margin. The flowers are irregular, bisexual and borne in terminal or lateral long-stalked cincinni which are enclosed in a large boat-shaped bract. The perianth consists of two whorls each of three segments, the outer whorl more or less equal and the inner whorl comprising two unequal lateral segments and the third elongated and arrow-shaped, folding around the style. There are five (rarely six) stamens with long, rigid filaments crowned by linear anthers with two locules. The ovary is inferior, comprising three fused carpels, with three locules, each with one to many ovules on axile placentas. The style is filiform with a three-lobed stigma. The fruit is a woody capsule dehiscing loculicidally by three valves, or a fleshy schizocarp. The seeds contain a straight embryo in endosperm and may or may not have an aril.

**Classification.** *Strelitzia*, *Ravenala* and *Phenakospermum* have perianth segments free, the ovary contains numerous ovules and the fruit is a capsule. In *Strelitzia* the flowers are very irregular and the seeds have an aril. In *Ravenala* and *Phenakospermum* the flowers are only slightly irregular; in the former there are six stamens and the seeds have an aril and the latter has five stamens and the seeds have no aril. *Heliconia* has only a single basal ovule in each locule, the fruit is a schizocarp splitting into three one-seeded parts, and the seeds are without an aril. *Phenakospermum* was once included within *Ravenala*, but is now considered taxonomically distinct as well as very disjunct in distribution. *Heliconia* may be either separated into its own family, the Heliconiaceae, or included in the Musaceae.

The Strelitziaceae is most nearly allied to the Musaceae and Lowiaceae, differing from the former in having leaves and bracts in two ranks, bisexual flowers and fruit not a berry. (Lowiaceae does not have an entry in this book, but comprises just one genus, *Orchidantha* (*Lowia*), with two species native to Malaysia and Borneo. They are more or less stemless rain forest herbs, the main

features dissimilar to those of the Musaceae and Strelitziaceae being bisexual flowers with a long hypanthium, much surpassing the ovary, and leaves with conspicuous transverse minor veins.)

**Economic uses.** All genera are more or less ornamental. *Ravenala* (traveller's tree) is a stately tree widely grown in the tropics. *Strelitzia* species (bird of paradise flower) are also common there and in temperate glasshouses. *Heliconia*, with a more or less banana-like habit and ecology, was little cultivated 30 years ago but now, very justifiably, appears ever more prominently in tropical gardens.       N.W.S.

## ZINGIBERACEAE
*Ginger, Cardamom and Turmeric*

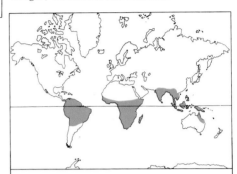

**Number of genera:** 49
**Number of species:** about 1,300
**Distribution:** tropical, chiefly Indomalaysia.
**Economic uses:** spices (eg ginger, cardamom, turmeric), perfumes, medicines, dyes and tropical and greenhouse ornamentals (eg *Hedychium*, *Costus*).

The Zingiberaceae is a distinctive family of perennial aromatic forest plants which yield spices (including ginger), dyes, perfumes, medicines, and a number of ornamental species cultivated for their showy flowers.

**Distribution.** The family is tropical, occurring chiefly in Indomalaysia.

**Diagnostic features.** All species have branched, underground, fleshy rhizomes and frequently possess tuberous roots. The aerial stems, when present, are invariably short, usually leafless, but sometimes quite leafy. The leaves emerge from the rhizomes as two distinct ranks and toward the base they consist of open or closed sheaths. The blades are fairly large with numerous, closely parallel, pinnate nerves diverging obliquely from the midrib. A distinctive long ligule is to be found at the junction of the sheath and blade. The inflorescence usually occurs as a dense head or cyme but can also be a raceme or consist of a solitary flower. The flowers are irregular and bisexual and their structure is unique and very complicated, the most distinctive part being a conspicuous two- or three-lobed lip (labellum) produced by the fusion of two staminodes. There is only one fertile stamen, corresponding to one of the

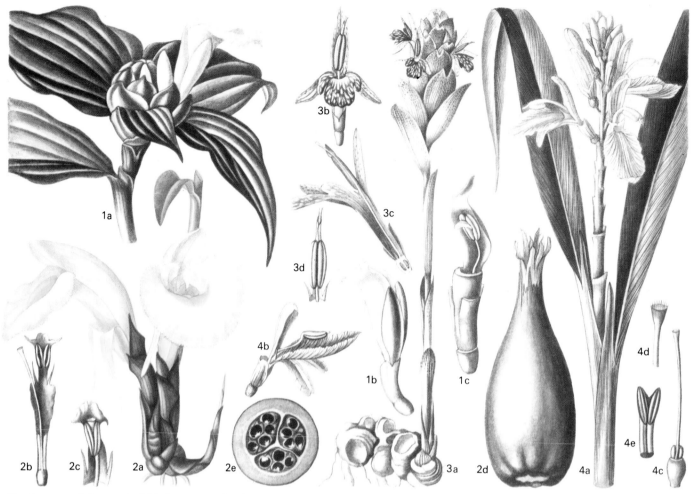

Zingiberaceae. 1 *Costus afer* (a) inflorescence showing upper leaves each with a basal ligule ($\times\frac{2}{3}$); (b) flower ($\times1$); (c) flower with perianth removed showing staminode, fertile stamen and hood-like stigma ($\times\frac{1}{2}$). 2 *Aframomum melegueta* var *minor* (a) habit ($\times\frac{2}{3}$); (b) ovary, style, stigma and petaloid fertile stamen with two anthers ($\times\frac{2}{3}$); (c) stigma cupped between anthers ($\times1$); (d) fruit ($\times\frac{2}{3}$); (e) cross section of fruit ($\times\frac{2}{3}$). 3 *Zingiber officinale* (ginger) (a) habit ($\times\frac{1}{2}$); (b) flower ($\times1$); (c) half flower ($\times1$); (d) anthers with extended connective folded around the style ($\times10$). 4 *Alpinia officinarium* (a) upper leaves and inflorescence ($\times\frac{2}{3}$); (b) flower ($\times1$); (c) gynoecium ($\times1\frac{1}{3}$); (d) stigma ($\times4$); (e) anthers ($\times1\frac{1}{3}$).

members of the inner whorl of stamens; the other two members may be present as staminodes flanking it on each side, or may be absent altogether. One "stamen" of the outer whorl is absent altogether and the other two staminodes form the labellum. The flower is enclosed by three outer perianth segments fused into a tubular calyx, and three inner segments which are petal-like, often showy and more or less united, with the posterior segment often enlarged; the whole is subtended by a bract which is often sheathing. The style is quite reduced and often so thin that it cannot support itself. It is fused to a groove along the length of the anther and only the stigma protrudes above it. The ovary is inferior, of three fused carpels with either three (occasionally two) locules with axile placentas or one locule with parietal (rarely basal) placentas. There are usually numerous ovules in each locule. The fruit is a brightly colored and sometimes very fleshy capsule. The seeds are large, rounded or angled, with copious endosperm: many are wrapped in a distinctive red aril.

**Classification.** The chief genera are *Zingiber* (80–90 species), *Costus* (150 species), *Alpinia* (250 species), *Curcuma* (70 species), *Kaemp-*

*feria* (55 species), *Hedychium* (50 species).

Together with the Musaceae, the Cannaceae, the Strelitziaceae and the Marantaceae, the Zingiberaceae forms an isolated order, the Zingiberales, which belongs to the superorder Commelinidae. The genera *Costus, Dimerocostus, Monocostus* and *Tapeinochilos* are separated by some authorities into the family Costaceae.

**Economic uses.** A large number of the principal genera have beautiful flowers and many of them are cultivated in the tropics, and as hothouse ornamentals in temperate countries. Particularly striking examples include various species of *Hedychium* (ginger lily), *Kaempferia, Costus* and *Roscoea*.

Many Zingiberaceae are rich in volatile oils and are widely used as condiments, herbs, dyes and medicinal plants. Perhaps the best known is ginger (*Zingiber officinale*). Other important products include the abir, a perfumed powder obtained from the rhizome of *Hedychium spicatum*; East Indian arrowroot, derived from the tubers of *Curcuma angustifolia*; *C. longa* (*C. domestica*), which yields turmeric, one of the main coloring and aromatic ingredients of curry powder and also used as a yellow dye;

zedoary, a spice, tonic and perfume made from the rhizomes of *C. zedoaria*; *Alpinia officinale* of Hainan and *A. galanga* of the Moluccas, which yield the medicinal and flavoring rhizome galangal; and *Elettaria cardomomum* from Indonesia, which yields the eastern spice cardomom. Several useful products are also derived from different species of *Aframomum*, including the spice Melegueta pepper (*Aframomum melegueta*).

C.J.H.

# CANNACEAE
### Queensland Arrowroot

The Cannaceae is a family containing a single genus, *Canna*, comprising large perennial herbs with spectacular flowers. *Canna edulis* yields purple or Queensland arrowroot, and several species are grown as greenhouse and tropical garden ornamentals.

**Distribution.** The family is tropical and subtropical, native to the West Indies and Central America.

**Diagnostic features.** The plants have a swollen underground tuberous rhizome, from which arise aerial stems bearing large, broad, pinnately veined leaves with a distinct

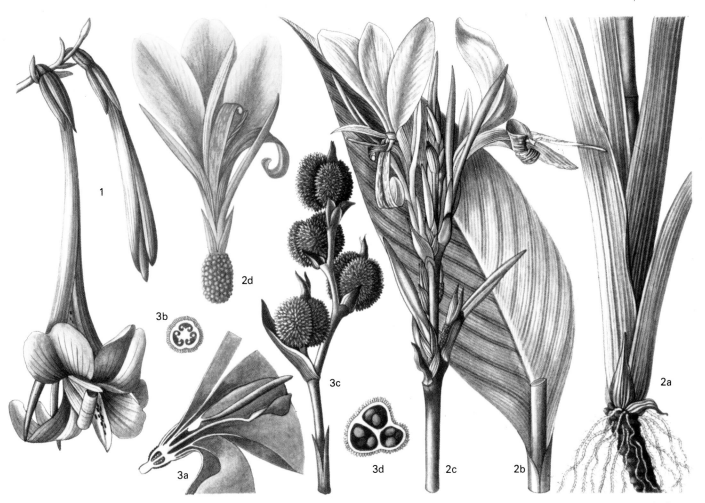

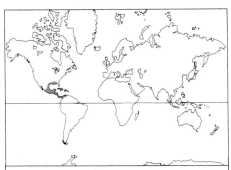

Cannaceae. 1 *Canna indifolia* tip of inflorescence ($\times\frac{2}{3}$). 2 *Canna glauca* (a) base of plant showing swollen rhizome and sheathing leaf bases ($\times\frac{2}{3}$); (b) leaf ($\times\frac{2}{3}$); (c) inflorescence ($\times\frac{2}{3}$); (d) flower; shown from base upwards are: worty inferior ovary, two green sepals, two orange lanceolate petals, outer staminode whorl of two broad wings plus curved labellum, inner staminode whorl of wing-like staminode plus slightly coiled staminode with coiled half-anther attached and central petaloid style with hairy stigmatic surface ($\times 1$). 3 *Canna generalis* (a) half section of flower base ($\times 1$); (b) cross section of ovary ($\times 2$); (c) worty fruits ($\times\frac{2}{3}$); (d) cross section of fruit ($\times\frac{2}{3}$).

**Number of genera:** 1
**Number of species:** 30–55
**Distribution:** tropical and subtropical in Central America and the W Indies.
**Economic uses:** purple or Queensland arrowroot grown commercially and numerous greenhouse and tropical garden ornamentals (cannas).

midrib. The petiole sheaths the stem, but there is no ligule.

The flowers are large, conspicuous, bisexual and borne in racemose inflorescences, each flower being subtended by a bract. The perianth comprises three free, imbricate sepals which are usually green, and three petals which are similar, but narrower and fused at the base, with one of them usually smaller than the others. The petals are united to the fused staminal column, which is composed of four to six, mainly sterile, petallike and brightly colored stamens, which form the most showy part of the flower. Essentially, the stamens form two series or whorls of which the outermost is composed of three petaloid staminodes, the largest of these (the labellum) being reflexed and rolled back on itself. The inner series consists of one or two staminodes and a free petaloid fertile stamen with half an anther joined to it along one edge. The ovary is inferior, of three fused carpels, and has three locules each containing two rows of numerous ovules on axile placentas. The single petaloid style is usually longer than the staminal tube. The fruit is a warty capsule containing many small seeds with straight embryos and very hard endosperm.

**Classification.** The variable numbers of species accredited to *Canna* reflect the differences in interpretation by different botanists. Criteria of value in classifying the genus include shape of leaves and perianth parts, length of the staminal tube, and numbers and shape of the staminodes.

This family belongs to the same order as the families of the banana (Musaceae), ginger (Zingiberaceae) and arrowroot (Marantaceae), with which it shares such features as irregular flowers, reduction in the number of functional stamens, inferior ovary and seeds with endosperm. It differs from the closely related Zingiberaceae in not having ligules.

**Economic uses.** *Canna edulis* is a species of considerable economic importance as the rhizomatous tubers are the source of a starch known as purple or Queensland arrowroot. It is grown as a food crop in the Pacific and parts of Asia, and on a commercial scale in Australia. The starch is easily digested and therefore suitable for incorporation in invalid and infant diets. The rhizomes of some other species, such as *C. bidentata*, are sometimes used as emergency foods, while those of *C. gigantea* and *C. speciosa* yield extracts with medicinal properties. A number of the species (notably *C. indica*) have been developed as ornamental plants for heated greenhouses in temperate zones or in tropical gardens.                          S.R.C.

Marantaceae. 1 *Calathea villosa* (a) leaf showing basal sheath, petiole and blade ( ×⅔); (b) inflorescence with flowers subtended by green bracts ( ×⅔).
2 *C. concolor* (a) flower comprising fused sepals, three irregular petals, two petaloid staminodes and one petaloid stamen with fertile anther and a
hooded style ( ×1); (b) upper part of flower opened ( ×1). 3 *Stromanthe sanguinea* (a) upper leaf and inflorescence ( ×⅔); (b) flower ( ×2); (c)
flower opened out ( ×3). 4 *Maranta arundinacea* (a) shoot with leaves and inflorescence ( ×⅔); (b) tuber ( ×⅔); (c) flower ( ×1); (d) petaloid
staminodes, dorsal view of fertile stamen and style ( ×1⅓); (e) cross section of unilocular ovary ( ×3); (f) fruit ( ×2).

# MARANTACEAE
*Arrowroot*

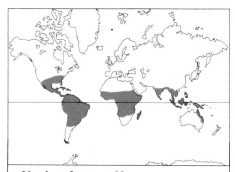

**Number of genera:** 30
**Number of species:** about 350
**Distribution:** mainly tropical, chiefly
America.
**Economic uses:** true arrowroot
(*Maranta arundinacea*), flowers and
tubers of *Calathea* species eaten,
leaves used for roofing and
basketwork locally, and many
greenhouse ornamentals (*Calathea*
and *Maranta*).

The Marantaceae, a smallish tropical family
of herbaceous perennials, includes several
useful species, such as the West Indian

arrowroot (*Maranta arundinacea*).
**Distribution.** Most of the 30 genera are
native to tropical America; seven are native to
Africa and six to Asia.
**Diagnostic features.** The plants usually have
underground rhizomes or tubers. The leaves
are arranged in two rows and the petioles
have sheathing bases. The leaf blade is
narrow or broad, with pinnate veins running
closely parallel from the midrib. The petiole
may be winged, and where it joins the blade it
is generally swollen into a pulvinus – a mass
of cells active during movements of the leaf
in response to stimuli.

The inflorescence is a spike or panicle
usually subtended and enclosed by spathe-
like bracts. The flowers, which are not very
conspicuous, are irregular and bisexual,
usually with an outer series of three free
sepals and an inner series of three distinctly
petaloid segments, more or less united into
a tube and irregularly three-lobed. The androecium is attached to the corolla and
consists of only one fertile stamen (usually
petaloid and with a single-celled anther),
other stamens being absent or variously
modified to form petaloid staminodes. The
ovary, of three fused carpels, is inferior and

has either three locules or one (the other two
aborting), each locule containing one basal,
erect ovule. There is a single style. The fruit is
either fleshy or a loculicidal capsule. The
seeds have an aril and abundant endosperm
surrounding a curved embryo.
**Classification.** The genera may be divided
into two tribes, the PHRYNIEAE (ovary with
three locules) and the MARANTEAE (ovary
with one locule). The former includes such
genera as *Calathea* (one staminode), *Marant-
ochloa* (two staminodes and deciduous
bracts) and *Phrynium* (two staminodes and
persistent bracts), and the latter contains
*Ischnosiphon* (persistent bracts), *Thalia* (one
outer staminode) and *Maranta* (two conspicuous outer staminodes).

The Marantaceae is closely related to the
Musaceae, Zingiberaceae, Cannaceae and
Strelitziaceae, these families sharing
sufficient vegetative and floral characters to
be placed in the same order (Zingiberales).
The Marantaceae is the most highly evolved
family of this group by virtue of the extreme
reduction in both stamens and carpels.
**Economic uses.** Economically the most
important genus is *Maranta*. West Indian
arrowroot or maranta starch is obtained by

grinding and washing the rhizomes of *Maranta arundinacea*. The species is cultivated commercially in the West Indies and tropical America as the starch, being readily digestible, is useful in special diets. The tough and durable leaves of *Calathea dis-* *color* are used to make waterproof baskets and those of *C. lutea* are used for roofing houses in the Caribbean and Central America. The flowers of two Mexican species, *C. macrosepala* and *C. violacea*, are cooked and eaten as a vegetable. The so-called topi- tambu, or sweet corn root, is the edible tuber of the West Indian species *C. allouia*. Some species of *Calathea* and *Maranta* are cultivated in temperate zones as greenhouse ornamentals and houseplants for their attractive foliage.                    S.R.C.

# ARECIDAE

## ARECALES

### PALMAE
*The Palm Family*

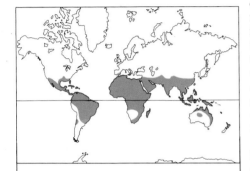

**Number of genera:** about 212
**Number of species:** about 2,780
**Distribution:** chiefly tropical, with some subtropical and a few temperate outliers.
**Economic uses:** most important products are coconuts, copra, dates, sago, palm oil, fibers (eg coir, raffia). Many are grown as ornamentals.

The Palmae (Arecaceae) is a natural and ancient group of plants. Amongst the monocotyledons only the Orchidaceae, Gramineae and Liliaceae have more genera and species, and in the moister tropics, which palms epitomize, only the orchids equal them in conspicuousness.

**Distribution.** Palms are mainly tropical and occur in all habitats from perhumid lowland rain forest to deserts and from mangrove swamps to high mountain thickets. There are a few subtropical and temperate outliers, including the European species *Chamaerops humilis* which attains latitude 44°N. All palms have a single apical bud and if that is killed by frost the stem dies. Few species have overcome this limitation. *Trachycarpus* reaches 2,400m (8,000ft) altitude at 32°N in the Himalayas, where the land is under snow from November to March, and *Serenoa* reaches 30°N in North America. The distribution of palms is strikingly disbalanced. Africa has 16 genera and 116 species. This contrasts with 29 genera and 132 species in the nearby but much smaller Indian Ocean islands (Mascarenes, Comoro islands and Madagascar). The New World has 64 genera and 857 species, mostly in South America. The eastern tropics are easily the richest,

with 97 genera and 1,385 species. The poverty of Africa (the whole continent has fewer palms than Singapore island) is believed to result from desiccation during the Pleistocene which greatly reduced the extent of moist habitats. It is noteworthy that nearly all palms are restricted to one of these four regions. Amongst species exceptions are the date palm (*Phoenix dactylifera*) and the coconut (*Cocos nucifera*), both widely cultivated, and two other species found in both Africa and the Indian Ocean islands. At the generic level, the raffia palms (*Raphia*) and oil palms (*Elaeis*) span the Atlantic and four genera are common to Africa and Asia (*Borassus, Calamus, Hyphaene* and *Phoenix*).
**Diagnostic features.** The life form of the palms is distinctive; it is characterized, for example, by the coconut, with its single unbranched trunk bearing a terminal tuft of feather-like leaves, with axillary inflorescences and a basal mass of slender, renewing roots. The family as a whole contains a great deal of diversity, but much less within the natural groups now recognized. The stem has no secondary thickening (the seedling builds up an inverted cone to the full width before growth commences and this width is then maintained) and is traversed by numerous small, separate vascular bundles, each with a hard fibrous sheath; these may be more or less evenly spaced (as in the coconut) or concentrated at the periphery (from which hard springy planks can then be made, though these trunks are as hard as steel to cut). The surface is often hoop-marked by leaf scars. Branching occurs very rarely and is in most cases truly dichotomous (*Hyphaene, Nypa*) which is a rare mode in flowering plants as a whole. The apical bud is usually well protected by leaf bases or spines, or the tissues may be poisonous. The vasculature, with major bundles in the stem center, is characteristic of monocotyledons, and is fundamentally different from the arrangement in dicotyledons.

Monocotyledon trees have several features in common: a few large leaves are formed one at a time at the stem tip, each with a broad sheathing base. The young, still-furled leaf stands in the crown center like a sword. The palm leaf, however, is more complex than that of other monocotyledons. The blade is plicate (folded like a fan) and the folds are either displayed more or less normally along both sides of an extended

rachis, as in the feather palms, or else arise crowded from a short central rib or costa, as in fan palms. A few species of both kinds have entire blades (eg *Licuala grandis, Verschaffeltia splendida*). The folds develop as the leaf grows in bud before it is unfurled and, as it expands from the bud, swellings (pulvini) develop which push the folded leaflets out to their mature position. The leaflets may contain one fold or several and be linear, fishtail-shaped, acute or praemorse (looking as if bitten off). Before final expansion of the leaf, superficial cells often divide to form hairs or scales which are often of great complexity and beauty. The loose surface layer so formed acts as a lubricant during the opening of the leaf. Large fan leaves are intermediate between the feather and fan types, and have a massive central costa (costapalmate). A more fundamental distinction, however, is between leaves that are V-folded (induplicate) and those that are Λ-folded (reduplicate). The former never possess an apical leaflet, the latter always have one – a distinction that arises early in development. Nearly all fan leaves are reduplicate. Leaves vary enormously in size; *Raphia fainifera* has the largest flowering plant leaf, sometimes over 20m (65ft) long.

Palm inflorescences are highly diverse, from huge panicles with many orders of branching (*Calamus*) or tree-like monopodia with some 250,000 flowers (talipot palm), down to simple spikes. They are usually lateral and borne in the crown or below it, but are occasionally terminal (*Corypha, Metroxylon*). Stems of the latter mode are monocarpic, storing up starch in the trunk and utilizing it in one gigantic burst of reproduction after which the whole plant dies. The caryotoid palms are intermediate. In most of them axillary inflorescences form from the stem tip downwards and the stem progressively dies.

The flowers may be bisexual or unisexual, the sexes then being borne either on the same plant (monoecious) or on separate plants (dioecious); some species have bisexual and unisexual flowers on the same plant (polygamous). Floral parts are normally in threes. There are three separate or connate sepals which may be imbricate, and three separate or connate petals which may be imbricate in female flowers and valvate in male flowers. The stamens are in two whorls each of three. The anthers have two locules and open by longitudinal slits. The ovary is

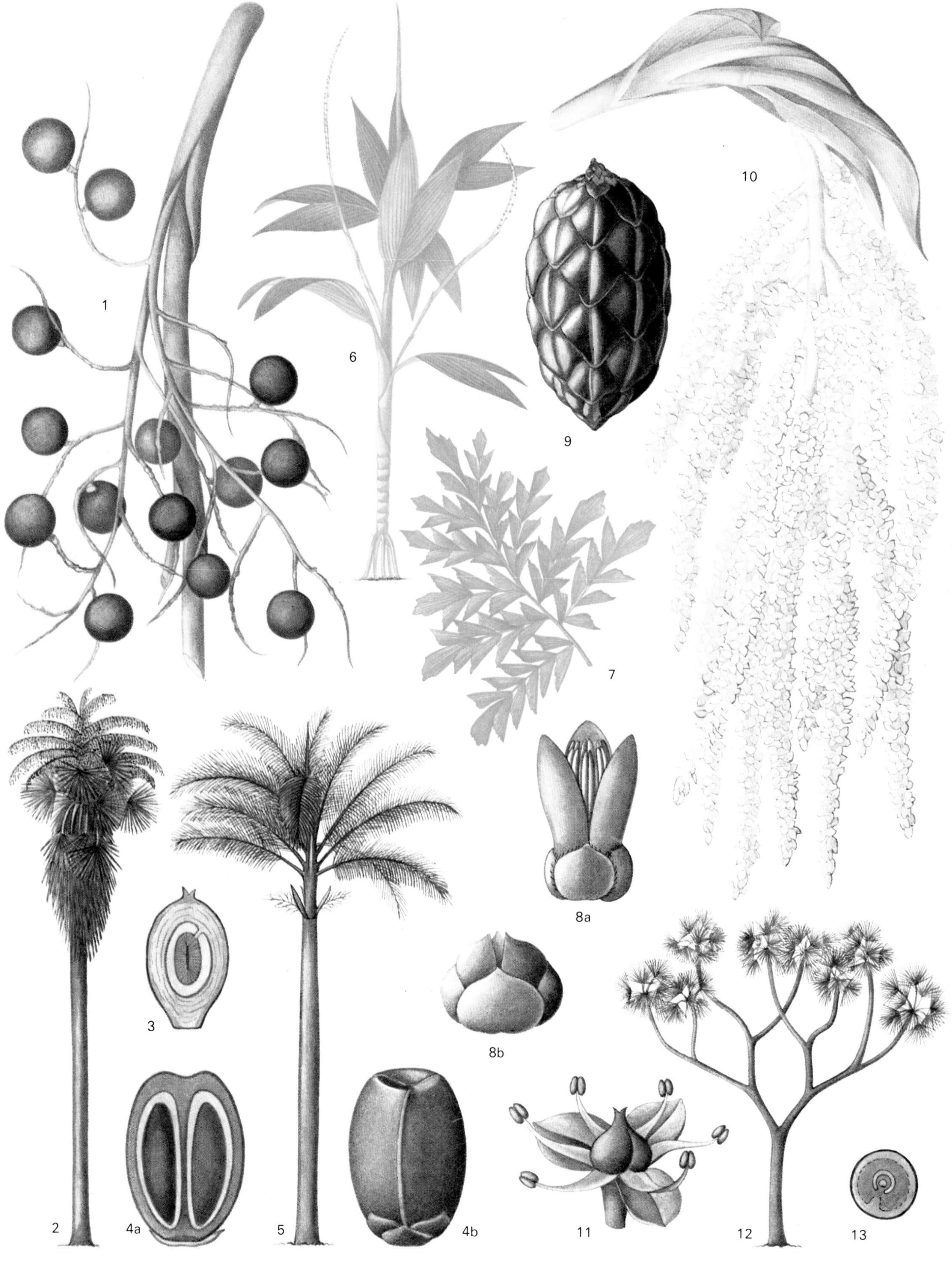
1
2
3
4a
4b
5
6
7
8a
8b
9
10
11
12
13

superior and comprises three carpels which are either free (apocarpous) or fused (syncarpous). In syncarpous ovaries there are either one or three locules. There are three (or sometimes one) erect or pendulous anatropous (rarely semi-anatropous or orthotropous) ovules in each locule. The ovary is vestigial or absent in male flowers. There are adaptations to pollination by wind (*Cocos, Elaeis, Phoenix*) or beetles (*Bactris, Johannesteijsmannia*) or other insects. Numerous stamens (eg *Phytelephas*) or other parts may represent modifications to permit insect feeding. The flowers commonly persist only one day or less. The pollen is monocolpate.

The fruits are mostly one-seeded berries or drupes with a vast range in size; *Lodoicea* (double coconut) has the world's largest seed. The fruit surface is most often smooth but may be warty or covered with beautiful, geometrically arranged scales. The mesocarp is fleshy, or dry and variously fibrous. The endocarp if distinct is mostly thin. Many fruits are brightly colored and virtually all are indehiscent. The storage tissue is endosperm which is oily or fatty rather than starchy; it is sometimes extremely hard in which case it is known as vegetable ivory and utilized for carvings or (formerly) for buttons. Germination is hypogeal (the basic mode in monocotyledons); often the cotyledon remains as a haustorium within the testa.

**Classification.** The palms, though conventionally considered to be a single family, could equally well be considered to be a group of families. Palms are difficult to work with, poorly collected, and intractable to the herbarium method. They have therefore been relatively neglected by taxonomists. They are supremely a group for the field botanist for whom they provide an exhilarating challenge.

Classic subdivisions of this great family do not closely reflect natural groups. Recently a revised subdivision into 15 natural groups belonging to five evolutionary lines has been proposed.

### CORYPHOID LINE

The three groups of this line comprise all the

Palmae. 1 *Livistonia rotundifolia* part of branch bearing indehiscent fruit ( ×⅔). 2 *Corypha umbrauculifera* habit showing massive terminal inflorescence. 3 *Elaeis guineensis* vertical section of the fruit containing a single seed. 4 *Arenga westerhoutii* (a) vertical section of fruit with two seeds ( ×⅔); (b) fruit ( ×⅔). 5 *Roystonea regia* habit. 6 *Chamaedorea geonomiformis* a small reed-like palm shown here producing male inflorescences. 7 *Caryota mitis* bipinnate leaf ( ×⅔). 8 *Caryota cumingii* (a) male flower with three sepals, three petals and three stamens ( ×6); (b) female flower ( ×5). 9 *Raphia vinifera* scaly fruit a distinctive feature of the Lepidocaryoid line ( ×⅔). 10 *Chamaedorea fragrans* male inflorescence ( ×⅔). 11 *Corypha umbrauculifera* bisexual flower ( ×6). 12 *Hyphaene thebiaca* habit showing the dichotomous branching, an unusual feature in the palms. 13 *Phoenix* sp cross section of ovary with a single locule and ovule ( ×6).

genera with induplicate leaves (except the caryotoid line), all those with palmate leaves (except two of the lepidocaryoids), and 14 of the 15 apocarpous genera, many of which also have perfect flowers.

CORYPHOID PALMS. This includes those palms, indubitably the most primitive, with much-branched inflorescences of bisexual apocarpous flowers and the most primitive wood vessels (the *Trithrinax* alliance). This large group (32 genera, 300 species) shows very diverse inflorescence and flower structure and syncarpy as well as apocarpy. The chromosome complement is a uniform n = 18. Many genera have brightly colored fruits apparently adapted to dispersal by birds.

PHOENICOID PALMS. This comprises the single genus *Phoenix* which is clearly related to the coryphoids but has pinnate, induplicate leaves (in which it is unique) and is dioecious, and somewhat dimorphic. The flowers are probably wind-pollinated in specialized inflorescences. *Phoenix* occurs in Africa and Asia, mainly in dry habitats though several species occur in swamps including two in mangrove forest; one species occurs in Crete.

BORASSOID PALMS. This comprises six genera and 56 species of more specialized palms which seem clearly related and derived from a coryphoid stock. They are dioecious with stout inflorescences and dimorphic flowers often sunken in pits in the axis. Leaves are large and costapalmate. Chromosome number is n = 18, 17 or 14. Fruits are fibrous or corky. The group is restricted to the Old World. *Borassus* is the least specialized genus and ranges most widely, from Africa and Madagascar to India, New Guinea and possibly Queensland. Other genera include *Latania* and *Lodoicea*.

### LEPIDOCARYOID LINE

This derives its name from the unique, highly distinctive scaly fruits. Nearly all climbing palms belong here, including rattan cane (*Calamus*), as well as *Raphia*, the fabulous *Pigafettia*, the sago palms (*Metroxylon*) and the important New World genera *Mauritia* and *Mauritiella* (the only palms with induplicate palmate leaves and which are therefore bifid at the apex). Nearly all are armed with spines on the trunk, leaf sheath, leaves and inflorescence. There is a wide diversity of inflorescence type but a uniform pattern of floral arrangement. The flowers are bisexual and the plants polygamous, monoecious or dioecious. The fruits commonly have a fleshy layer, sweet to devastatingly sour, probably an attractant for dispersing mammals and birds. Chromosome number is n = 14. This large line (22 genera, 664 species) is mainly confined between 25° north and south and to the wet tropics.

### NYPOID LINE

This comprises a single monotypic genus, *Nypa*, an inhabitant of salt water. This is the only non-coryphoid genus with an apocarpous gynoecium. The carpels have a vascular

structure different from that of all other palms. Chromosome number is n = 17. *Nypa* is the oldest fossil palm with records back to the Cenomanian (100–110 million years) and is one of the seven oldest angiosperms known. The present range is Far Eastern but fossils extend to Europe and southern England (Eocene, London Clay), indicating a distribution along the former Tethys Sea, to West Africa and to America. *Nypa* is highly distinctive and has no close relatives but is possibly closer to the arecoid line than to any other.

### CARYOTOID LINE

This line comprises three monoecious genera and 35 species. The leaves are morphologically induplicate in vernation but are anatomically reduplicate. Some are bipinnate (*Caryota*), others have pinnately nerved leaflets. The flowers are unisexual. In most species stems flower basipetally then die. The fruits are fleshy, and often contain irritant spicules but are nevertheless animal-dispersed. Chromosome number is n = 5.

### ARECOID LINE

This line, comprising nine separate groups and with 146 genera and 1,684 species, contains 68% of all the known palm genera and 60% of the species.

PSEUDOPHOENIX. This single Caribbean genus of four species is the most primitive. *Pseudophoenix* has much-branched inflorescences, bisexual flowers and fruits with one to three seeds.

COCOSOID PALMS. This comprises 28 genera and 583 species and contains the well-known oil palm (*Elaeis guineensis*) and coconut (*Cocos*). Twenty-four of the genera are South American (and include a few climbers), two more are found in the West Indies, one is in South Africa and *Cocos* itself is best considered to be of Melanesian origin. Many of the less specialized species are adapted to cooler, drier, more seasonal climates. This group has previously been considered to be a separate line but is in fact very close to the arecoids. The most conspicuous difference is the bony endocarp with three pores. The single large inflorescence bract is distinctive but is a feature shared with other arecoids. The flowers are dimorphic, in triads, and the chromosome number is n = 15 or 16.

ARECOID PALMS. In the restricted sense this is a widely distributed and numerous group (88 genera, 760 species). It is highly advanced with a single inflorescence bract and strikingly dimorphic flowers borne in triads. Chromosome number is n = 16 or 18. These palms are mainly confined to humid rain forests within the tropics. Eighteen natural alliances can be recognized. America has 10 genera, Africa only one, the Indian Ocean islands 19 genera and the eastern tropics 58, of which the *Clinostigma* alliance has shown exceptionally prolific evolution at genus level.

The other groups of the Arecoid line are the CEROXYLOID (America, Indian Ocean),

CHAMAEDOROID (America, Indian Ocean), IRIARTOID (America), PODOCOCOID (Africa), GEONOMOID (America) and PHYTELE-PHANTOID (America).

Palms contain, amongst their great diversity, many primitive traits of the monocotyledons and must be considered as an early monocotyledon stock.

The gynoecium in the more primitive palms shows much similarity to that of primitive Ranunculales and Magnoliales amongst the dicotyledons. It is stipitate, follicular, conduplicate, with an open ventral suture, laminar or submarginal placentation and the locular canal remains open. Amongst monocotyledons the palms have primitive flowers, and this is combined with woodiness.

The distribution of the five lines of palms, and especially of those genera exhibiting primitive traits, shows 13 of the 15 groups occurring in South America and Africa (if fossil *Nypa* is included), with only *Pseudophoenix* and the caryotoids missing. These two regions also have a concentration of primitive genera. The eastern tropics show numerical superiority but this is largely due to the considerable proliferation there of the advanced arecoids and the climbing lepidocaryoids *Calamus* and *Daemonorops*. Of particular interest are two groups of the arecoid line, the Chamaedoroid and Geonomoid palms, found in America and in the Indian Ocean islands.

Modern reconstruction of the past positions of the continents shows that South America and Africa were joined as a single land mass, West Gondwana, until the late Jurassic, when they were separated by the opening of the Atlantic. The climate remained warm throughout the Mesozoic and early Tertiary. The physical setting might well have been conducive to palm evolution. The present evidence suggests that palms probably evolved very early in the history of the monocotyledons. Floral resemblances between monocotyledons, such as the Alismataceae, and dicotyledons, such as the Nymphaeaceae, are best regarded as convergent evolution by reduction to the herbaceous condition and aquatic habitat. The sheer antiquity of palms, with the advanced *Nypa* (one of the seven oldest known flowering plants), raises the possibility that monocotyledons and dicotyledons have a separate origin (comparable to that of birds and mammals) from a proto-angiospermous stock and are not evolved one from the other.

**Economic uses.** Coconuts and dates are central to the economy of many producing countries. An important by-product of the coconut palm (*Cocos nucifera*) is copra, and oil is also extracted from the oil palm (*Elaeis guineensis*) and species of *Orbigyna*. A major source of carbohydrate for many people living in the tropics is sago, which is processed from the pith of palms of the genus *Metroxylon* (sago palm) and some species of other genera (eg *Arenga* and *Caryota*). Palm wine (toddy) is another useful product of many palms including *Borassus* and *Caryota*. This is evaporated to produce palm sugar (jaggery) or distilled to form the base for the liquor known as arrack. However, the term arrack is most often applied to the alcoholic spirit obtained from the sap of the coconut palm.

Apart from their obvious importance as a source of food, the palms produce a variety of useful fibers. These include coir, which comes from the husk of the coconut, raffia fiber obtained by stripping the surface of young leaflets of the genus *Raphia*, and piassava fiber, from the leaf sheaths or fibrous stems of South American *Leopoldinia piassaba* and *Attalea funifera*. *Caryota urens* gives a black bristle fiber (kitul fiber) used for ropes and broom heads. For canework, the much-used rattan cane is invaluable; this comes mainly from *Calamus* species, which also provides malacca cane, a stout cane used for walking sticks and baskets.

Waxes are obtained from *Copernicia* (carnauba wax) and *Ceroxylon*. Vegetable ivory, from the ivory nut palm and others, was once an important commodity, used for buttons and a general substitute for real ivory. The betel nut comes from *Areca catechu* and is widely used in India, Malaysia and tropical Africa.

Many palms, with their tall slender trunks and dense crown of attractive leaves, are ideal ornamental subjects. Notable examples are the Cuban royal palm (*Roystonea regia*), the Chinese windmill palm (*Trachycarpus fortunei*) and the coquitos palm (*Jubaea spectabilis* (*J. chilensis*)).

T.C.W.

# CYCLANTHALES

## CYCLANTHACEAE
*Panama Hat Plant*
The Cyclanthaceae is a small family of perennial stemless herbs or climbers chiefly notable for the leaves from which Panama hats are made.

**Distribution.** The family is native to the West Indies and Tropical America.

**Diagnostic features.** The plants are either rhizomatous with no aerial stems, or somewhat woody climbers; a few species are partly epiphytic. A feature of this family is the watery or milky juice in all tissues. The leaves are either two-ranked or spirally arranged and distinctly palm-like, being very deeply bilobed, and each lobe being further subdivided. The petiole is sheathing at the base.

The flowers are unisexual with both male and female on the same plant, densely crowded on to axillary spadixes which are enveloped by two or more conspicuous

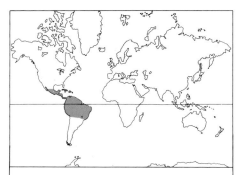

**Number of genera:** 11
**Number of species:** 180
**Number of species:** W Indies and tropical America.
**Economic uses:** the leaves are made into Panama hats and lesser products are mats, baskets, thatching and brooms.

(sometimes petaloid) non-persistent bracts. The arrangement of the flowers on the spadix is either spiral, the females single with about four male flowers around, or clusters of male and female flowers in superposed whorls. The male flowers have either a small, toothed, cup-shaped perianth or no perianth at all, and numerous stamens fused at the base with the filaments also adherent to the perianth when the latter is present. The female flowers have either no perianth or a perianth of four persistent, free or fused segments; sometimes those of several flowers become collectively fused, enlarged and hardened in the fruit. The female flower also possesses four short, or long and thread-like, staminodes. The ovary, of four fused carpels, is usually superior, and often embedded in the axis. It has one locule, containing numerous ovules attached either to one apical or to four parietal or apical placentas. The one to four spreading stigmas are either stalkless or mounted on a short style. The multiple, fleshy fruit has either separate or fused berries containing several seeds. The seed has a succulent testa and a minute embryo surrounded by copious endosperm.

**Classification.** The family may be split into two subfamilies on the basis of inflorescence and leaf characteristics.

CARLUDOVICOIDEAE. Male and female flowers in spirally arranged groups, spadix not screw-like, leaves cleft at the apex, fan-like or entire; *Carludovica, Schultesiophytum, Asplundia, Thoracocarpus, Evodianthus, Dicranopygium, Sphaeradenia, Stelestylis, Ludovia, Pseudoludovia*.

CYCLANTHOIDEAE. Male and female flowers in separate, alternate whorls, or sometimes part spirals, spadix screw-like, the leaves deeply cleft with a forked main rib; *Cyclanthus*.

The family is closely related to the Palmae, Pandanaceae and Araceae, and shows considerable evolutionary advances in the possession of much reduced unisexual flowers. It might be regarded as an advanced derivative

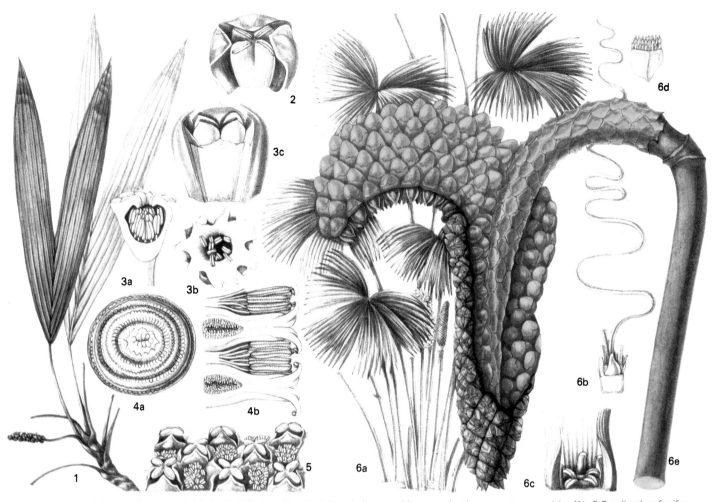

Cyclanthaceae. 1 *Asplundia vagans* habit ( ×⅔). 2 *Stelestylis stylaris* female flower with one perianth segment removed ( × 4½). 3 *Evodianthus funifer* (a) half male flower in bud ( × 4½); (b) male flower ( × 6); (c) young fruit ( × 4). 4 *Cyclanthus bipartitus* (a) tip of spadix with alternating whorls of male and female flowers ( × 1½); (b) section of spadix showing male and female flowers ( × 4). 5 *Sphaeradenia chiriquensis* portion of spadix with male and female flowers ( × 2). 6 *Carludovica rotundifolia* (a) spadices and fan-like leaves ( × 1½); (b) female flower ( × 2); (c) young fruit ( × 4½); (d) male flower ( × 2); (e) ripe composite fruit cut open to show the stalk (pink) into which bases of the berries (orange) fit ( ×⅔).

of the palm habit with the same degree of evolutionary specialization as the Araceae.

**Economic uses.** The family is economically important for *Carludovica palmata* (the Panama hat plant), from which young leaves are made into Panama hats. Ecuador alone exports over 1,000,000 of these hats per year. About six young leaves are required to make a single hat. Older coarser leaves are used for mat- and basket-making. The leave of *C. angustifolia* are used for thatching native huts in Peru and the leaves of *C. sarmentosa* are used to manufacture brooms in Guyana.

S.R.C.

# PANDANALES

## PANDANACEAE
*Screw Pines*

The Pandanaceae is a large family of tropical trees, shrubs and climbers.

**Distribution.** The family is distributed throughout the tropics and subtropics of the Old World and most members favor coastal or marshy areas.

**Diagnostic features.** The tall stems bear the annual scars of the leaf bases, are branched

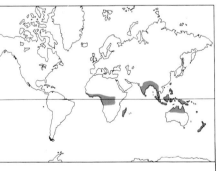

**Number of genera:** 3
**Number of species:** about 700
**Distribution:** Old World tropics and subtropics.
**Economic uses:** large edible starchy fruits, leaves used for thatching and basket-weaving, aerial roots for cordage, flowers used in perfumery and some ornamentals.

and are usually supported by aerial roots. The leaves are terminal in three ranks but as the stem is often twisted they appear to be spirally arranged. They are long and narrow, sometimes stiff and sword-like or in some cases almost grass-like.

The flowers are unisexual and arranged in a racemose spadix (except in *Sararanga*), subtended by a spathe which is sometimes brightly colored. Male and female are borne on separate plants and lack both calyx and corolla. In the male flower the numerous stamens are arranged in a raceme or umbel with free or fused filaments, and are sometimes represented as staminodes in the female inflorescence. The female flower consists of a superior ovary usually of many carpels in a ring, sometimes reduced to a row of carpels or a single carpel. There are one to many locules depending on the degree of fusion between the carpels; each locule has one to numerous anatropous ovules on basal or parietal placentas. The stigmas are sessile or almost so. The fruit is a berry or multilocular drupe, containing small seeds with fleshy endosperm and a minute embryo.

**Classification.** The trees and shrubs of the most important genus, *Pandanus* (about 600 species), bear cone-like fruits that resemble pineapples. Most species of *Freycinetia* (about 100 species) are climbing perennials with slender stems which bear roots that penetrate the supporting host. *Sararanga* (two species) differs from the other two

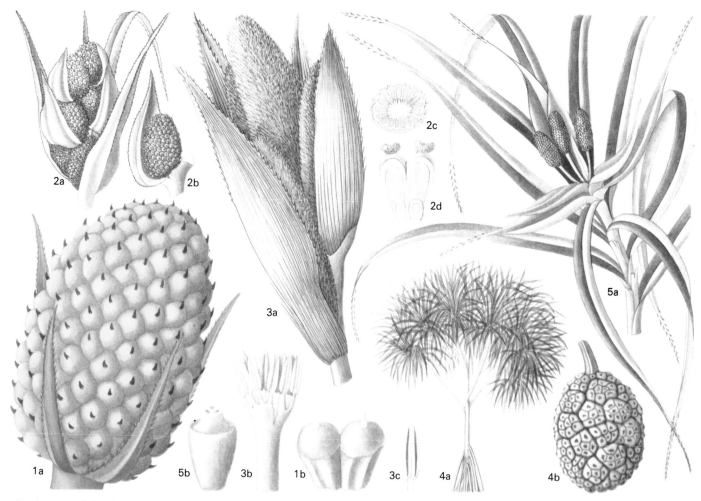

Pandanaceae. 1 *Pandanus minor* (a) cone-like fruit ($\times\frac{2}{3}$); (b) two segments of the fruit ($\times 1$). 2 *P. pygmaeus* (a) female inflorescences ($\times\frac{2}{3}$); (b) female inflorescence and subtending bract ($\times\frac{2}{3}$); (c) cross section of female inflorescence ($\times 1$); (d) female flowers showing sessile stigmas and with part of wall cut away to show basal ovules ($\times 4$). 3 *P. houlletii* (a) male inflorescences and bracts ($\times\frac{2}{3}$); (b) male flower comprising a whorl of stamens with fused filaments ($\times 6$); (c) dehiscing stamen ($\times 8$). 4 *P. kirkii* (a) habit showing aerial roots at base of trunk ($\times\frac{1}{50}$); (b) fruit ($\times\frac{1}{5}$). 5 *Freycinetia angustifolia* (a) flowering shoot with sword-like leaves ($\times\frac{2}{3}$); (b) ovary and base of style ($\times 16$).

genera in that it lacks the aerial roots, has a panicoid inflorescence and pediceled rather than sessile flowers. The fruit is a drupe.

The Pandanaceae is most closely related to the Cyclanthaceae, Palmae, Araceae and Lemnaceae though the relationships are distant.

**Economic uses.** Many species of the genus *Pandanus* (screw pines) are useful sources of food; *Pandanus leram* (Nicobar breadfruit) produces a large globular fruit which is usually boiled in water to form a mealy mass; *P. utilis* and *P. andamanensium* also have edible starchy fruits.

The leaves of the most common species, *P. odoratissimus*, are used for thatching and weaving, especially the thornless variety *laevis*. These leaves are dried, beaten to make them supple, soaked in water and then sun-bleached. Wide strips are used for thatching and matting and fine strips for hats and baskets. Fibers are made from the aerial roots and used for cordage and brushes. This same species is cultivated for its flowers which are used for the popular Indian perfume kewda attar. The fragrant white spathes of the male inflorescences are distilled in water and the vapor absorbed in sandalwood oil. In Malaya the fragrant leaves of *P. odorus*, which never flowers, are used in potpourris. Several species are ornamentals, notably *Freycinetia banksii*, a hothouse plant from New Zealand, which is commonly trained to grow around peat-covered pillars, and *Pandanus veitchii* which has glossy dark green leaves with a silvery white border.                M.C.D.

# ARALES

## LEMNACEAE
### Duckweeds

The Lemnaceae is a family of small or minute aquatic herbs which may be free-swimming, floating or submerged. They include the well-known duckweed (*Lemna*), which often forms a green carpet over stagnant water, and *Wolffia*, the smallest known flowering plant.

**Distribution.** The family is found in fresh water all over the world.

**Diagnostic features.** Modification of the vegetative body has been carried so far that the usual distinction between leaf and stem is no longer maintained and representatives of

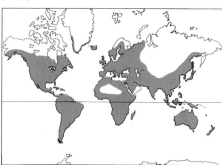

**Number of genera:** 6
**Number of species:** about 43
**Distribution:** cosmopolitan, in freshwater habitats.
**Economic uses:** food for fish and water fowl.

this family consist of undifferentiated fronds (thalluses) of various forms. Simple roots may be present or absent. Male and female flowers are on the same plant (rarely on separate plants) and are borne in pouches or sheaths. The inflorescence consists of one female and two male flowers, which are naked or surrounded by a spathe. The male flower has one or two stamens and anthers

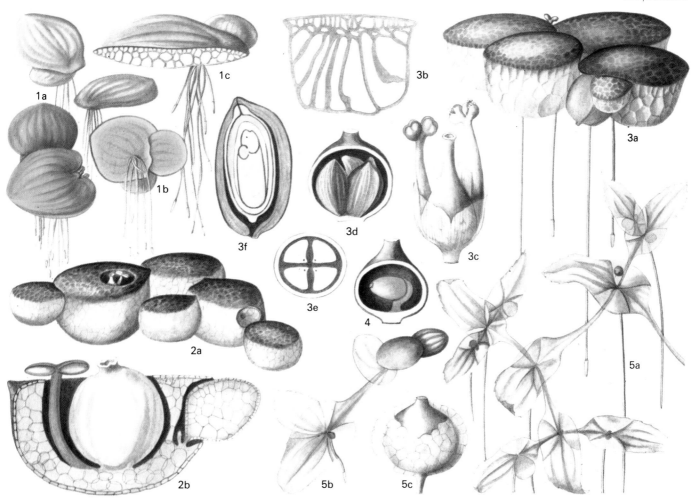

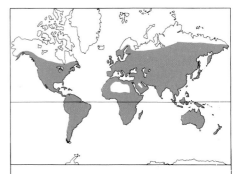

Lemnaceae. 1 *Lemna polyrhiza* (a) thallus with long adventitious roots and prominent root caps ($\times 2\frac{2}{3}$); (b) underside of thallus ($\times 2\frac{2}{3}$); (c) vertical section of thallus ($\times 5\frac{2}{3}$). 2 *Wolffia arrhiza* (a) floating thalli ($\times 26$); (b) section of thallus with budding pouch and male and female flowers in a single cavity ($\times 52$). 3 *Lemna gibba* (a) thalli ($\times 12$); (b) vertical section of thallus ($\times 8$); (c) inflorescence of two male flowers and one female flower enclosed in a spathe ($\times 20$); (d) half fruit ($\times 20$); (e) cross section of fruit ($\times 20$); (f) vertical section of seed ($\times 40$). 4 *L. minor* ovary with wall removed ($\times 20$). 5 *L. trisulca* (a) thalli with numerous side branches ($\times 2\frac{1}{3}$); (b) seedling ($\times 6$); (c) fruit ($\times 26$).

with one or two locules. The female flower has a short style and an ovary with one locule and one to seven ovules. The fruit is a utricle and the seeds are smooth or ribbed, with a straight embryo and either fleshy or no endosperm. Vegetative reproduction by buds or turions is known in the family.

**Classification.** Roots are present in *Spirodela* and *Lemna*, but absent in the other genera. *Spirodela*, with six cosmopolitan species, is distinguished from *Lemna* by dorsal and ventral scales on the fronds, several, sometimes many, roots and prominent nerves, usually seven to 15 on the broadly ovate thallus, the underside of which is often red-brown owing to the presence of brown pigment cells. *Lemna* (with 15 cosmopolitan species) has no thalloid scales and one root (sometimes no roots), only one to three nerves and no pigment cells. Roots are absent in the *Wolffia* group, and the genera are distinguished as follows. *Wolffia* and *Pseudowolffia* are floating plants and *Wolffiopsis* and *Wolffiella* are submerged. *Wolffia*, with 10 tropical and temperate species, has a floating globular and thickened thallus, whereas *Pseudowolffia*, with three species in North and central Africa, has

a thin flat floating thallus, the buds coming from a linear transverse slit in the thallus; in *Wolffia* this opening is circular. *Wolffiopsis* is represented by only one species, from tropical America and Africa, and is similar to *Wolffiella* which has eight species with a similar distribution, but *Wolffiopsis* has a symmetrical thallus and in *Wolffiella* the fronds are ligulate. *Wolffiella* is solitary or occurs in stellate colonies.

*Wolffia arrhiza* is the smallest known flowering plant; individuals are scarcely visible to the naked eye, and are generally seen as scum on the surface of water. About 12 flowering individuals could be accommodated on a single frond of *Lemna minor*.

The Lemnaceae is closely related to the Araceae. The adult structure of *Lemna* is very similar to that of the seedlings of the monotypic aquatic genus *Pistia* (Araceae), and it seems highly likely that *Lemna* evolved from *Pistia* by paedomorphosis (ie a process in which sexual maturity is attained at the juvenile stage).

**Economic uses.** No direct uses for this family are known, but they are important as a source of food for waterfowl and fish and are serious weeds of still water.                 S.A.H.

# ARACEAE
*The Aroids*

**Number of genera:** about 110
**Number of species:** about 2,000
**Distribution:** mainly tropics, with a few in temperate zones.
**Economic uses:** several species used locally and more widely for their edible swollen stems and a number are attractive ornamentals notably the arum lily.

The Araceae (the aroids) is a large family of mostly herbaceous plants, with great variety in vegetative habit. In the main they are

1

2c

2b

2a

3a

3b

3c

3d

3e

4

herbaceous with aerial stems or underground tubers or rhizomes, but there are a few woody members. The family includes a number of climbers and epiphytes as well as a floating water plant (*Pistia*).

**Distribution.** The family is pantropical, with a few species found in temperate zones. Some species are marsh plants.

**Diagnostic features.** Many members contain a watery or milky sap (latex) and raphides (calcium oxalate crystals). Many plants, eg those of the genus *Dieffenbachia*, are poisonous. The leaves are simple or compound, basal or produced on the aerial stems. The blades are often expanded with parallel, pinnate or palmate veins and are subtended by a petiole with a membranous sheathing base. The leaf of the fruit salad vine (*Monstera deliciosa*) develops large holes in the course of development.

The roots of all species are adventitious and most of the climbing and epiphytic forms develop two types, one of which is absorbent, growing downwards towards the soil, while the other, not influenced by gravity, grows away from the light and clasps firmly as it grows into crevices in the branches of the supporting tree. In some species (eg of *Philodendron*) the seed may be excreted by a bird on one of the upper branches of a tree where it may germinate, first producing clasping roots and later producing unbranched aerial roots which hang in the air as they grow downward towards the soil. Many aerial roots develop an outer water-absorbing tissue similar to the velamen of orchids. Some truly epiphytic species of *Anthurium* (eg *Anthurium gracile*), although producing both types of root, have no connection with the soil, and the absorbent roots obtain water and mineral salts from the humus which collects on the trunk of the tree on which they are growing.

The inflorescence is characteristic for the family, consisting of a large spathe (bract), often conspicuous and petaloid, subtending and sometimes enveloping a spadix of numerous small flowers which may be bisexual or unisexual. If the flowers are unisexual both types are usually borne on the same spadix with the male higher than the female flowers. The genus *Arisaema* is exceptional in having male and female spadixes borne on separate plants. The four to six small perianth segments, which may be fused to form a cup, are usually only present in bisexual flowers. The stamens vary in number from one to six and are often fused together as in *Colocasia*. Staminodes may be present in female flowers. The ovary is superior or embedded in the spadix, and comprises one to numerous carpels and one to numerous locules with one to numerous anatropous, amphitropous or orthotropous ovules on basal, apical, axile or parietal placentas. The style is variously shaped and sometimes absent, the stigma thus being sessile. The fruit is a berry, sometimes leathery, with one to many seeds, which are usually endospermic with a straight embryo, although a few species have no endosperm and a curved embryo.

The inflorescence of many aroids emits a nauseous, fetid odor which attracts carrion flies to effect pollination. *Arum* displays a highly evolved pollination mechanism. The spathe envelops the flowers and is lined with downward-projecting hairs past which the flies fall or crawl until they get to the bottom. The flies, which may be covered in pollen collected from the stamens of the male flowers of another *Arum* plant, crawl over the female flowers on the lower part of the spadix. As the lower hairs on the spathe wither, the flies crawl further up the spadix and become dusted with the pollen from the stamens of the male flowers which are now beginning to mature. Finally the upper hairs of the spathe wither, allowing the flies to escape and effect cross-pollination by being trapped in another inflorescence.

**Classification.** This very diverse family is difficult to classify and involves the use of internal as well as external features. Experts have variously grouped genera into a number of subfamilies or tribes. In one system emphasis is placed on the nature of the spathe and spadix, bisexual as opposed to unisexual flowers and the presence or absence of a perianth, while the other system groups genera according to the habit, the presence or absence of latex, leaf-form and details of floral structure. Genera lacking latex and with the spadix entirely covered with bisexual flowers include *Anthurium*, *Acorus* and *Monstera*, although the latter lacks a perianth. Laticiferous genera with the spadix entirely covered with unisexual flowers and possessing endospermic seeds include *Philodendron* and *Zantedeschia*. Genera such as *Arum*, *Amorphophallus*, *Arisaema*, *Alocasia*, *Xanthosoma* and most species of *Colocasia* all contain latex and possess unisexual flowers set on a spadix with a barren upper region.

The family is closely related to the Lemnaceae (duckweed family) and is considered to have been evolved either from liliaceous or palm-like ancestors.

**Economic uses.** Economically the family is of considerable importance, as the edible aroids of the genera *Colocasia*, *Xanthosoma*, *Alocasia*, *Amorphophallus* and *Cyrtosperma* are grown throughout the tropics and subtropics, where their starchy swollen tuberous corms are cultivated primarily as a subsistence food. However, the growing of *Colocasia* (taro, dasheen, eddo, cocoyam) and *Xanthosoma* (tanier, yautia) in some countries has reached a commercial scale. The taro or dasheen (*Colocasia esculenta*) is of Asian origin and consists of many varieties, some of which are adapted to upland and well-drained areas and others to lowland flooded conditions. The corms contain crystals of calcium oxalate which have to be destroyed by boiling or baking. The starch grains are small and easily digested, thus making it a suitable food for infants and invalids. Tanier (*Xanthosoma sagittifolium*, *X. atrovirens*, *X. violaceum*), of South American origin, is closely related to *Colocasia* but most varieties produce larger corms with coarse starch grains. The major food species of the other genera are *Alocasia indica* and *A. macrorrhiza*, *Amorphophallus campanulatus* and *Cyrtosperma chamissonis*, all of which are to be found mainly in Indonesia and the Pacific islands. The inflorescence of *Monstera* is sometimes used as food. Many genera contain species which are grown as ornamentals. perhaps the best-known being *Philodendron*, *Dracunculus* and the arum lily of florists – *Zantedeschia aethiopica*.

S.R.C.

Araceae. 1 *Philodendron verrucosum* epiphytic stem with clinging, adventitious roots ($\times\frac{1}{9}$). 2 *Pistia statiotes* (a) habit showing floating rosette of leaves ($\times\frac{2}{3}$); (b) inflorescence with subtending spathe ($\times 2$); (c) vertical section of inflorescence with a single basal female flower with a curved style and above a whorl of male flowers each with pairs of fused stamens ($\times 4$). 3 *Arum maculatum* (a) inflorescence comprising a large spathe enveloping the spadix ($\times\frac{2}{3}$); (b) leaf ($\times\frac{2}{3}$); (c) vertical section of inflorescence showing, from the base, female flowers, male flowers, rudimentary female flowers represented by hairs and the tip free of flowers ($\times\frac{2}{3}$); (d) fruiting spike ($\times\frac{2}{3}$); (e) ovary ($\times 4$). 4 *Anthurium andraeanum* var *lindenii* spathe and spadix with bisexual flowers along its whole length ($\times\frac{2}{3}$).

# LILIIDAE

## LILIALES

### PONTEDERIACEAE

*Water Hyacinth and Pickerel Weed*

The Pontederiaceae is a small family of freshwater aquatics that includes the water hyacinth (*Eichhornia*), which is probably the world's most serious aquatic weed.

**Distribution.** The family is pantropical. *Pontederia* (five species), *Reussia* (two species), *Zosterella* (two species) and the monotypic *Hydrothrix* and *Eurystemon* are confined to the New World. *Heteranthera* (10 species) and *Eichhornia* (seven species) occur in both the New and Old World, and *Monochoria* (five species) and the monotypic *Scholleropsis* are Old World genera.

**Diagnostic features.** Both annual and perennial species are represented and these can be submerged, free-floating or emergent. The stems may be rhizomes, or stolons, or erect

Pontederiaceae. 1 *Pontederia cordata* var *lancifolia* (a) leaf and inflorescence with subtending spathe ($\times\frac{2}{3}$); (b) flower ($\times 3$); (c) gynoecium ($\times 4$); (d) cross section of ovary ($\times 5$); (e) vertical section of ovary ($\times 5$). 2 *Heteranthera limosa* (a) leafy shoot and flowers ($\times\frac{2}{3}$); (b) flower with perianth removed showing two stamens with smaller anthers than the other ($\times 2$); (c) cross section of ovary ($\times 3$). 3 *Eichhornia paniculata* (a) base of plant with sheathing leaf bases and swollen petioles ($\times\frac{2}{3}$); (b) inflorescence, each flower with conspicuous green and white nectar guides on the upper petal ($\times\frac{2}{3}$); (c) flower opened out ($\times\frac{1}{3}$); (d) gynoecium ($\times 3$); (e) long stamen ($\times 4$); (f) short stamen ($\times 6$).

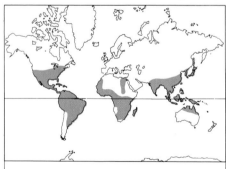

**Number of genera:** 9
**Number of species:** 34
**Distribution:** pantropical, in freshwater habitats.
**Economic uses:** aquatic ornamentals, edible leaves and weeds.

and unbranched. In many cases they are enveloped by sheathing leaf bases. In the free-floating species such as *Eichhornia crassipes*, buoyancy is provided by inflated petioles. The leaves are opposite or in whorls, with lanceolate to broadly ovate blades. In *Hydrothrix* they are reduced to hair-like pseudowhorls. In the perennial species fragmentation of stems and stolons is a common method of vegetative reproduction.

The inflorescence is a raceme or panicle subtended by a spathe-like sheath. The flowers are bisexual, regular or rarely irregular, blue, lilac, yellow or white and often showy. The perianth is usually composed of six segments. There are three or six stamens (rarely one). The ovary, of three fused carpels, is superior, comprising three locules with numerous ovules on axile placentas, or one locule with a single ovule. The style is long and the stigma entire or slightly lobed. The fruit is a capsule or nutlet. The seeds have a straight embryo and copious endosperm.

The floral morphology is exceptionally diverse for a small family. The genera *Eichhornia*, *Pontederia* and *Reussia* possess species with styles of three different lengths (tristylous) as well as species with styles all the same length (homostylous). In the tristylous species, flowers are of three distinct morphological types: with a long style and anthers at two levels below the stigma; with a medium style and one set of anthers above the stigma and one below; with a short style and anthers at two levels above the stigma. The pollen grains produced by the three levels of anthers differ in size. *Monochoria*, *Heteranthera* and *Scholleropsis* have members with dimorphic stamens, but their styles are of a uniform length. In *Hydrothrix* only one stamen is present. After flowering, species of *Eichhornia*, *Pontederia*, *Reussia* and *Monochoria* exhibit a downward curvature of the floral axis (hydrocarpy) and their fruits mature below the water surface.

Some members of the Pontederiaceae have interesting pollination systems. The aerial, showy flowers of the tristylous species are pollinated by insects and have conspicuous nectar guides on the perianth parts. There is a close association between North American populations of *Pontederia cordata* and a small solitary bee *Dufourea novae-angliae*. The flowering of the former coincides with the emergence of the bee and the bee appears to visit no other plant for nectar or pollen.

**Classification.** The Pontederiaceae, together with its most closely related family the Philydraceae, forms a peripheral subgroup within the Liliales.

**Economic uses.** Apart from *Eichhornia crassipes* and *Pontederia cordata* which are grown as aquatic ornamentals, the family is

Iridaceae. 1 *Crocus flavus* (a) habit ($\times\frac{2}{3}$); (b) capsule ($\times\frac{2}{3}$); (c) tip of trilobed style ($\times 4$). 2 *Crocus sp* flower with perianth opened ($\times\frac{2}{3}$). 3 *Gladiolus papilio* (a) inflorescence (spike) ($\times\frac{2}{3}$); (b) cross section of trilocular ovary ($\times 2\frac{2}{3}$). 4 *G. melleri* (a) half flower ($\times\frac{2}{3}$); (b) tip of style ($\times 3$). 5 *Iris laevigata* (a) apex of rhizome and leaf bases ($\times\frac{2}{3}$); (b) inflorescence with fully opened flower consisting of three reflexed "falls", three erect inner "standards" and three petaloid style branches behind the stamens ($\times\frac{2}{3}$); (c) standard ($\times 1$); (d) stamen ($\times 1$); (e) petaloid style branch ($\times 1$). 6 *I. germanica* cross section of ovary ($\times 2\frac{2}{3}$). 7 *I. foetidissima* dehiscing capsule ($\times\frac{2}{3}$).

not utilized by Man to any great extent. However, members of the family are of considerable economic importance as aquatic weeds, the free-floating *Eichhornia crassipes* being the most widespread and serious. Other members of the family are weeds of rice fields. These include the emergent aquatics *Heteranthera limosa* and *H. reniformis* (United States of America), *Reussia rotundifolia* and *Pontederia cordata* (South America), *Eichhornia natans* (Africa) and *Monochoria vaginalis* (Asia). The leaves of *Monochoria* species are also utilized in Asia as a green vegetable.

S.C.H.B.

## PHILYDRACEAE

The Philydraceae is a small family of perennial herbs native to Southeast Asia, New Guinea and Australia.

The plants possess erect aerial stems arising from an underground rhizome. The leaves are linear, and radical or clustered at the stem base. The flowers are solitary, bisexual and irregular, with a perianth of two whorls, each of two free segments. The single stamen has a bilocular anther inserted on a flattened filament. The ovary is superior and consists

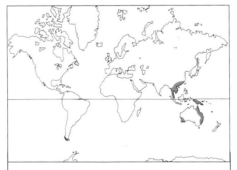

**Number of genera:** 4
**Number of species:** 6
**Distribution:** E and SE Asia, Malaysia, Australia.
**Economic uses:** none.

of three fused carpels forming a single locule with parietal placentation or three locules with axile placentation. There are numerous anatropous ovules in each locule. The style is simple. The fruit is a capsule containing numerous endospermic seeds with a straight embryo.

Three genera, *Philydrum*, *Philydrella* and *Orthothylax* are monotypic while the fourth, *Helmholtzia*, contains three species. The

Philydraceae is closely related to the Pontederiaceae and both families stand apart from the rest of the Liliales. The family has no known economic uses. S.R.C.

## IRIDACEAE
*The Iris Family*

The Iridaceae is a family of perennial herbs, including such horticulturally important genera as *Crocus*, *Freesia*, *Gladiolus* and *Iris*.
**Distribution.** The family has a worldwide distribution in both tropical and temperate regions, but South Africa, the eastern Mediterranean and Central and South America are especially rich in species.
**Diagnostic features.** Most of the Iridaceae are herbaceous and possess storage organs which are either corms (eg *Gladiolus*, *Iris*), rhizomes (eg many *Iris* species, *Sisyrinchium*) or more rarely bulbs (eg *Iris* of the Juno group). Some species are evergreen and usually possess a rhizome, which is sometimes very slender or compact and of little importance as a means of food storage. The leaves are usually narrow and linear, rather tough in texture and most commonly arranged in two ranks, often forming a flat "fan."

The structure of the inflorescence varies

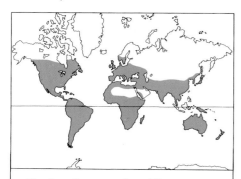

**Number of genera:** about 70
**Number of species:** about 1,800
**Distribution:** cosmopolitan.
**Economic uses:** numerous garden and indoor ornamentals (eg crocuses, freesias, gladioli and irises) a dye (saffron) and orris root.

considerably, but is usually terminal and cymose, the branches often reduced to one flower giving a spike or raceme (eg *Ixia*, *Gladiolus*). In some cases, for example in *Crocus*, the whole inflorescence may be reduced to a single flower which is practically sessile and therefore produced just above ground level.

The flowers show a remarkable degree of variability and some have become extremely modified from the basic shape. The whole family is characterized by having bisexual flowers which have six perianth segments, arranged either regularly (eg *Sisyrinchium*) or irregularly (eg *Gladiolus*) and either united in a perianth tube at the base (as in *Crocus*) or more or less free (eg *Moraea*). The perianth segments are in two whorls which may be more or less equal in size, giving a flower of completely regular shape, or arranged in two series of different appearance; the "bearded" irises are a good example of the latter, with the three outer segments deflexed or spreading horizontally and furnished with a patch of hairs (the "falls"), while the inner three segments are erect and glabrous (the "standards"). The stamens are three in number and arranged opposite the outer perianth segments. The anthers have two locules and nearly always dehisce extrorsely. The ovary is inferior (very rarely superior), comprising three fused carpels and three locules with axile placentas or more rarely one locule with three parietal placentas. The ovules are usually numerous, rarely one or few.

One of the most variable features is the style which in its simplest form is branched into three at the apex with terminal stigmatic surfaces (as in many *Crocus* species). However, in several genera (eg *Iris*, *Moraea*) the style has evolved into three flattened, often colorful, petal-like structures which are furnished with enlarged showy "crests" overtopping the stigmatic surface. In *Iris* these style branches curve outwards away from the axis of the flower and form, with three of the perianth segments, a protective

tunnel-like organ over each anther. Any pollinating insect is attracted by nectar at the base of the segments and by a colorful "signal patch" on the expanded blade of the segments and crawls along the tube, thus becoming dusted with pollen. Although most of the Iridaceae are certainly insect-pollinated, a few have adaptions to other methods such as bird-pollination (for example the scarlet flowers of *Rigidella*) or wind-pollination, as in *Dierama*, where the stems and pedicels are very long and slender. The fruit is a capsule and the seeds have a small embryo and copious endosperm.

**Classification.** The Iridaceae can be divided into 11 tribes, using such features as the type of rootstock and the degree of regularity or irregularity in the shape of the flower. The more important of these tribes range from the primitive, rhizomatous, regular-flowered SISYRINCHEAE (eg *Sisyrinchium*, *Libertia*) to the MARICEAE (eg *Cypella*), IRIDEAE (eg *Iris*), IXIEAE (eg *Ixia*, *Freesia*, *Lapeyrousia*), CROCEAE (eg *Crocus*, *Romulea*) and GLADIOLEAE (eg *Gladiolus*, *Tritonia*, *Acidanthera*) finally reaching the ANTHOLYZEAE (eg *Antholyza*, *Anomalesia*) which is considered to be the most advanced with extremely irregular flowers.

The family is related to the Liliaceae and Amaryllidaceae. The Liliaceae differs in normally having six stamens and a superior ovary.

**Economic uses.** The family is valued principally for its garden and indoor ornamentals and a great deal of work has gone into the production of cultivars and hybrids in such genera as *Gladiolus*, *Iris*, *Crocus*, *Freesia*, *Ixia*, *Sparaxis*, *Crocosmia* (*Montbretia*) and *Tigridia*. In the first two genera the hybridization has been carried on for well over a century and it is now virtually impossible to discover the origins of the hybrids.

Apart from their horticultural value, the Iridaceae have not been used to any extent for commercial purposes. Saffron, obtained from the scarlet style of *Crocus sativus*, was at one time used widely as a dye and flavoring agent in cooking; the plant was cultivated for this purpose throughout Europe and Asia. Orris root, from *Iris florentina*, is used in making perfumes and cosmetics.                    B.F.M.

# LILIACEAE
*The Lily Family*
The Liliaceae is one of the largest families of flowering plants and certainly one of the most important horticulturally, as it includes the lilies and numerous other outstandingly beautiful cultivated genera. Its chief economic representative is the onion.
**Distribution.** The family is cosmopolitan, although many of the smaller groups have a limited distribution in definite areas.
**Diagnostic features.** Most of the Liliaceae are herbs, and of these a large percentage have

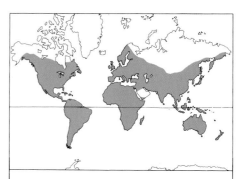

**Number of genera:** about 250
**Number of species:** about 3,500
**Distribution:** cosmopolitan.
**Economic uses:** many popular ornamentals (eg lilies, tulips and hyacinths), vegetables (eg onions, leeks, garlic and asparagus) and some medicinal uses.

swollen storage organs such as bulbs, corms, rhizomes or thick fleshy roots. However, some genera, for example *Aloe* and *Haworthia*, are evergreen succulents while a few, such as *Lapageria*, are evergreen climbers with woody stems. In Australia many of the genera have evolved in extreme xerophytic conditions and now bear little resemblance to any other members of the family. *Borya*, for example, produces tufts of needle-like leaves and dense heads of flowers not unlike those of an *Armeria*.

The leaf characters vary enormously within the family from basal and linear with parallel veins (eg *Ornithogalum*, *Endymion*, *Eremurus*, *Anthericum*) to cauline and broadly ellipsoid with net-veining (eg *Trillium*). In *Gloriosa*, tendrils are produced at the leaf tips while in *Asparagus* the leaves tips while in *Asparagus* the leaves are reduced to insignificant scales.

The flowers are usually regular and bisexual, and borne in a raceme, sometimes solitary (eg *Tulipa*) or more or less condensed into a cyme (eg *Hemerocallis*). The flowers of *Allium* are produced in umbel-like cymes which may be two- or three-flowered or may carry a very large number of flowers in a huge spherical head. There are usually six more or less equal perianth segments (rarely four or more than six) in two whorls

Liliaceae. 1 *Lapageria rosea* leafy shoot with solitary axillary flowers ($\times \frac{2}{3}$). 2 *Allium cyaneum* habit showing flower in umbel-like cymes ($\times \frac{2}{3}$). 3 *Calochortus uniflorus* habit showing basal bulb ($\times \frac{2}{3}$). 4 *Aloe jucunda* habit showing basal rosette of spined, fleshy leaves ($\times \frac{2}{3}$). 5 *Kniphofia triangularis* inflorescence ($\times \frac{2}{3}$). 6 *Lilium martagon* (a) inflorescence ($\times \frac{2}{3}$); (b) fruit—a capsule ($\times \frac{2}{3}$). 7 *L. canadense* half flower showing petaloid perianth segments, and superior ovary containing numerous ovules on axile placentas and crowned by a single style with a lobed stigma ($\times \frac{1}{2}$). 8 *Convallaria majalis* fruits—berries ($\times \frac{2}{3}$). 9 *Colchicum callicymbium* habit showing basal bulb, leaves beginning to emerge and flower with a six-lobed perianth, six yellow stamens and a trifid style ($\times \frac{2}{3}$).

of three, free (eg *Tulipa*) or united into a perianth tube (eg *Kniphofia*). The outer three segments are sometimes smaller than the inner and rather sepal-like, enclosing and protecting the three more showy inner segments during the bud stage (eg *Calochortus*). There are usually six stamens (rarely three or up to 12), always arranged opposite the perianth segments; the anthers have two locules usually with latrorse dehiscence. The ovary, of three fused carpels, is superior (rarely inferior or semi-inferior), usually with three locules and axile placentas, rarely with one locule and parietal placentas. The ovules are usually numerous (rarely solitary) and arranged in two rows in each locule. The styles are entire or divided, rarely free. The fruit is either a dry capsule or, less frequently, a fleshy berry. The seeds have a straight or curved embryo and abundant endosperm. Many of the Liliaceae are insect-pollinated, the attraction being in the form of honey secreted by the ovary or by exposed nectaries at the base of the perianth segments, these being well displayed in most species of *Fritillaria*.

**Classification.** Different authorities have recognized between 12 and 28 tribes within the family. It is thus difficult to recommend any particular system since it is obvious that a great deal of critical work is required in order to clarify the relationships within the family. Certain groups have, however, remained fairly uniform throughout the various treatments, for example the important horticultural groups TULIPEAE, containing *Lilium, Tulipa, Nomocharis, Erythronium,* and *Calochortus,* and SCILLEAE, containing *Scilla, Muscari, Chionodoxa, Eucomis, Ornithogalum* (chincherincheree), *Camassia* and *Puschkinia.*

The genus *Allium* and several related genera such as *Tulbaghia, Agapanthus, Brodiaea* and *Triteleia* are placed by some taxonomists in a separate family the Alliaceae, and by others in the Amaryllidaceae, but in this book are retained in the Liliaceae. The main differences between the various classifications of the family have centered round the inclusion or exclusion of *Allium* and its relatives, of *Yucca* and *Sansevieria* which are considered here to belong to the Agavaceae, and of *Ophiopogon* and *Liriope* which are liliaceous but have in the past been placed in the Haemodoraceae.

The Liliaceae is closely related to two other major families of petaloid monocotyledons, the Iridaceae and Amaryllidaceae. The Iridaceae, however, usually has flowers with three stamens and an inferior ovary, while the latter combines six stamens with an inferior ovary.

**Economic uses.** The main use of members of the Liliaceae is in horticultural display, for many of the genera are of extreme beauty. The most popular plants are probably the tulips. There has been a great deal of hybridization in the genus *Tulipa*, especially

in the last two centuries, and a vast range of large showy hybrids are now available, but some of the species have been known as garden plants since the middle of the 16th century. *Lilium* contains perhaps the most beautiful of species in the whole family while many other genera such as *Scilla, Muscari, Hyacinthus, Erythronium, Agapanthus, Colchicum, Kniphofia, Aloe, Hemerocallis, Hosta, Convallaria* and *Gloriosa* are widely known to gardeners.

*Allium* is the only important food plant genus with *Allium cepa* (onion, shallot), *A. porrum* (leek), *A. sativum* (garlic) and *A. schoenoprasum* (chives) being the best known. *Asparagus officinalis* is cultivated for its tender young shoots. Some of the Liliaceae have been used medicinally, such as *Aloe* (bitter aloes), *Urginea* (squill), *Veratrum* (hellebore powder) and *Colchicum,* the seed and corms of which yield the alkaloid colchicine. Apart from its medical uses colchicine is used in plant genetics to induce polyploidy.                                    B.F.M.

## AMARYLLIDACEAE
*The Daffodil Family*

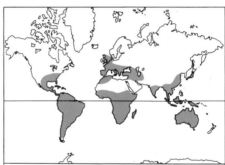

**Number of genera:** about 75
**Number of species:** about 1,100
**Distribution:** mainly warm temperate and subtropical.
**Economic uses:** many garden and indoor ornamentals, notably the daffodils and snowdrops.

The Amaryllidaceae is a large, horticulturally important family which includes the daffodils (*Narcissus*) and snowdrops (*Galanthus*) and many other popular greenhouse and house plant subjects, including the "amaryllis" (*Hippeastrum*).

**Distribution.** The family is mainly warm temperate, subtropical and tropical in distribution, although the genera *Narcissus* and *Galanthus* occur as far north as Britain.

**Diagnostic features.** Most members of the Amaryllidaceae possess a bulbous rootstock, although some have rhizomes or a bulb with a short rhizome attached to the base. They are mostly deciduous, but a few (eg *Clivia*) are evergreen. The leaves are more or less linear, growing from the base of the plant. The inflorescence consists of a scape carrying an umbel, although in many cases it is reduced to only a few flowers or to a

solitary flower (eg in many *Narcissus* species). In some genera (eg *Sternbergia*) the scape is much reduced and does not appear above ground level. The ovary is subterranean, at the base of a long perianth tube, until the fruiting stage when the scape elongates and pushes the developing capsule to the surface. This feature is often associated with autumnal species that flower at the beginning of a rainy season after drought.

The same habit has evolved independently in the Liliaceae (eg *Colchicum*) and Iridaceae (eg *Crocus*, where there are many autumn-flowering species). It has the advantage that plants can make full use of a short wet season, the flowers being produced early, often before the leaves, so the capsules and foliage have the rest of the growing period in which to mature before the onset of dry conditions once more. The subterranean ovary is protected below ground from any severe frosts that occur during the winter. Other Amaryllidaceae which exhibit this adaptation and which also occur in the cooler areas of the range of distribution are, for example, *Haylockia* in temperate South America and *Gethyllis* in South Africa. The majority of Amaryllidaceae, however, occur in warm regions and their flowers are carried on a completely aerial scape with a spathe or spathes subtending the inflorescence.

The flowers are bisexual and regular, and the showy perianths are made up in two whorls of three segments, either free from each other (eg *Galanthus, Leucojum, Amaryllis, Nerine*) or joined into a tube (eg *Crinum, Zephyranthes, Sternbergia, Cyrtanthus, Stenomesson*). In many genera there is a corona present, this being very obvious in *Narcissus*, where it appears as a distinct trumpet or cup, depending on the species. The origin of this corona varies from one genus to another; in *Narcissus* it is an appendage of the perianth segments, but in *Pancratium* it is made up of the expanded and fused filaments of the stamens. The stamens are six in number, arranged in two whorls of three, either more or less free, fused to the perianth tube, or joined together by their filaments into a staminal cup. The ovary of three fused carpels is inferior, containing three (or rarely one) locules with axile (rarely parietal) placentas. The ovules are anatropous and usually numerous in each locule. The style is slender, with a capitate or three-lobed stigma, and the fruit is either a loculicidally dehiscing capsule or a fleshy berry. The seeds have a small straight embryo and fleshy endosperm.

**Classification.** The family as defined in this book (ie excluding genera with a superior ovary such as *Allium*) can be divided into two broad groups, depending primarily on the absence or presence of a corona. The former state is thought to be more primitive and includes such genera as *Galanthus, Amaryllis, Crinum* and *Zephyranthes*. The more advanced genera, possessing a corona,

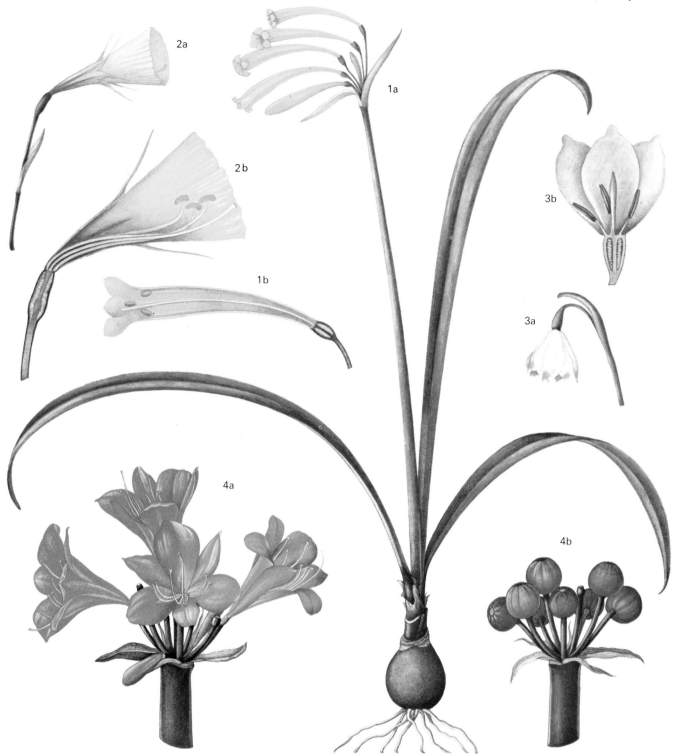

Amaryllidaceae. 1 *Cyrtanthus* sp (a) habit showing basal bulb, linear leaves, inflorescence borne on a leafless stalk (scape) and flowers with a tubular perianth ($\times\frac{2}{3}$); (b) half flower ($\times 1\frac{1}{3}$). 2 *Narcissus bulbocodium* var *citrinus* (a) flower showing subtending spathe, and perianth with finely-toothed lobes and large tubular corona ($\times\frac{2}{3}$); (b) half flower ($\times 1\frac{1}{3}$). 3 *Leucojum vernum* (a) flower with two whorls of free perianth segments ($\times\frac{2}{3}$); (b) half flower ($\times 1\frac{1}{3}$). 4 *Clivia miniata* (a) inflorescence ($\times\frac{1}{2}$); (b) fruits—berries ($\times\frac{1}{2}$).

include *Narcissus*, *Pancratium* and *Hymenocallis*.

There is considerable disagreement as to which genera should be included in the Amaryllidaceae. The current tendency is to retain the genus *Allium* and its relatives (eg *Brodiaea*, *Agapanthus*) in the Liliaceae, but some authorities include them in the Amaryllidaceae, while others recognize a separate family, the Alliaceae. The problem with

*Allium* and its relatives is that they resemble members of the Amaryllidaceae in having an umbellate inflorescence, but have a superior ovary, as do the Liliaceae. If the taxonomist considers the form of inflorescence to be the most important feature of the plant, then they are classified in the former, but if he gives priority to the possession of a superior ovary then they must be in the Liliaceae. Others consider that the combination of

these characters justifies the formation of the Alliaceae. *Allium* and all its relatives are treated in this book as members of the Liliaceae.

*Agave* and *Vellozia* have also been considered as members of the Amaryllidaceae, but are here placed in the Agavaceae and Velloziaceae respectively. *Hypoxis*, *Curculigo* and five other genera (here included in Amaryllidaceae) have sometimes been made into a separate family, the Hypoxidaceae; and *Alstroemeria, Bomarea, Leontochir* and *Schickendantzia* have been made into the Alstroemeriaceae.

The Amaryllidaceae is closely related to the Liliaceae and Iridaceae.

**Economic uses.** The family is important horticulturally, but not economically important in any other way, although alkaloids exist in *Galanthus* and *Narcissus* and these may be of medical interest. The fleshy fruits of the South African *Gethyllis* are edible, but although considered a delicacy they are consumed only on a very small scale and are not cultivated for this purpose. The genus *Narcissus* includes the daffodils and is probably the most important of the family for garden purposes. Many thousands of daffodil varieties have been raised over a period of at least 400 years, resulting in a large and valuable industry. The Netherlands is especially important in this respect, exporting large quantities of daffodils all over the world. In addition to the many hybrids there are a considerable number of wild *Narcissus* species that are of value, especially in rock gardening.

*Amaryllis belladonna* and *Nerine bowdenii*, both from South Africa, are highly valued in gardens for their showy, late-autumn flowers. The hybrid *Crinum, C. × powellii*, is another hardy plant popular in temperate gardens, and the South American *Zephyranthes candida*, which has crocus-shaped flowers, is often planted in warm borders for a September display. *Sternbergia* is a small genus from the Mediterranean region, whose species all have yellow goblet-shaped flowers; *S. lutea*, an autumn-flowering plant, is especially popular. *Galanthus* and *Leucojum*, the snowdrops and snowflakes, are closely related genera, usually producing white flowers in early spring, although a few *Leucojum* species produce leafless inflorescences in the autumn. Most of the other genera are best treated as greenhouse subjects, the most important being *Hippeastrum* ("amaryllis"), the huge-flowered hybrids of which flower in winter, *Nerine, Stenomesson, Clivia, Cyrtanthus, Doryanthes, Haemanthus, Crinum, Pamianthe, Eucharis, Hymenocallis, Phaedranassa, Habranthus, Sprekelia* and *Vallota*.

B.F.M.

## AGAVACEAE
*Sisal Hemp, Pulque and Dragon Tree*

The Agavaceae is a family of rhizomatous, woody, sometimes climbing plants. Many species produce valuable fibers, such as sisal hemp, and *Agave americana* is the source of the Mexican drink pulque.

**Distribution.** Members of the family are found throughout the tropics and subtropics, particularly in arid and semiarid locations.

**Diagnostic features.** The leaves are usually crowded at the base of the stem and are stiff,

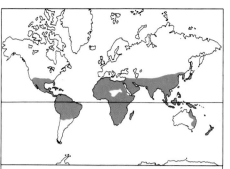

**Number of genera:** 20
**Number of species:** about 700
**Distribution:** tropics and subtropics mainly arid regions.
**Economic uses:** source of fibers (eg sisal hemp) alcoholic beverages (pulque and mezcal) and limited use as ornamentals.

often fleshy, narrow and sharp-pointed. They are entire or have prickles at the margin and may be as much as 3m (10ft) in length. Many species are succulents.

The flowers are borne in racemes or panicles and are regular or slightly irregular and usually bisexual; when unisexual, male and female flowers may be borne on separate plants or together with bisexual flowers on the same plant. The perianth is composed of two similar petal-like whorls, each in three parts, united into a long or short tube. There are six stamens inserted on the perianth tube or at the base of the segments. The filaments are thread-like or somewhat thickened towards the base, and the anthers have two locules which open by longitudinal slits. The ovary, of three fused carpels, is superior or inferior, and comprises three locules, with axile placentation and with solitary to numerous ovules in each locule. The style is single and slender. The fruit is a capsule or berry, containing one to numerous seeds with a small embryo surrounded by copious fleshy endosperm.

**Classification.** The most important genera are *Agave* (about 300 species), *Dracaena* (150 species), *Sanseverinia* (60 species), *Yucca* (40 species), *Furcraea* (20 species), *Cordyline* (15 species) and *Phormium* (two species).

At one time the members of this family were distributed in the Liliaceae and the Amaryllidaceae. They represent the xerophytic, fibrous-leaved members of both these families and thus form a rather heterogeneous group. They possess common features of morphology and inflorescence and grow in similar environments, but the individuals may be more closely related phylogenetically to members of the Liliaceae or the Amaryllidaceae than to each other.

**Economic uses.** The Agavaceae is a family of considerable economic importance. A number of species are the source of strong, durable fibers used for cordage, matting, fishing nets etc. Examples include *Agave sisalana* and *A. fourcroydes* (sisal hemp), *A.*

*heteracantha* (istle fiber), *Sanseverinia zeylanica* (bowstring hemp) and *Phormium tenax* (New Zealand flax). A red resin known as dragon's blood is obtained from *Dracaena cinnabari* and *D. draco* (the dragon tree). The fermented sap of *Agave americana* is the source of pulque, a national Mexican drink which either is consumed immediately, without aging, or is distilled to yield the spirit mezcal (mescal).

S.R.C.

## XANTHORRHOEACEAE
*Grass Trees*

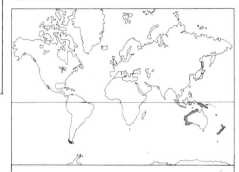

**Number of genera:** 8
**Number of species:** about 66
**Distribution:** Australia, New Caledonia, New Zealand.
**Economic uses:** *Xanthorrhoea* species yield a gum used for varnishes.

The Xanthorrhoeaceae is a family of stout woody rhizomatous perennials, often with tall, few-branched stems. Many are xerophytes.

**Distribution.** The family is found in Australia, New Caledonia, New Guinea and New Zealand.

**Diagnostic features.** The leaves are simple, linear, usually sheathing. The flowers are regular, often dry and papery, bisexual or unisexual (male or female then borne on separate plants), and usually borne in spikes, panicles or heads. The perianth is in two whorls of three segments. The stamens are in two whorls of three, the inner whorl often attached to base of the inner perianth segments and the outer whorl usually free. The anthers have two locules with introrse or latrorse dehiscence. The ovary, of three fused carpels, is superior, with either three locules and axile placentation or one locule and basal placentation; there are one to few ovules per locule. The styles are free or more or less connate. The fruit is a capsule or rarely a one-seeded nut; the seeds have a hard endosperm and a straight embryo.

**Classification.** The eight genera can be simply distinguished by their perianth characters. The perianth may be very small, as in *Chamaexeros* and *Acanthocarpus*, or with the outer parts glumaceous and the inner scarious, as in *Xanthorrhoea* and *Dasypogon*, or finally the perianth parts may be rigid and sometimes colored, as in *Kingia, Baxteria, Calectasia* and *Lomandra*.

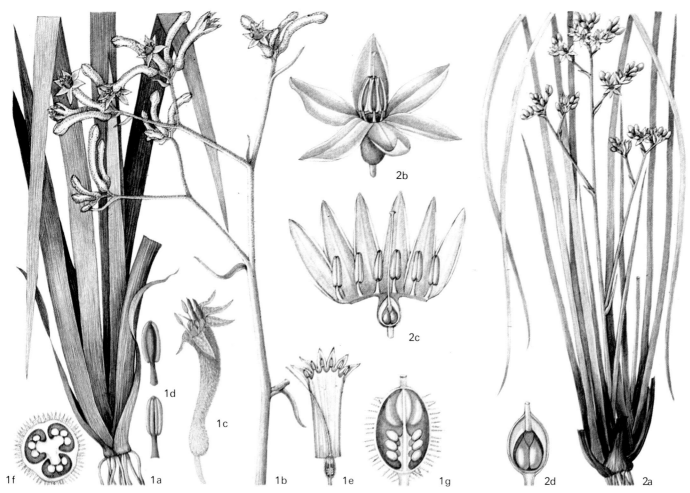

Haemodoraceae. 1 *Anigozanthos flavidus* (a) habit ($\times\frac{2}{3}$); (b) inflorescence ($\times\frac{2}{3}$); (c) flower showing curved green perianth tube and six stamens ($\times 1$); (d) stamen front (lower) and back view (upper) ($\times 3$); (e) flower dissected showing epipetalous stamens ($\times\frac{2}{3}$); (f) cross section of trilocular ovary with three axile placentas ($\times 6$); (g) vertical section of ovary ($\times 6$). 2 *Phlebocarya ciliata* (a) habit ($\times\frac{2}{3}$); (b) flower showing perianth in two whorls and six stamens ($\times 6$); (c) flower dissected showing epipetalous stamens ($\times 6$); (d) vertical section of ovary ($\times 14$).

The family is closely related to the Liliaceae and Agavaceae, and has also been associated with the Juncaceae.

**Economic uses.** *Xanthorrhoea* species (grass trees) yield a gum used in making varnishes.
S.A.H.

## VELLOZIACEAE

The Velloziaceae is a small family of fibrous shrubby plants.

**Distribution.** The family is native to arid regions of South America, tropical and subtropical Africa, and Madagascar.

**Diagnostic features.** The stems are woody and dichotomously branched, with narrow, drought-resistant leaves, usually clustered at the stem tips. When the leaves fall off, their thick, fibrous bases remain attached to the stem, making it appear thicker than it is. The base of the stem is covered with a dense mat of adventitious roots which can rapidly absorb any available water.

The flowers are bisexual, regular and solitary. They have a petaloid perianth composed of two series each of three segments, which may be free or united at the base. The stamens are either arranged in two whorls of three, or in six bundles in species where they are numerous. The gynoecium

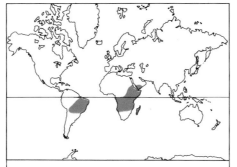

**Number of genera:** 4
**Number of species:** about 300
**Distribution:** arid regions of S America, tropical and subtropical Africa, and Madagascar.
**Economic uses:** none.

consists of three carpels fused into an inferior ovary with three locules and numerous ovules on axile placentas. The single style terminates either in a flattened head or in three distinct lobes. The fruit is a dry or hard capsule containing numerous seeds each with a small embryo surrounded by hard endosperm.

**Classification.** The four genera are *Vellozia* (100 species in South America and Africa

and one species in Arabia), *Barbacenia* (140 species in South Africa), *Xerophyta* (55 species in Africa, Madagascar and South America) and *Barbaceniopsis* (two species in South America). Features used to distinguish genera include the degree of fusion of the perianth and the numbers of stamens.

The family is related to the Haemodoraceae and the Taccaceae.

**Economic uses.** None are known.
S.R.C.

## HAEMODORACEAE
*Kangaroo Paw*

The Haemodoraceae is a family of tropical and warm temperate herbs, a few of which are cultivated as ornamentals.

**Distribution.** The family occurs in South Africa, Australasia (excluding New Zealand) and North and tropical America.

**Diagnostic features.** The plants are herbs with fibrous roots, tubers, rhizomes or stolons, and linear leaves arising from the ground and sheathing one another at the base. The leaves are hairy or glabrous, and have closely parallel or fan-like veins. The flowers are bisexual, regular or slightly irregular, and are borne in cymes, racemes or panicles, the whole often covered with a thick layer of woolly hair. These hairs may

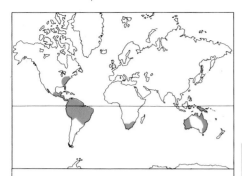

**Number of genera:** 17
**Number of species:** about 100
**Distribution:** Australasia (excluding New Zealand), S Africa, N and tropical America.
**Economic uses:** *Anigozanthos* species and a few others cultivated as ornamentals.

look like small feathers under a hand lens. The perianth is persistent, in one or two whorls each of three segments, and may form a straight or curved tube. When in two whorls the outer segments more or less cover the inner. There are three or six stamens with free filaments inserted on the inner perianth lobes, each anther having two locules which open by longitudinal slits. The ovary, of three fused carpels, is superior or inferior, with three locules, each locule containing one to many ovules on axile placentas. The style is usually filiform, with a capitate stigma. The fruit is a capsule, opening by three valves. The seeds have a small embryo and much endosperm.

**Classification.** The family can be divided into two tribes:

HAEMODOREAE. Perianth in two whorls, tube short or absent, stamens three or six. *Lanaria* (one species, South Africa), *Phlebocarya* (three species, Australia), *Schiekia* (one species, America), *Pyrrorhiza* (one species, Venezuela), *Wachendorfia* (25 species, South Africa), *Lachnanthes* (one species, North America), *Dilatris* (three species, South Africa), *Xiphidium* (one or two species, America), *Hagenbachia* (one species, Brazil), *Barberetta* (one species, South Africa), *Haemodorum* (20 species, Australia).

CONOSTYLEAE. Perianth in one whorl, tube often long and curved, stamens six, flowers always woolly. *Lophiola* (two species, North America), *Tribonanthes* (five species, Australia), *Conostylis* (22 species, Australia), *Blancoa* (one species, Australia), *Anigozanthos* (10 species, Australia), *Macropidia* (one species, Australia).

The family is related to the Liliaceae, in which some botanists place it.

**Economic uses.** Several species of *Anigozanthos* (kangaroo paw) are cultivated. They have racemes or spikes of tubular flowers clad in woolly hair, and are much prized in Australian gardens; they also make fine pot plants. *Anigozanthos manglesii* is the State emblem of Western Australia, with

orange to yellow, red and green six-lobed flowers. Other colorful species include *A. flavidus*, *A. rufus* and *A. pulcherrimus*. Other occasionally cultivated genera include *Conostylis*, *Macropidia*, *Blancoa* and *Lanaria*.

B.M.

# TACCACEAE
*East Indian and Hawaiian Arrowroot*

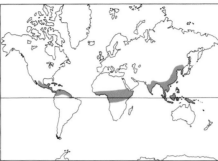

**Number of genera:** 2
**Number of species:** 31
**Distribution:** pantropical and China.
**Economic uses:** East Indian arrowroot.

The Taccaceae is a small family of perennial tropical herbs with tubers or creeping rhizomes, some of which are a source of food in the Pacific area.

**Distribution.** *Tacca* is pantropical; *Schizocapsa* is native to China.

**Diagnostic features.** The leaves are radical, large, broad, entire or deeply lobed, often with long stalks. The flowers are bisexual, regular, subtended by broad or elongated bracts, and borne in umbellate inflorescences. The bracts may collectively form a whorl or involucre below the flowers. The perianth is made up of two series, each of three segments, all fused together into a short, six-lobed, cup-shaped tube. The six stamens are attached by short filaments to the perianth tube. The ovary is inferior and consists of three fused carpels forming a single locule with three parietal placentas bearing numerous ovules. The style is short and terminates in three reflexed stigmas which are often petaloid. The fruit is usually a berry (rarely a capsule), containing numerous seeds, each with a minute embryo surrounded by copious endosperm.

**Classification.** The two genera, *Tacca* (30 species) and *Schizocapsa* (one species) can be distinguished on the nature of the leaves and fruit. In *Tacca* the leaves are either entire or much divided, and the fruit is a berry, while in *Schizocapsa* the leaves are entire and the fruit is a capsule.

The family is related to the Velloziaceae and Haemodoraceae, sharing with them such features as vegetative habit and bisexual flowers with a tubular perianth, six stamens and seeds with copious endosperm.

**Economic uses.** Economically the family is important for *Tacca pinnatifida*, whose

rhizomatous tubers yield a starch known as East Indian arrowroot. It is used in the Pacific islands and Africa for bread-making and as a starch in laundry work. The rhizomes of *T. hawaiiensis* (Hawaiian arrowroot) are used similarly. The leaves of some other species such as *T. fatsiifolia* and *T. palmata* are used medicinally to treat a variety of external and internal disorders.

S.R.C.

# STEMONACEAE

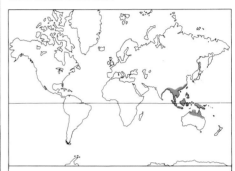

**Number of genera:** 3
**Number of species:** 30
**Distribution:** E Asia, Indomalaysia, N Australia, E N America.
**Economic uses:** *Stemona tuberosa* yields an insecticide.

The Stemonaceae is a small family of perennial erect or climbing herbs with rhizomes or tubers.

**Distribution.** The family is distributed throughout eastern Asia, Indomalaysia, and northern Australia, with some *Croomia* species in eastern North America.

**Diagnostic features.** The leaves are alternate, opposite or in whorls. The blade has a number of parallel main veins which are cross-linked by numerous transverse parallel veins. The flowers are borne solitary or in cymes or racemes in the axils of the leaves. They are regular and usually bisexual with a perianth of two series, each of two sepaloid or petaloid segments. There are four free stamens, and in *Stemona* the connective is extended well beyond the anther lobes. The ovary of two or three fused carpels is superior to semi-inferior and consists of a single locule containing few to many ovules on basal or apical placentas. It is surmounted by two or three sessile stigmas in *Croomia* and *Stichoneuron*, and a single sessile or subsessile stigma in *Stemona*. The fruit is an ovoid capsule, opening by two valves. It contains seeds which are often attached to the placenta by long stalks. The seed contains a small embryo surrounded by copious endosperm.

**Classification.** The three genera are *Stemona* (25 species), *Croomia* (three species) and *Stichoneuron* (two species). Some botanists place *Croomia* and *Stichoneuron* into a separate family (the Croomiaceae) because of several floral differences, including size of

Cyanastraceae. 1 *Cyanastrum cordifolium* (a) habit showing basal corm and heart-shaped leaves ($\times\frac{2}{3}$); (b) part of flower showing stamens with anthers free ($\times$2); (c) stamen ($\times$4). 2 *Cyanella lutea* (a) habit showing basal corm with fibrous covering and linear leaves ($\times\frac{2}{3}$); (b) flower showing six stamens: five smaller with anthers united, and one large and free ($\times 1\frac{1}{3}$); (c) large anther ($\times$2); (d) small anther dehiscing by apical pores ($\times$2); (e) gynoecium ($\times$2); (f) cross section of ovary ($\times$3); (g) fruit—a fleshy capsule ($\times 2\frac{2}{3}$). 3 *Conanthera campanulata* (a) leafy shoot and inflorescence ($\times\frac{2}{3}$); (b) part of flower with anthers arranged in a cone ($\times 2\frac{2}{3}$); (c) anther with extended connective ($\times$4).

the perianth segments and the nature of the placentation.

The Stemonaceae shows a number of vegetative and floral similarities with the Dioscoreaceae. It is probably most closely related to the Liliaceae.

**Economic uses.** The family is not economically important but the roots of *Stemona tuberosa* have been used for insecticidal purposes.                                    S.R.C.

## CYANASTRACEAE

The Cyanastraceae (Tecophilaeaceae) is a family of perennial herbs with corms or tubers.

**Distribution.** Of the six genera three (*Conanthera*, *Zephyra*, and *Tecophilaea*) are native to Chile, one (*Odontostomum*) is native to California, and the remaining two (*Cyanastrum* and *Cyanella*) are native to central and southern Africa, respectively.

**Diagnostic features.** The plants perennate by means of a fibrous, tunicated corm or a thick, flattened, disk-shaped tuber. The leaves are generally either alternately arranged at the base of the flowering stem, or arise directly from the underground stem. They are variable in shape, but usually heart-

**Number of genera:** 6
**Number of species:** about 22
**Distribution:** Chile, California, central and southern Africa.
**Economic uses:** none.

shaped, disk-shaped, oval or long and narrow, and usually smooth and glabrous. The flowers are bisexual and regular, arranged in simple or branching racemes, and subtended by large or small, often membranous bracts. There are two perianth whorls, each of three segments. These are either free or fused together at the base into a short tube. The apices of the segments are either spreading or bent downwards. The

stamens are in two whorls of three, attached to the perianth segments. In some species, one of the whorls is sterile, in the form of staminodes. Instead of the more usual longitudinal dehiscence, the anthers of most species of this family liberate their pollen from a terminal pore on each locule. Another somewhat unusual feature is that the connective is often extended at both ends, the basal part being swollen or spurred. The ovary is semi-inferior and consists of three fused carpels which form three locules, with numerous ovules arranged in double rows on axile placentas in each locule. The ovary is surmounted by a thread-like and slender style, terminating in a three-lobed stigma. The fruit is a capsule containing numerous seeds, each with a large embryo, surrounded by fleshy endosperm.

**Classification.** The genera can initially be divided into two groups on the basis of details of the stamens. *Conanthera* (anthers converging into a cone), *Odontostomum* and *Cyanastrum* (anthers free) all have six equal stamens. *Cyanella*, *Zephyra*, and *Tecophilaea* have either dissimilar stamens or some stamens modified as staminodes.

This family shows a number of similarities

to the Liliaceae, particularly in relation to the floral structure (both families have regular flowers, the perianth in two whorls, the stamens usually six in number and the ovary with three locules and axile placentation). The tendency for reduction in some of the stamens and the semi-inferior ovary are indications of advancement and the Cyanastraceae might therefore be regarded as intermediate between the Liliaceae and the Iridaceae.

**Economic uses.** None are known.

S.R.C.

## SMILACACEAE
*Smilax and Sarsaparilla*

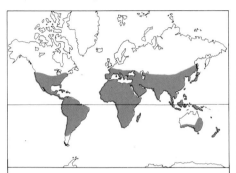

**Number of genera:** 4
**Number of species:** about 375
**Distribution:** mainly tropical and subtropical with a few temperate.
**Economic uses:** various *Smilax* species have medicinal use as tonics (sarsaparilla) and stimulants.

The Smilacaceae is a family of mainly climbing shrubs, the vast majority of which are in a single genus, *Smilax*.

**Distribution.** The family is mainly tropical and subtropical, but extends into temperate regions.

**Diagnostic features.** The prickly aerial stems arise from a rhizome and bear rather leathery, three-nerved leaves which, atypically for monocotyledons, display reticulate (net-like) venation between the main veins. The leaves are opposite or alternate, and the leaf sheaths may develop into tendrils, as in *Smilax*. These and the prickly hooks on the stem assist in the climbing habit. The flowers are regular, usually unisexual (bisexual in *Rhipogonum*) and arranged in axillary racemes, spikes or umbels. The male and female flowers are borne on separate plants. The perianth consists of two whorls each of three segments, free or rarely fused into a tube. There are usually two whorls each of three stamens (nine stamens in *Pseudosmilax* and three in *Heterosmilax*), with unilocular anthers. The female flower possesses staminodes. The ovary, of three fused carpels, is superior, consisting of three locules with one or two pendulous, orthotropous or semi-anatropous ovules per locule. The fruit is a berry containing one to three seeds with a small embryo and hard endosperm.

**Classification.** The four genera are *Smilax* (about 350 species), *Heterosmilax* (15 species), *Rhipogonum* (seven species) and *Pseudosmilax* (two species).

This family is closely related to the Liliaceae but differs in the leaf characters and in having male and female flowers on separate plants.

**Economic uses.** Various species of *Smilax* are sources of commercial sarsaparilla, used for treating rheumatism and other ailments. Other species have medicinal uses, the best known being *Smilax china* (China root) whose dried rhizome yields an extract with stimulant qualities. A New Zealand species of *Rhipogonum* (*R. scandens*) is sometimes used as a substitute for sarsaparilla.  S.R.C.

## DIOSCOREACEAE
*Yams*

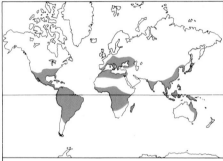

**Number of genera:** 6
**Number of species:** about 630
**Distribution:** pantropical with some temperate species.
**Economic uses:** staple food (yams) and diosgenin which has potential use in oral contraceptives.

The Dioscoreaceae is a family of mainly tropical climbers whose best-known members are the yams.

**Distribution.** The family is distributed throughout the tropics and some temperate regions.

**Diagnostic features.** All species, except for a few dwarf shrubs, are perennial herbaceous or shrubby climbers with well-developed tubers or rhizomes at their bases. The stems are unable to support their own weight for any great height and climb by twining on a support. The leaves are alternate and usually simply cordate, but are sometimes lobed as well.

The flowers are regular, small, inconspicuous and bisexual or unisexual (male and female then being borne on separate plants). They are usually axillary and borne in panicles, spikes or racemes, and have six perianth segments in two whorls, usually fused into a short bell-shaped tube at the base. There are two whorls of three stamens attached to the perianth lobes; one row is sometimes reduced to staminodes or absent. The ovary, of three fused carpels, is inferior, containing three locules, each with two (rarely numerous) ovules on axile placentas

(rarely a single locule with parietal placentation). There are three styles or a single style with three stigmas. The fruit is a capsule, a berry or a samara, and the seeds are usually winged, with endosperm and a small embryo.

**Classification.** Three genera (*Trichopus*, *Stenomeris* and *Avetra*) have bisexual flowers; *Dioscorea*, *Rajania* and *Tamus* have unisexual flowers. *Trichopus* (southern India) comprises a single dwarf shrub species whose fruit is a berry. *Stenomeris* (two species, western Malaysia) and *Avetra* (one species, Madagascar) are climbers with samara fruits, the former having numerous ovules per locule (all the other genera have two). *Dioscorea* (600 species, pantropical), *Rajania* (25 species, West Indies) and *Tamus* (five species, Canaries, Madeira, Europe and Mediterranean) are all climbers with capsule, samara and berry fruits respectively. The family is most closely related to the Liliaceae.

**Economic uses.** The only economically important genus is *Dioscorea*, the yams, the tubers of about 60 species being cultivated as a subsistence crop in three main centers, Southeast Asia, West Africa and Central and South America. Some species are the source of diosgenin, a steroidal sapogenin developed in recent years for its use in oral contraceptives.  C.J.H.

# ORCHIDALES

## BURMANNIACEAE

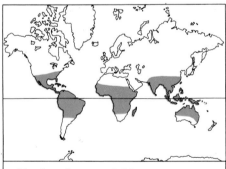

**Number of genera:** 5–16
**Number of species:** about 125
**Distribution:** pantropical.
**Economic uses:** none.

The Burmanniaceae is a small family of mainly tropical forest herbs, most of which are colorless saprophytes.

**Distribution.** The family is found throughout the tropics, especially in the tropical forests of Brazil, equatorial Africa and Southeast Asia.

**Diagnostic fatures.** Most species are small, annual or perennial herbs with slender, upright, unbranched stems produced from rhizomes or tubers. Leaves are often absent; when present, they are crowded near the base of the stem and usually long and narrow with the typically parallel veins of monocotyle-

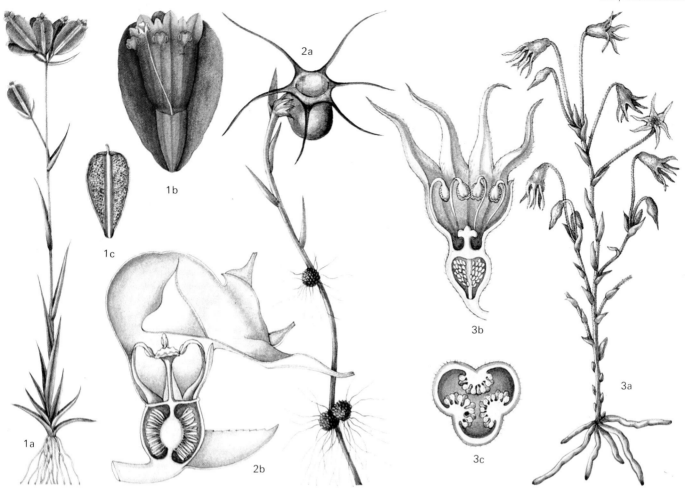

Burmanniaceae. 1 *Burmannia coelestis* (a) habit showing the slender, upright, unbranched stem crowned by an inflorescence of winged flowers (×1); (b) part of perianth opened out showing three sessile anthers and style with three stigmas (×4); (c) vertical section of ovary (×4). 2 *Afrothismia* sp nov (a) habit (×⅔); (b) half flower showing inferior ovary containing ovules on axile placentas (×8). 3 *Haplothismia exannulata* (a) habit (×⅔); (b) vertical section of flower; (c) cross section of unilocular ovary with ovules on three parietal placentas (×6).

donous plants. Those higher up the stem are very much smaller or scale-like and are arranged alternately. In many species the vegetative parts are colorless, in which case the plant feeds saprophytically on decaying materials in the soil and needs no chlorophyll for photosynthesis.

The flowers are usually carried at the tip of the upright stem as a raceme or cyme. Less often the flowers may be solitary. They are regular (rarely irregular), bisexual (rarely unisexual) and white, bright blue or rarely yellow. The perianth has six segments in two whorls which are fused at their bases into a three-winged tube. The outermost whorl encloses and protects the inner whorl in the bud. The innermost perianth segments carry three or six stamens without filaments. The ovary is inferior, of three fused carpels, forming either three locules with axile placentation or one locule with three parietal placentas. There are numerous minute ovules in each locule. The style has three stigmas.

The fruit is a single capsule dehiscing by three longitudinal slits. The perianth usually persists to give the capsule a three-winged appearance. The seeds have little or no endosperm and the embryo is very small.

**Classification.** The earliest classifications divided the family into three tribes: BURMANNIEAE with flowers possessing three stamens, and containing the important genus *Burmannia*; THISMIEAE with flowers with six stamens and an irregular perianth, and containing the genus *Thismia*; CORSIEAE with flowers with six stamens and a regular perianth. Some authorities have elevated all three tribes to family rank, while others have separated out only the last and given it family rank as the Corsiaceae, a family then containing two genera, *Corsia* (flowers bisexual) and *Arachnites* (flowers unisexual). The number of genera recognized by different taxonomists within these three tribes has varied widely. The Burmanniaceae is closely related to the Orchidaceae.

**Economic uses.** None are known.

B.N.B

# ORCHIDACEAE
## The Orchid Family

The Orchidaceae is a very large family distributed throughout the world. Prized for their spectacularly beautiful flowers, orchids are cultivated with sometimes fanatical devotion, and enormous numbers of new hybrids are produced, often commanding

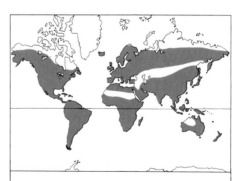

**Number of genera:** about 750
**Number of species:** about 18,000
**Distribution:** cosmopolitan.
**Economic uses:** basis of a huge floriculture industry and source of flavoring essence (vanillin).

high prices. Many species, however, are in danger of extinction through the destruction of natural habitats.

**Distribution.** The family is worldwide, with representatives in every country excepting a few isolated islands and Antarctica. Their ecological range is equally wide, with plants occurring in all but the most extreme environments such as the sea, the driest

deserts and the tops of the coldest mountains. Orchid species with surprisingly similar flowers can be found from the damp mountainsides of Norway and New Zealand to the dry savannahs of tropical Africa and South America, and from the coastal strands of the South Pacific to rocky ledges in the Himalayas.

**Diagnostic features.** Although the orchid family is very large, it shows less vegetative and major floral diversity than many much smaller families. Orchids have several characteristics, mostly connected with their flowers, which distinguish them from plants of other families.

Firstly, the seed from which a plant develops is very small, just a few cells forming a more or less undifferentiated embryo. The fruit (capsule) from a single fertilized flower spike may produce vast numbers of these seeds; figures of a million or more having been recorded from many species. In order to germinate, these minute seeds require the help of a fungus in a special symbiotic relationship; even so, it can take an inordinate time for a flowering-size orchid plant to appear. The ripening of the seed in an orchid capsule usually takes from two to 18 months; the germination period is similar, but it can take a further four years for the life-cycle to be complete. Under laboratory conditions this can be reduced to about three years, and if the ripening process is omitted by sowing green seeds or even excised embryos, it can be reduced by a further year.

For successful growth in the wild, and in cultivation, orchid plants must maintain their symbiosis with various species of fungi, the hyphae of which can be found in all their vegetative tissue, but mainly concentrated in the roots. About half of the world's total of species are terrestrial, with "normal" roots, but most of the remainder, all tropical or subtropical, are epiphytes, usually with long, tangled and dangling, dull gray or white, velamen-covered, aerial roots with a green, densely meristematic apex which absorbs moisture and the minimal amounts of nutrients required. These epiphytes generally have thick ligulate leaves with ligules, although many other types occur; all minimize the loss of water and encourage maximum photosynthetic activity. The extreme environmental conditions experienced by many epiphytes and lithophytes (plants growing on rocks and stones) have led to

actively photosynthesizing tissue being developed in such organs as the stems, roots and even the flowers of several species.

Many tropical and subtropical orchids, both terrestrial and epiphytic, possess special water and nutrient storage organs called pseudobulbs. These vary, depending on the species, from almost imperceptibly swollen stems to apple-like, very hard and shiny green organs from which the leaves arise. Sometimes a plant has only one pseudobulb, or they may form a closely packed cluster. With some tropical species, the pseudobulbs can be spaced out along a creeping or climbing rhizome. Pseudobulbs vary in size from mere pinheads in an Australian species to thick cylinders up to 3m (10ft) tall in an Asian species. Pseudobulbs are always on or above soil level, but in several groups of terrestrial orchids from the cooler, temperate regions there are similar organs developed on the roots in the soil. The swollen roots, called tubers, have given rise to the name "orchid"; the testiculate appearance of paired tubers led the Greeks to call them *orchis*, their word for testicle.

Although all of the world's orchids can be considered as either terrestrials or epiphytes there are some variations so extreme as to warrant a mention. The terrestrial habit is so very pronounced in a small group of Australian species that they spend their entire life-cycle underground, and apparently never surface even to flower. The nutrition of these plants is saprophytic, and there are many other examples of saprophytism in both terrestrial and epiphytic orchids. Some of the epiphytic saprophytes are so well adapted to the conditions of the humid tropics that they are better considered as very rapid growing climbers or scramblers. There are helophytic (bog) orchids but as yet no truly halophytic (salt-tolerant), aquatic or parasitic species have been recorded although some species are insectivorous-like, developing a bucket-type trap into which the pollinating insects are lured as part of the pollination mechanism.

The leaves of the Orchidaceae are alternate, often in two rows, rarely opposite, sometimes reduced to scales. They are simple, often fleshy, and sheathing at the base, the sheath usually enclosing the stem.

It is undoubtedly the flowers which most characterize orchids and serve to distinguish them absolutely from any other plants. They are nearly always bisexual and borne solitary or in spikes, racemes or panicles. Basically they conform to an irregular pattern with three sepals and three petals. The sepals are usually similar but the two laterals or the single dorsal may be elongated or bear a longitudinal crest. The petals, however, are usually dissimilar with the two laterals being quite distinct from the dorsal. It is this dorsal petal, termed the labellum or lip, which gives the orchid flower its characteristic appearance. The two lateral petals and the

sepals frequently are very similar morphologically, in which case they are termed the tepals, but whether or not the petals and sepals are distinguishable the labellum is always quite different. It can be two-, three- or four-lobed, or so divided as to resemble closely a lichen such as *Usnea*. It can be much smaller than the other perianth segments but is usually very much larger. Its surface can be ornamented in many ways with plates, hairs, calluses or keels and frequently it bears the most bizarre color combinations. Orchid flowers vary in color from white, yellow and green through enamel blues to deepest, dullest rusty maroons and purples: the petals and sepals may be of relatively uniform color, but the labellum generally has markedly distinct colors. It may be produced backwards to form a spur which can vary from the smallest sac to a long tube up to 30cm (12in) long; in some species the spur is bilobed.

Lying opposite the labellum in the floral whorl are the sexual organs. In orchids these are always united to form a single structure termed the column, which in its simplest, basic form is surmounted by the anthers with the receptive stigmatic surface just underneath them, usually separated by a flap of sterile tissue called the rostellum. There are one or two stamens with bilocular anthers and three stigmas, one of which is usually sterile and transformed into the rostellum. The many thousands of variations of columnar structure provide the basis for much of the technical classification of orchids. In evolutionary terms it is the columnar structure that has determined the particular pollinating agent of each species, and hence the great range of species found today.

The pollen is not an amorphous powder, as in most flowering plants, but is aggregated into two, four, six or eight waxy, horny or mealy pollen masses, called pollinia, and these are variously attached to the column apex, and usually are stalked. The simplest form of pollinating mechanism in orchids is the accidental removal, by a bee, of one or more or all of the pollinia which attach themselves to its head or thorax during its nectar search at the base of the flower or in the spur. Bees, wasps, flies, ants, beetles, hummingbirds, bats and frogs have all been observed as the pollinating agents of orchids; the underground species mentioned below probably adding mites or even snails or slugs to this list. The attachment of the pollinia to the insect is helped by a variety of additional mechanisms such as a quick-setting glue on the pollinial stalks or an explosive device that can project the pollinia up to 60cm (2ft) from the flower, and by the development of features, both visual and olfactory, that attract the insect to the right position on the right flower. A well-known mechanism is termed pseudocopulation. Although these mechanisms by and large involve specific orchid species and pollinat-

Orchidaceae. 1 *Bulbophyllum barbigerum* habit showing the swollen pseudobulbs at the base of the leaves ( ×⅔). 2 *Dendrobium pulchellum* habit ( ×⅔). 3 *Saphronitis coccinea* habit ( ×⅔). 4 *Oncidium tigrinum* (a) flowers borne in a wiry raceme ( ×⅔; (b) column (united sexual organs) ( ×⅓). 5 *Paphiopedilum concolor* (a) an evergreen orchid lacking pseudobulbs ( ×⅔); (b) column side view ( ×1); (c) column front view ( ×1). 6 *Coelogyne parishii* (a) aerial shoot with large pseudobulbs ( ×⅔); (b) column ( ×1⅓).

ing agent relationships, cross-fertilization can occur, and in orchids there are more naturally occurring hybrids than in all the other plant families put together. This "promiscuity" extends to hybrids between species in two different genera, and under the controlled conditions of the plant-breeding establishment up to 20 species and five genera can be combined into one plant. New man-made hybrids, many of them multi-generic, are entering the commercial world at the rate of about 150 each month.

Another characteristic feature of orchid flowers is the great range of scents produced, presumably as an integral part of the pollination mechanism. These vary from the smell of rotting carrion through sickly sweet vanilla-like odors to unquestionably very pleasing perfumes, often from quite medi-ocre inflorescences. Sometimes the un-pollinated flowers, for example those of certain New Guinean species, live for nine months or even more, in contrast with the two or three hours' life of those of some tropical American species.

Orchid fruits, like orchid flowers, are also widely variable, ranging from single, long, narrow "pods" to clusters of many small grape-like capsules. The ovaries from which the fruits are derived are inferior and usually have one locule with three parietal placentas but occasionally there are three locules with axile placentation. There are numerous ovules in each locule. In every case the capsules open, when fully ripe, laterally by three or six longitudinal slits.

**Classification.** The classification of the orchid family is rather technical and largely based on characteristics of the column and, at the higher levels, the nature of the pollinia. It does not integrate closely with the life-forms and geographical or ecological range of species but, coincidentally, appears to correlate with the most commonly agreed evolutionary development of the group. The usually accepted classification divides the family into three subfamilies, six tribes, about 80 subtribes and about 750 genera. The smallest subfamily (the APOSTASIOID-EAE) contains one tribe which comprises one subtribe consisting of two genera, *Apostasia* and *Neuwiedia*, to give a total of about 20 species. The next subfamily, the CYPRI-PEDIOIDEAE, has a single tribe comprising four subtribes and about 120 species of

lady's slipper orchids grouped into five genera, those most commonly encountered being *Cypripedium* and *Paphiopedilum*. Many botanists treat these two subfamilies as separate families, the Apostasiaceae and Cypripediaceae, respectively, because of the unusual structure of the flowers.

The third subfamily, the ORCHIDOIDEAE, contains over 99% of the family's species and consists of four tribes, the ORCHIDEAE, NEOTTIEAE, EPIDENDREAE and the VANDEAE. The genera in these tribes can vary from one to 1,200 species or more; many genera are rarely seen whereas others, from the smallest to the largest, are well-known to botanists, ecologists and growers and breeders.

The Orchideae contains mainly terrestrial orchids similar in general appearance to those found throughout Europe; commonly encountered genera are *Aceras, Anacamptis, Coeloglossum, Dactylorhiza, Disa, Gymna-denia, Habenaria, Himantoglossum, Ophrys, Orchis* and *Satyrium*.

Many of the Neottieae are saprophytic and most are terrestrial but some are better considered as epiphytes. Common genera are *Anoectochilus, Cephalanthera, Goodyera, Listera, Neottia* and *Spiranthes*.

The Epidendreae contains the majority of tropical epiphytic orchids but there are some terrestrial species. Genera commonly encountered include *Bletia, Brassia, Bulbo-phyllum, Cattleya, Coelogyne, Dendro-bium, Epidendrum, Eria, Masdevallia, Pleïone, Polystachya, Sophronitis, Vanilla* and *Zygo-petalum*.

The Vandeae are also predominantly epiphytic but include many genera which are basically terrestrial. Among the frequently encountered genera are *Angraecum, Cata-setum, Cymbidium, Eulophia, Lycaste, Maxil-laria, Miltonia, Odontoglossum, Oncidium, Phalaenopsis, Stanhopea* and *Vanda*.

The relationship of the Orchidaceae to other families is not at all clear. The Burmanniaceae, Liliaceae and Amarylli-daceae (especially the genus *Hypoxis* and its relatives) have been suggested, but probably the Orchidaceae has most in common with the Burmanniaceae, especially the segregate family, the Corsiaceae.

**Economic uses.** Excepting the flavoring essence vanillin obtained from *Vanilla plani-folia* and the questionably nutritive "salep" from the tubers of certain species, the orchids have little direct economic importance other than as the basis for a vast floricultural industry. The legends surrounding the early discoveries, importations, sale, cultivation and breeding of choice orchids are among the classics of botanical and horticultural literature. The facts are no less remarkable; the privations suffered by the privately sponsored explorers, the vast losses of plants sustained on the long journeys back to Europe, the fabulous prices realized at auctions for fresh importations are well documented. Today, many orchid expe-

ditions are still being organized and there are many problems involved with success-fully establishing exotic species in culti-vation.

The orchid growing and breeding industry today is based in the United States of America, but it is also a major export earner in such countries as Singapore, Hawaii, Australia, Thailand, England, Holland and Western Germany. Orchid plants are culti-vated in the controlled environment of greenhouses in the temperate regions of the world but in the subtropics and tropics they are grown out of doors in the same way as other plants. They are increasingly grown as room or "house" plants and in Germany, for example, special indoor greenhouses called "orchidaria" are used. Nearly all the most popular cultivated orchids (those of the genera *Aërides, Brassia, Cattleya, Coelo-gyne, Cymbidium, Dendrobium, Epiden-drum, Laelia, Miltonia, Odontoglossum, Oncidium, Paphiopedilum, Phalaenopsis* and *Vanda* and their intergeneric hybrid de-rivatives) are epiphytes and require a special potting compost. This is ideally a mixture of osmunda fiber and sphagnum moss, but with difficulties in obtaining sufficient supplies of these materials natural and man-made substitutes are used today. These include fir bark, dried bracken, peat and shredded plastic waste. The terrestrial species, such as those of *Calanthe* and *Cypripedium*, require a loam rich mixture.

Orchid cultivation has progressed greatly since imported plants were first grown in Europe over 200 years ago. Recent research has shown that they were grown first of all in China nearly 1,000 years ago. A recent development has been the growth of societies devoted entirely to the growing and exhibit-ing of orchids. There are over 400 local orchid societies in the world.

In many ways the continued survival of certain orchid species is enhanced by their popularity in cultivation, but, on the other hand, the bizarre, attractive and unusual appearance of many plants leads to their destruction by continued collecting and flower picking. However, it is the ever-increasing destruction of suitable orchid habitats for urban development, and their alteration by drainage and other agricultural procedures, industrial pollution and military operations that is responsible for placing probably a quarter or even a third of the 18,000 species in danger of extinction.

Fortunately Man has realized the prob-able consequences of his actions and today throughout the world orchids are now among the best-protected plants. Legislation prohibiting their removal and sale or export, the establishment of nature reserves and the application of new technology, such as meristem propagation, to the raising of large numbers of threatened species are some of the ways in which orchids are being con-served. P.F.H.

Orchidaceae (continued) 7 *Neottia nidus-avis* (the bird's nest orchid) a colorless, saprophytic plant ($\times\frac{2}{3}$). 8 *Anoectochilus roxburghii* leaf and flowering spike ($\times\frac{2}{3}$). 9. *Ophrys bertolonii* habit ($\times\frac{2}{3}$). 10 *Orchis purpurea* leaves and inflorescence ($\times\frac{2}{3}$). 11 *Corybas bicalcarata* habit ($\times\frac{2}{3}$). 12 *Cypripedium calceolus* flower section ($\times 1$). 13 *Apostasia nuda* (a) flower ($\times 4$); (b) detail of stamen ($\times 8$). 14 *Disa hamatopetala* spike ($\times\frac{2}{3}$). 15 *Cypripedium irapeanum* flower and leaves ($\times\frac{2}{3}$). 16 *Dactylorhiza fuchsii* flower (a) side view ($\times 2\frac{2}{3}$) and (b) front view ($\times 2$).

# INDEX

# ACKNOWLEDGEMENTS

The Publishers acknowledge the following reference sources:

*Annals of the Royal Botanic Garden, Calcutta* (1887–present). Calcutta.

Bailey, L. H. (1949). *Manual of Cultivated Plants*. Revised edition. New York.

Bailey, L. H. & Bailey, E. Z. (1976). Revised by the staff of the Liberty Hyde Bailey Hortorium. *Hortus Third*. New York and London.

Baillon, H. E. (1886–1895). *Histoire des Plantes*. Paris.

Bean, W. J. (1970–1976). *Trees and Shrubs hardy in the British Isles*, ed. 8, edited by G. Taylor and D. L. Clarke. London.

Bentham, G. & Hooker, J. D. (1862–1883). *Genera Plantarum* 1–3. London.

*Botanical Register* (1863–1942). London.

Chittenden, F. J. (ed.) (1951). *Dictionary of Gardening*, 1–4. Oxford.

Cronquist, A. (1968). *The Evolution and Classification of Flowering Plants*. London.

Oalla Torre, C. G. de & Harms, H. (1900–1907). *Genera Siphonogamarum*. Leipzig.

Davis, P. H. & Cullen, J. C. (1965, 1978). *The Identification of Flowering Plant Families*, ed. 1 (1965), 2 (to be published). Edinburgh (ed. 1); Cambridge (ed. 2).

Engler, H. G. A. (ed.) (1900–1953). *Das Pflanzenreich. Regni vegetabilis Conspectus*, nos 1–107. Berlin.

Engler, H. G. A. (1964). *Syllabus der Pflanzenfamilien*, ed. 6, by H. Melchior, vol 2. Berlin.

Engler, H. G. A. & Prantl, K. A. E. (1887–1915). *Die natürlichen Pflanzenfamilien*, ed. 1. Leipzig.

Engler, H. G. A. & Prantl, K. A. E. (1925–present). *Die natürlichen Pflanzenfamilien*, ed. 2. Leipzig and Berlin.

Graf, A. B. (1963). *Exotica 3*. Roehrs Company, Rutherford, N.J.

Harrison, S. G., Masefield, G. B. & Wallis, M. (1969). *The Oxford Book of Food Plants*. Oxford.

Hegi, G. (1906–1931). *Illustrierte Flora von Mittel-Europa*, ed. 1, 1–7. München.

Hegi, G. (1936–present). *Illustrierte Flora von Mittel-Europa*, ed. 2, 1→. München.

Hegi, G. (1966–present). *Illustrierte Flora von Mittel-Europa*, ed. 3, 1→. München.

Hill, A. F. (1952). *Economic Botany*, ed. 2. New York, Toronto and London.

Hooker, W. J. (1836–1854). *Icones Plantarum*. London.

Howes, F. N. (1974). *A Dictionary of Useful and Everyday Plants and their Common Names*. Cambridge.

Hutchinson, J. (1926, 1934). *The Families of Flowering Plants*, 1–2. London.

Hutchinson, J. (1959). *The Families of Flowering Plants*, ed. 2, 1–2. Oxford.

Hutchinson, J. (1964, 1967). *The Genera of Flowering Plants (Angiospermae)*, 1 (1964), 2 (1967). Oxford.

Irvine, F. R. (1969). *West African Crops*. Oxford.

Janick, J. *et al.* (1974). *Plant Science*, ed. 2. San Francisco.

*Journal of Botany, British and Foreign* (1863–1942). London.

Lawrence, G. H. M. (1951). *Taxonomy of Vascular Plants*. New York.

Marloth, R. (1913–1932). *The Flora of South Africa*. Capetown.

Martius, C. F. P. von (1840–1906). *Flora Brasiliensis*. München, Wien, Leipzig.

Porter, C. L. (1967). *Taxonomy of Flowering Plants*. San Francisco and London.

Purseglove, J. W. (1968). *Tropical Crops. Dicotyledons*. London.

Purseglove, J. W. (1972). *Tropical Crops. Monocotyledons*. London.

Radford, A. E., Dickison, W. C., Massey, J. R. and Bell, C. R. (1974). *Vascular Plant Systematics*. New York.

Rehder, A. (1940). *Manual of Cultivated Trees and Shrubs*, ed. 2. New York. (Reprint 1956).

Rendle, A. B. (1904, 1938). *The Classification of Flowering Plants*, ed. 1, 1–2. Cambridge.

Rendle, A. B. (1930). *The Classification of Flowering Plants*, ed. 2, 1. Cambridge.

Ross-Craig, S. (1948–1973). *Drawings of British Plants*. London.

Simmonds, N. W. (ed.) (1976). *Evolution of Crop Plants*. London and New York.

Soó, R. von (1963). *Fejlödéstörténeti Növényrendszertan*. Budapest.

Stebbins, G. L. (1974). *Flowering Plants. Evolution above the Species Level*. London.

Swift, L. H. (1974). *Botanical Classifications*. Connecticut.

Synge, P. M. (ed.) (1956). *Supplement to the Dictionary of Gardening*. Oxford.

Synge, P. M. (ed.) (1969). *Supplement to the Dictionary of Gardening*, ed. 2. Oxford.

Takhtajan, A. (1959). *Die Evolution der Angiospermen*. Jena.

Takhtajan, A. (1969). *Flowering Plants. Origin and Dispersal*. Translated by C. Jeffrey. Edinburgh.

*The Botanical Magazine*. (1793→). London.

Thorne, R. F. (1974a). The "Amentiferae" or Hamamelidae as an artificial group: a summary statement. *Brittonia*, 25: 395–405.

Thorne, R. F. (1974b). A Phylogenetic Classification of the Annoniflorae. *Aliso*, 8: 147–209.

Thorne, R. F. (1976). A Phylogenetic Classification of the Angiospermae. *Evolutionary Biology*, 9: 35–106.

*Transactions of the Linnean Society of London*. (1791–1875). London.

*Transactions of the Linnean Society of London (Botany)* (1875→). London.

Tutin, T. G., Heywood, V. H. *et al. Flora Europaea*, 1 (1964), 2 (1968), 3 (1972), 4 (1976). Cambridge.

*Urania Pflanzenreich. Höhere Pflanzen*, 1 (1971), 2 (1973). Leipzig, Jena and Berlin.

Wettstein, R. (1933–1935). *Handbuch der Systematischen Botanik*, ed. 4. Leipzig–Wien.

Willis, J. C. (1973). *A Dictionary of the Flowering Plants and Ferns*, ed. 8, revised by H. K. Airy Shaw. Cambridge.